Carbon Dioxide Mineralization and Utilization

Pen-Chi Chiang · Shu-Yuan Pan

Carbon Dioxide Mineralization and Utilization

Springer

Pen-Chi Chiang
Graduate Institute of Environmental Engineering
National Taiwan University
Taipei
Taiwan

Shu-Yuan Pan
Graduate Institute of Environmental Engineering
National Taiwan University
Taipei
Taiwan

ISBN 978-981-10-3267-7 ISBN 978-981-10-3268-4 (eBook)
DOI 10.1007/978-981-10-3268-4

Library of Congress Control Number: 2016963187

Printed on acid-free paper

This Springer imprint is published by Springer Nature
The registered company is Springer Nature Singapore Pte Ltd.
The registered company address is: 152 Beach Road, #22-06/08 Gateway East, Singapore 189721, Singapore

Preface

Carbon dioxide (CO_2) mineralization and utilization is an important technology wherein CO_2 is captured and stored for utilization instead of being released into the atmosphere. CO_2 mineralization and utilization demonstrated in the waste-to-resource supply chain can "reduce carbon dependency, promote resource and energy efficiency, and lessen environmental quality degradation," thereby reducing the environmental risks and increasing the economic benefits towards sustainable development goals. This book provides comprehensive information on CO_2 mineralization and utilization using alkaline wastes via accelerated carbonation technology from theoretical and practical considerations, presented in 20 chapters. Engineers, scientists, government officers, and project managers will consider this book as an essential reference on CO_2 mineralization and utilization.

In this book, the concept of carbon cycle from the thermodynamic point of view was first introduced. The principles, applications, and environmental impact assessment of carbon capture and storage technologies also are illustrated in Part I. Among the carbon capture and utilization processes, CO_2 mineralization via accelerated carbonation technology is especially focused in Part II. Throughout the carbonation process, huge amounts of CO_2 and alkaline wastes generated from industries can be reclaimed and reused. From the theoretical consideration, the process chemistry, reaction kinetics, mass transfer, and system analysis for accelerated carbonation are systematically presented. On the other hand, from the practical consideration, the analytical methods and the application of accelerated carbonation are introduced as well. In Part III, it then explores the utilization of carbonated products as green materials such as supplementary cementitious materials and high value-added chemicals. Key performance indicators for evaluating the function and properties of carbonated products are developed. Lastly, an integral approach for waste treatment and resource recovery is proposed to establish a waste-to-resource supply chain towards a circular economy system. It discusses the challenges, barriers, and strategies of integrated air pollution control at industry in detail, and then illustrates the importance and significance of establishing waste-to-resource green supply chain. Furthermore, the carbonation system is critically assessed and optimized from aspects of engineering, environmental, and economic analysis.

Reduction in CO_2 emission in industries and/or power plants should be a portfolio option. Integrated alkaline waste treatment with CO_2 mineralization and utilization is an attractive approach to achieving direct and indirect reduction in greenhouse gas (GHG) emissions in industries. The accelerated carbonation can not only stabilize alkaline wastes but also fix CO_2 in flue gas from industries as a safe and stable carbonate precipitate. On the other hand, the amount of CO_2 reduction by carbonation could be certified as emission reduction credits, in conjunction with the joint implementation (JI), emission trading scheme (ETS), and clean development mechanism (CDM) issued by the Kyoto Protocol. Therefore, it suggests that the establishment of a waste-to-resource supply chain should provide a method of overcoming the barriers of energy demand, waste management, and GHG emissions to achieve a circular economy system, under which the "win-win" philosophy demonstrating green economy and healthy environment can be coexisted.

We gratefully acknowledge Prof. Liang-Shih Fan at the Ohio University for his thoughtful comments and invaluable support during the development of this book. We also deeply appreciate Prof. Chung-Sung Tan, Prof. Young Ku, Prof. E.-E. Chang, and Prof. Yi-Hung Chen, who have contributed to this book. Special thanks go to Dr. Pen-Tai Chiang, Dr. Mengyao Gao, Shelley Yang, and Ming-I Chen for their hard work and patience. Moreover, we are wholeheartedly grateful to Dr. Kinjal J. Shah, Andrew Chiang, Teresa Wang, Michael Du, Elena Blair, and Michael Lin, who reviewed one or more chapters of this book and provided valuable suggestions and comments. Our sincerest appreciation also goes to all the laboratory group members, Silu Pei, Tai-Chun Chung, Jeffrey Chen, Chen-Hsiang Hung, Kuan-Wei Chen, and Tse-Lun Chen for their hard work on developing carbon mineralization and utilization technologies.

Furthermore, we would like to express our gratitude to Mark Goedkoop, who kindly granted permission to use their photographs and valuable information in this book. Over the years, Ministry of Science and Technology (Taiwan), along with several industrial partners including China Steel Corp., Tung-Ho Steel Enterprise Corp., Formosa Petrochemical Corp., and Cheng Loong Corp., provided funding in support of our research grants for the development and deployment of carbon mineralization and utilization technologies. Much of the results reported in this book are based on the aforementioned efforts. Our thanks also go to Xiao-Li Pei for her assistance in the preproduction of this book.

Taipei, Taiwan

Pen-Chi Chiang
Shu-Yuan Pan

Contents

Part IV Valorization of Carbonation Product as Green Materials

Chapter 1
Introduction

Abstract Since the industrial revolution in 1750, human activities have resulted in a 40% increase in the atmospheric concentration of CO_2, thereby leading to rapid global warming. To mitigate the global warming and climate change caused by huge anthropogenic CO_2 emissions, different strategies, action plans, and economic instruments have been proposed and implemented around the world. In this chapter, the significance and importance of climate change and global warming are illustrated. An overview of several important formal meetings of the United Nations Framework Convention on Climate Change (UNFCCC) Parties, i.e., Conferences of the Parties (COP), is provided to reveal key milestones in dealing with global greenhouse gas emissions. One such method uses accelerated carbonation of alkaline wastes to capture and utilize CO_2, the theoretical and practical considerations of which are presented in 19 Chapters in this book.

1.1 Climate Change and Global Warming: Significance and Importance

Greenhouse gases (GHGs) are gases in the atmosphere that can absorb and emit radiation within the thermal infrared range, thereby leading to the greenhouse effect. Without GHG, the average temperature of Earth's surface would be approximately 0 °F (−18 °C), rather than present average of 59 °F (15 °C) [1, 2]. Of the gases affecting the ambient temperature of the Earth, the following are most interesting because they are known as long-lived greenhouse gases (LLGHGs):

- Carbon dioxide (CO_2)
- Methane (CH_4)
- Nitrous oxide (N_2O)
- Chlorofluorocarbons (CFCs)
- Hydrochlorofluorocarbons (HCFCs)
- Hydrofluorocarbons (HFCs)
- Perfluorocarbons (PFCs)
- Sulfur hexafluoride (SF_6)

P.-C. Chiang and S.-Y. Pan, *Carbon Dioxide Mineralization and Utilization*, DOI 10.1007/978-981-10-3268-4_1

The most abundant GHGs in the atmosphere of the Earth are water vapor (H_2O), CO_2, CH_4, N_2O, O_3, and CFCs. These gases can be discharged into the atmosphere by natural and anthropogenic sources. However, since the beginning of the industrial revolution, human activities have produced a 40% increase in the atmospheric concentration of CO_2, from 280 ppm in 1750 to 400 ppm in 2015. The rapid increase of CO_2 concentration in the atmosphere has spurred worldwide concerns of global climate change from government, industrial, and academic groups. Anthropogenic emissions of CO_2 mainly come from combustion of carbon-based fossil fuels (such as coal, oil, and natural gas), along with deforestation, soil erosion, and animal agriculture. It is noted that the major anthropogenic GHGs are CO_2, CH_4, N_2O, SF_6, HFCs, and PFCs, which are regulated under the international Kyoto Protocol treaty. The global warming potential (GWP) depends on both the efficiency of the molecule as a GHG and its atmospheric lifetime. CO_2 is defined to have a GWP of one over all time period. For instance, methane has an atmospheric lifetime of 12 ± 3 years, resulting in a GWP value of 72 over a timescale of 20 years [3].

1.1.1 Kyoto Protocol in 1997

The Kyoto Protocol is an international treaty signed in 1997, which extends the 1992 United Nations Framework Convention on Climate Change (UNFCCC). The Kyoto Protocol was adopted in Kyoto (Japan), and originally aimed to attain, by 2012, a reduction of global GHG emissions at least 5% less than the observed levels in 1990. A total of six GHGs, including CO_2, CH_4, N_2O, HFCs, PFCs, and SF_6, were regulated in the Kyoto Protocol, which came into effect in 2005. As a result of the Kyoto Protocol, the European Union (EU) issued a global reduction aim of GHG levels by 8%. The Protocol defines three flexibility mechanisms to meet the emission limitation commitment for the Annex I Parties, which include international emissions trading (IET), the clean development mechanism (CDM), and joint implementation (JI). The economic basis for providing this flexibility is that the marginal cost of reducing emissions differs among countries [4].

To negotiate the Kyoto Protocol for establishing legally binding obligations of reducing GHG emissions for developed countries, the United Nations Climate Change Conferences (UNCCC) are held annually in the framework of the UNFCCC. They serve as the formal meeting of the UNFCCC Parties, i.e., Conferences of the Parties (COP), which assess the progress in dealing with climate change. The first UNCCC (COP 1) was held at Berlin, Germany, in 1995. From 2011, the COP meetings have also been used to negotiate the Paris Agreement, as part of the Durban platform activities (adopted at COP 17 in 2011), until its conclusion in 2015.

1.1.2 Cancún Agreement (COP 16) in 2010

The 2010 UNFCCC, officially referred as the 16th session of the Conference of the Parties (COP 16), was held at Cancún, Mexico, in 2010. The agreement includes voluntary pledges made by 76 countries to control GHG emissions. At the time of the agreement, these countries were collectively responsible for 85% of annual global CO_2 emission. The most significant outcome was the agreement for a "Green Climate Fund (GCF)" and a "Climate Technology Centre," adopted by the states' parties. The GCF aimed to distribute US$100 billion per year by 2020 to assist poorer countries in financing emission reductions and adapting to climate change. It also asked rich countries to reduce their GHG emissions as pledged in the Copenhagen Accord and planed to reduce the emissions for developing countries. However, at that time, the funding of the GCF was not agreed upon.

1.1.3 Durban Agreement (COP 17) in 2011

COP 17 meeting was held at Durban, South Africa, in 2011. In this meeting, the implementation of carbon capture and storage (CCS) technologies was regarded as eligible for clean development mechanism (CDM) projects and activities. However, the geological storage of CO_2 demonstrated around the world still faces many uncertainties and risks, such as accidental leakage of CO_2, environmental impacts, and public acceptance. On the other hand, carbon capture, utilization, and storage (CCUS) have recently received global attention as a viable option for reducing CO_2 emissions from industries and/or power plants [5–8]. In this meeting, the creation of the GCF was also discussed.

1.1.4 Paris Agreement (COP 21) in 2015

The COP 21 meeting was held at Paris (France) in 2015. Negotiations resulted in the adoption of the Paris Agreement, which represented a consensus of the representatives of the 196 parties, to govern climate change reduction measures starting from 2020. The agreement will become legally binding only if at least 55 counties, which together produce at least 55% of the global GHG emissions, ratify the agreement [9]. The agreement ended the work of the Durban platform which was established during COP 17. The expected key result of COP 21 was highlighted by the below statement:

> Holding the increase in the global average temperature to well below 2 °C above pre-industrial levels and to pursue efforts to limit the temperature increase to 1.5 °C above pre-industrial levels, recognizing that this would significantly reduce the risks and impacts of climate change.

The agreement also called for "zero net anthropogenic GHG emissions" to be reached by 2050. Prior to the conference, a total of 146 national climate panels each publicly presented draft national climate contributions, called intended nationally determined contributions (INDCs), which was estimated to limit global warming to 2.7 °C by 2100. For instance, the EU suggested the INDC should set a binding target for at least a 40% domestic reduction in GHG emissions by 2030, compared to 1990 [10]. It also suggested that the regulated GHGs by EU members should include CO_2, CH_4, N_2O, HFCs, PFCs, SF_6, and NF_3.

1.2 Mitigation and Adaptation

The ocean is the major short-term sink in nature because of the imbalance between CO_2 concentrations in the ocean and the atmosphere. Although the natural sink is very important, offering −0.5 °C of temperature reduction following an overshoot [11], the major application of anthropogenic sinks, such as carbon capture utilization and storage (CCUS) and rapid reforestation, is also required to achieve a plateau at 2 °C. Without technologies that remove CO_2 from the atmosphere, the 350 CO_2 ppm target is out of reach in the twenty-first century [11].

To mitigate rapid global warming and adapt to the climate change caused by huge anthropogenic CO_2 emissions, different strategies and tools from various aspects have been proposed and implemented. Action plans and practical technologies have been executed to pursue scientific solutions for overcoming the challenges of global warming [12, 13]. According to the international energy agency (IEA) report, the strategies for reducing CO_2 emissions include the following: (1) improving overall energy efficiency, (2) implementing carbon capture and storage (CCS) technologies, and (3) utilizing renewable energy and material recycling [14]. Among the above strategies, the CCS technologies could reduce CO_2 emissions by 9–50% in industrial sectors, compared to the present level, by 2050 and could mitigate cumulative global climate change by 15–55% in 2100 [15].

Putting a price on carbon emission can also help shift the burden of the environmental damage back to those who can reduce it. There are two types of carbon pricing instruments that can be utilized to accelerate the CO_2 emission reduction: (1) emissions trading systems (ETS) and (2) carbon taxes. The choice of carbon pricing tools depends on national and economic circumstances. The ETS is sometimes referred to as a cap-and-trade system. It caps the total level of greenhouse gas emissions and allows industries with low emissions to sell their extra allowances to larger emitters. By creating a platform of supply and demand for emission allowances, an ETS can effectively establish a market price for GHG emissions, ensuring that the emitters will be kept within their pre-allocated carbon budget. Conversely, a carbon tax directly sets a price on carbon by defining a tax rate on the GHG emissions (or the carbon content) of fossil fuels. It is different from an ETS because the emission reduction outcome of a carbon tax is not predefined but the carbon price is.

1.3 Structure and Contents of This Book

This book provides comprehensive information on CO_2 capture and utilization using alkaline wastes via accelerated carbonation technology from theoretical and practical considerations, presented in the following 19 chapters. This book should be beneficial to readers who take scientific and practical interests in the current and future *accelerated carbonation technology* for CO_2 mineralization and utilization. Engineers, scientists, government officers, and project managers will find this book an essential reference on CO_2 mineralization and utilization.

In Part I, a broad review on challenges and opportunities for global warming issues is provided, including post-combustion carbon capture, storage and utilization (Chap. 2), CO_2 mineralization and utilization via accelerated carbonation (Chap. 3), and environmental impact assessment (EIA) and carbon capture and storage (CCS) guidance (Chap. 4).

In Part II, the integrated waste treatment via ex situ accelerated carbonation is systematically presented, in terms of theories and principles (Chap. 5), analytical methods for carbonation material (Chap. 6), mechanisms and modelling (Chap. 7), practices and applications (Chap. 8). Chapter 9 covers the system analysis methodology, including response surface methodology (RSM), life cycle assessment (LCA), cost–benefit analysis (CBA), and 3E (Engineering, Environmental, and Economic) triangle model.

In Part III, various types of feedstock for CO_2 mineralization are illustrated, including natural silicate and carbonate minerals (Chap. 10), iron and steel slags (Chap. 11), fly ash, bottom ash, and dust (Chap. 12) and paper industry, construction, and mining process wastes (Chap. 13).

In Part IV, the valorization of carbonization product as green materials is discussed, including utilization of carbonation product as green materials (Chap. 14), supplementary cementitious materials (SCMs) in cement mortar (Chap. 15), and aggregates and other high-value products (Chap. 16).

In Part V, the concepts of integral approach for waste treatment and resource recovery are illustrated. First, the carbon capture with flue gas purification (e.g., SO_x, NO_x, and particulate matter) via process integration and intensification is provided in Chap. 17. After that, the importance and significance of waste-to-resource (WTR) supply chain are discussed, in terms of barriers, challenges, strategies, and action plans (Chap. 18). Following that the principles of system optimization, such as (1) mathematical programming approach, (2) graphical presentation for optimization, and (3) comprehensive performance evaluation, are introduced to demonstrate the best available technology (Chap. 19). Moreover, several demonstration and action plans around the world are reviewed. Finally, the prospective and perspective on the strategies toward zero waste for sustainability are provided in Chap. 20.

References

1. NASA GISS (2016) Science briefs: greenhouse gases: refining the role of carbon dioxide. NASA GISS
2. Karl T, Trenberth K (2003) Modern global climate change. Science 302(5651):1719–1723
3. IPCC (2007) IPCC fourth assessment report (AR4). Climate change 2007: the physical science basis. Intergovernmental Panel on Climate Change, Cambridge
4. Toth FL et al (2001) Where should the response take place? The relationship between domestic mitigation and the use of international mechanisms. 10. Decision-making frameworks
5. Birat JP (2010) Global technology roadmap for CCS in industry: steel sectorial report, 5th edn. ArcelorMittal Global R and D, Maizières-lès-Metz, France
6. Chiu P-C, Ku Y (2012) Chemical looping process—a novel technology for inherent CO_2 capture. Aerosol Air Qual Res 12:1421–1432. doi:10.4209/aaqr.2012.08.0215
7. Yu C-H, Huang C-H, Tan C-S (2012) A review of CO_2 capture by absorption and adsorption. Aerosol Air Qual Res 12:745–769. doi:10.4209/aaqr.2012.05.0132
8. Pan S-Y, Chiang A, Chang E-E, Lin Y-P, Kim H, Chiang P-C (2015) An innovative approach to integrated carbon mineralization and waste utilization: a review. Aerosol Air Qual Res 15:1072–1091. doi:10.4209/aaqr.2014.10.02
9. UNFCCC (2015) Historic paris agreement on climate change—195 nations set path to keep temperature rise well below 2 degrees celsius. UN Climate Change Newsroom, 12 Dec 2015
10. Intended Nationally Determined Contribution of the EU and its Member States (2015). EU
11. Paltsev S, Reilly J, Sokolov A (2013) What GHG concentration targets are reachable in this century?. Massachusetts Institute of Technology, MA, USA
12. IPCC (2007) Climate change 2007: mitigation of climate change
13. Yang H, Xu Z, Fan M, Gupta R, Slimane RB, Bland AE, Wright I (2008) Progress in carbon dioxide separation and capture: a review. J Environ Sci 20(1):14–27. doi:10.1016/s1001-0742(08)60002-9
14. International Energy Agency (IEA) (2014) Energy technology perspectives 2014—harnessing electricity's potential. IEA, France
15. International Energy Agency (IEA) (2012) Energy technology perspectives 2012: pathway to a clean energy system. International Energy Agency, France

Part I
Global Warming Issues: Challenges and Opportunities

Chapter 2
Post-combustion Carbon Capture, Storage, and Utilization

Abstract Deployment of carbon capture, storage, and utilization technologies worldwide from the CO_2 emission point source is a strategy that has been proposed to address the challenge of climate change and global warming. As a viable option for reducing CO_2 emissions, moving carbon capture and storage technology to incorporate "utilization" (carbon capture and utilization) has received dramatically global attention. This chapter provides an overview of various types of carbon capture, storage, and utilization technologies. After that, one of the carbon capture and utilization technologies, i.e., microalgae pond systems, is illustrated in detail. The principles of microalgae open pond systems, key parameters affecting productivity, and economic considerations of operating open ponds are systematically illustrated.

2.1 Significance and Importance

Human activities resulted in emissions of four long-lived greenhouse gases (GHGs): CO_2, CH_4, N_2O, and halocarbons, of which CO_2 is the most important anthropogenic GHG due to its solitary responsibility for about two-thirds of the enhanced greenhouse effect [1]. Meanwhile, rapid economic growth in developing countries such as China and India is driving worldwide energy demand and usage day by day. At the same time, it has been predicted that fossil fuels will remain the dominant energy source around the world for at least another 20 years to fulfill such energy demands [2]. As CO_2 keeps accumulating in the atmosphere after generating from power plant, concerns about serious and irreversible damage, such as rising water level and species extinction, are being raised regarding its influence on climate change. Consequently, it is clear that effective control of CO_2 emissions is required to achieve the goal of global CO_2 concentration below 550 ppm over next the 100 years [3].

P.-C. Chiang and S.-Y. Pan, *Carbon Dioxide Mineralization and Utilization*, DOI 10.1007/978-981-10-3268-4_2

2.1.1 Strategies on Global CO_2 Mitigation

It is noted that the increased global average CO_2 concentration in the atmosphere is likely to cause further warming and induce many changes in the global climate system. To reduce CO_2 in the atmosphere, five strategies can be considerable [4]:

- Strategy 1: Reducing the amount of CO_2 producer sources (reducing energy intensity)
- Strategy 2: Using CO_2 (or reducing carbon intensity)
- Strategy 3: Capturing and storing of CO_2
- Strategy 4: Switching to less carbon-intensive fuels from conventional fuels
- Strategy 5: Increasing the use of renewable energies.

The third strategy involves the development of innovative, available, and cost-effective carbon capture and storage (CCS) technologies because of a 50-year estimate for the continued widespread burning of fossil fuels, the goal of reaching a 500 ppm atmospheric CO_2 concentration plateau, and the lag time needed for the development and implementation of new carbonless sources of energy [4, 5]. In general, the carbon capture technologies can be classified into three categories:

- Precombustion capture
- Post-combustion capture
- Oxy-fuel combustion

2.1.2 Transition from Storage to Utilization

As a viable option for reducing CO_2 emissions, moving CCS technology to incorporate "utilization" [i.e., carbon capture and utilization (CCU)] has received dramatically global attention. Europe (in particular Germany), the USA, and Australia are well advanced in research and development of carbon capture, utilization, and storage (CCUS) technologies [6, 7]. The utilization routes of the captured CO_2 include enhanced fuel recovery (i.e., enhanced oil recovery and enhanced gas recovery), biological conversion (i.e., algae), food industry, chemicals (i.e., fertilizer and liquid fuel), refrigerant, inerting agents, fire suppression, plastics, and even mineralization as carbonates (i.e., precipitated calcium carbonates (PCC) and construction materials). The benefits of CCU technologies include the following: [6, 8].

- Potentially reduce annual CO_2 emissions by at least 3.7 Gt, equivalent to about 10% of the world's current annual emissions
- Value-added products that create green jobs and economic benefits and help offset the abatement cost.

As a result, CCUS technologies are key strategies to attenuate the impacts of global warming during the transition period for developing sustainable energy technologies. CCUS is considered to be a critical strategy in the pathway to a sustainable energy system, contributing ~14% of reductions in global CO_2 emissions by 2050 [9, 10]. However, none of the CCUS technologies alone can provide a short—to medium-term solution to reduce CO_2 emissions at a level necessary to stabilize current concentrations. Rather, a portfolio solution must be identified to achieve the most effective CO_2 reduction while minimizing social and economic costs.

2.1.3 Concept of Carbon Capture, Utilization, and Storage (CCUS)

Figure 2.1 shows the major CCUS technologies, including CO_2 capture, storage (sequestration), utilization (direct use), and conversion into chemicals and/or fuels.

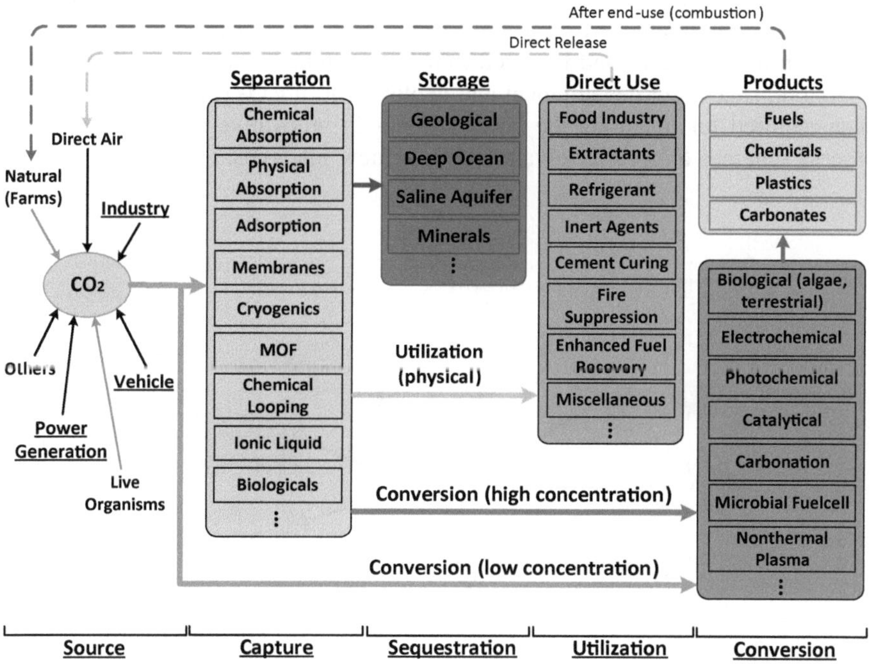

Fig. 2.1 Concepts of post-combustion carbon capture, utilization, and storage (CCUS) technologies

For example, the CCS technologies can effectively capture CO_2 from emission sources, transport it, and then store it at suitable and permanent geological sites.

2.2 Post-combustion Carbon Capture and Storage

As the first step for CO_2 capture, dilute CO_2 in flue gas from industries and/or conventional power plants should be separated and concentrated to a high purity in a cost-effective manner with low energy consumptions. After capture, the CO_2 can be stored into geological or saline formations to ensure long-term sequestration. Also, the concentrated CO_2 stream can be directly utilized or converted into carbon-based materials, such as fuels and chemicals.

2.2.1 Post-combustion CO_2 Capture Technologies

Figure 2.2 shows various approaches to post-combustion CO_2 capture from flue gas or air. Although various CO_2 capture technologies are available, only a few processes have been deployed on a large scale due to significant mass transfer limitations in the processes and the need to treat a significant amount of flue gas [11]. Therefore, successful development and deployment of CO_2 capture processes have been required for obtaining the breakthroughs in innovative reactor concepts and process schemes as well as advancement for new materials.

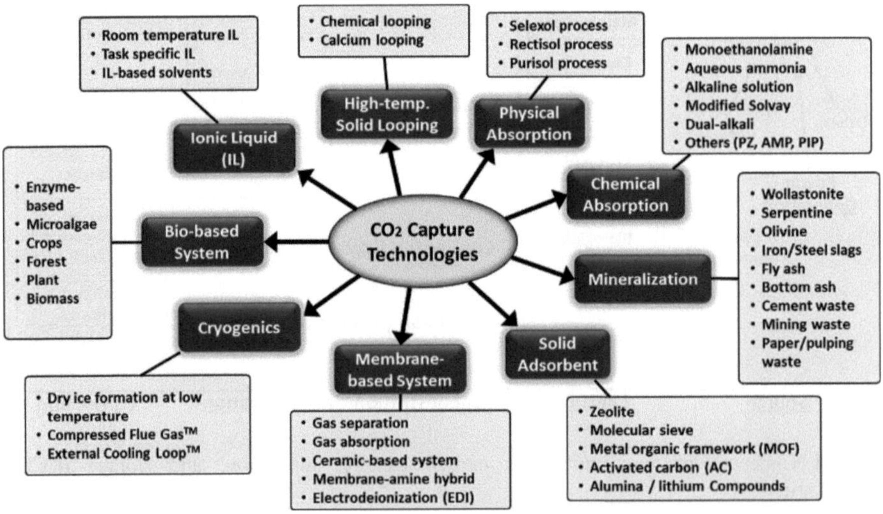

Fig. 2.2 Different approaches to CO_2 capture technologies

In order to achieve the above goals, the dilute CO_2 in flue gas (or air) can be concentrated via various technologies:

- Chemical absorption: alkaline solutions such as NaCl, ammonium solution, and monoethanolamine [11, 12]
- Physical adsorption: zeolite [13], activated carbon [14], and metal-organic frameworks [15]
- Selective membrane separation [16, 17]
- Cryogenic techniques [18]
- Ionic liquid absorption process [19]
- High-temperature solid looping processes: calcium looping [20] and chemical looping [21].

The aforementioned capture technologies can concentrate the dilute CO_2 in flue gas to nearly pure CO_2. After that, it should take the sequential storage or utilization into consideration. Other approaches of post-combustion capture are integrated with CO_2 utilization including the following:

- Mineral carbonation: natural ores and/or solid wastes [22]
- Biological method: microalgae and enzyme-based processes [23].

Some of the CO_2 capture technologies, such as biological method and mineral carbonation, are related to direct conversion and utilization of CO_2 because the physico-chemical property of CO_2 is changed after capture process. Therefore, no additional CO_2 storage site is required with the capture plant. These two approaches, i.e., biological method and mineral carbonation, are illustrated in detail in the following Sect. 2.4 and Chap. 3, respectively.

2.2.1.1 Absorption and Adsorption Processes

Table 2.1 presents the comparison of the post-combustion CO_2 capture technologies by absorption process, such as using aqueous absorbents and ionic liquid (IL). Meanwhile, chemical absorption via aqueous alkanolamine solutions is regarded as the most applicable technology for CO_2 capture by 2030 [24]. It can be accomplished in two stages: (1) CO_2 absorption using an absorbent or solvent, and followed by (2) desorption using pressure, temperature, or electric swing. However, several technological issues, including equipment corrosion, energy consumption in regeneration, and absorber volume, should be critically assessed in using alkanolamine aqueous solutions as absorbents. Thus, a modification and intensification of the absorption process should be considered to enhance the mass transfer between CO_2 gas and solution, for example, a high-gravity rotating packed-bed reactor [25, 26]. In addition, appropriate absorbent genomes, such as using piperazine with diethylenetriamine [27], piperazine with diethylene glycol [28], and NaOH solution [12], are needed to achieve high CO_2 capture efficiency and low regeneration energy.

Table 2.1 Merits and demerits of various physical and chemical absorption processes for post-combustion CO_2 capture

Process description/chemical components		Advantages	Disadvantages
Physical	• Selexol process • Rectisol process • Purisol process	• Low vapor pressure and toxicity (Selexol) • Low corrosion (Rectisol) • Low energy consumption (Purisol)	• Low absorption capacity • Limited refractory life (Selexol) • High capital and operating costs (Rectisol)
Chemical	• Alkanolamine solution (MEA, DEA, and MDEA) • Sterically hindered amine (AMP) • Promoter (PZ, PIP)	• High absorption capacity • Low operating pressure and temperature • Suitable for retrofitting of the existing power plant	• Severe equipment corrosion rate • High energy consumption in regeneration • Large absorber volume required • Amine degradation by SO_2, NO_2, and O_2
	• Ionic liquid (IL)	• Low vapor pressure • Non-toxicity • Good thermal stability • High polarity	• High viscosity • High energy requirement for regeneration • High unit costs

Table 2.2 presents the post-combustion CO_2 capture technologies by adsorption process using solid adsorbent and metal-organic frameworks (MOFs). The adsorption processes exhibit a lower CO_2 adsorption capacity, compared to chemical absorption processes.

2.2.1.2 Chemical Looping Process

Chemical looping process (CLP) is an advanced combustion process, where CO_2 is inherently separated from the other flue gas components. Oxygen-carrier materials, such as Fe-, Cu-, Ni-, Mn-, and Co-based metal oxides, are frequently used in the transfer of oxygen from combustion air to the fuel. In this case, direct contact between fuel and air can be avoided, thereby resulting in near 100% CO_2 in flue gas. Figure 2.3 shows the schematic diagram of CLP for power generation. Oxygen-carrier material (Me_xO_y) is reduced by carbonaceous fuel, such as coal and CH_4, to generate H_2O and CO_2 in the fuel reactor, as described in Eq. (2.1).

$$C_nH_{2m} + (2n+m)Me_xO_y \rightarrow nCO_2 + mH_2O + (2n+m)Me_xO_{y-1} \qquad (2.1)$$

Table 2.2 Post-combustion CO_2 capture processes via physical and chemical adsorption

Process description		Advantages	Disadvantages
Physical	• Activated carbon (AC) • Zeolite • Mesoporous silica (MS) • Metal-organic frameworks (MOFs)	• Wide availability and low cost • High thermal stability (AC) • Low sensitivity to moisture (AC) • High pore size and tunable pore size (MS and MOFs)	• Low CO_2 adsorption capacity • Low CO_2 selectivity • Slow adsorption kinetics • Thermal, chemical, and mechanical stability in cycling
Chemical	• Amine-based adsorbent • Alkali earth metal adsorbent • Lithium-based adsorbent	• Exothermic reaction • High adsorption capacity • Low cost in natural minerals	• Deactivation of synthesis adsorbent • Low CO_2 selectivity • Serious diffusion resistance
	• Alkaline solid waste (steelmaking slag, ashes, etc)	• Thermodynamically stable product • High availability of wastes • Reuse product in a variety of application • Decreased leaching of heavy metal trace elements from the wastes	• Low CO_2 adsorption capacity • Slow adsorption kinetics and mass transfer • High energy consumption in crushing

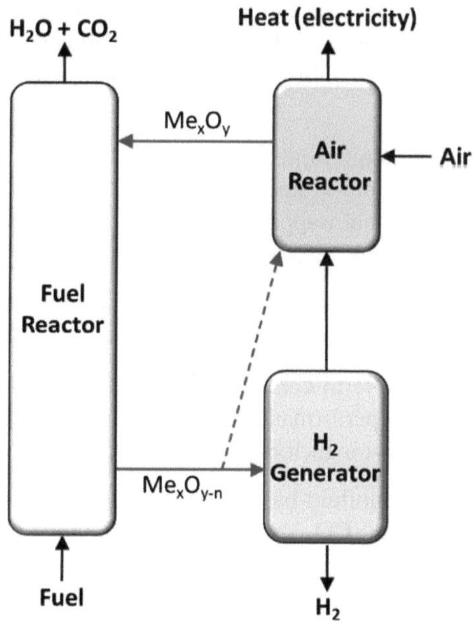

Fig. 2.3 Schematic diagram of chemical looping process (CLP) for power generation with high-purity CO_2 production

The reduced metal oxide (Me_xO_{y-n}) is then oxidized to Me_xO_y by oxygen in the air reactor, as expressed in Eq. (2.2), which is an exothermic reaction that provides heat to external facilities, such as electricity generator.

$$2Me_xO_{y-1} + O_2 \rightarrow 2Me_xO_y \tag{2.2}$$

In order to optimize CO_2 production, several challenges in commercialization of CLP still need to be addressed: (1) development of a low-cost oxygen carrier with high reactivity and recyclability; and (2) optimization of reactor design and operating condition. The reactivity of Ni- and Cu-based oxygen carriers is great for chemical looping process; however, their development is still limited due to their high fabrication cost [21]. Regarding the scale of CO_2 capture capacity, CLP has been successfully demonstrated in actual operation in a size of 0.3–1 MW and should be ready to further scale up to the size range of 1–10 MW [29].

2.2.2 Carbon Storage Technologies

The concentrated CO_2 can be pressurized and stored in geological formations underground, so-called geological CO_2 storage, such as

- Deep ocean [30]
- Saline aquifers [31, 32]
- Unminable coal beds [33]
- Depleted oil/gas reservoirs [34]

With widespread geographic distribution, CO_2 (high purity after capture process) injection into confined geological formations can offer a potentially large storage capacity [35]. In general, four major mechanisms for geological CO_2 storage are as follows:

- Dynamic fluid trapping
- Dissolution trapping
- Residual trapping
- Mineral trapping

From the technological point of view, the risks of geological CO_2 storage may include the following [36]:

- Long-term containment risks: potential leakage and induced seismicity
- Site performance risks: improper analysis of wellbore, near-wellbore, and reservoir factors

A nanofluid-based supercritical CO_2 technique could be effectively used for geological CO_2 storage since nanofluids can not only enhance homogeneous CO_2 transport in reservoir but also mitigate the adverse effects of stratigraphic heterogeneity on migration and accumulation of CO_2 plume [32]. Regarding the reaction mechanism, extensive studies have been carried out to evaluate changes in host rock properties when exposed to CO_2 [37]. A large-scale commercial storage should be conducted with potential exploration of geothermal resource in deep-seated hot rocks. Successful CO_2 sequestration in geological formations, on the one hand, requires a cross-disciplinary effort in geochemical, mineralogical, and petrophysical properties of reservoir and seal materials from the microscopic to the macroscopic scales. On the other hand, prediction and regular monitoring of storage capacity, CO_2 migration and phase state, and long-term reservoir behavior are crucial to ensure the safety, security and sustainability of CO_2 sequestration.

2.2.3 *Large-Scale Demonstration Plans*

The "utilization" of CO_2 through physical, chemical, or enhanced biological methods should result in effective reductions in CO_2 emissions toward a sustainable carbon cycle. With the development of innovative utilizationprocesses, direct and indirect environmental benefits are expected regarding (1) the reduction of both CO_2 emissions and fossil fuel extractions, and (2) the substitution of chemicals [2]. However, the widespread deployment of CO_2 capture, storage, and utilization (CCSU) projects has not been achieved due to several barriers including the following:

- High capital investment
- Uncertainties in policies, regulations, and technical performance
- Public acceptance
- Concerns about human health and safety
- Environmental risks.

Many countries, such as Norway, Canada, the USA, Germany, and Australia, are advanced in research and development of CCSU technologies [7]. In this section, two large-scale demonstrations in Norway (i.e., CO_2 capture by amine absorption process) and in Canada (i.e., CO_2/SO_2 capture and H_2SO_4 production) are provided and illustrated.

2.2.3.1 Carbon Capture by Amine in Norway (2012)

In Norway, the carbon tax mechanism has been implemented since 1991, thereby leading to the successful development of the first large-scale carbon capture project in the world. As shown in Fig. 2.4, it not only takes CO_2 from a power station but also extracts CO_2 from the natural gas coming from the offshore Sleipner field, owned by Statoil. The efficiency of a modern gas power plant in Norway is typically 59%, such as the one on the southwest coast at Karsto. However, the efficiency would drop to around 50%, if the CO_2 capture unit (including transportation of the CO_2) is installed. Usually, it requires energy to drive the CO_2 capture and

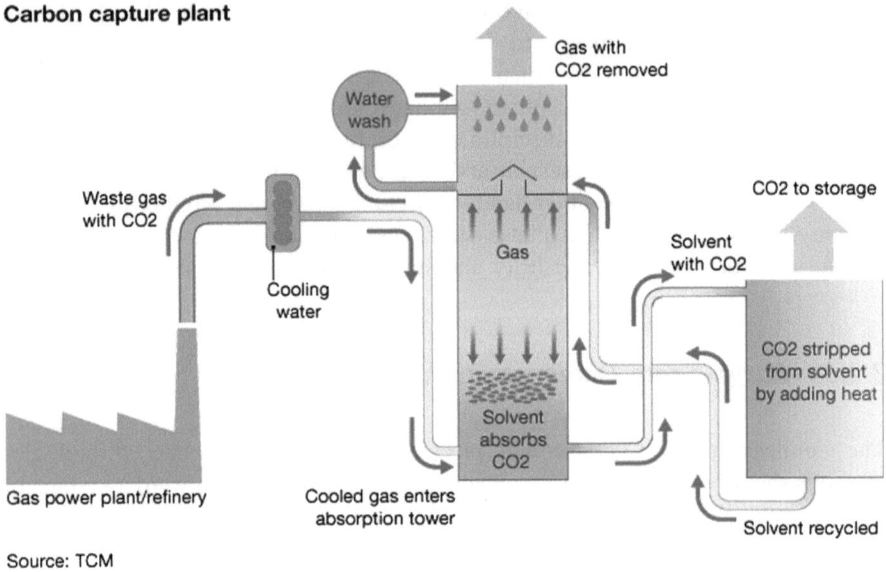

Fig. 2.4 Schematic diagram of the carbon capture plant by amine in Norway. Reproduced from Ref. [38] by permission of Technology Centre Mongstad (TCM)

storage processes. Therefore, the power station must burn more fuel to produce the same amount of electricity.

In Dong Energy's power station, the CO_2 concentration in the flue gas is about 3%, while the pipe from the cracker at Statoil's Mongstad refinery comes in at around 13%. After absorption process, approximately 90% of CO_2 in the flue gas can be captured. The liquid surface in the absorption tower is estimated to be the same as eight soccer fields.

2.2.3.2 Sask Power Plant in Canada (2015): CO_2/SO_2 Capture and H_2SO_4 Production

At Canada, the Boundary Dam power station (Unit 3), operated by the SaskPower team, has integrated with a carbon capture plant. The project transformed the aging Unit #3 at Boundary Dam power tation near Estevan, Saskatchewan, into a reliable, long-term producer of up to 115 megawatts (MW) of base load electricity. The captured CO_2 could be compressed and permanently stored underground ($\sim$3.4 km deep), or be liquidized and transported by pipeline to nearby oil company for further utilization, e.g., enhanced oil recovery.

In addition to CO_2, other by-products could be produced from the project. For instance, SO_2 can be captured and converted to sulfuric acid (H_2SO_4) for industrial use. Fly ash, another by-product of coal combustion, could also be sold for use in ready-mix concrete, precast structures, and concrete products. The designed CO_2

and SO_2 emission reduction are 90 and 100%, respectively. In this case, one million tonnes of CO_2 could be captured each year, equivalent of taking more than 250,000 cars off Saskatchewan roads annually.

2.3 Carbon Utilization and Valorization

The most valuable perspective in CO_2 utilization and valorization is not the amount of CO_2 used, since the fixed CO_2 would be reformed within a short time after a CO_2-made chemical is used. Rather, it is the introduction of innovative technologies for cleaner production that could reduce the use of materials and energy. Carbon utilization and valorisation can achieve a real carbon cycle toward environmental sustainability. After CO_2 capture process, the concentrated CO_2 can be either utilized directly, or converted and transformed into other products, as shown in Fig. 2.5.

In this section, two approaches to utilizing concentrated CO_2, i.e., (1) direct utilization and (2) conversion and transformation, are briefly illustrated.

2.3.1 *Direct Utilization of Concentrated CO_2*

Direct use of CO_2 involves phase changes (i.e., states of matter) between gas, liquid, solid, and supercritical fluid. In this case, the concentrated CO_2 can be directly applied in various fields such as [39].

- Food industry
- Soft drinks
- Fire extinguishers
- Extractants
- Solvents for reactions, separation, synthesis, and modification of material: supercritical CO_2.

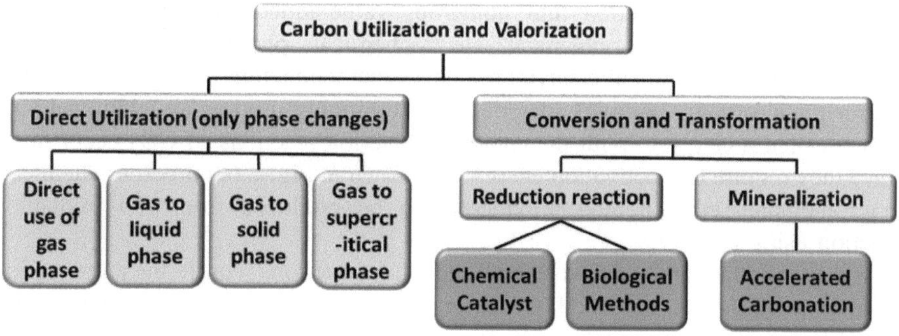

Fig. 2.5 Various approaches to CO_2 utilization and valorization toward sustainable carbon cycle

The economics of the CO_2 utilization technology reply on the quality of CO_2 output (e.g., purity, temperature, and pressure) and the capture processes involved. Currently, the market scales of such applications are still small, which cannot contribute to a huge impact on the overall CO_2 emission mitigation. Moreover, there is a large difference of predicted CO_2 price between the studies. The most acceptable estimate is that the CO_2 price in 2050 might be in the range of 100–400 US \$ per ton CO_2 [40].

2.3.2 *CO_2 Conversion and Transformation*

Conversion of the captured CO_2 to useful products is an essential strategy toward a sustainable carbon cycle. As CO_2 is a thermodynamically stable compound, conversion of CO_2 typically goes through a catalytic process with additional work input, e.g., renewable energy source. Among the carbon-species compounds, the CO_2 molecule exhibits the highest oxidation state (4+). For this reason, CO_2 conversion and transformation can be realized by.

- Reduction reaction [41]: to a negative-going oxidation state.
- Mineralization [42]: to a lower Gibbs free energy, compared to gaseous CO_2.

Extensive efforts have been underway to increase the CO_2 conversion efficiency under various novel processes. The CO_2 reduction reaction can be achieved by several different approaches, such as electrocatalytic [43, 44], photocatalytic [41], and biological [23] methods. It can not only reduce CO_2 accumulations in the atmosphere but also use CO_2 for furnishing chemicals and energy products. In the following content, the reduction reaction of CO_2 via either chemical catalyst or biological method is provided. The mineralization of CO_2 will be discussed in detail in Chap. 3.

2.3.2.1 Chemical and Catalytic Reaction

The CO_2 can be converted into numerous value-added chemicals via several methods, such as thermochemical, electrochemical, photoelectrochemical, and photocatalytic. In chemical catalytic systems, semiconducting (e.g., TiO_2 and CdS) and/or metal-organic complex materials are commonly used as an additional substance called a catalyst. With the catalyst, reduction reactions of CO_2 occur faster since they require less activation energy. As shown in Fig. 2.6, numerous chemicals (e.g., methane, ethanol, and polymers) can be produced from CO_2 reduction reaction using catalyst.

It is noted that some of the catalytic reduction reactions require the participation of hydrogen (H_2) or proton (H^+). For this reason, the H_2 (or proton) can be produced from water using exclusively renewable energy, which has been considered

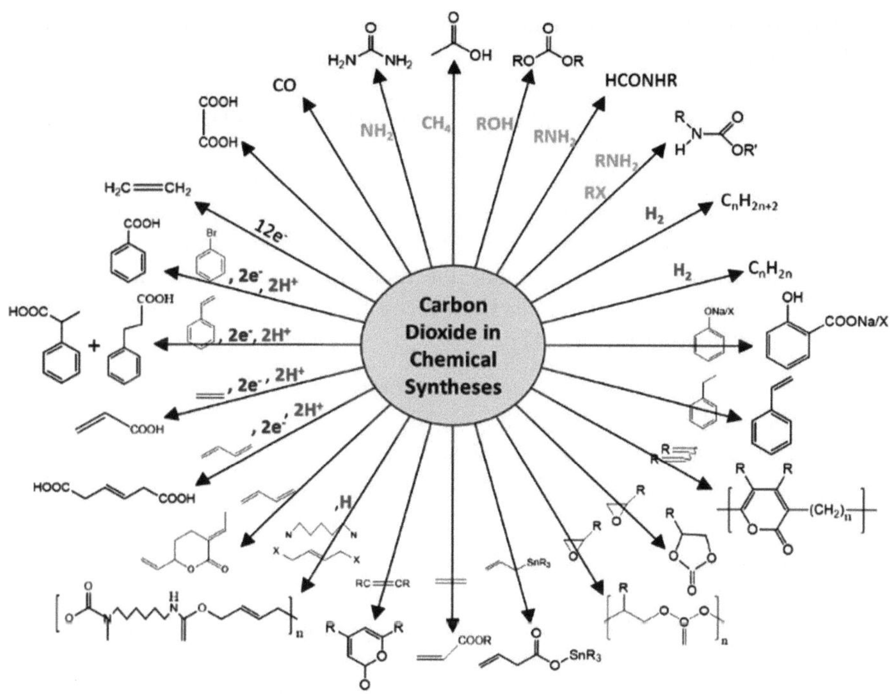

Fig. 2.6 Plentifully potential uses of CO_2 as chemicals through various conversion technologies

as an alternative to store intermittent energy. In electrochemical process, the protons and electrons could be generated from water in an anode compartment using electricity from renewable energy [45], as shown in Eq. (2.3):

$$H_2O + \text{energy} \rightarrow 2H^+ + 2e^- + 1/2O_2 \uparrow \quad (2.3)$$

The concept of using CO_2 and H_2 to produce chemical fuels, such as methane, methanol, and dimethyl ether (DME), is so-called the production of sustainable organic fuel for transport (SOFT). The CO_2 and H_2 can be converted into liquid hydrocarbon through a combination of the reverse water–gas shift (RWGS) reaction and Fischer–Tropsch (FT) process, as shown in Eqs. (2.4) and (2.5), respectively.

$$CO_2 + H_2 \rightarrow CO + H_2O \quad \Delta H^0_{298} = +42\,\text{kJ/mol} \quad (2.4)$$

$$(2n+1)\,H_2 + n\,CO \rightarrow C_nH_{2n+2} + n\,H_2O \quad (2.5)$$

where n is typically ranged from 10 to 20. The FT synthesis involves a series of chemical reaction that produce a variety of hydrocarbons (C_nH_{2n+2}).

On the other hand, the CO_2 and H_2 can be converted into, for instance, methanol via Eq. (2.6):

$$CO_2 + 3H_2 \rightarrow CH_3OH + H_2O \quad \Delta H^0_{298} = -49\,\text{kJ/mol} \tag{2.6}$$

The key to successful SOFT production replies on three elements: (1) a cheap source of CO_2, which depends on the types of CO_2 capture process; (2) a cheap source of H_2, which may be combined with renewable energy utilization; and (3) an effective and robust catalyst to initiate the reduction reaction. Regarding the catalytic systems, they typically suffer from low efficiency of CO_2 reduction reaction due to several limiting factors [41]:

- Fast electron–hole (e^--h^+) recombination rates
- Complicated backward reactions
- Low CO_2 affinity of the photocatalyst

2.3.2.2 Enhanced Biological Fixation

Another approach to CO_2 reduction reaction is via the biological methods using fast-growing biomass, called enhanced biological fixation. It corresponds to the production of aquatic and/or terrestrial biomass under non-natural photosynthetic conditions. For instance, direct use of CO_2 by "microalgae" has attracted great interest since microalgae can not only consume CO_2 but also be converted to biochemicals or biofuels. Microalgal cells are sunlight-driven cell factories that can convert CO_2 into raw materials for producing biofuels (or solar fuels), animal food chemicals, and high-value bioactive compounds such as docosahexaenoic acid [46]. The reaction can be achieved by two separate mechanisms:

- Light-dependent set: Solar light (uv) energy is used to excite electrons for reducing the coenzyme $NADP^+$ to NADPH and creating the high-energy molecule ATP.
- Light-independent set: The reduced molecules are utilized to convert CO_2 to organic compounds that can be used as a source of energy by algae.

Due to its carbon neutral property, the production of bio-based chemicals in biorefineries using biomass is a key opportunity for sustainable green growth. Based on a cradle-to-grave life cycle assessment [47], bioproducts could provide reductions of 39–86% in greenhouse gas emissions, compared to their fossil counterparts. Furthermore, the reaction products, such as biochemicals, may offer a high added value, providing an opportunity to cover the costs of biomass production (e.g., algal). Biotechnologies are expected to contribute to 2.7% of GDP in 2030 within the OECD region and make the largest economic contribution in industry and primary production [48]. As a result, the biological process for CO_2 utilization and transformation should be feasible from an economic standpoint and acceptable from an energy perspective.

2.4 Case Study: Microalgae Pond for CO_2 Capture and Utilization

Algal technology offers great potential to combat the global energy and CO_2 crisis and malnutrition, while also generating high value-added products. Microalgae have been recognized as an alternative feedstock to energy conversion and/or chemicals production. They are fast-growing and ubiquitous photosynthetic organisms, which are rich in protein and chemical compounds, and can be converted to biodiesel fuel using a variety of different methods. Microalgae have been recognized as a third-generation source of biofuels not only because they use CO_2 from the atmosphere, as indicated in Eq. (2.7), but also due to their high lipid content per biomass compared to other plants [49].

$$6CO_2 + 12H_2O \overset{light}{\rightarrow} C_6H_{12}O_6 + 6\ O_6 \tag{2.7}$$

Microalgae are superior to terrestrial energy crops, in terms of biomass and biodiesel yield [50]:

- Microalgae in photobioreactors: ~150 Mt/ (ha-year)
- Microalgae in open ponds: 50–70 Mt/ (ha-year)
- Switch grass (terrestrial): ~13 Mt/ (ha-year)
- Corn (terrestrial): ~9 Mt/ (ha-year)
- Soybeans (terrestrial): ~3 Mt/ (ha-year).

Microalgae can be used as a feedstock for manufacturing a number of products:

- Bioenergy: biodiesel, biogas, bioethanol, etc.
- Non-energy bioproducts: carbohydrates, pigments, protein, biomaterials, etc.
- Animal feed.

As a result, microalgae could provide huge market potential for biodiesel as well as other valuable biochemicals. However, since the productivity (or yields) of the bioproducts has certain stoichiometric and thermodynamic constrains (fundamental principles of biochemistry), the maximum theoretical energy conversion of the full sunlight spectrum into organic matter is about 10% [49]. Aside from stoichiometric and thermodynamic constrains, algal cultivation is considered as one of the technological barriers. It is noted that algal cultivation accounts for one-third of the total cost involved in the algal biofuel production process [51].

In the following section, the types of open pond systems and the key parameters affecting productivity for the microalgae cultivation as well as the process economics are discussed.

2.4.1 *Types of Open Pond Systems*

For algal cultivation, the open ponds are easier to construct and operate than most closed systems, thereby resulting in low production costs and low operating costs. Four major open pond systems can be used for algal cultivation:

- Shallow big ponds
- Tanks
- Circular ponds
- Raceway ponds.

Since each of them exhibits its own characteristic features, the selection of an appropriate open ponds for cultivation relies on the types of algal species, location of plant installation, local climatic conditions, and cost of water and lands [52]. Figure 2.7 shows the schematic diagram of material flows in an open pond system. Microalgae require the same basic element inputs as plants, including light, CO_2, water, and inorganic nutrients (such as P compounds). As a result, environmental factors such as light intensity, temperature, pCO_2, PO_2, pH, and salinity play a crucial role in microalgae productivity. During algal growth, the chlorophyll-α content, algae cell density, DO, and pH of the solution can grow rapidly.

In general, open ponds can offer greater CO_2 fixation capacity than tubular reactors due to their greater culture volume per area [53]:

- Open ponds: 100–300 L-CO_2/m^2
- Tubular reactors (1–5-cm tubes): 8–40 L-CO_2/m^2.

Production of microalgae in raceway ponds is considered the most promising technology for algal cultivation, especially in the large scale [51]. Typically, the surface-to-volume ratio of raceway ponds (i.e., the depth of pond) is approximately 5–10 m^{-1} [54]. The shallow configuration is required for raceway ponds to prevent

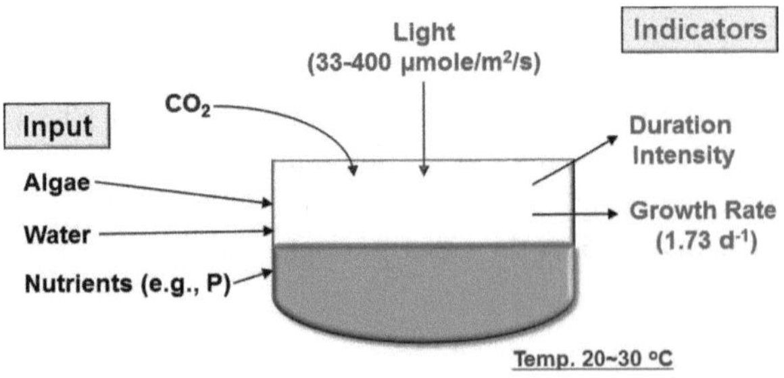

Fig. 2.7 Schematic diagram of material flows in an open pond system

light limitation inside the culture, thereby resulting in a linear land footprint when it comes to scale-up. In addition, typical process hurdles for raceway ponds still may include the following: [55].

- High risk of culture contamination from strains of bacteria or other outside organisms
- Lack of temperature control (or environment condition control due to weather)
- Inefficient stirring mechanisms
- Poor gas–liquid mass transfer
- Low final biomass concentrations incurring high harvesting costs

2.4.2 *Key Parameters Affecting Productivity*

Numerous external and internal factors will significantly affect algal growth and productivity in the open pond systems, including the following:

- Environmental factors: location of the cultivation system, rainfall, solar radiation, etc.
- Engineering factors: pond depth, CO_2 delivery system, methods of mixing, power consumption, etc.
- Biological factors: light, pH, oxygen accumulation, salinity, algal predators, etc.

Table 2.3 presents the aforementioned factors affecting algal growth, biomass accumulation, and production for most algae in open ponds. Normally, the most challenging and important factors discussed in photobioreactor design and operation are sufficient mixing (stirring), rapid carbon utilization, and the accumulation of photosynthetically produced oxygen [53]. Therefore, controlling pH, conductivity, and O_2 concentration of the culture is important to obtain a high biomass concentration and productivity [56]. It was noted that an excess of both pH and dissolved O_2 in ponds would significantly reduce biomass concentration and productivity, especially in mid summer, while no inhibition of growth by excess irradiance and temperature is expected [56].

In the following section, several important factors, including (1) light (radiation), (2) temperature, (3) mixing, (4) CO_2 delivery and cultural pH control, (5) accumulation of dissolved oxygen, and (6) salinity, are discussed.

2.4.2.1 Light (Radiation)

Light (or photolight uv) is essential for algal growth since the growth of all algal species is highly dependent on solar radiation of the pond or water body. It suggests that maximum algal growth rate should be obtained at the light saturation (irradiation) point. Beyond the light saturation point, algal growth would be inhibited because of the photoinhibition mechanism. The photoinhibition is a series of

Table 2.3 Significant factors affecting algal growth, biomass accumulation, and production for most of algae in open pond system. Reprinted by permission from Taiwan Association for Aerosol Research: Ref. [80], copyright 2016

Categories	Factors	Influence (features)	Rule of thumb	Reference
Environmental (location)	Irradiance (light intensity)	Algae growth	Minimum of 4.65 kWh m^{-2} d^{-1}	[57]
	Rainfall	Chance of culture dilution	Not more than 1 m per year	[58]
	Land slope	Earth-moving costs during pond construction	Not more than 5%	[59]
	Contiguous area	Ensure a commercial scale (i.e., 3.8 dam^3 y^{-1} of oil feedstock)	Minimum 3.2–10.5 km^2	[58]
Engineering	Pond depth	Temperature, light utilization efficiency, mixing, and power consumption of mixing, etc.	Typical 15–20 cm (raceway)	[23]
			Typical 20–30 cm	[60]
			Productivity increased 134–200% in depth of 40 cm (compared to 20 cm)	[61]
	Mixing	Light utilization efficiency, oxygen removal rate, etc.	Kept between 5 and 30 cm s^{-1}	[51]
	CO_2 delivery system	Affected by diffusor	At least 65 μmol L^{-1}	[53]
	Power consumption	Affected by water head and types of paddle wheel	Typical 1.5–8.4 W m^{-3}	[62]
Biological	Light (wavelength, photoperiod)	Autotrophic growth and photosynthetic activity	400–500 nm (for chlorophyll a, chlorophyll b, and carotenoids)	[63]
	Biomass concentration (areal density)	Pigmentation to block light	Up to 0.66 g L^{-1}	[64]
	pH	CO_2 concentration in the culture	Optimal in the range of 7–8, no more than 8.5	[53, 65]
	Oxygen accumulation	Results in photooxidation and respiration of cell	Not more than 25 mg L^{-1}	[56]
	Salinity	Growth and cell composition of microalgae	Range between 22 and 28 mS cm^{-1}	[56]
	SO_2	May not directly inhibit growth but increase acidity	Must below 100 mg/L	[23]
	C:N ratio	Composition in algal cells	Range from 6 to 8	[66]
	Temperature	Cellular chemical composition, uptake of nutrients and CO_2, growth rates for every species of algae	Minimal temperature around 18 °C	[67]
			Different optimal temperatures ranging from 24 to 42 °C	[68]
	Nutrient/media (N, P, S)	Growth and composition of benthic algae. Formation of microalgae ($CH_{1.7}O_{0.4}N_{0.15}P_{0.0094}$)	N: 4–8% per dry weight basis algae P: 0.1% per dry weight basis algae S: 0.5% per dry weight basis algae	[60]

complex reactions that inhibit different activities of photosystem II (or water–plastoquinone oxidoreductase). So far, these is no consensus on the real mechanisms of photoinhibition. The photoinhibition can occur at all light intensities, and the rate constant of photoinhibition is directly proportional to light intensity. From the aspect of engineering design, the depth of open ponds is directly dependent on irradiance spectra. The appropriate depth of open pond system can be determined by measuring the irradiance attenuation coefficient as a function of wavelength for each strain [63].

2.4.2.2 Temperature

Temperature maintenance in raceway ponds is challenging and important work. Productivity of microalgae increases as the pond temperature increases up to an optimum temperature. Above the optimum temperature, algal respiration and photorespiration increase, thereby reducing overall productivity. It suggests that the optimum ranges of temperature and light irradiance for growth are 20–30 °C and 33–400 $\mu mol\ m^{-2}\ s^{-1}$, respectively, for algae species such as [69].

- Most of green algae
  - *Botryococcus*
  - *Chlorella*
  - *Chlamydomonas*
  - *Haematococcus*
  - *Neochloris*
  - *Nannochloropsis*
  - *Scenedesmus*
  - *Spirogyra*
  - *Ulva*
- A few species of brown algae
- A few species of red algae
- A few species of blue-green algae

Similarly, the Algae Raceway Integrated Design (ARID) also suggests that diurnal and seasonal temperature fluctuations should be minimized. The temperature should maintain within the optimal range of 15–30 °C.

2.4.2.3 Mixing

Mixing, especially vertical mixing, is considered the most significant factor affecting the performance of raceway ponds since it is largely related to light utilization efficiency. The importance and significance of sufficient mixing in raceway ponds includes the following [51, 54, 70]:

- Periodic exposure of cells to sunlight
- Keeping cells in suspension
- Availability of the nutrient to algal cells
- Removal of photosynthetically generated dissolved oxygen (DO) to avoid photooxidation and photoinhibition by respiration, thereby enhancing light utilization efficiency

Typically, mixing accounts for ~69% of total utility costs [71]. Therefore, the mixing intensity should be carefully adjusted according to the environmental conditions. For instance, the mixing velocity should be reduced during the night and even in winter time to avoid excess biomass loss in the absence of light.

2.4.2.4 CO_2 Delivery and Effect of Culture PH Value Control

Microalgae can be successfully grown in various types of liquid phases, such as (piggy) wastewater and deep-sea water. Typical, the carbon-to-nitrogen (C:N) ratio in the algal cells ranges from 6 to 8 [66]. When wastewater is used for algae cultivation, additional CO_2 delivery is imperative for maintaining a sufficient carbon source because the C:N ratio of wastewater is generally 3. In high-rate ponds, a supply of concentrated CO_2 greater than 5% is needed to sustain algal growth [72]. CO_2 delivery largely depends on culture pH and the factors affecting mass transfer rates, such as type of sparger, mixing intensity, liquid velocity, and gas–liquid contact time. The rate of CO_2 absorption in the solution increases with an increase in the pH of the solution. For most algae species, the optimal pH ranges from 7 to 8 to maintain sufficient bicarbonate species in the solution [65].

In open ponds, the dissolved CO_2 after recarbonation tends to be desorbed to the atmosphere. It was estimated that the mass transfer coefficient (K_L) for CO_2 release through the surface of a 100-m^2 pond should be about 0.10 m/h [53]. For the actively growing algal cells, a sufficient CO_2 mass transfer rate should be at a pH value of 8 [73]. In some cases, however, maintaining a pH in the range of 9.5–10.5 (such as *Spirulina* sp.) is necessary to minimize the chances of contamination with other microalgae [56, 66]. As a result, open pond systems should be operated at higher pH and lower alkalinity than tubular reactors.

2.4.2.5 Accumulation of Dissolved Oxygen (DO)

Oxygen (O_2) is a by-product generated from the algal culture in the course of photosynthesis. Stoichiometric analysis revealed that 1.9 g of O_2 would be generated per gram of the algal biomass synthesis [66]. However, a high concentration of dissolved oxygen (DO) during algal cultivation would severely damage the algal cells by photooxidation and inhibit photosynthesis via respiration. The accumulation of DO in the culture thus will lead to a reduction in biomass productivity. It is necessary to maintain sufficient mass transfer by sparging, even when no carbon is required. The energy required for increasing mass transfer rate and reducing DO concentrations could be compensated by the increased biomass and its associated potential energy yields.

Removal of accumulated DO from microalgae system is a more critical design criterion than other unit processes, such as carbon supply. It is noted that an increase in DO concentration greater than 25 mg/L might exhibit a negative impact on the biomass productivity [56]. At the maximal rate of photosynthesis, a 1-cm tubular reactor would accumulate 8–10 mg O_2/L/min, which may easily result in the O_2 concentrations over 100 mg/L [53]. In contrast, the DO in open ponds rises much more slowly as a consequence of the much greater volume per unit surface area, where the maximum DO concentration is typically 25–40 mg/L [53, 66]. It was estimated that the maximum DO levels for an open pond with a 100-m^2 surface and a 20-cm depth should be 14.5 mg/L at a mixing velocity of 30 cm/s, or 19.0 mg/L at a mixing velocity of 3.7 cm/s [53].

2.4.2.6 Salinity

In water chemistry, salinity is defined as the saltiness or dissolved salt content of a water body. It is a thermodynamic state variable (along with temperature and pressure), which governs physical characteristics of a water body such as the density and heat capacity. The salinity of the solution would affect the growth and cell composition of microalgae via several mechanisms, including the following: [74].

- Osmotic changes
- Ion (salt) stress
- Changes of the cellular ionic ratios (membrane selective ion permeability)

In open ponds, fluctuation in the salinity of the culture due to evaporation, rain, and precipitation is a common issue, especially with brackish or saline water [51]. The simplest way to solve the fluctuation issue is to add extra freshwater or salt as necessary. Daily refilling with freshwater in open ponds could maintain the conductivity of the culture ranging 22–28 mS cm^{-1} [56].

Aside from consecutive addition of freshwater, appropriate water treatment and/or separation processes for the culture could be employed to maintain the salinity of the culture. Available treatment technologies for the culture desalting include membrane separation [75]. However, the main challenge in membrane process is the precipitation of calcium salts, especially in calcium-laden water, thereby resulting in loss of alkalinity and other minerals (such as phosphorus and iron) [76].

2.4.3 *Economic Considerations*

For microalgae technology, the optimal control of pH, conductivity, and O_2 concentration in the culture is extremely important to obtain a high biomass concentration and productivity. A comprehensive performance evaluation (CPE) among the unit processes is critical to achieve high engineering performance while maintaining low cost and environmental impacts. The most challenging work in assessing the economics of microalgae process are (1) the cost of the CO_2 supply and (2) the uncertain nature of downstream processing [77].

The high cost of CO_2 capture and transportation is the major obstacle for algal biomass production. Therefore, converting CO_2 into a bicarbonate/carbonate aqueous solution is preferred since it can be easily transported in a water pipeline under normal pressure [78]. To achieve the above goal, efficient conversion of CO_2 from a flue gas into aqueous bicarbonate/carbonate solution should be developed, for example,

- Carbonate–bicarbonate buffer [65]
- CO_2 hydrate [79]
- Electrochemical membrane process [75]
- High-gravity rotating packed-bed reactor

On the other hand, regarding the downstream processing, biofuels (e.g., algae-to-fuel technology) should be produced simultaneously with value-added coproducts from the perspective of green economy and green design [49]. Furthermore, integration of an energy-efficient ex situ water treatment process for the culture solution with existing open ponds might be an alternative. In this case, provision of sufficient CO_2 concentration and removal of excess O_2, (in-)organic acid, and salinity from the culture could be simultaneously achieved.

References

1. Nyambura MG, Mugera GW, Felicia PL, Gathura NP (2011) Carbonation of brine impacted fractionated coal fly ash: implications for CO_2 sequestration. J Environ Manage 92(3):655–664. doi:10.1016/j.jenvman.2010.10.008
2. Aresta M (2010) Carbon dioxide: utilization options to reduce its accumulation in the atmosphere. In: Aresta M (ed) Carbon dioxide as chemical feedstock. Wiley-VCH, p 13
3. Fernandez Bertos M, Simons SJ, Hills CD, Carey PJ (2004) A review of accelerated carbonation technology in the treatment of cement-based materials and sequestration of CO_2. J Hazard Mater 112(3):193–205. doi:10.1016/j.jhazmat.2004.04.019
4. IPCC (2005) IPCC special report on carbon dioxide capture and storage, Intergovernmental Panel on Climate Change, Cambridge. ISBN-13 978-0-521-86643-9
5. Pacala S, Socolow R (2004) Stabilization wedges: solving the climate problem for the next 50 years with current technologies. Science 305(5686):968–972
6. CSLF (2011) InFocus: what is carbon utilization? Carbon Sequestration Leadership Forum (CSLF)
7. Styring P, Jansen D, Coninck Hd, Reith H, Armstrong K (2011) Carbon capture and utilization in the green economy: using CO_2 to manufacture fuel, chemical and materials
8. Pan S-Y, Du MA, Huang IT, Liu IH, Chang EE, Chiang P-C (2015) Strategies on implementation of waste-to-energy (WTE) supply chain for circular economy system: a review. J Clean Prod 108:409–421. doi:10.1016/j.jclepro.2015.06.124

9. IEA (2014) Tracking clean energy progress 2014—energy technology perspectives 2014 excerpt IEA input to the clean energy ministerial. International Energy Agency, France
10. IEA (2014) Energy technology perspectives 2014—harnessing electricity's potential. International Energy Agency, France
11. Yu C-H, Huang C-H, Tan C-S (2012) A review of CO_2 capture by absorption and adsorption. Aerosol Air Qual Res 12:745–769. doi:10.4209/aaqr.2012.05.0132
12. Lin C-C, Chu C-R (2015) Feasibility of carbon dioxide absorption by NaOH solution in a rotating packed bed with blade packings. Int J Greenhouse Gas Control 42:117–123. doi:10.1016/j.ijggc.2015.07.035
13. Lee SC, Hsieh CC, Chen CH, Chen YS (2013) CO_2 adsorption by Y-type zeolite impregnated with amines in Indoor air. Aerosol Air Qual Res 13:360–366. doi:10.4209/aaqr.2012.05.0134
14. Chen LC, Peng PY, Lin LF, Yang TCK, Huang CM (2014) Facile preparation of nitrogen-doped activated carbon for carbon dioxide adsorption. Aerosol Air Qual Res 14 (3):916–927. doi:10.4209/aaqr.2013.03.0089
15. Ganesh M, Hemalatha P, Peng MM, Cha WS, Jang HT (2014) Zr-fumarate MOF a novel CO_2-adsorbing material: synthesis and characterization. Aerosol Air Qual Res 14(6):1605–1612. doi:10.4209/aaqr.2013.11.0337
16. Lin YF, Chen CH, Tung KL, Wei TY, Lu SY, Chang KS (2013) Mesoporous fluorocarbon-modified silica aerogel membranes enabling long-term continuous CO_2 capture with large absorption flux enhancements. ChemSusChem 6(3):437–442. doi:10.1002/cssc.201200837
17. Ramasubramanian K, Zhao Y, Winston Ho WS (2013) CO_2 capture and H_2 purification: prospects for CO_2-selective membrane processes. AIChE J 59(4):1033–1045. doi:10.1002/aic.14078
18. Wang B, Gan ZH (2014) Feasibility analysis of cryocooler based small scale CO_2 cryogenic capture. Comment on "Energy analysis of the cryogenic CO_2 process based on Stirling coolers" Song CF, Kitamura Y, Li SH [Energy 2014; 65: 580–89]. Energy 68:1000–1003. doi:10.1016/j.energy.2014.02.032
19. Zhang X, Zhang X, Dong H, Zhao Z, Zhang S, Huang Y (2012) Carbon capture with ionic liquids: overview and progress. Energy Environ Sci 5(5):6668. doi:10.1039/c2ee21152a
20. Chang MH, Huang CM, Liu WH, Chen WC, Cheng JY, Chen W, Wen TW, Ouyang S, Shen CH, Hsu HW (2013) Design and experimental investigation of calcium looping process for 3-kWth and 1.9-MWth facilities. Chem Eng Technol 36(9):1525–1532. doi:10.1002/ceat.201300081
21. Chiu P-C, Ku Y (2012) Chemical looping process—a novel technology for Inherent CO_2 capture. Aerosol Air Qual. Res 12:1421–1432. doi:10.4209/aaqr.2012.08.0215
22. Olivares-Marín M, Maroto-Valer MM (2012) Development of adsorbents for CO_2 capture from waste materials: a review. Greenhouse Gases: Sci Technol 2(1):20–35. doi:10.1002/ghg.45
23. Klinthong W, Yang YH, Huang CH, Tan CS (2015) A review: microalgae and their applications in CO_2 capture and renewable energy. Aerosol Air Qual Res 15(2):712–742. doi:10.4209/aaqr.2014.11.0299
24. Rochelle GT (2009) Amine scrubbing for CO_2 capture. Science 325(5948):1652–1654. doi:10.1126/science.1176731
25. Tan C, Chen J (2006) Absorption of carbon dioxide with piperazine and its mixtures in a rotating packed bed. Sep Purif Technol 49(2):174–180. doi:10.1016/j.seppur.2005.10.001
26. Lin CC, Lin YH, Tan CS (2010) Evaluation of alkanolamine solutions for carbon dioxide removal in cross-flow rotating packed beds. J Hazard Mater 175(1–3):344–351. doi:10.1016/j.jhazmat.2009.10.009
27. Yu C-H, Lin Y-X, Tan C-S (2014) Effects of inorganic salts on absorption of CO_2 and O_2 for absorbents containing diethylenetriamine and piperazine. Int J Greenhouse Gas Control 30:118–124. doi:10.1016/j.ijggc.2014.09.005

28. Yu C-H, Wu T-W, Tan C-S (2013) CO_2 capture by piperazine mixed with non-aqueous solvent diethylene glycol in a rotating packed bed. Int J Greenhouse Gas Control 19:503–509. doi:10.1016/j.ijggc.2013.10.014
29. Abanades JC, Arias B, Lyngfelt A, Mattisson T, Wiley DE, Li H, Ho MT, Mangano E, Brandani S (2015) Emerging CO_2 capture systems. Int J Greenhouse Gas Control 40:126–166. doi:10.1016/j.ijggc.2015.04.018
30. Phelps JJC, Blackford JC, Holt JT, Polton JA (2015) Modelling large-scale CO_2 leakages in the North Sea. Int J Greenhouse Gas Control 38:210–220. doi:10.1016/j.ijggc.2014.10.013
31. Soong Y, Howard BH, Hedges SW, Haljasmaa I, Warzinski RP, Irdi G, McLendon TR (2014) CO_2 sequestration in saline formation. Aerosol Air Qual Res 14(2):522–532. doi:10.4209/aaqr.2013.06.0195
32. Yang DX, Wang S, Zhang Y (2014) Analysis of CO_2 migration during nanofluid-based supercritical CO_2 geological storage in saline aquifers. Aerosol Air Qual Res 14(5):1411–1417. doi:10.4209/aaqr.2013.09.0292
33. Kieke D, Imbus S, Cohen K, Galas C, Gasperikova E, Pickles W, Silver E (2009) The CO_2 capture project phase 2 (CCP2) storage program: progress in geological assurance in unmineable coal beds. Energy Procedia 1(1):79–86. doi:10.1016/j.egypro.2009.01.013
34. Olajire AA (2014) Review of ASP EOR (alkaline surfactant polymer enhanced oil recovery) technology in the petroleum industry: prospects and challenges. Energy 77:963–982. doi:10.1016/j.energy.2014.09.005
35. IPCC (2007) Climate change 2007: mitigation of climate change
36. Pawar RJ, Bromhal GS, Carey JW, Foxall W, Korre A, Ringrose PS, Tucker O, Watson MN, White JA (2015) Recent advances in risk assessment and risk management of geologic CO_2 storage. Int J Greenhouse Gas Control 40:292–311. doi:10.1016/j.ijggc.2015.06.014
37. Birkholzer JT, Oldenburg CM, Zhou Q (2015) CO_2 migration and pressure evolution in deep saline aquifers. Int J Greenhouse Gas Control 40:203–220. doi:10.1016/j.ijggc.2015.03.022
38. Black R (2012) Norway aims for carbon leadership. BBC News, 11 May 2012
39. Huang C-H, Tan C-S (2014) A review: CO_2 utilization. Aerosol Air Qual Res 14:480–499. doi:10.4209/aaqr.2013.10.0326
40. Hoel M, Greaker M, Ground C, Rasmussen I (2009) Climate policy: costs and design. TemaNord, Copenhagen, Denmark
41. Wang WN, Soulis J, Yang YJ, Biswas P (2014) Comparison of CO_2 photoreduction systems: a review. Aerosol Air Qual Res 14(2):533–549. doi:10.4209/aaqr.203.09.0283
42. Duan Y, Zhang K, Li XS, King DL, Li B, Zhao L, Xiao Y (2014) Ab initio thermodynamic study of the CO_2 capture properties of M_2CO_3 (M = Na, K)- and $CaCO_3$-promoted MgO sorbents towards forming double salts. Aerosol Air Qual Res 14(2):470–479. doi:10.4209/aaqr.2013.05.0178
43. Yan Y, Gu J, Bocarsly AB (2014) Hydrogen bonded pyridine dimer: a possible intermediate in the electrocatalytic reduction of carbon dioxide to methanol. Aerosol Air Qual Res 14(2):515–521. doi:10.4209/aaqr.2013.06.0227
44. Angamuthu R, Byers P, Lutz M, Spek AL, Bouwman E (2010) Electrocatalytic CO_2 conversion to oxalate by a copper complex. Science 327:313–315. doi:10.1126/science.1177981
45. Ganesh I (2014) Conversion of carbon dioxide into methanol—a potential liquid fuel: fundamental challenges and opportunities (a review). Renew Sustain Energy Rev 31:221–257. doi:10.1016/j.rser.2013.11.045
46. Mwangi JK, Lee WJ, Whang LM, Wu TS, Chen WH, Chang JS, Chen CY, Chen CL (2015) Microalgae oil: algae cultivation and harvest, algae residue torrefaction and diesel engine emissions tests. Aerosol Air Qual Res 15(1):81–98. doi:10.4209/aaqr.2014.10.0268
47. Adom F, Dunn JB, Han J, Sather N (2014) Life-cycle fossil energy consumption and greenhouse gas emissions of bioderived chemicals and their conventional counterparts. Environ Sci Technol 48(24):14624–14631. doi:10.1021/es503766e
48. OECD (2009) The bioeconomy to 2030: designing a policy agenda. Organization for Economic Co-operation and Development

49. Trivedi J, Aila M, Bangwal DP, Kaul S, Garg MO (2015) Algae based biorefinery—how to make sense? Renew Sustain Energy Rev 47:295–307. doi:10.1016/j.rser.2015.03.052
50. Adesanya VO, Cadena E, Scott SA, Smith AG (2014) Life cycle assessment on microalgal biodiesel production using a hybrid cultivation system. Bioresour Technol 163:343–355. doi:10.1016/j.biortech.2014.04.051
51. Kumar K, Mishra SK, Shrivastav A, Park MS, Yang J-W (2015) Recent trends in the mass cultivation of algae in raceway ponds. Renew Sustain Energy Rev 51:875–885. doi:10.1016/j.rser.2015.06.033
52. Borowitzka M (1999) Commercial production of microalgae: ponds, tanks, tubes and fermenters. J Biotechnol 70:313–321
53. Weissman JC, Goebel RP, Benemann JR (1988) Photobioreactor design Mixing, carbon utilization, and oxygen accumulation. Biotechnol Bioeng 31:336–344
54. Chiaramonti D, Prussi M, Casini D, Tredici MR, Rodolfi L, Bassi N, Zittelli GC, Bondioli P (2013) Review of energy balance in raceway ponds for microalgae cultivation: re-thinking a traditional system is possible. Appl Energy 102:101–111. doi:10.1016/j.apenergy.2012.07.040
55. Posten C (2009) Design principles of photo-bioreactors for cultivation of microalgae. Eng Life Sci 9(3):165–177. doi:10.1002/elsc.200900003
56. Jimenez C (2003) Relationship between physicochemical variables and productivity in open ponds for the production of spirulina: a predictive model of algal yield. Aquaculture 221(1–4):331–345. doi:10.1016/s0044-8486(03)00123-6
57. Benemann J, Goebel R, Augenstein D, Weissman J (1982) Microalgae as a source of liquid fuels. U.S. Department of Energy, Office of Energy Research, Washington, DC
58. USDOE (2010) National algal biofuels technology roadmap. U.S. Department of Energy, Office of Energy Efficiency and Renewable Energy. http://www1.eere.energy.gov/biomass/pdfs/algal_biofuels_roadmap.pdf
59. Bennett MC, Turn SQ, Chan WY (2014) A methodology to assess open pond, phototrophic, algae production potential: a Hawaii case study. Biomass Bioenergy 66:168–175. doi:10.1016/j.biombioe.2014.03.016
60. Greenwell HC, Laurens LM, Shields RJ, Lovitt RW, Flynn KJ (2010) Placing microalgae on the biofuels priority list: a review of the technological challenges. J R Soc Interface 7(46):703–726. doi:10.1098/rsif.2009.0322
61. Sutherland DL, Turnbull MH, Craggs RJ (2014) Increased pond depth improves algal productivity and nutrient removal in wastewater treatment high rate algal ponds. Water Res 53:271–281. doi:10.1016/j.watres.2014.01.025
62. Mendoza JL, Granados MR, de Godos I, Acien FG, Molina E, Heaven S, Banks CJ (2013) Oxygen transfer and evolution in microalgal culture in open raceways. Bioresour Technol 137:188–195. doi:10.1016/j.biortech.2013.03.127
63. Murphy TE, Kapili BJ, Detweiler AM, Bebout BM, Prufert Bebout LE (2015) Vertical distribution of algal productivity in open pond raceways. Algal Res 11:334–342. doi:10.1016/j.algal.2015.07.003
64. Kumar K, Sirasale A, Das D (2013) Use of image analysis tool for the development of light distribution pattern inside the photobioreactor for the algal cultivation. Bioresour Technol 143:88–95. doi:10.1016/j.biortech.2013.05.117
65. Gonzalez-Lopez CV, Acien Fernandez FG, Fernandez-Sevilla JM, Sanchez Fernandez JF, Molina Grima E (2012) Development of a process for efficient use of CO_2 from flue gases in the production of photosynthetic microorganisms. Biotechnol Bioeng 109(7):1637–1650. doi:10.1002/bit.24446
66. Kumar K, Dasgupta CN, Das D (2014) Cell growth kinetics of chlorella sorokiniana and nutritional values of its biomass. Bioresour Technol 167:358–366. doi:10.1016/j.biortech.2014.05.118
67. Richmond A (1986) Outdoor mass cultures of microalgae. In: Richmond A (ed) Handbook of algal mass culture. CRC Press, Boca Raton, pp 285–330

68. Vonshak A, Tomaselli L (2000) Arthrospira (Spirulina): systematics and ecophysioiogy. In: Whitton BA, Potts M (eds) The ecology of cyanobacteria, pp 505–522. doi:10.1007/0-306-46855-7_18
69. Singh SP, Singh P (2015) Effect of temperature and light on the growth of algae species: a review. Renew Sustain Energy Rev 50:431–444. doi:10.1016/j.rser.2015.05.024
70. Prussi M, Buffi M, Casini D, Chiaramonti D, Martelli F, Carnevale M, Tredici MR, Rodolfi L (2014) Experimental and numerical investigations of mixing in raceway ponds for algae cultivation. Biomass Bioenergy 67:390–400. doi:10.1016/j.biombioe.2014.05.024
71. Hreiz R, Sialve B, Morchain J, Escudié R, Steyer J-P, Guiraud P (2014) Experimental and numerical investigation of hydrodynamics in raceway reactors used for algaculture. Chem Eng J 250:230–239. doi:10.1016/j.cej.2014.03.027
72. Putt R, Singh M, Chinnasamy S, Das KC (2011) An efficient system for carbonation of high-rate algae pond water to enhance CO_2 mass transfer. Bioresour Technol 102(3):3240–3245. doi:10.1016/j.biortech.2010.11.029
73. de Godos I, Mendoza JL, Acien FG, Molina E, Banks CJ, Heaven S, Rogalla F (2014) Evaluation of carbon dioxide mass transfer in raceway reactors for microalgae culture using flue gases. Bioresour Technol 153:307–314. doi:10.1016/j.biortech.2013.11.087
74. Fon Sing S, Isdepsky A, Borowitzka MA, Lewis DM (2014) Pilot-scale continuous recycling of growth medium for the mass culture of a Halotolerant Tetraselmis sp. in raceway ponds under increasing salinity: a novel protocol for commercial microalgal biomass production. Bioresour Technol 161:47–54. doi:10.1016/j.biortech.2014.03.010
75. Datta S, Henry MP, Lin YJ, Fracaro AT, Millard CS, Snyder SW, Stiles RL, Shah J, Yuan J, Wesoloski L, Dorner RW, Carlson WM (2013) Electrochemical CO_2 capture using resin-wafer electrodeionization. Ind Eng Chem Res 52(43):15177–15186. doi:10.1021/ie402538d
76. Shimamatsu H (2004) Mass production of spirulina, an edible microalga. Hydrobiologia 512 (1–3):39–44
77. Williams PJlB, Laurens LML (2010) Microalgae as biodiesel & biomass feedstocks: Review & analysis of the biochemistry, energetics & economics. Energy Environ Sci 3(5):554. doi:10.1039/b924978h
78. Chi Z, O'Fallon JV, Chen S (2011) Bicarbonate produced from carbon capture for algae culture. Trends Biotechnol 29(11):537–541. doi:10.1016/j.tibtech.2011.06.006
79. Nakano S, Chang KH, Shijima A, Miyamoto H, Sato Y, Noto Y, Ha JY, Sakamoto M (2014) A usage of CO_2 hydrate: convenient method to increase CO_2 concentration in culturing algae. Bioresour Technol 172:444–448. doi:10.1016/j.biortech.2014.09.019
80. Li P, Pan SY, Pei S, Lin YPJ, Chiang, PC (2016) Challenges and perspectives on carbon fixation and utilization technologies: an overview. Aerosol and Air Quality Research 16:1327–1344. doi:10.4209/aaqr.2015.12.0698

Chapter 3
CO_2 Mineralization and Utilization via Accelerated Carbonation

Abstract In recent years, CO_2 emission control in a large-scaled industrial process has also drawn lots of attention due to climate change and global warming issues. For instance, to establish a sustainable resource cycle, an integrated multiwaste treatment via carbonation process has been proposed using CO_2 in flue gas as a chemical to stabilize active components in alkaline solid wastes. In this chapter, the basic information regarding CO_2 and carbon-related species is illustrated in terms of thermodynamics and process chemistry. In addition, the types of CO_2 mineralization via carbonation are summarized and reviewed.

3.1 Thermodynamics of Carbon Dioxide

Over the past 200 years, the amount of global CO_2 emissions has dramatically increased due to rapid growth in energy demand, population, and economic development. An increase of greenhouse gas in the atmosphere enhances global warming has been recently proved by scientists [1]. Over the past 250 years, the concentration of CO_2 in the atmosphere increased from 275 to 400 ppm, which results in an increase of the global mean surface temperature by 0.075 ± 0.013 °C per decade from 1901 to 2012 [2]. Therefore, proper CO_2 emission control in a large-scaled industrial process has also drawn lots of attention due to climate change and global warming issues.

In this section, basic information regarding the formation of heat for C-related species, Gibbs Free Energy for carbon cycle, and physicochemical properties of CO_2 is illustrated.

3.1.1 Gibbs Free Energy

In thermodynamics, the Gibbs free energy is a thermodynamic potential that measures the "usefulness" or process-initiating work obtainable from a thermodynamic

P.-C. Chiang and S.-Y. Pan, *Carbon Dioxide Mineralization and Utilization*, DOI 10.1007/978-981-10-3268-4_3

system at a constant temperature and pressure (isothermal, isobaric). Gibbs free energy (*G*) is a state function which is related to three variables, e.g., enthalpy (*H*), temperature (*T*), and entropy (*S*), as shown in Eq. (3.1):

$$G = H - T * S \tag{3.1}$$

At the molecular scale, the entropy (*S*) of a substance can be related to the number of ways the total energy can be distributed among all the particles. In this case, each possible distribution of the energy characterizes a "microstate" of the system. In other words, entropy is a measure of the number of microstates that are accessible at a given total energy. The above interpretation can be developed by Boltzmann equation, as expressed in Eq. (3.2):

$$S = k_B \times \ln(W) \tag{3.2}$$

where k_B is Boltzmann's constant (i.e., 1.381×10^{-23} J/K) and *W* is number of microstates. A number of factors contribute to the entropy of a substance, such as the physical state, temperature, molecular size, intermolecular forces and dissolution, and mixing.

3.1.2 Formation of Heat for C-Related Species

One of the promising approaches to CO_2 mitigation technologies is CO_2 mineralization, which has been considered an environmentally friendly alternative for both CO_2 capture and storage because it can not only reduce the CO_2 emission but also convert CO_2 to stable precipitates. Figure 3.1 shows the standard molar free energy of formation for several carbon-related substances at 298 K. CO_2 mineralization can be achieved by accelerated carbonation reaction, also known as carbon capture and utilization by mineralization (CCUM), which has been proven thermodynamically practical for enhancing the natural weathering process [3]. The energy of formation for gaseous CO_2 is approximately −400 kJ mol^{-1}, and it can decrease to around −1100 kJ mol^{-1} as solid $CaCO_3$ precipitates [4].

3.1.3 Physico-chemical Properties of CO_2

CO_2 is a colorless and odorless gas vital to life on earth. Table 3.1 presents the thermodynamic state variables of CO_2, comparable to water (H_2O). CO_2 has a critical point at 31.06 °C and 73.8 bars (1070 lb/in^2) and a critical density of 0.469 g/cm^3. In addition, diffusion of CO_2 in water is approximately 10,000 times lower than in air. The binding of the solvent with CO_2 occurs at high pressure, and a

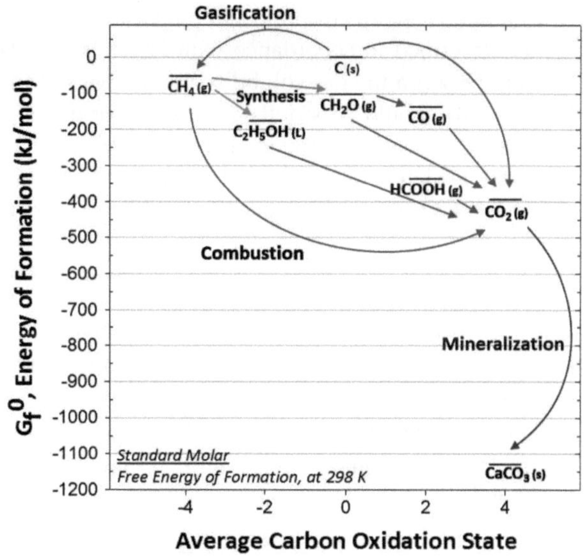

Fig. 3.1 Standard molar free energy of formation for several carbon-related substances at 298 K

reduction in pressure releases the gas [5]. At standard temperature and pressure, the density of CO_2 is around 1.98 kg/m^3, about 1.67 times that of air.

Table 3.1 Thermodynamic state variables of carbon dioxide (CO_2) and water (H_2O)

Property[a]	Unit	Value	
		CO_2	H_2O
Density	kg/m^3	1.778	995.65
Specific inner energy	kJ/kg	–	125.73
Specific enthalpy	kJ/kg	510.09	125.83
Specific entropy	kJ/kg/K	2.753	0.4368
Specific isobar heat capacity (cp)	kJ/kg/K	0.856	4.1800
Specific isochor heat capacity (cv)	kJ/kg/K	0.662	4.1175
Isobar coefficient of thermal expansion	1/K	3.352×10^{-3}	–
Heat conductance	W/(m K)	1.703×10^{-2}	–
Thermal conductivity	W/(m K)	–	0.6155
Dynamic viscosity	kg/m/s	1.517×10^{-5}	7.973×10^{-4}
Kinematic viscosity	m^2/s	8.530×10^{-6}	8.008×10^{-7}
Thermal diffusivity	m^2/s	1.132×10^{-5}	–
Prandtl number	–	0.7624	–
Schmidt number	–	0.7550	–
Coefficient of compressibility Z	–	0.9952	–
Speed of sound	m/s	270.69	1512.07

[a]Calculated at 1.013 bar and 30 °C by Peace Software (Web site) from http://www.peacesoftware.de/einigewerte/co2_e.html

CO_2 is a weak electrophile and physically soluble in water (or a solvent), which can be described in accordance with Henry's law. Henry's law indicates the relationship between the gas solubility in pore water and the partial pressure of the gas. Henry's law should be strictly valid only for gases that can be infinitely diluted in solution. CO_2 can be produced from a reversible reaction form of carbonic acid (H_2CO_3), which is a weak acid since its ionization in water is incomplete. The amount of CO_2 dissolution in water can be expressed by Henry's law in Eq. (3.3):

$$C'_{CO_2} = H'_{CO_2} \times P_{CO_2} \tag{3.3}$$

where C' is the concentration of CO_2 dissolved in aqueous solution (M); H'_{CO_2} is Henry's constant for CO_2 (i.e., $10^{-1.46}$ M atm^{-1} at 25 °C); and P_{CO_2} is the partial pressure of CO_2 in the gas phase (atm).

The dissociation of gaseous CO_2 into carbonic acid (H_2CO_3) is shown in Eq. (3.4). The hydration equilibrium constant (K_h) of H_2CO_3 at 25 °C in pure water is about 1.7×10^{-3}.

$$CO_{2(g)} + H_2O_{(aq)} \rightarrow H_2CO_{3(aq)} \tag{3.4}$$

In aqueous solution, H_2CO_3 includes dissolved CO_2 (denoted as $CO_{2(aq)}$) and effective H_2CO_3 (denoted as $H_2CO^*_{3(aq)}$), as shown in Eq. (3.5). Typically, the concentration of $H_2CO_3^*$ is much lower than the concentration of CO_2. Therefore, the H_2CO_3 has two acid dissociation constants (K_a).

$$H_2CO_{3(aq)} = CO_{2(aq)} + H_2CO^*_{3(aq)} \tag{3.5}$$

The dissociation of the carbonic ions into HCO_3^- and carbonate CO_3^{2-} can be described, as shown in Eqs. (3.6) and (3.7). The first one for the dissociation of carbonic ion into the bicarbonate (also called hydrogen carbonate) ion at 25 °C is 2.5×10^{-4} mol/L, corresponding to $pK_{a1} = 3.6$, as shown in Eq. (3.6).

$$H_2CO_{3(aq)} \rightarrow H_{(aq)} + HCO_{3^-(aq)} \tag{3.6}$$

The second one for the dissociation of carbonic ion into the bicarbonate ion at 25 °C is 4.47×10^{-7} mol/L, corresponding to $pK_{a1} = 6.35$, as shown in Eq. (3.7).

$$H_2CO^*_{3(aq)} \rightarrow H^+_{(aq)} + HCO_{3^-(aq)} \tag{3.7}$$

The formed bicarbonate (HCO_3^-) ion can be dissociated into carbonate (CO_3^{2-}) ion, as expressed in Eq. (3.8), where the constant for the dissociation at 25 °C is 4.69×10^{-11} mol/L, corresponding to $pK_{a2} = 10.33$.

$$HCO_{3(aq)^-} \rightarrow H^+_{(aq)} + CO^{2-}_{3(aq)} \tag{3.8}$$

The relative concentrations of CO_2, H_2CO_3, and the deprotonated forms bicarbonate (HCO_3^-) and carbonate (CO_3^{2-}) depend on the pH value of the solution. At a low pH (~4), the production of H_2CO_3 dominates, at a mid pH (~8) HCO_3^- dominates, and at a high pH (~12) CO_3^{2-} dominates. As a result, the mole balances of the carbonic acid system can be expressed as Eq. (3.9):

$$C_T = \left[H_2CO^*_{3(aq)}\right] + \left[HCO^-_{3(aq)}\right] + \left[CO^{2-}_{3(aq)}\right] \tag{3.9}$$

where C_T is the total inorganic carbon (TIC) concentration (M).

3.2 CO_2 Mineralization via Carbonation

Accelerated carbonation, considered as a chemical adsorption reaction, can be referred as "mineral sequestration" or "mineral carbonation." The accelerated carbonation is to mimic the natural weathering processes, where CO_2 reacts with metal-oxide-bearing materials to form stable and insoluble carbonates. Among the carbon capture, utilization, and storage (CCUS) technologies, accelerated carbonation of natural minerals (or industrial alkaline solid wastes) is attractive for the mitigation of CO_2 emissions in the coming decades because the gaseous CO_2 is fixed as carbonate precipitation and rarely released after mineralization. Moreover, since carbonation is an exothermal reaction, energy consumption and costs may be reduced by its inherent properties [6].

As shown in Fig. 3.1, the CO_2 generated from combustion of fossil fuels or carbon-related chemicals can be directly fixed as mineral carbonates. Since the carbonates are naturally occurring minerals and possess the lowest free energy of formation, the carbonation products can be permanently fixed and stored over geologic periods of time. In other words, it could ensure the issues of safe and stable storage, thereby being negligible release of CO_2 to the environment.

In the accelerated carbonation process, the gaseous CO_2 is dissolved into solution to form carbonate ions. Calcium oxide or magnesium oxide is the most favorable metal oxide in reacting with CO_2. The carbonate ions are reacted with alkaline earth metal oxide (e.g., CaO and MgO) and then converted into carbonate precipitates in the presence of aqueous environments. In all cases, accelerated carbonation must provide base ions, such as monovalent sodium and potassium, or divalent calcium and magnesium ions to neutralize the carbonic acid [7]. Other carbonate-forming elements, such as iron metals, are not practical due to their unique and precious features [8]. Figure 3.2 shows the potential pathways of carbonate ions (CO_3^{2-}) in reaction with alkaline solid particles, including (1) direct conversion inside the solid particle, (2) crystallization on the surface of the particle, (3) precipitation in bulk solution, and then (4) attachment on the surface of the particle.

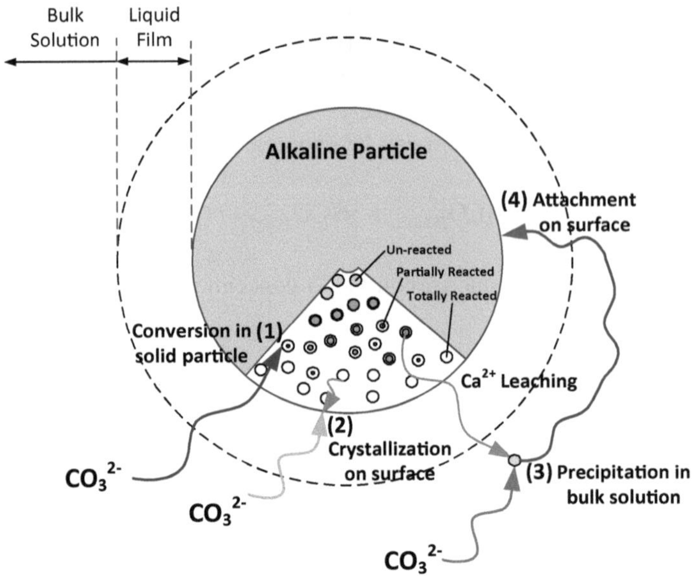

Fig. 3.2 Pathways of carbonate ions (CO_3^{2-}) in reaction with alkaline solid particles

Typically, accelerated carbonation can be accomplished by either of the following processes [9–11]:

- In situ carbonation [12, 13]: "underground" in geologic formation. The concentrated CO_2 is transported to underground igneous rocks (typically basalt) and is permanently fixed within the hosting rocks as solid carbonates, such as $CaCO_3$ and $MgCO_3$.
- Ex situ carbonation [7, 14]: "aboveground" in a chemical processing plant. A source of calcium silicate feedstock (natural ores or alkaline solid wastes) is carbonated aboveground, such as
  - Carbonation at industrial sites
  - Biologically mediated carbonation
  - Carbon mineralization in industrial reactors

3.2.1 In Situ Carbonation

In situ (mineral) carbonation is considered as the worldwide development of storage projects through basalt or peridotite minerals. Concentrated CO_2 is injected into silicate rocks to promote carbonate formation underground. This can be regarded as one type of CO_2 storage method. However, several challenges in the large-scale deployment of in situ carbonation are found [11]:

- Identification of appropriate sites where natural characteristics (e.g., geothermal gradient) are specifically favorable to carbonation process
- Simulation of longer reaction time periods on both the CO_2 solubility and Si or Mg solubility as a function of temperature and pressure
- Decrease in the energy requirement for artificial methods of reaction enhancement
- Slow pace and reaction kinetics of the process

In situ carbonation can offer economic advantages over ex situ carbonation, but may not be an economically viable option for small-to-medium CO_2 emission sources. Conversely, the ex situ carbonation process may be attractive in industries that lack of geological storage and must proceed quickly to near completion. A great number of large-scale industrial wastes that are available close to the CO_2 emission source can be considered as the feedstock for ex situ carbonation. Therefore, several environmental benefits can be achieved by ex situ carbonation although ex situ carbonation cannot compete with in situ carbonation (geological storage), in terms of potential capacity and cost of CO_2 sequestration.

3.2.2 *Ex Situ Accelerated Carbonation*

Appropriate materials for ex situ accelerated carbonation are generally rich in metal oxides, including calcium, magnesium, aluminum, iron, and manganese oxides. These appropriate feedstock sources include [15–23] the following:

- Natural ores: serpentine and wollastonite.
- Alkaline solid wastes: iron- and steel-making slag, combustion residues, cement/concrete wastes, and fly ashes.

Accelerated carbonation using "natural ores" may create environmental impacts due to the massive mineral requirements and associated scale of mining [7]. As a result, the size of the mining operation is considered as the most significant economic and environmental barrier for large-scale carbonation process using natural minerals. Conversely, accelerated carbonation using "alkaline solid wastes" can offer several advantages, from the 3E (engineering, environmental and economic) aspects [6, 10, 24–27]:

- Engineering aspect:
  - It offers great sequestration capacity due to the high availability of deposits at low material cost.
  - Carbonation products, such as $CaCO_3$ and $MgCO_3$, are thermodynamically stable under ambient conditions, that is, in the absence of acidification.

   - Products may be beneficially reused in a variety of application, such as construction materials.
   - It can reduce the amount of free CaO and its associated hydration expansion in service.

- Environmental aspect:
   - It can prevent globally available industrial alkaline solid wastes from landfill.
   - It creates alternative materials to reduce the need to transport suitable natural sands, or the energy required to produce manufactured aggregates.
   - Carbonation reduces environmental impacts due to decreased leaching of heavy metal trace elements from residues and stabilizing the wastes.

- Economic aspect:
   - Energy consumption and costs may be reduced by its inherent properties (i.e., it is an exothermal reaction).
   - It could neutralize the pH of the solution because of the formed carbonate precipitations, if alkaline wastewater is used as the liquid agents.
   - It is a cost-effective process because no transport at sites within steelworks is required.

Although ex situ carbonation is not economically viable so far, relevant research and tests are active because the raw materials required for carbonation are globally abundant. Basically, two branches for ex situ carbonation have been developed:

- Direct carbonation: reaction occurs in one single step.
- Indirect carbonation: the mineral has to first be refined, and then the refined mineral is carbonated.

Table 3.2 presents various types of methods to ex situ carbonation using alkaline solid wastes. From the feedstock point of view, the achievable CO_2 capture capacity of iron and steel slags, such as basic oxygen furnace slag (BOFS), argon oxygen decarburization slag (AODS), and continuous casting slag (CCS), is relatively higher than that of other solid wastes such as fly ash (FA). The reaction temperature would affect not only the leaching kinetics of calcium ions from solid wastes but also the rates of CO_2 dissolution and carbonate precipitation. It was observed that the appropriate operating temperature for carbonation should be set at the range between 60 and 80 °C, in the cases of aqueous carbonation using SS and FA [28–30].

Figure 3.3 illustrates the conceptual diagram of ex situ accelerated carbonation using industrial wastes, including flue gas (CO_2), wastewater, and alkaline solid wastes. Through the process, the gaseous CO_2 in flue gas could be fixed as solid carbonates, while the wastewater could be neutralized to a pH value of 6–7. In addition, both the chemical and physical properties of solid wastes could be

Table 3.2 Performance evaluation of ex situ carbonation using alkaline solid wastes

Types of wastes[a]	Methods	Reactor	Process characteristics	Performance[b]	References
CKD	Indirect carbonation	Batch flask	Extraction by NH_4NO_3/CH_3COONH_4 (1 M) Without adjusting pH during carbonation S/L ratio = 50 g/L, CO_2 flow rate = 200 mL/min	• CC: 180 kg CO_2/t-waste • Production: 420 kg $CaCO_3$/t-waste • Vaterite purity > 98% (Spherical)	[31]
CKD, SDA, and CDSA	Indirect carbonation	Batch flask	Solvent leaching by $NaHCO_3$ (0.5 M) for 24 h Primarily amorphous $CaCO_3$ formation	• CC: 101–123 kg CO_2/t-waste • CE: 77–93%	[32]
FA (MSWI)	Indirect carbonation	Batch flask	Two-step carbonation using ammonia pH of solution reached 6.95 within 5 min Formation of $CaCO_3$, NaCl, $CaSO_4$ precipitates	• CO_2 Uptake: 59% • CE: 57%	[33]
SS	Indirect carbonation (pH swing)	Batch flask	Dissolution using NH_4SO_4 at pH 8.2–8.3; particle size: 75–150 μm; S/L ratio = 15 g/L; 1 h; 65 °C E_a = 42.0 kJ/mol	• CE: 74%	[34, 35]
PS				• CE: 67%	
BFS				• CE: 59%	
RG	Direct gas–solid carbonation	Autoclave	Mixed with NH_4OH (1 M) for 5 mL/g-waste Production of both $CaCO_3$ and $FeCO_3$ Operated at 70 bar with a particle size < 45 μm	• CE = 41% • $CaCO_3$ purity = 25% • $FeCO_3$ purity = 19%	[36]
AODS	Direct aqueous carbonation	Autoclave	Operated at 30 bar and 180 °C for 120 min Optimum S/L ratio = 25–250 g/L for D_p of 46 μm Reduction in basicity and heavy metal leeching	• CC = 260 kg CO_2/t-waste	[37]
CCS				• CC = 310 kg CO_2/t-waste	

(continued)

Table 3.2 (continued)

Types of wastes[a]	Methods	Reactor	Process characteristics	Performance[b]	References
Coal FA (Victorian brown coal)	Direct aqueous carbonation	Autoclave	Operated at 40 °C and 3 MPa with 60-rpm stirring Optimum L/S ratios = 0.2–0.3 (w/w) Increase in heat for faster rate of CO_2 transfer	• CC = 27.1 kg CO_2/t-waste • CE = 14% • E_a = 34.5 kJ/mol	[28]
BOFS	Direct aqueous carbonation	Rotating packed bed	Integrated with metalworking wastewater Optimum L/S ratio = 20 mL/g at 65 °C for 30 min CaO_f and $Ca(OH)_2$ in BOFS was eliminated	• CC = 277–290 kg CO_2/t-waste • CE = 91–94%	[38]

[a] *CKD* Cement kiln dust; *SDA* Spray dryer absorber ash; argon oxygen decarburization slag (AODS); *CDSA* Circulating dry scrubber ash; *CCS* Continuous casting slag; *RG* Red gypsum

[b] *CC* CO_2 capture capacity (kg CO_2 per ton of waste); *CE* Carbonation efficiency (%); E_a Activated energy

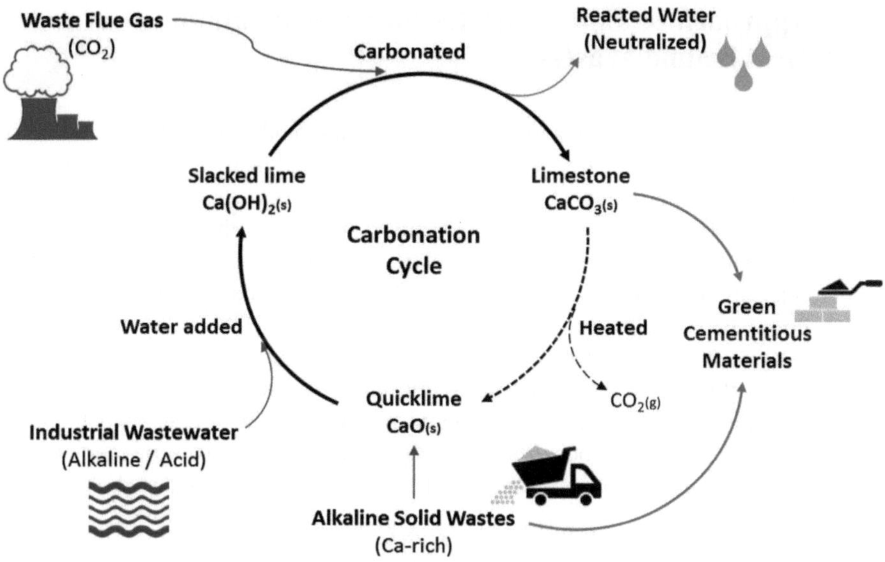

Fig. 3.3 Innovative approach to integrating CO_2 mineralization and waste utilization through accelerated carbonation

upgraded after carbonation [39], which facilitate its reuse in a variety of applications as green construction materials (e.g., supplementary cementitious materials). Carbonation is an effective way to improve the durability of concrete/mortar since relatively insoluble $CaCO_3$ is formed from the soluble $Ca(OH)_2$ in the concrete/mortar [40]. Moreover, accelerated carbonation could efficiently immobilize heavy metals, such as Pb, Cd, and Cr, leaching from the alkaline solid waste by the formation of metal carbonates. For both APC fly ash and bottom ash as examples [41–44], the leaching of heavy metals including Pb, Cr, Zn, Cu, and Mo could be significantly reduced upon carbonation. Cd and Pb have a strong affinity with $CaCO_3$ product and also form complexes with Fe and Al (hydr-)oxides [45]. Similarly, Sb could also be immobilized by combining with other processes, such as sorbent adding, during carbonation reaction [41].

From the carbonation method point of view, the CO_2 capture efficiency using direct carbonation was found to be superior to that using indirect carbonation. In other words, a higher CO_2 removal capacity and rate can be achieved by direct carbonation, compared to indirect carbonation. However, the purity of the produced $CaCO_3$ precipitate from indirect carbonation was higher than that of using direct carbonation, which could provide a higher value-added product for utilization. Thus, the utilization of the carbonated products should be implemented to couple the CO_2 reduction and waste utilization in industry.

3.3 Approach to Enhancing Ex Situ Carbonation for Alkaline Wastes

To establish a sustainable resource cycle, an integrated approach to multiwaste treatment via accelerated carbonation has been proposed using CO_2 in flue gas as a chemical to stabilize active components in alkaline solid wastes. From the economic point of view, ex situ carbonation can reduce the treatment cost of wastewater and increase the added value of alkaline solid. Moreover, the amount of CO_2 reduction could be considered as certified emission reduction (CER) credits, which could be used in the emission trading scheme (ETS) under the clean development mechanism (CDM) issued by the Kyoto Protocol.

Alkaline solid wastes can be used to sequester great amount of CO_2, especially if the wastes are generated nearby the source of CO_2, for achieving both the environmental and economic benefits. To achieve a cost-effective manner, several strategies on ex situ carbonation for alkaline solid wastes are suggested as follows:

- **Appropriate Carbonation Efficiency**: A carbonation conversion of higher than 85% for solid wastes should be acceptable to achieve waste stabilization and CO_2 fixation [4]. Also the feedstock mineral must be near the CO_2 emission point source to minimize transportation costs.
- **Enhanced Energy Recovery**: The procedures of feedstock crushing (in the case of steel slag), process heating, and slurry stirring are normally energy-intensive processes. It needs to be compensated by the exothermic carbonation process to make the process economically viable in an industrial context [9, 46–48].
- **Integrated Wastes Treatment**: Aqueous accelerated carbonation of alkaline solid wastes is suggested as a link to utilize wastewater for large-scale application [49].
- **Diversified Product Utilization**: The carbonated (or treated) solid wastes can be converted into several high value-added materials, such as glass ceramics (with red mud from the sintering process) and precipitated calcium carbonate (PCC).
- **Accelerated Pilot-Scale Demonstrations**: Significant technological breakthroughs should be needed before deployment can be considered. This includes reactor design, waste-to-resource supply chain, and system optimization from the 3E (Engineering, Environmental, and Economic) aspects [50].

References

1. Cubasch U, Wuebbles D, Chen D, Facchini MC, Frame D, Mahowald N, Winther J-G (2013) Introduction. In: Stocker TF, Qin D, Plattner G-K, Tignor M, Allen SK, Boschung J, Nauels A, Xia Y, Bex V, Midgley PM (eds) Climate change 2013: the physical science basis. Contribution of working group I to the fifth assessment report of the intergovernmental panel on climate change. Cambridge University Press, Cambridge

2. Hartmann DL, Klein Tank AMG, Rusticucci M, Alexander LV, Brönnimann S, Charabi Y, Dentener FJ, Dlugokencky EJ, Easterling DR, Kaplan A, Soden BJ, Thorne PW, Wild M, Zhai PM (2013) Observations: atmosphere and surface. In: Stocker TF, Qin D, Plattner G-K, Tignor M, Allen SK, Boschung J, Nauels A, Xia Y, Bex V, Midgley PM (eds) Climate change 2013: the physical science basis. Contribution of working group I to the fifth assessment report of the intergovernmental panel on climate change. Cambridge University Press, Cambridge
3. Lackner KS, Wendt CH, Butt DP, Joyce EL, Sharp DH (1995) Carbon dioxide disposal in carbonate minerals. Los Alamos National Laboratory, Los Alamos
4. Pan S-Y, Chiang A, Chang E-E, Lin Y-P, Kim H, Chiang P-C (2015) An innovative approach to integrated carbon mineralization and waste utilization: A review. Aerosol Air Qual Res 15:1072–1091. doi:10.4209/aaqr.2014.10.02
5. Klemeš J, Cockerill T, Bulatov I, Shackley S, Gough C (2007) Engineering feasibility of carbon dioxide capture and storage. In: Shackley S, Gough C (eds) Carbon capture and its storage: an integrated assessment. Ashgate, England
6. Lackner KS (2003) A guide to CO_2 sequestration. Science 300(5626):1677–1678
7. Gerdemann SJ, O'Connor WK, Dahlin DC, Penner LR, Rush H (2007) Ex situ aqueous mineral carbonation. Environ Sci Technol 41(7):2587–2593
8. Lackner KS (2002) Carbonate chemistry for sequestering fossil carbon. Annu Rev Energy Env 27(1):193–232. doi:10.1146/annurev.energy.27.122001.083433
9. IEA (2013) Mineralisation—carbonation and enhanced weathering. International Energy Agency
10. Pan S-Y, Chang EE, Chiang P-C (2012) CO_2 capture by accelerated carbonation of alkaline wastes: a review on its principles and applications. Aerosol Air Qual Res 12:770–791. doi:10.4209/aaqr.2012.06.0149
11. Sanna A, Uibu M, Caramanna G, Kuusik R, Maroto-Valer MM (2014) A review of mineral carbonation technologies to sequester CO_2. Chem Soc Rev 43(23):8049–8080. doi:10.1039/c4cs00035h
12. Kelemen PB, Matter J (2008) In situ carbonation of peridotite for CO_2 storage. In: The National Academy of Sciences of the United States of America, pp 17295–17300
13. Muriithi GN, Petrik LF, Fatoba O, Gitari WM, Doucet FJ, Nel J, Nyale SM, Chuks PE (2013) Comparison of CO_2 capture by ex-situ accelerated carbonation and in in-situ naturally weathered coal fly ash. J Environ Manage 127:212–220. doi:10.1016/j.jenvman.2013.05.027
14. Pan SY, Chiang PC, Chen YH, Tan CS, Chang EE (2013) Ex Situ CO_2 capture by carbonation of steelmaking slag coupled with metalworking wastewater in a rotating packed bed. Environ Sci Technol 47(7):3308–3315. doi:10.1021/es304975y
15. Olajire AA (2013) A review of mineral carbonation technology in sequestration of CO_2. J Petrol Sci Eng 109:364–392. doi:10.1016/j.petrol.2013.03.013
16. Eloneva S, Said A, Fogelholm C-J, Zevenhoven R (2012) Preliminary assessment of a method utilizing carbon dioxide and steelmaking slags to produce precipitated calcium carbonate. Appl Energy 90(1):329–334. doi:10.1016/j.apenergy.2011.05.045
17. Nduagu E, Romão I, Fagerlund J, Zevenhoven R (2013) Performance assessment of producing $Mg(OH)_2$ for CO_2 mineral sequestration. Appl Energy 106:116–126. doi:10.1016/j.apenergy.2013.01.049
18. Pan S-Y, Chiang P-C, Chen Y-H, Tan C-S, Chang EE (2014) Kinetics of carbonation reaction of basic oxygen furnace slags in a rotating packed bed using the surface coverage model: maximization of carbonation conversion. Appl Energy 113:267–276. doi:10.1016/j.apenergy.2013.07.035
19. Said A, Mattila HP, Jarvinen M, Zevenhoven R (2013) Production of precipitated calcium carbonate (PCC) from steelmaking slag for fixation of CO_2. Appl Energy 112:765–771. doi:10.1016/j.apenergy.2012.12.042
20. Sanna A, Dri M, Hall MR, Maroto-Valer M (2012) Waste materials for carbon capture and storage by mineralisation (CCSM)—A UK perspective. Appl Energy 99:545–554. doi:10.1016/j.apenergy.2012.06.049

21. Teir S, Eloneva S, Fogelholm C, Zevenhoven R (2009) Fixation of carbon dioxide by producing hydromagnesite from serpentinite. Appl Energy 86(2):214–218. doi:10.1016/j.apenergy.2008.03.013
22. Renforth P, Washbourne CL, Taylder J, Manning DA (2011) Silicate production and availability for mineral carbonation. Environ Sci Technol 45(6):2035–2041. doi:10.1021/es103241w
23. Huntzinger DN, Gierke JS, Kawatra SK, Eisele TC, Sutter LL (2009) Carbon dioxide sequestration in cement kiln dust through mineral carbonation. Environ Sci Technol 43 (6):1986–1992
24. Gunning PJ (2011) Accelerated carbonation of hazardous waste. University of Greenwich, UK
25. Lim M, Han GC, Ahn JW, You KS (2010) Environmental remediation and conversion of carbon dioxide (CO(2)) into useful green products by accelerated carbonation technology. Int J Environ Res Public Health 7(1):203–228. doi:10.3390/ijerph7010203
26. Bonenfant D, Kharoune L, Sauve S, Hausler R, Niquette P, Mimeault M, Kharoune M (2009) Molecular analysis of carbon dioxide adsorption processes on steel slag oxides. Int J Greenhouse Gas Control 3(1):20–28. doi:10.1016/j.ijggc.2008.06.001
27. Monkman S, Shao Y, Shi C (2009) Carbonated ladle slag fines for carbon uptake and sand substitute. J Mater Civ Eng 21:657–665. doi:10.1061//asce/0899-1561/2009/21:11/657
28. Ukwattage NL, Ranjith PG, Yellishetty M, Bui HH, Xu T (2014) A laboratory-scale study of the aqueous mineral carbonation of coal fly ash for CO_2 sequestration. J Clean Prod. doi:10.1016/j.jclepro.2014.03.005
29. Chang EE, Pan SY, Chen YH, Tan CS, Chiang PC (2012) Accelerated carbonation of steelmaking slags in a high-gravity rotating packed bed. J Hazard Mater 227–228:97–106. doi:10.1016/j.jhazmat.2012.05.021
30. Chang EE, Wang Y-C, Pan S-Y, Chen Y-H, Chiang P-C (2012) CO_2 capture by using blended hydraulic slag cement via a slurry reactor. Aerosol Air Qual Res 12:1433–1443. doi:10.4209/aaqr.2012.08.0210
31. Jo H, Park S-H, Jang Y-N, Chae S-C, Lee P-K, Jo HY (2014) Metal extraction and indirect mineral carbonation of waste cement material using ammonium salt solutions. Chem Eng J 254:313–323. doi:10.1016/j.cej.2014.05.129
32. Noack CW, Dzombak DA, Nakles DV, Hawthorne SB, Heebink LV, Dando N, Gershenzon M, Ghosh RS (2014) Comparison of alkaline industrial wastes for aqueous mineral carbon sequestration through a parallel reactivity study. Waste Manag. doi:10.1016/j.wasman.2014.03.009
33. Jung S, Wang LP, Dodbiba G, Fujita T (2014) Two-step accelerated mineral carbonation and decomposition analysis for the reduction of CO(2) emission in the eco-industrial parks. J Environ Sci (China) 26(7):1411–1422. doi:10.1016/j.jes.2014.05.006
34. Dri M, Sanna A, Maroto-Valer MM (2014) Mineral carbonation from metal wastes: effect of solid to liquid ratio on the efficiency and characterization of carbonated products. Appl Energy 113:515–523. doi:10.1016/j.apenergy.2013.07.064
35. Dri M, Sanna A, Maroto-Valer MM (2013) Dissolution of steel slag and recycled concrete aggregate in ammonium bisulphate for CO_2 mineral carbonation. Fuel Process Technol 113:114–122. doi:10.1016/j.fuproc.2013.03.034
36. Azdarpour A, Asadullah M, Junin R, Manan M, Hamidi H, Mohammadian E (2014) Direct carbonation of red gypsum to produce solid carbonates. Fuel Process Technol 126:429–434. doi:10.1016/j.fuproc.2014.05.028
37. Santos RM, Van Bouwel J, Vandevelde E, Mertens G, Elsen J, Van Gerven T (2013) Accelerated mineral carbonation of stainless steel slags for CO_2 storage and waste valorization: effect of process parameters on geochemical properties. Int J Greenhouse Gas Control 17:32–45. doi:10.1016/j.ijggc.2013.04.004
38. Chang EE, Chen T-L, Pan S-Y, Chen Y-H, Chiang P-C (2013) Kinetic modeling on CO_2 capture using basic oxygen furnace slag coupled with cold-rolling wastewater in a rotating packed bed. J Hazard Mater 260:937–946. doi:10.1016/j.jhazmat.2013.06.052

39. Fernandez Bertos M, Li X, Simons SJR, Hills CD, Carey PJ (2004) Investigation of accelerated carbonation for the stabilisation of MSW incinerator ashes and the sequestration of CO_2. Green Chem 6(8):428. doi:10.1039/b401872a
40. Tsuyoshi S, Etsuo S, Minoru M, Nobuaki O (2010) Carbonation of γ-Ca2SiO4 and the mechanism of vaterite formation. J Adv Concr Technol 8(3):273–280
41. Cappai G, Cara S, Muntoni A, Piredda M (2012) Application of accelerated carbonation on MSW combustion APC residues for metal immobilization and CO_2 sequestration. J Hazard Mater 207–208:159–164. doi:10.1016/j.jhazmat.2011.04.013
42. Fernandez Bertos M, Simons SJ, Hills CD, Carey PJ (2004) A review of accelerated carbonation technology in the treatment of cement-based materials and sequestration of CO_2. J Hazard Mater 112(3):193–205. doi:10.1016/j.jhazmat.2004.04.019
43. Arickx S, Van Gerven T, Vandecasteele C (2006) Accelerated carbonation for treatment of MSWI bottom ash. J Hazard Mater 137(1):235–243. doi:10.1016/j.jhazmat.2006.01.059
44. Santos RM, Mertens G, Salman M, Cizer O, Van Gerven T (2013) Comparative study of ageing, heat treatment and accelerated carbonation for stabilization of municipal solid waste incineration bottom ash in view of reducing regulated heavy metal/metalloid leaching. J Environ Manage 128:807–821. doi:10.1016/j.jenvman.2013.06.033
45. Rendek E, Ducom G, Germain P (2006) Carbon dioxide sequestration in municipal solid waste incinerator (MSWI) bottom ash. J Hazard Mater 128(1):73–79. doi:10.1016/j.jhazmat.2005.07.033
46. Pan SY, Chiang PC, Chen YH, Chen CD, Lin HY, Chang EE (2013) Systematic approach to determination of maximum achievable capture capacity via leaching and carbonation processes for alkaline steelmaking wastes in a rotating packed bed. Environ Sci Technol 47 (23):13677–13685. doi:10.1021/es403323x
47. Huijgen WJJ, Ruijg GJ, Comans RNJ, Witkamp GJ (2006) Energy consumption and net CO_2 sequestration of aqueous mineral carbonation. Ind Eng Chem Res 45(26):9184–9194
48. Huijgen W, Comans R, Witkamp G (2007) Cost evaluation of CO_2 sequestration by aqueous mineral carbonation. Energy Convers Manag 48(7):1923–1935. doi:10.1016/j.enconman.2007.01.035
49. Pan SY, Chen YH, Chen CD, Shen AL, Lin M, Chiang PC (2015) High-gravity carbonation process for enhancing CO_2 fixation and utilization exemplified by the steelmaking industry. Environ Sci Technol 49(20):12380–12387. doi:10.1021/acs.est.5b02210
50. Pan S-Y, Lorente Lafuente AM, Chiang P-C (2016) Engineering, environmental and economic performance evaluation of high-gravity carbonation process for carbon capture and utilization. Appl Energy 170:269–277. doi:10.1016/j.apenergy.2016.02.103

Chapter 4
Environmental Impact Assessment and CCS Guidance

Abstract Avoiding the adverse impacts of carbon capture and storage activities on the environment and human health would require careful site selection, effective regulatory oversight, and appropriate monitoring program. The strategic environmental assessment and environmental impact assessment are procedural tools for evaluating and assessing possible environmental effects of a policy or certain project. This chapter provides the overview of the principles and methodology for strategic environmental assessment and environmental impact assessment for carbon capture and storage activity. The guidelines of carbon capture and storage activities to accomplish strategic environmental assessment or environmental impact assessment are also discussed and illustrated.

4.1 Strategic Environmental Assessment (SEA)

Strategic environmental assessment (SEA) methodology is a widely recognized and useful tool when structuring environmental aspects and performing the environmental impact assessment of large projects. It was laid out in Directive 2001/42/EC on the assessment of the effects of certain plans and programs on the environment [1], which has been transposed into legislation, i.e., Statutory Instrument 2004 No. 1633 [2]. An SEA is a systematic process for evaluating environmental consequences of proposed policies, plans, and programs to ensure the consequences are fully understood and appropriately addressed from the earliest stages of decision making [3–5]. As a result, the SEA should be transparent and suitable for communication with stakeholders that show a greater interest in the concept. Most practitioners consider SEA as a decision-aiding process rather than a decision-making process [5].

Governmental programs which contain decisions on the appointment of possible locations or routes or the consideration of alternative locations or routes for CCS activities are expected to be SEA-obligated [6]. In general, the SEA is undertaken at an earlier stage in the decision-making process than an environmental impactassessment (EIA) to ensure that environmental considerations are properly integrated into this stage of the decision-making process. In most cases around the

P.-C. Chiang and S.-Y. Pan, *Carbon Dioxide Mineralization and Utilization*, DOI 10.1007/978-981-10-3268-4_4

world such as the Netherlands [7], the permit for deploying a CCS project requires an EIA procedure. The purpose of the EIA is to clarify the potential effects of a CCS project on the environment, economy, natural resources, and society. More details regarding the EIA procedure are illustrated in Sect. 4.2.

4.1.1 Methodology and Framework

The objectives of SEA are to broadly present the preferred environmental, ecological, social, and economic outcomes to minimize detrimental effects of a project or activity. The SEA Directive does not have a list of plans or programs similar to the EIA [4]. Since the plans and programs proposed by the private sector may have considerable environmental impacts, the voluntary use of SEA methodology is encouraged. So far, the SEA has mainly been used for the evaluation of public infrastructure policies, programs, and plans [5, 6]. The methodology of the SEA basically comprises of five phases:

- Phase 1: Screening
  - Determination of obligation
  - Identification of key factors
  - Judgment by competent authority
- Phase 2: Scoping
  - Public notification
  - Public consultation
  - Determination of system boundary
- Phase 3: Formulation, analysis, and valuation
  - Formulation of environmental report
  - Analysis of current state (business-as-usual) of the environment
  - Description of alternatives to the plan
  - Publication of preliminary plan and environmental report
- Phase 4: Assessment
  - Public consultation
  - Decision making and action plans
  - Determination of key performance indicators
- Phase 5: Evaluation
  - Performance check
  - Evaluation of environmental impacts

Within the framework, the available baseline information could be collected. Although general guidelines exist, SEA guidelines and the best practice framework do not exist for applying to a CCS program. In the case of developing a CCS program, all activities that will be carried out to construct, operate, and close a CCS facility and all factors that could be affected by the above activities should be considered in the SEA [3]. Accordingly, the scope and boundary of the SEA would be determined. In the scoping stage, several strategies could be considered for future developments [5]:

- To widen the scope of assessment
- To expand the consultation requirements
- To increase the consideration of sustainability issues and health impacts

Afterward, the environmental impacts of different alternatives would be analyzed and quantified. In the following valuation step, preferable technical alternatives would be identified and weighted. From the results of the valuation, the best and worst alternatives with respect to the CCS program can be identified for the decision-making procedure.

4.1.2 Screening Key Aspects and Available Information

Since SEA is a key tool in sustainable development strategic decision making, it is usually applied at an earlier stage in a CCS development than EIA. The screening stage is the first step of the SEA to define the key aspects and issues for proposing an SEA work plan. Typically, a screening matrix with an overview of the status of knowledge in different environmental areas would be developed. It is suggested that SEA should be used when strategic-level decisions are being made for alternatives options and preferred options [5].

An essential part of the SEA process is to identify the current baseline of environmental conditions as a "business-as-usual" scenario. Most of the environmental issues relating to CCS are with the engineering aspect, e.g., ensuring the CO_2 remains in the storage reservoir for hundreds to thousands of years without significant leakage or seepage [5]. These two issues of engineering and environment are inevitably entwined. In other words, it is only with sufficient knowledge of the existing conditions that the key issues may be properly identified and addressed through the assessment. Therefore, it should be addressed from a strategic environmental perspective through the creation of minimum national or international standards and requirements for site selection, including [5] the following:

- Geology
  - Seal thickness and integrity
  - Fluid compatibility
  - Geochemical reaction
- Reservoir Property Assessment
- Well

- Disposal well selection
- Design (cementing, materials, corrosion) and modelling
- Monitoring

- Others
  - Surrounding environment conditions
  - Failure of wells and pipelines
  - Lateral migration potential

To determine suitable areas that might be acceptable for CCS, the above components should be carefully considered in the SEA. In addition, the minimum standards need to be used in tandem with good operational and monitoring procedures [5]. Since each individual CCS project has different characteristics, the generic standards would have limitations in the level of achievable protection.

4.1.3 Technical Description of Alternatives for CCS

CCS involves three stages: (1) capture and concentration, (2) transport, and (3) storage. Although the three distinctive steps of CCS are formally separated activities, an integrated approach to combining the separate EIA procedures for CCS activities into one procedure, or at least to provide close linkage between them, should be considered. For CCS, the capture of CO_2 from industries and/or power plants using fossil fuel combustion can be done by separating CO_2 from the flue gas either prior to fuel combustion or post-combustion. There are a range of CCS technologies at different stages of development. Therefore, in the SEA, characteristics and processes of CO_2 generation sources and various technical alternatives for CCS should be comprehensively described.

The components of the CCS system are briefly illustrated in the following content. More general information regarding carbon capture, utilization, and storage technologies can be referred to in Chaps. 2 and 3 in this book.

4.1.3.1 CO_2 Capture

Currently, several technologies for capturing, transporting, and storing CO_2 from coal-fired power plants are available in the literature. For CO_2 capture, four main technical alternatives could be considered:

- Precombustion capture
- Post-combustion capture
- Oxy-fuel combustion capture
- Industrial separation from natural gas processing, ammonia production, etc.

4.1.3.2 CO_2 Transport

Normally, CO_2 is captured as a gas and needs to be compressed (or cooled) for the transport process. For CO_2 transport, various alternatives could be considered:

- Pipeline (i.e., onshore and offshore)
- Shipping (i.e., offshore)
- Truck
- Railway

From the economical feasibility point of view, large-scale transport options are shipping and pipeline, while truck and train are possible means of transport for small-scale projects in the start-up phase of a CCS program. Typically, pipeline is the best alternative for transporting large quantities of CO_2 onshore, e.g., >1 Mt/year [3], which is a commercial technology. Tankers would generally only be functional for smaller volumes, e.g., ~1 Mt/year [5].

4.1.3.3 CO_2 Storage

For CO_2 storage, the following alternatives could be considered:

- Enhanced oil recovery (EOR) or enhanced gas recovery (EGR)
- Enhanced coal bed methane recovery (ECBM)
- Saline reservoirs
- Depleted hydrocarbon reservoirs
- Ocean storage (e.g., dissolution type or lake type)
- Mineral carbonation

The captured CO_2 can be stored both in onshore terrestrial geological formations and in offshore subseabed geological formations. In addition, although other options (such as ocean storage) exist, they are either in early phases of development or demonstration phases.

4.2 Environmental Impact Assessment (EIA)

Table 4.1 presents the comparison of EIA and SEA for CCS activities. Similar to SEA, the concept of EIA refers to the examination, analysis, and assessment of the proposed activities with a view to ensure environmental, social, and economic integrity for achieving long-term sustainable development. In particular for certain CCS activities, the comprehensive environmental and socioeconomic impact assessments should be thoroughly performed from a life cycle approach. In other words, with specific relation to CCS, an EIA would be conducted to a particular CCS project, while an SEA would examine CCS opportunities and policy on a regional basis (e.g., country wide).

Table 4.1 Comparison of environmental impact assessment (EIA) and strategic environmental assessment (SEA) for CCS activities

Category	EIA	SEA
Feature	Usually reactive to a proposed CCS development proposal	Proactive and informs CCS development proposals
Assessment contents	The effect of a proposed CCS development on the environment	• The effect of CCS policy, plans, or programs on the wider environment • The effect of the environment on the CCS development needs and opportunities
Target	A specific proposed CCS project	Areas, regions, or sectors of CCS development
System boundary	A well-defined beginning and end	A continuing process aimed at providing information at the right time
Impacts assessment	Direct impacts and benefits of a proposed CCS project	Cumulative CCS impacts and identifies implications and issues for a sustainable development
Focus	• Mitigation of CCS impacts and possible CO_2 leakages • Specific impacts of a proposed CCS project	Maintaining a chosen level of environmental quality
Perspective	A narrow site-specific perspective and a high level of detail	• A wide global perspective and a low level of detail to provide a vision and overall framework. • Provides a review of cumulative global effects of CCS

Courtesy of [5]

With the EIA procedure, the relevant information on environmental impacts required for various administrative decisions is gathered in a single report: the environmental impact statement (EIS). The EIS report should represent the knowledge base on environmental impacts due to the activity and is used as reference work in the decision making process [6].

Currently, CCS projects are not specifically mentioned in EIA around the world since the CCS relevant technologies are relatively new and under development [5]. However, in some cases (such as in the EU), CCS projects may be constrained by existing legislation. To ensure capture of a CCS development may be to amend EIA legislation in national countries by suggesting that the CCS projects are specifically required to be subject to an EIA [5]. Although legislation usually refers to guidelines for conducting EIAs, in many cases, it does not specifically require the use of the guidelines.

4.2.1 Methodology and Framework

A variety of frameworks for EIA procedures can be found in international guidelines, the European Union, and core countries, but they are fundamentally similar.

In many cases, some elements of the good practice of EIA are not actually required by law [5]. Det Norske Veritas (DNV) Ltd has proposed the best practice of the EIA procedure, which is based on International Finance Corporation guidelines combined with best practices identified from countries where DNV operate. With regard to compliance with CCS best practices, EIA frameworks may require amendments to ensure that the minimum requirements for acceptance by mechanism, such as clean development mechanism (CDM) and joint implementation (JI), can be achieved. The roles of CCS in the CDM are discussed in Sect. 4.3.4.

Generally, the EIA is used to safeguard environmental interests in the face of normally highly positive economic and socially beneficial impacts. The suggested stages for conducting EIAs are as following:

- (Stage 1) screening and scoping
- (Stage 2) analysis of alternatives
- (Stage 3) project descriptions
- (Stage 4) review on environmental baseline and legislation
- (Stage 5) impact assessment
- (Stage 6) environmental management plan for impact mitigation
- (Stage 7) environmental monitoring plan
- (Stage 8) reporting and review
- (Stage 9) project implementation and operations

Any possible risks or uncertainties that could cause the CCS project to be abandoned should be identified in the EIA. The risks and uncertainties can be determined via various approaches.

This can provide an overall insight into the environmental burdens of the entire CCS chain, thereby being able to streamline various decision-making procedures. Since the designs of CCS chain networks consist of multiscale concerns, a sound decision-making framework at material, process, and supply chain levels is required. Various approaches could be applied to achieve the goal, such as a hierarchical and multiscale framework to minimize investment, operating costs, and material costs, as shown in Fig. 4.1 [8]. In all cases, the best available techniques should be applied to ensure high-level protection for the environment and for communities [9].

Currently, a wide variety of methods have been used in the EIA, such as

- Life cycle assessment (LCA) [6]: quantification of the environmental impacts
- Environmental risk assessment (ERA) [10]: identification of potential hazards of a proposal to manage uncertainty
- Acoustics models [11]: calculation of the sound propagation in ocean
- Geodetic deformation analysis [12]: determination of the trend of movements (displacements) for all the common points in a monitoring network
- Water discharge analysis [13]: identification of thermal and waste substances during water discharge
- Other forms of surveys: ecological, archeological, geo-hydrological analyses

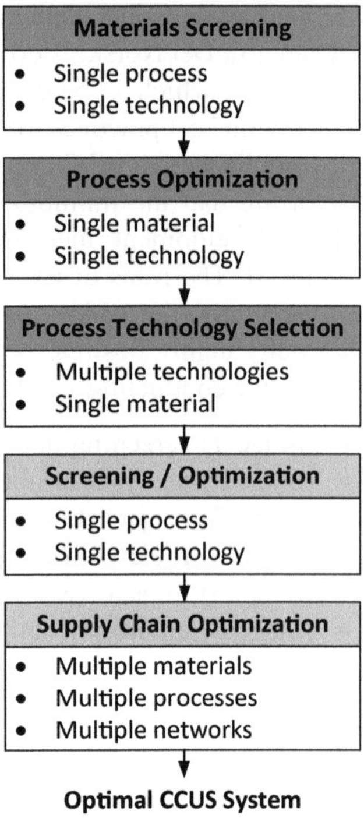

Fig. 4.1 Framework of multiscale systems engineering for CO_2 capture, utilization, and sequestration (CCUS)

In an EIA, different decision-making procedures for obtaining permits or exemptions are incorporated into a single procedure [7]. The EIA can be influenced by third parties by requesting additional and/or challenging information. Therefore, the possibility of public participation could play a key role in the public perception and rules of acceptation of CCS. It suggests that the following requirements and guidelines for a CCS project should be incorporated in the environmental assessment to avoid a significant release of CO_2 [5]:

- An integrated environmental, social, and health impact assessment (EHSIA) approach
  - Identification of environment resources
  - Requirement of operator commitment for monitoring
  - Provision of handing long-term liability
  - Consideration of storage performance assessment (SPA) as an inherit part
- A risk-based source-pathway-receptor approach
  - Identification of project with high risks of early closure
  - Evaluation of a carbon balance across the entire project life cycle
  - Provision of clear regulatory guidance on the play-off in priorities between local pollution concerns and climate change concerns

4.2.2 Environmental and Natural Resource Aspect

To include the environment in the decision-making process on permits and investments of the involved parties, EIA is usually introduced to quantify the environmental impacts of specific activities. The purpose of the EIA is to evaluate different alternatives and find the best option, in terms of environmental benefits, for a certain project.

For the purpose of permanent storage of CO_2, the environmental impact of a CCS project highly depends on (1) the characteristics of underground geological formation, (2) overpressure issues of the reservoir, and (3) lithologies adjacent to the storage reservoir [14]. The potential environmental impacts that may arise from CCS activities include the following:

- Air emissions: particulate matter, nitrogen oxides, sulfur oxides, dust, mercury, polycyclic aromatic hydrocarbons, etc.
- Water use associated with current CCS technologies: enhanced oil recovery (EOR)
- (Ground) water pollution: lubricant for drilling operations
- Solid waste generation: during drilling operations
- Noise: disturbing levels of noise
- Human health and safety: health of population
- Biodiversity: impact on ecosystems and habitats
- Geology: landscape, soils, and underground space.

On the other hand, the potential natural resource impact that may be caused from the CCS activities includes the following:

- Resources and raw materials: natural asset, energy source
- Waste utilization: management hierarchy

4.2.3 Socioeconomic Aspect

From an economic point of view, the cost of geological storage of CO_2 is highly site-specific, depending on factors such as (1) the location of the project (onshore or offshore), (2) the number of wells for injection, and (3) the depth of the storage formation. However, in all cases, the costs for storage (including monitoring) typically are in the range 0.6–8.3 US$/t-$CO_2$ stored [15]. The potential socioeconomic impacts that may result from the CCS activities include the following:

- Traffic and transport: travel and transport on communities
- Economy and skills
- Archeology and cultural heritage: heritage resources, historic building, archeological features.

4.3 CO_2 Capture and Storage (CCS) Guideline

Geological storage of CO_2 has drawn extensive attention around the world from a concept of limited interest to one that is quite widely regarded as a potentially important mitigation option, as shown in Fig. 4.2. It is noted that the density of CO_2 will increase with depth. Until at about 800 m deep or greater, the injected CO_2 will be in a dense supercritical state [15].

4.3.1 Challenges

The existing challenges for widely deploying CCS activities as a CO_2 emissions control option [16] could be categorized into four aspects:

- Institutional barriers:
  - Building public understanding, awareness, and acceptance.
- Regulatory barriers:
  - Establishing an adequate legal and regulatory framework to support broad CCS deployment, including dealing with long-term liability;

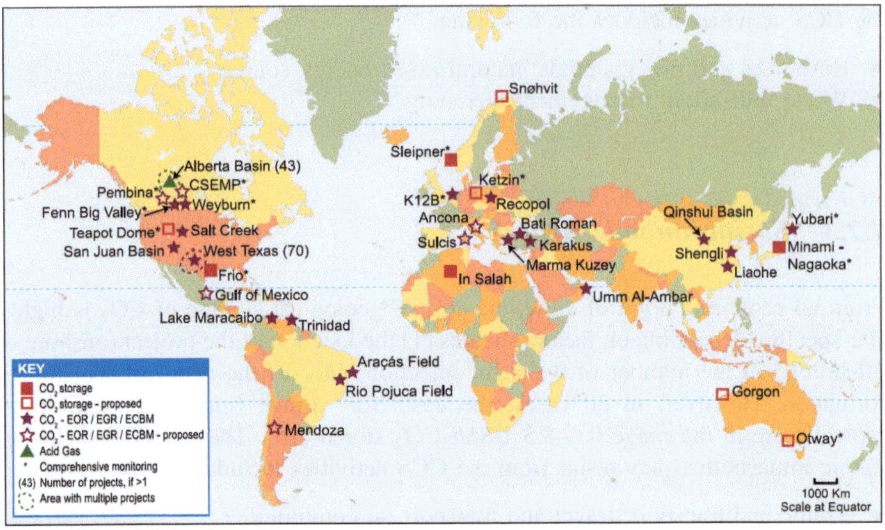

Fig. 4.2 Location of geological storage sites where CCS activities are planned or under way (courtesy of Special Report of the Intergovernmental Panel on Climate Change IPCC, Carbon Dioxide Capture and Storage, © Intergovernmental Panel on Climate Change 2005, published by Cambridge University Press ISBN 9780521866439 [15])

- Technological barriers:
  - Addressing the cost and energy penalty of capture;
  - Proving CO_2 storage permanence;
  - Verifying that sufficient storage capacity exists;
  - Developing best practices for the life cycle of a CCS project, from site selection through to site closure and post-closure monitoring.
- Financial barriers:
  - Global need for significant financial investments to bring numerous commercial-scale demonstration projects online in the near future;

4.3.2 Risk of CO_2 Release

The CO_2 storage site is the key area of risk in the CCS chain. It is noted that the risk of CO_2 release into the atmosphere during the phases of injection and storage exists. In general, the injection phase has a relatively limited period of operation, e.g., ~50 years. Based on experience with the oil and gas industries, the risk of release of significant CO_2 is estimated to be 10^{-3} per reservoir per annum [5]. One of the major reasons is due to the corrosion of injection equipment, which could be controlled to less than 2.5 μm/pa by using polyethylene. However, more experience from CCS trials should be collected to confirm the available information from oil and gas injection wells. Other possible CO_2 release pathways include the following:

- Failure of abandoned wells and/or wellbore
- Diffusion flow through caprock via faults or by buoyancy through permeable zones
- Dissolution and transport of CO_2 charges waters by groundwater flow (most important leakage mechanism from aquifers)

Scientific knowledge and industrial experience can serve as a basis for appropriate risk management. Mapping of the reservoir and surrounding area should be a critical component to reduce the risk of CO_2 release from the storage site. In addition, risk management would need to incorporate the results of the storage performance assessment (SPA), which could be as an inherent part of EIA, as shown in Fig. 4.3. Furthermore, both a better understanding of the impact of impurities and the development of a suitable modelling technique are essential to predict the short- and long-term fate of stored CO_2 in a variety of geological formations [17].

4.3.3 Monitoring Program

Since it is possible that the stored CO_2 could leak or seep out of a reservoir, it is necessary to monitor the CO_2 via a wide range of techniques, such as 2D and 3D

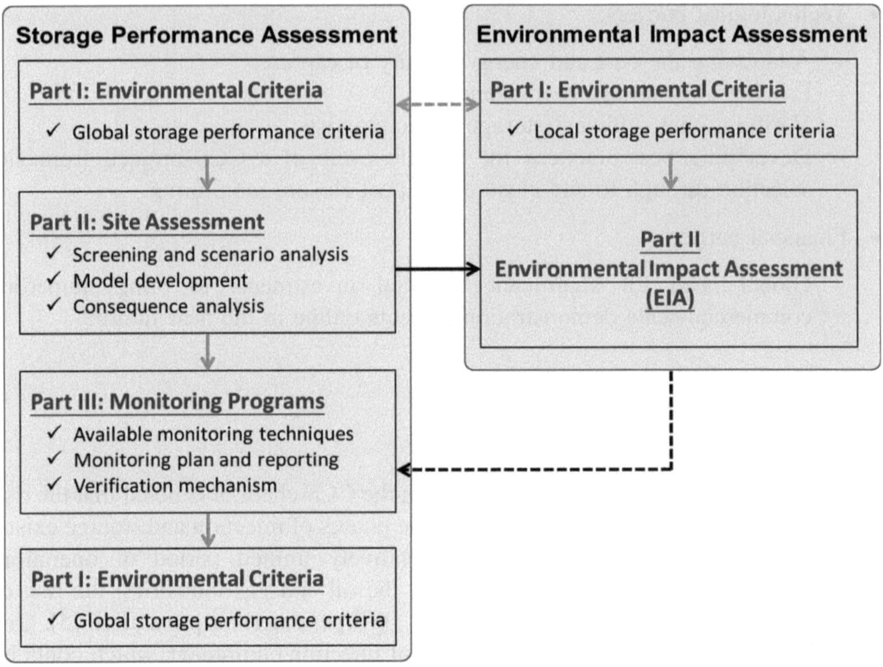

Fig. 4.3 Procedure of storage performance assessment (SPA) for EIA of a CCS project

seismic reflection surveys. Table 4.2 presents an overview of measurement, monitoring, and verification technologies for the CCS activity. Many monitoring techniques are mature but require further research and development. One of the main reasons is due to a lack of awareness of business opportunities, although industrial-scale projects usually have programs to develop and evaluate monitoring techniques. Therefore, a draft of monitoring methodology should be proposed for individual CCS projects.

4.3.4 CCS in Clean Development Mechanism (CDM)

4.3.4.1 Kyoto Protocol and Its Role in CCS Activity

Although the Kyoto mechanisms had been guaranteed by 2012, the experience in the development of these mechanisms in the Kyoto Protocol can be referred to. There are three mechanisms, i.e., (1) clean development mechanism (CDM), (2) joint implementation (JI), and (3) international emission trading (IET), included in the Kyoto Protocol:

Table 4.2 Measurement, monitoring, and verification technologies for the CCS activity (adapted from [5])

Techniques	Detection method	Technology readiness
Time lapse 4D multicomponent seismic	Acoustic	Well known
Cross well seismic tomography	Acoustic	Well known
Vertical seismic profiling	Acoustic	Well known
Down hole microseismic	Acoustic	Developmental
Electrical resistance tomography	Electrical	Developmental
Electromagnetic induction tomography	Electrical	Prototype
Soil gas sampling	Chemical	Well known
Noble gas tracing	Chemical	Early testing
Other gas tracing	Chemical	Early testing
Well head detectors	Chemical	Prototype
Brine sampling	Chemical	Well known
Subsurface and surface tilt meters	Physical	Developmental
Airborne hyper-spectral imaging	Optical	Developmental
Space-based monitoring	Microwave	Proposed

- Clean development mechanism (CDM): An incentive for companies in industrialized countries to invest in eligible emission reduction projects in developing countries.
- Joint implementation (JI): An incentive for companies in industrialized countries to reduce emissions through cooperative efforts, where a JI project may be involved.
- International emission trading (IET): Industrialized countries are allowed to meet their commitment via buying or selling excess emission credits among themselves.

Both the CDM and JI schemes are related to the project level, where introducing EIA as a set standard could be possible. In contrast, IET is related to a trading system at the international level, where EIA might possibly become a decision tool in it. Therefore, it is not anticipated that a CCS project would be accepted under the CDM and/or JI schemes without an EIA [5].

Appropriate amendments to CCS project activities should be applied under the CDM. It is noted that the CDM registry should be used to ensure the accurate accounting of the issuance, holding, transfer, acquisition, and cancellation of certified emission reductions (CERs) from CCS project activities under the CDM. In the project design, the project participants should clearly document the liability obligations arising from the proposed CCS project activity or its geological storage site [9].

According to the suggestion by the United Nations Framework Convention on Climate Change (UNFCCC), three major phases are recommended in proximity to the proposed CCS project or activity [9], including (1) characterization of suitable geological storage sites; (2) risk and safety assessment; and (3) monitoring and numerical modeling. Brief illustrations regarding CCS in the CDM can be found as follows:

4.3.4.2 Characterization of Suitable Geological Storage Site

Under the proposed conditions of use, geological storage sites should only be used to store carbon dioxide if there is no significant risk of seepage, environmental impact, and/or human health concerns. Typically, the geological storage site should not be located in international waters. Geological storage of CO_2 should be carefully evaluated to select a suitable storage site under the CDM. Available evidence (e.g., data, analysis and history matching) should be provided to indicate a complete and permanent storage of CO_2 [9], including

- Characterization of the geological storage site architecture and surrounding domains, such as
  - Structure of the geological containment
  - Areal and vertical extent of the site
  - Cap rock formation(s)
  - Overburden
  - Secondary containment zones
  - Fracture system
  - Fluid distribution and physical properties
  - Injection formation (associated with storage capacity)
- Characterization of dynamic behavior, sensitivity characterization, and risk assessment.
- Establishment of a site development and management plan (site preparation, well construction, injection rates, operating and maintenance programs, etc.).

As a result, a wide range of data and information should be collected and used in performing the characterization and selection of a suitable geological storage site. The timing and management of the closure phase of the CCS activity, including site closure and related activities, should also be addressed.

4.3.4.3 Risk and Safety Assessment

To assess the integrity of the geological storage site and potential impacts on human health and ecosystems, a thorough and comprehensive risk and safety assessment should be carried out. In this phase, the risk and safety assessment should be used to reveal the environmental and socioeconomic impact assessments of the sequestration activity. Therefore, the entire CCS chain, such as surrounding environments, should be taken into consideration and assessment. Also, it can be used to determine operational data for the application of development and management plans for the storage site. As a result, for example, the appropriate maximums of injection pressure that will not compromise the confining cap rock formation(s) and the overburden could be set.

Several key components of risks and effects should be especially considered in conducting risk and safety assessment [9, 15]:

- Containment failure: This results in emissions of greenhouse gases from above-ground installations and/or seepage from subsurface installations, thereby causing potential effects on (1) underground sources of drinking water, (2) the chemical properties of seawater, (3) human health, and (4) ecosystems.
- Continuous slow seepage from a geological storage site: It might arise due to seepage (1) along injection wells or abandoned wells; (2) along a fault or fracture; (3) through the cap rock formation; and/or (4) across faults and ineffective confining layers.
- Sudden mass release of CO_2 from surface CCS installations: it might arise due to pipeline rupture.
- Potential induced seismicity or other geological impacts.
- Other potential consequences for the environment, local ecosystems, property, and public health.

With risk and safety assessment, it could be used to help prioritize locations and approaches for enhanced monitoring activities [9]. Typically, a risk assessment comprises of four steps:

- Step 1: Hazard characterization:
  - Potential hazards resulting from the CCS activity
  - Potential seepage pathways from the geological storage site
  - Critical parameters affecting potential seepage and its magnitude
  - Sensitivity to various assumptions
- Step 2: exposure assessment:
  - Characteristics of surrounding populations and ecosystems
  - Potential fate and behavior of any seeped CO_2
- Step 3: effects assessment:
  - Sensitivity of species, communities or habitats linked to potential seepage events identified during the hazard characterization
  - Effects of elevated CO_2 concentrations in the atmosphere, biosphere, and hydrosphere
- Step 4: risk characterization:
  - Safety and integrity of storage site in the short-, medium-, and long-term scale
  - Risk assessment of seepage under the proposed conditions of use in development and management plan

Furthermore, a contingency plan for large incidents (such as seepage) should be prepared with all the necessary plans, including availability of (1) a team, (2) trained personnel, (3) materials and equipment, and (4) financial means to mitigate the adverse impacts of the large incidents.

4.3.4.4 Monitoring and Numerical Modeling

Monitoring of CCS project activities is essential to meet the following goals [9]:

- To determine the reductions in anthropogenic emissions by sources of GHGs that have occurred as a result of the registered CCS project activity.
- To provide assurance of the environmental integrity and safety of the geological storage site.
- To ensure that good site management is taking place, taking account of the proposed conditions of use set out in the site development and management plan.

In this phase, the monitoring task forces should be carried out to meet the following four objectives [9]:

- To ensure that the injected CO_2 is well contained within the storage site, as well as the project boundary.
- To confirm that injected CO_2 is behaving as predicted to minimize the risk of any seepage or other adverse impacts.
- To detect and estimate the flux rate and total mass of CO_2 from any seepage.
- To determine whether timely and appropriate remedial measures have been carried out in the event of seepage.

Typically, monitoring of the geological storage site begins before injection activities commenced to ensure adequate time for the collection of any required baseline data. The parameters and information that are monitored and collected should be transparently specified. The location and frequency of the application of different monitoring techniques during the operational phase, closure phase, and post-closure phase should also be determined. At an appropriate frequency, several key items of monitoring techniques and measurement targets include [9] the following:

- Geological, geochemical, and geomechanical parameters, such as fluid pressures, displaced fluid characteristics, fluxes, and microseismicity
- CO_2 stream and its composition at various points in the entire CCS chain
- Temperature and pressure at the top and bottom of the injection well(s) and observation well(s)
- Parameters in overburdened and surrounding domains of storage site, e.g., groundwater properties and soil gas measurements

- Detection of corrosion or degradation of the transport and injection facilities
- Effectiveness of any remedial measures taken in the event of seepage

To improve the accuracy and/or completeness of data and information, the numerical models used to characterize the storage site should be periodically updated by conducting new simulations using the monitored data and information. This could assist in adjusting the event of significant deviations between the observed and predicted behaviors. Therefore, it could confirm that no future seepage can be expected from the geological storage site.

References

1. Directive 2001/42/EC of The European Parliament and of The Council (2001) Off J Eur Commun 197(30)
2. The Environmental Assessment of Plans and Programmes Regulations 2004 (2004) UK
3. Eriksson S, Andersson A, Strand K, Svensson R (2006) Strategic environmental assessment of CO_2 capture, transport and storage: a report within the CO_2 free power plant project. Vattenfall Research and Development AB, Sweden
4. DECC (2009) Strategic environmental assessment for a framework for the development of clean coal-post adoption statement. Department of Energy & Climate Change, UK
5. Vendrig M, Purcell M, Melia K, Archer R, Harris P, Flach T (2007) Environmental assessment for CO_2 capture and storage. Det Norske Veritas (DNV) Ltd., UK
6. Faaij A, Turkenburg W (2010) The screening and scoping of environmental impact assessment and strategic environmental assessment of carbon dioxide capture & storage in the Netherlands. Environ Impact Assess Rev 28(6):392–414
7. Koornneef J, Faaij A, Turkenburg W (2009) Environmental impact assessment of carbon capture & storage in the Netherlands. Cato
8. Hasan MMF, First EL, Boukouvala F, Floudas CA (2015) A multi-scale framework for CO_2 capture, utilization, and sequestration: CCUS and CCU. Comput Chem Eng 81:2–21. doi:10.1016/j.compchemeng.2015.04.034
9. UNFCCC (2011) Conference of the parties serving as the meeting of the parties to the kyoto protocol: seventh session. United Nations Framework Convention on Climate Change
10. Hyett D (2010) Environmental risk assessment in environmental impact assessment: optional or mandatory. In: 30th Annual meeting of the international association for impact assessment conferences: the role of impact assessment in transitioning to the green economy, 6–11 Apr 2010, Switzerland, pp 1–6
11. Spiga I (2015) Ocean acoustics modelling for environmental impact assessments. North American Electric Reliability Corporation, USA
12. Setan H, Singh R (2001) Deformation analysis of a geodetic monitoring network. Geomatica 55(3)
13. Okadera T, Watanabe M, Xu K (2006) Analysis of water demand and water pollutant discharge using a regional input–output table: an application to the City of Chongqing, upstream of the Three Gorges Dam in China. Ecol Econ 58(2):221–237. doi:http://dx.doi.org/10.1016/j.ecolecon.2005.07.005
14. Barros N, Oliveira GM, Lemos de Sousa MJ (2012) Oliveira, Gisela. Environmental Impact Assessment of Carbon Capture and Sequestration.pdf. Paper presented at the 32nd annual meeting of the international association for impact assessment, Porto, Portugal

15. IPCC (2005) IPCC special report on carbon dioxide capture and storage. Intergovernmental Panel on Climate Change. Cambridge. doi:ISBN-13 978-0-521-86643-9
16. CSLF (2011) inFocus: what is carbon utilization? Carbon Sequestration Leadership Forum (CSLF)
17. National Resource Canada (2006) Canada's CCS technology roadmap. Canmet Energy Technology Centre

Part II
Fundamentals of Accelerated Carbonation

Chapter 5
Principles of Accelerated Carbonation Reaction

Abstract Industrial alkaline solid wastes are ideal accelerated carbonation materials due to their availability and low cost. These materials are generally rich in calcium content and often associated with CO_2 point source emissions so no mining is needed and the consumption of raw materials is avoidable. This chapter provides the principles and definitions of accelerated carbonation reaction using alkaline solid wastes. Two types of carbonation processes, i.e., direct and indirect carbonation, are briefly discussed from the theoretical considerations, including process chemistry and key performance indicators. The performance and application of both direct and indirect carbonation processes can be found in detail in Chap. 8.

5.1 Principles and Definitions

5.1.1 Theoretical Considerations

Accelerated carbonation was first proposed by Seifritz [1], which involved alkaline materials reacting with high-purity CO_2. The concept behind accelerated carbonation processes is to mimic natural weathering, where gaseous CO_2 reacts with metal-oxide-bearing materials in the presence of moisture. In this case, the reaction can be accelerated to a timescale of a few minutes or hours. After accelerated carbonation, stable and insoluble carbonates will be formed, where CaO and MgO are the most favorable metal oxides in reacting with CO_2 [2]. In principle, the affinity of oxides for carbonate formation depends on the following:

- chemisorption strength to CO_2 gas (especially for gas–solid interface);
- number of basic sites at the surface (especially for gas–solid interface);
- solubility product constant, i.e., K_{sp} (in aqueous carbonation reaction); and
- total content in solid particle.

P.-C. Chiang and S.-Y. Pan, *Carbon Dioxide Mineralization and Utilization*, DOI 10.1007/978-981-10-3268-4_5

Therefore, the affinity of oxides varies as follows: basic oxides (CaO, MgO) > amphoteric oxides (Al_2O_3, Cr_2O_3, TiO_2, MnO, iron oxides) > acidic oxides (SiO_2). From a thermodynamic point of view, both alkaline earth metals (e.g., Ca and Mg) and alkali metals (e.g., Na and K) can be carbonated [3, 4]. However, alkali carbonates and/or bicarbonates are soluble in water, which will result in releasing CO_2 back into the atmosphere and therefore considered to be unsuitable for the long-term storage of CO_2. Consequently, the capacity of CO_2 fixation by these alkaline residues depends directly on the proportion of binary oxide (CaO and MgO) and/or hydroxide ($Ca(OH)_2$ and $Mg(OH)_2$) content in the matrix. Moreover, CaO offers more potential for chemisorptions of CO_2 than MgO due to its basic characteristics [5]. Furthermore, a number of other metals, such as Mn, Fe, Co, Ni, Cu, and Zn, are impractical for carbonation due to their unique and precious features for recovery and utilization.

Accelerated carbonation can be classified into two main types: mineral carbonation (as discussed in this chapter) and alkaline solid waste carbonation. To provide a significant amount of CO_2 fixation, large amounts of cheap raw materials are required as feedstock for carbonation. Alkaline (industrial) solid wastes are ideal feedstocks for accelerated carbonation which is an exothermic reaction.

- Carbonation products are thermodynamically stable, which can be beneficially reused in a variety of application.
- Flue gas CO_2 can be used as a reactant, and the solid wastes are often associated with CO_2 point source emissions.
- It can neutralize the pH of the solution.
- It decreases leaching of heavy metals and trace elements.
- No on-site transport is required.

Figure 5.1 shows a scheme of material fluxes associated with ex situ carbonation of alkaline solid wastes to produce green construction materials. Accelerated carbonation is considered an effective approach to simultaneously fixing CO_2 and eliminating the contents of free-CaO and $Ca(OH)_2$ in solid residues [6, 7]. Also, the qualities of alkaline solid wastes can be improved to the point where they are qualified for civil engineering applications.

Accelerated carbonation of alkaline solid wastes can be carried out via two different approaches, as shown in Fig. 5.2:

- Direct carbonation: The reaction occurs in a single route step.
- Indirect carbonation: Alkaline earth metal is first extracted from the mineral matrix and subsequently carbonated.

Accelerated carbonation has been progressively evaluated to determine the CO_2 capture capacity of alkaline solid wastes via direct carbonation [9, 10] and indirect carbonation for precipitated calcium carbonate production [11, 12]. It is noted that carbonation can result in lowering of pH, affecting the solubility, and leaching of metals which are mobilized at high pH and fixed at low pH [13]. Carbonate precipitates are thermodynamically stable and could theoretically fix CO_2 permanently

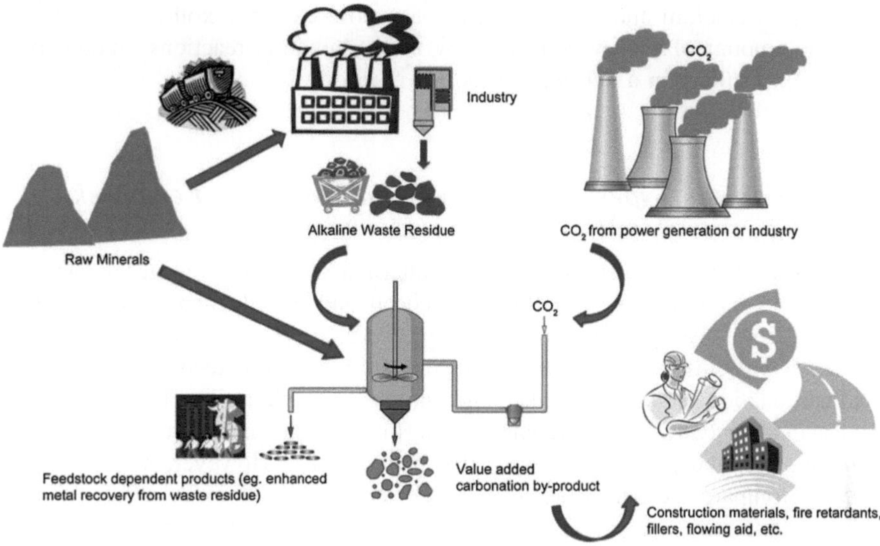

Fig. 5.1 Material fluxes associated with ex situ carbonation of alkaline solid wastes to produce green construction materials. Reprinted by permission from Macmillan Publishers Ltd: Ref. [8], copyright 2012

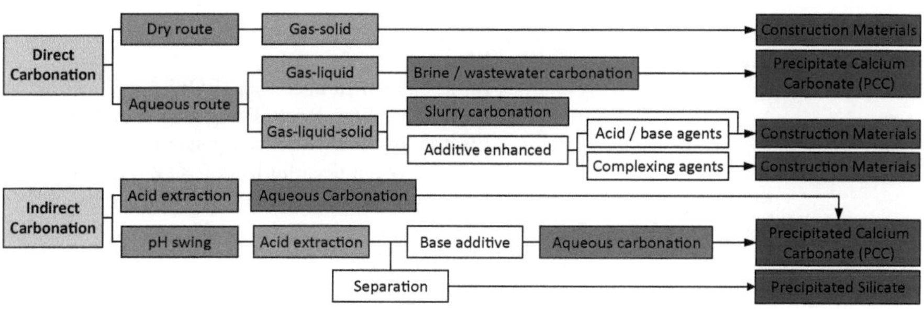

Fig. 5.2 Various process routes of accelerated carbonation for CO_2 mineralization

under ambient conditions, since they have a lower energy state than their reactants (CO_2 and silicates). However, they can be readily dissolved in the presence of strong acids in ambient conditions. Therefore, a risk of CO_2 release into the atmosphere is a concern if the carbonate precipitates make contact with acids, such as acid rain (pH 5–7).

A study on the Gibbs free energy of the carbonation reactions, calculated by Outokumpu HSC 5.1, indicated that the carbonation of calcium ions proceeds at temperatures over 45 °C, while the carbonation of magnesium ions should be possible only at temperatures over 144 °C [14]. In addition, the dissolution reactions of calcium and magnesium release more energy (exothermic, $\Delta H < 0$) than the amount of energy consumed by the carbonation reactions (endothermic, $\Delta H > 0$), resulting in a net exothermic net reaction.

5.1.2 Various Types of Alkaline Solid Wastes as Feedstock

Alkaline wastes can be suitably used as alternative feedstock in accelerated carbonation to natural ores since they are cheap and usually generated in a large quantity from nearby CO_2 emission points (e.g., power plants and industries). Table 5.1 presents the examples of alkaline solid wastes as feedstock for accelerated carbonation. Suitable alkaline solid wastes for accelerated carbonation include iron/steel slags (CaO-rich materials), ashes (fine particle size), cement wastes (fine particle size), and paper mill waste (fine particle size). However, it is difficult to directly compare the performance of accelerated carbonation for different wastes since each waste has its own unique set of advantages and disadvantages. Moreover, different approaches and processes are developed and performed, resulting in various forms of results.

Table 5.1 Example of alkaline solid wastes as feedstock for accelerated carbonation

Waste group	Example
Iron and steel slags	Blast furnace slag (BFS) Basic oxygen furnace slag (BOFS) Electric arc furnace oxidizing slag (EAFOS) Electric arc furnace reducing slag (EAFRS) Argon oxygen decarburization slag (AODS) Ladle furnace slag (LDS)
Air pollution control (APC) residues	Cyclone dust Cloth-bag dust Municipal solid waste incinerator (MSWI) fly ash Coal fly ash
Bottom ash (from furnace or incinerator)	MSWI bottom ash Boiler ash Coal slag Oil shale ash
Cement wastes	Cement kiln dust (CKD) Cement bypass dust (CBD) Construction and demolition waste Cement/concrete waste Blended hydraulic slag cement (BHC)
Mining and mineral processing wastes	Asbestos tailings Copper tailings (copper-nickel-PEG) Red mud (Bauxite residue)
Paper mill wastes	Lime kiln residues (calcium mud or lime mud) Green liquor dreg Paper sludge

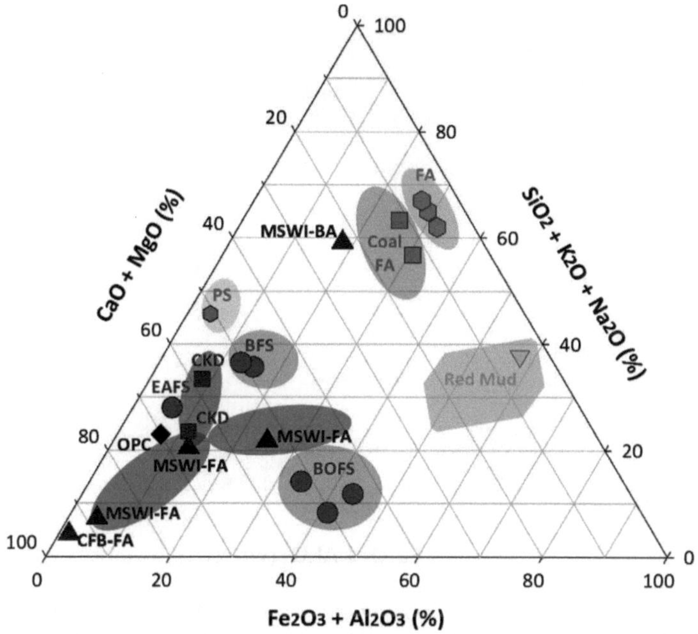

Fig. 5.3 Normalized phase diagram of (CaO + MgO) − (SiO_2 + Na_2O + K_2O) − (Al_2O_3 + Fe_2O_3) for various types of alkaline wastes. BFS (blast furnace slag); BOFS (basic oxygen furnace slag); PS (phosphorus slag); FA (fly ash); CKD (cement kiln dust); OPC (ordinary Portland cement); MSWI-FA (municipal solid waste incinerator fly ash); MSWI-BA: (municipal solid waste incinerator bottom ash); CFB-FA (circulate fluidized boiler bed fly ash). Reprinted by permission from Taiwan Association for Aerosol Research: Ref. [17], copyright 2016

Figure 5.3 shows the relationship of CO_2 capture capacity (in terms of CaO and MgO contents), hardness (in terms of Fe_2O_3 and Al_2O_3 contents), and pozzolanic property (in terms of SiO_2, K_2O and Na_2O contents) for different alkaline solid wastes. Alkaline solid wastes are chemically unstable with high calcium oxide content, e.g., iron and steel slag (30–60% CaO), APC residues (Ca content up to 35%), mining waste (Ca content ~5%), cement waste (30–50% CaO), municipal solid waste incinerator bottom ash (10–50% CaO), and coal fly ash (5–60% CaO). Ordinary portland cement (OPC) is a hydraulic product, while blast furnace slag (BFS) and fly-ash (FA) are, respectively, latent-hydraulic and pozzolanic byproducts [15]. Conversely, electric arc furnace slag (EAFS) is neither hydraulic nor pozzolanic due to its lack of tricalcium silicates and amorphous SiO_2 content [16]. In the presence of water, the calcium-bearing components in alkaline solid wastes could be hydrated and reacted with CO_2 in a high-pH solution (i.e., >10) to form calcium carbonates.

Typically, the amounts of CaO and SiO_2 contents in basic oxygen furnace slag (BOFS) increase with the decrease in particle size, while the Fe_2O_3 fraction decreases [18, 19]. Figure 5.4 illustrates the relationship of CO_2 capture capacity

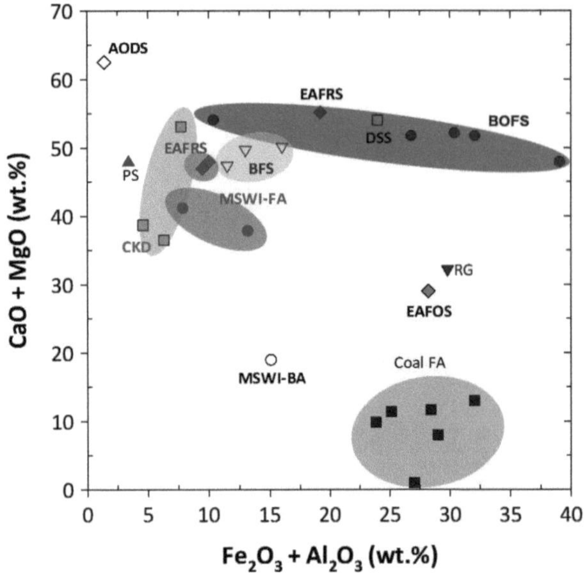

Fig. 5.4 Relationship of CO_2 capture capacity (i.e., CaO and MgO contents) and hardness (i.e., Fe_2O_3 and Al_2O_3 contents) for different solid wastes. Acronym: basic oxygen furnace slag (BOFS); electric arc furnace oxidizing slag (EAFOS); electric arc furnace reducing slag (EAFRS); argon oxygen decarburization slag (AODS); municipal solid wastes incinerator (MSWI); fly ash (FA); cement kiln dust (CKD); blast furnace slag (BFS), phosphorus slag (PS); steel slag (SS); red gypsum (RG); desulfurized slag (DSS)

(in terms of CaO and MgO contents) and hardness (in terms of Fe_2O_3 and Al_2O_3 contents) for various types of alkaline solid wastes. The contents of CaO and MgO in iron and steel slags are found to be relatively higher than those in fly ash (FA) or bottom ash (BA) . However, some steel slags such as basic oxygen furnace slag (BOFS) encounter hurdles in the need of energy-intensive process for material grinding due to their hard property (i.e., high contents of Fe_2O_3 and Al_2O_3). This would make it challenging as viable sinks for CO_2. Conversely, FA is relatively attractive since it is a fine powder. In other words, the costs for transportation, extraction, and crushing are minimal.

5.2 Types of Accelerated Carbonation Using Alkaline Solid Wastes

Alkaline solid wastes tend to be chemically less stable than geologically derived minerals. Therefore, carbonation of these solid wastes does not generally require the extraction of reactive ions from the solid matrix due to the alkaline-containing components as the mainly reactive phase.

5.2.1 Direct Carbonation

Direct carbonation can be divided into two types: (1) gas–solid (dry) carbonation, generally operated at a liquid-to-solid ratio of less than 0.2; and (2) aqueous (wet) carbonation, operated at a liquid-to-solid ratio of more than 5.

5.2.1.1 Gas–Solid (Dry) Carbonation

The direct gas–solid carbonation process was first developed by Lackner et al. [20], which consisted of converting silicate or metal oxides directly to carbonates using gaseous or supercritical CO_2. The gas–solid carbonation is the simplest approach to mineralization, as described in Eq (5.1):

$$\mathrm{Ca/Mg-silicate_{(s)}+CO_{2\,(g)}\rightarrow (Ca/Mg)CO_{3\,(s)}+SiO_{2\,(s)}} \tag{5.1}$$

In some cases, the process chemistry of gas–solid (dry) carbonation also can be expressed, in terms of various reacting minerals, as follows:

$$\mathrm{CaO_{(s)}+CO_{2(g)}\rightarrow CaCO_{3(s)}},\quad \Delta H=-179\,\mathrm{kJ/mol\,CO_2} \tag{5.2}$$

$$\mathrm{MgO_{(s)}+CO_{2(g)}\rightarrow MgCO_{3(s)}},\quad \Delta H=-118\,\mathrm{kJ/mol\,CO_2} \tag{5.3}$$

$$\mathrm{CaSiO_{3(s)}+CO_{2(g)}\rightarrow CaCO_{3(s)}+SiO_{2(aq)}},\quad \Delta H=-90\,\mathrm{kJ/mol\,CO_2} \tag{5.4}$$

$$\mathrm{Ca_2SiO_{4(s)}+2\,CO_{2(g)}\rightarrow 2\,CaCO_{3(s)}+SiO_{2(s)}},\quad \Delta H=-44\,\mathrm{kcal/mole\,CO_2} \tag{5.5}$$

$$\mathrm{Mg_2SiO_{4(s)}+2CO_{2(g)}\rightarrow 2MgCO_{3(s)}+SiO_{2(aq)}},\quad \Delta H=-44.5\,\mathrm{kJ/mol\,CO_2} \tag{5.6}$$

$$\begin{aligned}&\mathrm{Mg_3Si_2O_5(OH)_{4(s)}+3CO_{2(g)}\rightarrow 3MgCO_{3(s)}+2\,SiO_{2(aq)}+2H_2O_{(l)}}\\&\Delta H=-21\,\mathrm{kJ/mol\,CO_2}\end{aligned} \tag{5.7}$$

However, the challenges of the gas–solid dry process include the following: (1) the slow reaction kinetics at ambient pressure and temperature and (2) the significant energy requirement [21]. For instance, by exposing serpentine (100 μm) to a CO_2 pressure of 340 bar at 500 °C for 2 h, the highest carbonation conversion (referred to the definition in Chap. 6) was approximately 25% [20]. The reaction can be slightly accelerated by either [2]

- pretreating feedstock (such as grinding process and thermal activation) to increase the reactive surface area

  or
- possessing carbonation up to 500 °C.

Nevertheless, those treatments and processes are very energy intensive; therefore, the environmental benefits of carbonation might be easily offset. Moreover, the low capture efficiency is not currently viable on the industrial scale. Since dry carbonation exhibits a long reaction with low carbonation conversion, more efforts in the literature have focused on the performance of aqueous (wet) carbonation.

5.2.1.2 Aqueous (Wet) Carbonation

The addition of water to the carbonation (referred as aqueous carbonation) can significantly increase the reaction kinetics due to the mobilization of ions in the reaction of carbonic acid with alkaline metals. Figure 5.5 shows a typical direct aqueous carbonation for CO_2 fixation and product utilization as construction materials. The carbonation of alkaline solid wastes is carried out with the direct contact of flue gas from a stack, in the presence of water (typically using tap water). The wastewater generated in the same industries also can be used for the process to avoid the consumption of freshwater resources. After carbonation, the reacted slurry is then separated into liquid solution and carbonated solid wastes. The separated liquid solution can be heated by a heat exchanger with flue gas and recirculated into

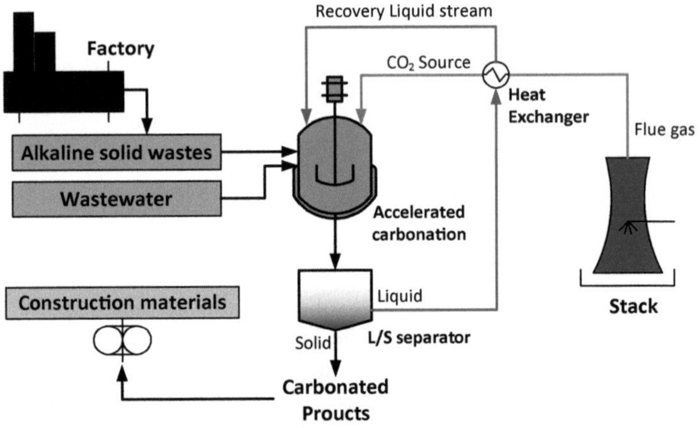

Fig. 5.5 Schematic diagram of ex situ direct aqueous carbonation (one-step carbonation) for CO_2 fixation and construction material production

the reactor for the next carbonation. Although no excessive heat is required in aqueous carbonation, the liquid solution (or slurry) can be moderately heated to about 60–80 °C to achieve a higher carbonation conversion, compared to that at ambient temperature [22, 23].

The aqueous carbonation reaction of calcium and magnesium silicates can be described as Eq. (5.8). The actual process chemistry of aqueous carbonation is complicated because most of the CaO content in solid wastes is not present in pure form. Typically, the CaO content is combined with silicates or other complex oxide phases. Therefore, not only CaO and $Ca(OH)_2$ but also other hydrate compounds may react with CO_2. Also, calcium silicate hydrates (C–S–H) can react with CO_2 to form $CaCO_3$ (e.g., calcite, vaterite, and aragonite) precipitates and amorphous silica gels [24, 25].

$$(\mathrm{Ca,Mg})_x\mathrm{Si}_y\mathrm{O}_{x+2y+z}\mathrm{H}_{2z(s)} + x\,\mathrm{CO}_{2(g)} \rightarrow x(\mathrm{Ca,Mg})\mathrm{CO}_{3\,(s)} + y\,\mathrm{SiO}_{2(s)} + z\,\mathrm{H_2O}_{(l)} \tag{5.8}$$

The aqueous carbonation occurred mainly through three reaction steps:

- Step 1: the leaching of CO_2-reactive metal ions (such as calcium and magnesium) from a solid matrix;
- Step 2: the contemporary dissolution of gaseous CO_2 into a liquid phase, converting carbonic acid to carbonate and/or bicarbonate ions; and
- Step 3: the consequent nucleation and precipitation of carbonates.

The first step of direct aqueous carbonation was believed to be the leaching of calcium ions from solid particles into solution. The second step should be the dissolution of gaseous CO_2 into the solution, as shown in Eq. (5.9):

$$\mathrm{CO}_{2\,(g)} + \mathrm{H_2O}_{(l)} \rightarrow \mathrm{H_2CO}^*_{3(aq)} \rightarrow \mathrm{H_2CO}_{3(aq)} + \mathrm{HCO}^-_{3\,(aq)} + \mathrm{CO}^{2-}_{3\,(aq)} \tag{5.9}$$

Since ${CO_3}^{2-}$ dominates at a high pH (e.g., >10), accelerated carbonation should be favored at a high pH due to the availability of carbonate ions (Eq. 5.10). It is noted that the balance between dissolution and precipitation is dependent on the kinetics and solubility of the feedstock present and possible products.

$$(\mathrm{Ca,Mg})^{2+}_{(aq)} + \mathrm{CO}^{2-}_{3\,(aq)} \rightarrow (\mathrm{Ca,Mg})\mathrm{CO}_{3(s)} \tag{5.10}$$

The mechanism of accelerated carbonation using alkaline solid wastes can be elucidated by various advanced techniques, such as quantitative X-ray diffraction (QXRD) analysis via either the relative intensity ratio method [26, 27] or Rietveld refinement [28]. For instance, Ca_2SiO_4 and $Ca_3Mg(SiO_4)_2$ could be the main CO_2-reacting components in EAFS [29]. The γ-C_2S component in the steel slag displayed higher reactivity to CO_2 than β-C_2S [30]. Similarly, the components of $Ca_2(HSiO_4)(OH)$, $CaSiO_3$, and $Ca_2(Fe, Al)_2O_5$ in steel slag were found to be mainly involved in the carbonation reaction [18].

$$\begin{aligned} & CaSiO_{3(s)} + CO_{2(g)} + 2H_2O_{(l)} \rightarrow CaCO_{3(s)} + H_4SiO_{4(aq)}, \\ & \Delta H = -75\,kJ/mol\,CO_2 \end{aligned} \quad (5.11)$$

Other cases for aqueous carbonation of (Mg, Fe)–silicate can be expressed as the following process chemistry:

$$\begin{aligned} & Mg_2SiO_{4(s)} + 2\,CO_{2(g)} + 2\,H_2O_{(l)} \rightarrow 2MgCO_{3(s)} + H_4SiO_{4(aq)}, \\ & \Delta H = -80\,kJ/mol\,CO_2 \end{aligned} \quad (5.12)$$

$$\begin{aligned} & 2Mg_2SiO_{4(s)} + CO_{2(g)} + 2\,H_2O_{(l)} \rightarrow MgCO_{3(s)} + Mg_3Si_2O_5(OH)_{4(s)}, \\ & \Delta H = -157\,kJ/mol\,CO_2 \end{aligned} \quad (5.13)$$

$$\begin{aligned} & Mg_3Si_2O_5(OH)_{4\,(s)} + 3\,CO_{2(g)} + 2\,H_2O_{(l)} \rightarrow 3\,MgCO_{3(s)} + 2H_4SiO_{4\,(aq)} + 2\,H_2O_{(l)}, \\ & \Delta H = -37\,kJ/mol\,CO_2 \end{aligned} \quad (5.14)$$

$$Fe_2SiO_{4(s)} + 2CO_{2(g)} + 2H_2O_{(l)} \rightarrow 2FeCO_{3(s)} + H_4SiO_{4(aq)} \quad (5.15)$$

The challenge in CO_2 fixation by carbonation of alkaline wastes is to accelerate the reaction and exploit the heat of reaction to minimize energy and material losses [8]. Another challenge encountered is that the dissolution of calcium species in alkaline solid waste was favored at the low pH; however, not favored for the precipitation of calcium carbonate.

5.2.2 *Indirect Carbonation*

Indirect carbonation is a multistep reaction, which involves several steps: (1) extraction of metal ions from alkaline solid wastes, (2) liquid–solid separation, and (3) carbonation of the filter solution, as shown in Fig. 5.6.

5.2.2.1 Acid Extraction

The indirect carbonation process was originally developed using acetic acid for the extraction of calcium ions from a solid particle, as expressed in Eq. (5.16) [11]:

$$CaSiO_{3\,(s)} + 2\,CH_3COOH_{(aq)} \rightarrow Ca^{2+}_{(aq)} + 2\,CH_3COO^-_{(aq)} + SiO_{2(s)} + H_2O_{(aq)} \quad (5.16)$$

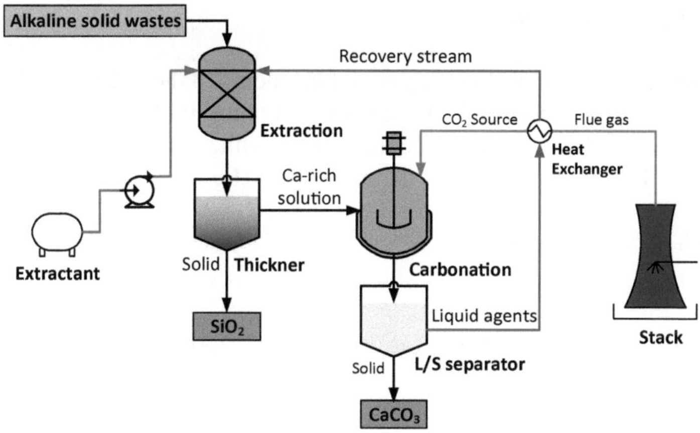

Fig. 5.6 Schematic diagram of ex situ indirect carbonation (multistep carbonation) for CO_2 fixation and $CaCO_3$ production

Calcium ions are extracted, for example, using acetic acid (CH_3COOH) from mineral crystals of $CaSiO_3$. The extracted solution is filtered through a fiber membrane to separate the mother solution (i.e., rich in calcium ion) and extracted solids (i.e., calcium-depleted SiO_2 particles). After that, CO_2 is introduced into the filtered solution, which can form a nearly pure calcium carbonate product for commercial use, as shown in Eq. (5.17) [11]:

$$Ca^{2+}_{(aq)} + 2\,CH_3COO^-_{(aq)} + CO_{2(g)} + H_2O_{(aq)} \rightarrow CaCO_{3\,(s)} \downarrow + 2\,CH_3COOH_{(aq)} \tag{5.17}$$

After carbonation, the end product is usually a pure carbonate (i.e., $CaCO_3$ and $MgCO_3$) because most of the oxides and hydroxides from the material have been extracted, followed by direct carbonation of the oxides and hydroxides with the CO_2. In addition, it is observed that the acetic acid can be recovered and recycled in this step for use in the next extraction. However, the acidity of the solution will reduce the efficiency of carbonation conversion.

According to the water chemistry, the optimum pH for aqueous carbonation should be above 10, while the dissolution of solid wastes (e.g., calcium leaching) occurs under low pH conditions. To increase the pH of the solution before carbonation, a pH-swing concept was proposed: The extraction step was done under acidic conditions, while the carbonation step was carried out in basic conditions [31]. In this case, (1) the addition of acidic or basic reagents and (2) the removal of acidic reagents by heating the solution are required.

5.2.2.2 Base Extraction

Aside from acid extraction, Kodama et al. [32] modified the indirect carbonation by a recyclable reaction solution using ammonium chloride (NH_4Cl). It is capable of swinging the pH of the solution spontaneously, while extracting the alkaline earth metal (Eq. 5.18) and precipitating the carbonates (Eq. 5.19):

$$4\,NH_4Cl_{(aq)} + 2\,CaO \cdot SiO_{2(s)} \rightarrow 2\,CaCl_{2\,(aq)} + SiO_{2(s)} \downarrow\ + 4\,NH_{3(aq)} + 2\,H_2O_{(l)} \tag{5.18}$$

$$4\,NH_{3(aq)} + 2\,CO_{2(aq)} + 2\,CaCl_{2\,(aq)} + 4\,H_2O_{(l)} \rightarrow 2CaCO_{3\,(s)} \downarrow\ + 4\,NH_4Cl_{(aq)} \tag{5.19}$$

In this modified process, an alkaline earth metal can be selectively extracted from solid wastes in an acidic condition using a weak base/strong acid solution of NH_4Cl solution. As the reaction proceeds, the solution exhibits alkalinity due to the generation of ammonia, which is beneficial to subsequent CO_2 absorption.

5.3 Process Chemistry

Calcium leaching and carbonation in alkaline solid wastes would occur simultaneously, for example, in the case of steelmaking slag [33]. From the view point of process chemistry, the challenge of accelerated carbonation using alkaline solid wastes is that the dissolution of calcium species in alkaline solid waste is favored at low pH, while low pH is not favored for the sequential precipitation of calcium carbonate. The balance between dissolution and precipitation is dependent on the kinetics and solubility of the present feedstock and possible products, where carbonates are probably the most important species. Therefore, finding a balanced operating condition between these two mechanisms is essential for optimizing the overall carbonation process.

In the following sections, the process chemistry of the major steps, i.e., (1) metal ion leaching in solution, (2) CO_2 dissolution in solution, and (3) formation of carbonate precipitates, for carbonation is illustrated in detail.

5.3.1 Metal Ion Leaching in Solution

In alkaline solid wastes, CaO and MgO are rarely present in pure form. Instead, the alkaline-metal oxides are primarily locked into a silicate, aluminate, or ferrite phase. In the course of the leaching process, the rapid and strong increase in pH (>10) is observed due to the dissolution of calcium-bearing components in the solid residues, e.g., lime and larnite, as shown in Eqs. (5.20) and (5.21), respectively.

$$CaO_{(s)}(\text{lime}) + H_2O_{(l)} \rightarrow Ca^{2+}_{(aq)} + 2\,OH^-_{(aq)} \tag{5.20}$$

$$Ca_2SiO_{4(s)}(\text{larnite}) + 2\,H_2O_{(l)} \rightarrow 2\,Ca^{2+}_{(aq)} + H_2SiO_{4(s)} + 2\,OH^-_{(aq)} \tag{5.21}$$

During the leaching of calcium (or magnesium) in solid particles, the properties of the solids may change considerably. It was observed that coarse, hard, or granular feed solids should be disintegrated into pulp or mush when their soluble content is removed [34]. In addition, carbonation may reduce the leaching of alkaline earth metals (except Mg) by the conversion of a calcium phase, such as portlandite, ettringite, and Ca–Fe–silicates, into calcite, which could contain traces of Ba and Sr [33].

5.3.2 CO_2 Dissolution in Solution

CO_2 can be physically absorbed in water (or a solvent) in accordance with Henry's law, as expressed in Eq. (5.22). Henry's law states that the relationship between the gas solubility in pure water and the partial pressure of the gas is strictly valid only for gases that can be infinitely diluted in solution. The higher partial pressure of CO_2 will result in a large amount of dissolved CO_2.

$$C'_{CO_2} = H'_{CO_2} \times P_{CO_2} \tag{5.22}$$

where C' is the concentration of CO_2 dissolved in aqueous solution (M); H'_{CO_2} is Henry's constant for CO_2 (e.g., $10^{-1.468}$ M atm^{-1} at 298 K); and P_{CO_2} is the partial pressure of CO_2 in the gas phase (atm). Henry's constant, a function of temperature, can be modified by Eq. (5.23) [35]:

$$H'_{CO_2,T} = H'_{CO_2,298K} \times \exp[C \times (1/T - 1/298)] \tag{5.23}$$

where C is the constant for all gases (2400 K for CO_2) and T is the temperature (K). Table 5.2 presents the different values of Henry's constant for CO_2 under different temperatures.

CO_2 could be dissolved from the atmosphere in available water, and carbonic acid can be created with a pH around 5.6, as shown in Eq. (5.24). The process is pH-dependent because of the dissociation of carbonic acid ($H_2CO_3^*$) into carbonate (CO_3^{2-}) and bicarbonate (HCO_3^-) ions, as shown in Eqs. (5.25) and (5.26), respectively:

Table 5.2 Henry's constant for carbon dioxide under different temperatures

Temp (°C)	Temp (K)	X^{a}	Henry's constant (M atm^{-1})
25	298	−1.468	0.034
30	303	−1.526	0.030
35	308	−1.582	0.026
40	313	−1.636	0.023
45	318	−1.689	0.020
50	323	−1.739	0.018
55	328	−1.788	0.016
60	333	−1.836	0.015
65	338	−1.882	0.013
70	343	−1.93	0.012
75	348	−1.971	0.011
80	353	−2.013	0.010

[a]Henry's constant = 10^{X}

$$CO_{2(g)} + H_2O_{(aq)} \rightarrow H_2CO^{*}_{3(aq)} \tag{5.24}$$

$$H_2CO^{*}_{3(aq)} \rightarrow H^{+}_{(aq)} + HCO^{-}_{3(aq)} \tag{5.25}$$

$$HCO^{-}_{3(aq)} \rightarrow H^{+}_{(aq)} + CO^{2-}_{3(aq)} \tag{5.26}$$

The corresponding equilibrium constants for Eqs. (5.25) and (5.26) can be expressed Eqs. (5.27) and (5.28), respectively:

$$K_a = \frac{[H^{+}][HCO_3^{-}]}{[H_2CO_3]} \tag{5.27}$$

$$K_b = \frac{[H^{+}][CO_3^{2-}]}{[HCO_3^{-}]} \tag{5.28}$$

where $K_a = 10^{-6.3}$ and $K_b = 10^{-10.3}$ at 25 °C.

Figure 5.7 shows the mole balance and equilibrium conditions for carbonation of alkaline solid wastes. Before ions equilibrium, the total inorganic carbon (TIC) concentration dynamically changes depending on several parameters, such as the pH of solution and the pressure of headspace CO_2 gas. At equilibrium, the mole balances of the carbonic acid system can be expressed as

$$C_T = \left[H_2CO^{*}_{3(aq)}\right] + \left[HCO^{-}_{3(aq)}\right] + \left[CO^{2-}_{3(aq)}\right] \tag{5.29}$$

where C_T is the TIC concentration (M).

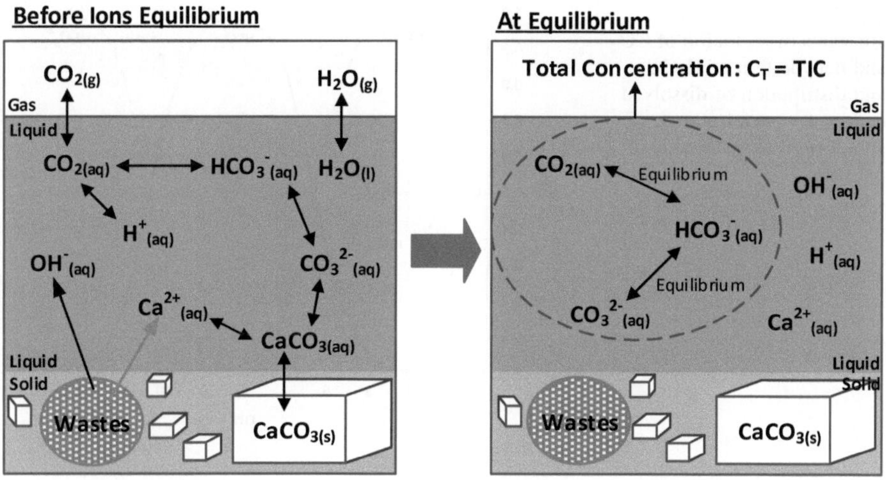

Fig. 5.7 Mole balance and equilibrium conditions for carbonation of alkaline solid wastes

By substitution of Eqs. (5.27) and (5.28) into Eq. (5.29), an explicit formula for $[H_2CO^*_{3(aq)}]$, $[HCO_3{}^-{}_{(aq)}]$ and $[CO_3{}^{2-}{}_{(aq)}]$ can be obtained as

$$\left[\mathrm{H_2CO^*_{3\,(aq)}}\right] = \alpha_0 C_\mathrm{T}, \quad \alpha_0 = \frac{[\mathrm{H}^+]^2}{[\mathrm{H}^+]^2 + K_a[\mathrm{H}^+] + K_\mathrm{a}K_\mathrm{b}} \tag{5.30}$$

$$\left[\mathrm{HCO^-_{3\,(aq)}}\right] = \alpha_1\, C_\mathrm{T}, \quad \alpha_1 = \frac{K_\mathrm{a}[\mathrm{H}^+]}{[\mathrm{H}^+]^2 + K_\mathrm{a}[\mathrm{H}^+] + K_\mathrm{a}K_\mathrm{b}} \tag{5.31}$$

$$\left[\mathrm{CO^{2-}_{3\,(aq)}}\right] = \alpha_2\, C_\mathrm{T}, \quad \alpha_2 = \frac{K_\mathrm{a}K_\mathrm{b}}{[\mathrm{H}^+]^2 + K_\mathrm{a}[\mathrm{H}^+] + K_\mathrm{a}K_\mathrm{b}} \tag{5.32}$$

Figure 5.8 shows the fraction (α_i) of each carbon species present, which is dependent on the pH of the solution. At a low pH ($\sim$4), the production of H_2CO_3 dominates; at a mid-pH ($\sim$8), $HCO_3{}^-$ dominates; and at a high pH ($\sim$12), $CO_3{}^{2-}$ dominates [35]. Therefore, accelerated carbonation is favored at a basic pH due to the availability of carbonate ions. In the case of pH at 6–7, the carbonation reaction should be retarded due to insufficient activity of the carbonate ions.

5.3.3 *Formation of Carbonate Precipitates: Nucleation and Growth*

Table 5.3 presents the solubility product constant (K_{sp}) of the relevant species that might participate in the carbonation reaction. The descending order for solubility of carbonates is as follows: $Mg^{2+} > Ca^{2+} > Zn^{2+} > Fe^{2+} > Cd^{2+} > Pb^{2+}$.

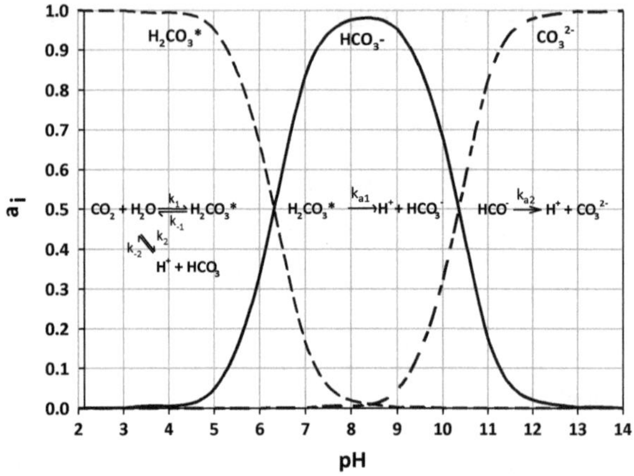

Fig. 5.8 Reaction rate constants of hydration of CO_2 and dehydration of H_2CO_3 and distribution of dissolved carbonate species at equilibrium as a function of pH value

Table 5.3 Solubility product constants (K_{sp}) of various carbonate precipitates near 25 °C*

Categories	Substance	Formula	K_{sp}	Log K_{sp}
Iron oxide	(Hematite)	α-Fe_2O_3	–	−42.7
Hydroxide	Calcium hydroxide	$Ca(OH)_2$	5.5×10^{-6}	−5.19
	Magnesium hydroxide	$Mg(OH)_2$	1.8×10^{-11}	−11.1
	Cadmium hydroxide	$Cd(OH)_2$	2.5×10^{-14}	–
	Lead(II) hydroxide	$Pb(OH)_2$	1.2×10^{-15}	–
	Iron(II) hydroxide	$Fe(OH)_2$	8.0×10^{-16}	−15.1
	Iron(III) hydroxide	$Fe(OH)_3$	4.0×10^{-38}	–
Carbonate	Magnesium carbonate	$MgCO_3$	3.5×10^{-8}	−7.46
	Calcium carbonate (calcite)	$CaCO_3$	4.5×10^{-9}	−8.35
	Calcium carbonate (aragonite)	$CaCO_3$	4.5×10^{-9}	−8.22
	Zinc carbonate	$ZnCO_3$	1.4×10^{-11}	–
	Iron(II) carbonate	$FeCO_3$	3.2×10^{-11}	−10.7
	Cadmium carbonate	$CdCO_3$	5.2×10^{-12}	–
	Lead(II) carbonate	$PbCO_3$	7.4×10^{-14}	–
	(Dolomite)	$CaMg(CO_3)_2$	–	−1.70
Sulfate	(Gypsum)	$CaSO_4 \cdot 2H_2O$	–	−4.62
Silicate	(Wollastonite)	$CaO \cdot SiO_2$	–	6.82
	(Larnite)	Ca_2SiO_4	–	37.65
	(Anorthite)	$CaAl_2Si_2O_8$	–	25.31
	(Gehlenite)	$Ca_2Al_2SiO_7$	–	55.23

*Available data from [35–37]

Typically, the theoretical K_{sp} of $CaCO_3$, as shown in Eq. (5.33), ranges from 3.7×10^{-9} to 8.7×10^{-9} at 25 °C [35], with 4.47×10^{-9} being widely used for calculation.

$$K_{sp} = [Ca^{2+}][CO_3^{2-}] \tag{5.33}$$

where $[Ca^{2+}]$ and $[CO_3^{2-}]$ are the activities of calcium and carbonate ions (M) in the solution, respectively.

The supersaturation ratio (S) of $[Ca^{2+}]$ and $[CO_3^{2-}]$ in the interface can be expressed as shown in Eq. (5.34). A low supersaturation ratio is disadvantageous for the nucleation of $CaCO_3$ [38].

$$S = \frac{[Ca^{2+}][CO_3^{2-}]}{K_{sp}} \tag{5.34}$$

Contact between Ca^{2+} ions and CO_2 leads to $CaCO_3$ precipitation, which is almost insoluble in water at pH levels above 9 (the solubility of $CaCO_3$ is 0.15 mmol/L at 25 °C). The carbonation formulation can be simply described by Eqs. (5.35) and (5.36):

$$Ca^{2+}_{(aq)} + CO^{2-}_{3\,(aq)} \rightarrow CaCO_{3(nuclei)} \tag{5.35}$$

$$CaCO_{3(nuclei)} \rightarrow CaCO_{3\,(s)} \tag{5.36}$$

Calcite and aragonite ($CaCO_3$), dolomite ($CaMg(CO_3)_2$), and magnesite ($MgCO_3$) are four of the most prevalent carbonate minerals found in Earth's crust. The crystal volume of calcium carbonate is approximately 11.7% more than that of calcium hydroxide [39]. There are three different polymorphs of calcium carbonate, i.e., calcite, aragonite, and vaterite, as shown in Fig. 5.9. Calcite typically has a rhombohedral, prismatic, or scalenohedral crystal type, while aragonite is typically in needle-like crystals with a high aspect ratio (length-to-width ratio). In nature, calcium carbonate occurs most commonly in hexagonal form (as calcite) but also occurs in orthorhombic form (aragonite) [34].

Figure 5.10 shows the structures and Ca–O coordination of calcite and vaterite. $CaCO_3$ polymorphs such as calcite and vaterite (known as μ $CaCO_3$) belong to the same structural family. However, unlike the calcite, vaterite is thought to be metastable [41]. It suggests that a preferential formation of vaterite should be in the carbonation of C–S–H with Ca/Si ratio of >0.75 [42]. Moreover, the lower $[Ca^{2+}]/[CO_3^{2-}]$ ratio also favors the formation of vaterite [43]. It also was noted that the formation mechanism of calcite, vaterite, and aragonite depends on the crystallographic structure of original 1.1 nm tobermorite [24].

Dissolution or crystallization of precipitates does not occur instantaneously. It suggests that some characteristic time, often longer than the time constant for the overall process, is needed to achieve a new equilibrium [45]. Calcite growth rates at 25 °C have been observed to be first order, second order, and higher (n = 3–3.9), indicating that calcite growth can be interpreted by adsorption, surface diffusion,

Fig. 5.9 Different crystal shapes of calcium carbonate: **a** amorphous calcium carbonate; **b** low-Mg calcite with 6% Mg; **c** high-Mg calcite with 20% Mg; **d** aragonite; **e** monohydrocalcite; **f** vaterite. Reprinted by permission from Macmillan Publishers Ltd: Ref. [40], copyright 2017

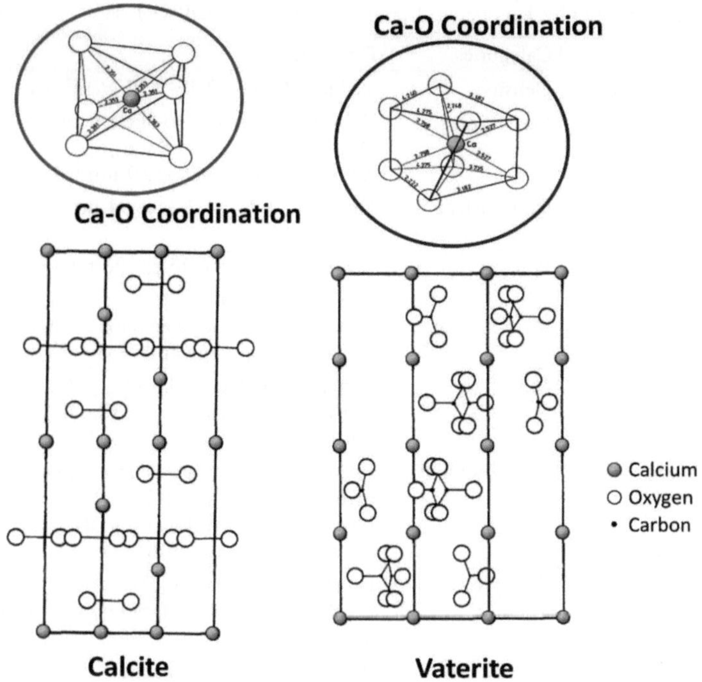

Fig. 5.10 Structures and Ca–O coordination of calcite and vaterite. Reprinted by permission from Macmillan Publishers Ltd: Ref. [44], copyright 1994

and polynucleation [46]. The pore structures of solid wastes are thought to become finer with carbonation, thus leading to a solid of lower porosity, lower tortuosity, and lower pore area with calcite infilling the pore space after carbonation [47, 48].

5.4 Changes in Physico-chemical Properties of Solid Wastes

After accelerated carbonation, the physico-chemical properties of alkaline solid wastes will significantly change. The use of carbonated products is closely related to their physico-chemical characteristics. As a result, characterization of materials before and after reaction is crucial for their subsequent utilization. The characterization of a material is usually divided into three parts:

- Physical properties: bulk density, morphology (shape), fitness, solubility, surface area, porosity, etc.

Table 5.4 Changes in physico-chemical properties of carbonated steel slag

Process	Type	Categories	Descriptions
Direct	Physical	Density	Reduce: 3.56 → 2.47–3.27 g/cm^3 [6]
		Particle size	Become coarser due to agglomeration [47] D [0.1]: 4.4 → 2.5–3.5 μm [6] D [0.9]: 31.8 → 28.4–29.9 μm [6]
		BET surface area	Increased: 1.73 → 7.22–8.94 m^2/g [6]
		Porosity	Becomes lesser due to calcite filling in the pores [49, 50]
		Microstructure	Product as compact continuous coatings on surface, a more regular morphology [6]
	Chemical	Mineralogy	Formation: $CaCO_3$, $Ca_5(SiO_4)_2CO_3$ Consumption: hydrogarnet, $Ca(OH)_2$ [9] $CaCO_3$: 3.2% → 12.6–17.0% [6]
		Free-CaO and $Ca(OH)_2$	Reduced from 6 to 0.3% for less than 75-μm-sized particles in ladle slag [9] Free CaO (5.1 wt%) and $Ca(OH)_2$ (1.8 wt%) completely eliminated [6]
		Heavy metal leaching	Leachable Cr reduced [51] Leaching of metals reduced by orders of magnitude [33]
Indirect	Physical	Particle size	PCC: spherical agglomerates of 10–20 μm [52]
		Particle shape	Cubic crystals (calcite) at 40–80 °C Plate- and needle-shaped (aragonite) at 90 °C [32]
	Chemical	Mineralogy	Calcite was formed at low temp (40–80 °C) and aragonite at around 90 °C on the surface of slag [32]

- Chemical properties: oxide contents, heavy metal leaching amounts, hazardous components, pozzolanic and cementitious properties, etc.
- Mineralogical properties: mineralogy, crystalline, phase fraction, grain size, defect, etc.

Table 5.4 presents the changes in physico-chemical properties of steel slag after carbonation, with the details illustrated in the following Sects. 4.1 (for direct carbonation) and 4.2 (for indirect carbonation). The changes in physico-chemical properties of solid wastes after carbonation include density, particle size, surface area, microstructure, chemical composition, and mineralogy.

5.4.1 Direct Carbonation

After the direct carbonation, significant changes in physical properties of solid wastes can be observed [6, 47], such as the following:

- a decrease in average density and overall particle size;
- an increase in BET surface area;
- finer pore structures with lower porosity, tortuosity, and pore area; and
- changes in surface topography and particle shape.

Since the crystal volume of $CaCO_3$ is about 11.7% greater than that of $Ca(OH)_2$ [39], the microstructure of solid wastes should be modified to some degree after carbonation. In general, carbonate precipitation using alkaline solid wastes (such as steel slags) as feedstock is mainly $CaCO_3$, in the form of calcite [2, 53]. The product of $CaCO_3$ calcite with regular morphology can form as compact continuous coatings on the surface of solid wastes during direct carbonation. At the same time, parts of the $CaCO_3$ product, with a size of 1–3 μm, appear to be isolated and separated [54].

In the case of ladle furnace slag (LFS), the formation of spurrite [$Ca_5(SiO_4)_2CO_3$] was observed by the reaction of hydrogarnet and calcium hydroxide [9]. For carbonated LFS, the extractable CaO content decreased to 0.2–0.3%, corresponding to a reduction of 95% in the extractable CaO content compared to the fresh LFS [9]. Similarly, in the case of steel slag, the contents of free-CaO and $Ca(OH)_2$ were completely eliminated after carbonation [6]. Furthermore, the leaching potential of heavy metals, such as Cr and Hg, from solid wastes can be reduced by the carbonation process.

5.4.2 *Indirect Carbonation*

For the indirect carbonation process, the precipitated calcium carbonate (PCC) is considered as high value-added products. The structures of calcium carbonate precipitates include the following:

- calcite: hexagonal β-$CaCO_3$;
- aragonite: orthorhombic λ-$CaCO_3$; and
- vaterite: μ-$CaCO_3$.

Calcite is the most stable polymorph of calcium carbonate. The potential use of PCC relies on the purity and variety of the crystal structures. The favored forms of PCC crystalline in most industrial applications [55] are as follows:

- rhombohedral calcite,
- scalenohedral calcite, and
- orthorhombic acicular aragonite.

The morphology of $CaCO_3$ particles can be controlled to meet the required specification for industrial applications [32]. Typically, the aragonite form of $CaCO_3$ can be prepared by precipitation at temperatures above 85 °C, while the vaterite $CaCO_3$ can be formed by precipitation at 60 °C [56]. The aragonite form will change to the calcite form at 380–470 °C [57]. In the case of using ammonium chloride (NH_4Cl) as an extractant, as the reaction temperature increases, the fraction of rhombohedral calcite decreases while the plate-shaped or needle-shaped particles of aragonite are generated [32]. Using acetic acid (CH_3COOH) as an extractant, spherical agglomerates of calcite crystals with a size of 10–20 μm can be formed by indirect carbonation of granulated blast furnace slag (<500 μm in size) [52].

According to thermodynamic equilibrium calculations of indirect carbonation using 33.3% acetic acid, almost all compounds (including Ca, Mg, Al, Si, Fe, and S) except Ti are expected to be dissolved in the solution. The leaching of Ca^{2+} ions is an exothermic reaction at 50 °C and feasible at temperatures lower than 156 °C, while the extraction of Mg^{2+} ions from $MgSiO_3$ proceeds at temperatures lower than 123 °C [11]. A large part of the metals could form acetates, such as magnesium acetate, calcium acetate, and iron (II) acetate [11]. On the other hand, the dissolution of metal ions from steel slags in acetic acid was much faster than from wollastonite [58]. The dissolved silicon from steel slags will form a gel at temperatures higher than 70 °C, which can be easily removed by mechanical filtration [11, 59]. To remove other dissolved ions such as iron, aluminum, and magnesium, other separation techniques are needed.

5.5 Challenges in Accelerated Carbonation Reaction

Aqueous accelerated carbonation of alkaline wastes combines the treatment of industrial wastes that are readily available near a CO_2 emission point. Thus, it could be part of an integrated approach to reducing CO_2 emissions for industrial plants. Although a great amount of alkaline waste (such as steel slags and ashes) is available for CO_2 mineralization, the costs of accelerated carbonation are too high for large-scale industrial deployment. Due to the inherent characters of carbonation reaction, a well-designed process with efficient mass transfer among gas–liquid–solid phases (high CO_2 fixation capacity with low energy consumption) is required.

From the technical point of view, the challenges in direct carbonation process include the following:

- slow rate of mass transfer between gaseous CO_2 and liquid solution phase;
- losses of material and energy (i.e., heat of reaction);
- influence of impurity in flue gas, such as sulfur dioxide (SO_2) and particulate matter (PM), on carbonation efficiency;
- high energy consumption of overall processes due to feedstock/material processing and reactor use (e.g., high temperature and pressure); and
- product separation and its subsequent utilization.

For indirect carbonation, the challenges in the process include the following [14, 58]:

- The high energy consumption for evaporation of the aqueous solution.
- The large variations in free energy resulting from the necessary formation of intermediate products.
- Other elements, such as heavy metals, may also leach out during the extraction step, thus leading to impure carbonate precipitate.

To overcome the above challenges, several innovative solutions are raised and discussed in Chap. 8.

References

1. Seifritz W (1990) CO_2 Disposal by means of silicates. Nature 345(6275):486–486
2. Pan S-Y, Chang EE, Chiang P-C (2012) CO_2 capture by accelerated carbonation of alkaline wastes: a review on its principles and applications. Aerosol Air Qual Res 12:770–791. doi:10.4209/aaqr.2012.06.0149
3. Huijgen WJJ, Comans RNJ (2003) Carbon dioxide sequestration by mineral carbonation literature review. Energy research Centre of the Netherlands
4. Huijgen WJJ, Comans RNJ (2005) Mineral CO_2 sequestration by carbonation of industrial residues: literature review and selection of residue
5. Bonenfant D, Kharoune L, Sauve S, Hausler R, Niquette P, Mimeault M, Kharoune M (2009) Molecular analysis of carbon dioxide adsorption processes on steel slag oxides. Int J Greenh Gas Control 3(1):20–28. doi:10.1016/j.ijggc.2008.06.001
6. Pan SY, Chiang PC, Chen YH, Tan CS, Chang EE (2013) Ex situ CO_2 capture by carbonation of steelmaking slag coupled with metalworking wastewater in a rotating packed bed. Environ Sci Technol 47(7):3308–3315. doi:10.1021/es304975y
7. Pan SY, Chen YH, Chen CD, Shen AL, Lin M, Chiang PC (2015) High-gravity carbonation process for enhancing CO_2 fixation and utilization exemplified by the steelmaking industry. Environ Sci Technol 49(20):12380–12387. doi:10.1021/acs.est.5b02210
8. Bobicki ER, Liu Q, Xu Z, Zeng H (2012) Carbon capture and storage using alkaline industrial wastes. Prog Energy Combust Sci 38(2):302–320. doi:10.1016/j.pecs.2011.11.002
9. Monkman S, Shao Y, Shi C (2009) Carbonated ladle slag fines for carbon uptake and sand substitute. J Mater Civ Eng 21:657–665. doi:10.1061//asce/0899-1561/2009/21:11/657
10. Bodor M, Santos R, Gerven T, Vlad M (2013) Recent developments and perspectives on the treatment of industrial wastes by mineral carbonation—a review. Open Eng 3(4). doi:10.2478/s13531-013-0115-8
11. Teir S, Eloneva S, Fogelholm C-J, Zevenhoven R (2007) Dissolution of steelmaking slags in acetic acid for precipitated calcium carbonate production. Energy 32(4):528–539. doi:10.1016/j.energy.2006.06.023
12. Eloneva S, Said A, Fogelholm C-J, Zevenhoven R (2012) Preliminary assessment of a method utilizing carbon dioxide and steelmaking slags to produce precipitated calcium carbonate. Appl Energy 90(1):329–334. doi:10.1016/j.apenergy.2011.05.045
13. Costa G, Baciocchi R, Polettini A, Pomi R, Hills CD, Carey PJ (2007) Current status and perspectives of accelerated carbonation processes on municipal waste combustion residues. Environ Monit Assess 135(1–3):55–75. doi:10.1007/s10661-007-9704-4
14. Teir S (2008) Fixation of carbon dioxide by producing carbonates from minerals and steelmaking slags. Helsinki University of Technology

15. Gruyaert E, Van den Heede P, De Belie N (2013) Carbonation of slag concrete: effect of the cement replacement level and curing on the carbonation coefficient—effect of carbonation on the pore structure. Cem Concr Compos 35(1):39–48. doi:10.1016/j.cemconcomp.2012.08.024
16. Muhmood L, Vitta S, Venkateswaran D (2009) Cementitious and pozzolanic behavior of electric arc furnace steel slags. Cem Concr Res 39(2):102–109. doi:10.1016/j.cemconres.2008.11.002
17. Li P, Pan SY, Pei S, Lin YPJ, Chiang PC (2016) Challenges and perspectives on carbon fixation and utilization technologies: an overview. Aerosol Air Qual Res 16(6):1327–1344. doi:10.4209/aaqr.2015.12.0698
18. Pan SY, Chiang PC, Chen YH, Chen CD, Lin HY, Chang EE (2013) Systematic approach to determination of maximum achievable capture capacity via leaching and carbonation processes for alkaline steelmaking wastes in a rotating packed bed. Environ Sci Technol 47 (23):13677–13685. doi:10.1021/es403323x
19. Zhang T, Yu Q, Wei J, Li J, Zhang P (2011) Preparation of high performance blended cements and reclamation of iron concentrate from basic oxygen furnace steel slag. Resour Conserv Recycl 56(1):48–55. doi:10.1016/j.resconrec.2011.09.003
20. Lackner KS, Wendt CH, Butt DP, Joyce EL, Sharp DH (1995) Carbon dioxide disposal in carbonate minerals. Los Alamos National Laboratory, Los Alamos, NM, USA
21. IPCC (2005) IPCC special report on carbon dioxide capture and storage. Intergovernmental Panel on Climate Change, Cambridge. ISBN 13 978-0-521-86643-9
22. Chang EE, Pan SY, Chen YH, Tan CS, Chiang PC (2012) Accelerated carbonation of steelmaking slags in a high-gravity rotating packed bed. J Hazard Mater 227–228:97–106. doi:10.1016/j.jhazmat.2012.05.021
23. Chang EE, Chen T-L, Pan S-Y, Chen Y-H, Chiang P-C (2013) Kinetic modeling on CO_2 capture using basic oxygen furnace slag coupled with cold-rolling wastewater in a rotating packed bed. J Hazard Mater 260:937–946. doi:10.1016/j.jhazmat.2013.06.052
24. Tsuyoshi S, Etsuo S, Minoru M, Nobuaki O (2010) Carbonation of γ-Ca_2SiO_4 and the mechanism of vaterite formation. J Adv Concr Technol 8(3):273–280
25. Villain G, Thiery M, Platret G (2007) Measurement methods of carbonation profiles in concrete: thermogravimetry, chemical analysis and gammadensimetry. Cem Concr Res 37 (8):1182–1192. doi:10.1016/j.cemconres.2007.04.015
26. Alègre R (1965) Généralisation de la méthode d' addition pour I' analyse quantitative par diffraction. X Bullentin de la Société Française de Minéralogie et de Cristallographie 88:569–574
27. Hosseini T, Selomulya C, Haque N, Zhang L (2015) Investigating the effect of the Mg2+/Ca2 + molar ratio on the carbonate speciation during the mild mineral carbonation process at atmospheric pressure. Energ Fuel 29(11):7483–7496. doi:10.1021/acs.energyfuels.5b01609
28. Rietveld HM (1969) A profile refinement method for nuclear and magnetic structures. J Appl Crystallogr 2:65–71
29. Uibu M, Kuusik R, Andreas L, Kirsimäe K (2011) The CO_2-binding by Ca-Mg-silicates in direct aqueous carbonation of oil shale ash and steel slag. Energy Procedia 4:925–932. doi:10.1016/j.egypro.2011.01.138
30. Ghouleh Z, Guthrie RIL, Shao Y (2015) High-strength KOBM steel slag binder activated by carbonation. Constr Build Mater 99:175–183. doi:10.1016/j.conbuildmat.2015.09.028
31. Park A, Fan L (2004) mineral sequestration: physically activated dissolution of serpentine and pH swing process. Chem Eng Sci 59(22–23):5241–5247. doi:10.1016/j.ces.2004.09.008
32. Kodama S, Nishimoto T, Yamamoto N, Yogo K, Yamada K (2008) Development of a new pH-swing CO_2 mineralization process with a recyclable reaction solution. Energy 33(5): 776–784. doi:10.1016/j.energy.2008.01.005
33. Huijgen WJJ, Comans RNJ (2006) Carbonation of steel slag for CO2 sequestration: leaching of products and reaction mechanisms. Environ Sci Technol 40(8):2790–2796
34. McCabe ML, Smith JC, Harriott P (2005) Leaching and extraction. In: Unit operations of chemical engineering, 7 edn. McGraw Hill, p 764
35. Morel FMM, Hering JG (1993) Principles and applications of aquatic chemistry. Wiley

36. Astrup T, Mosbaek H, Christensen TH (2006) Assessment of long-term leaching from waste incineration air-pollution-control residues. Waste Manag 26(8):803–814. doi:10.1016/j.wasman.2005.12.008
37. Chemistry 112 (2006) Solubility product constants near 25 °C. http://bilbo.chm.uri.edu/CHM112/tables/KspTable.htm
38. Wang M (2004) Controlling factors and mechanism of preparing needlelike $CaCO_3$ under high-gravity environment. Powder Technol 142(2–3):166–174. doi:10.1016/j.powtec.2004.05.003
39. Ishida T, Maekawa K (2000) Modeling of pH profile in pore water based on mass transport and chemical equilibrium theory. In: Proceedings of JSCE
40. Blue CR, Giuffre A, Mergelsberg S, Han N, De Yoreo JJ, Dove PM (2017) Chemical and physical controls on the transformation of amorphous calcium carbonate into crystalline $CaCO_3$ polymorphs. Geochim Cosmochim Ac 196:179–196. doi:10.1016/j.gca.2016.09.004
41. Santos A, Ajbary M, Morales-Florez V, Kherbeche A, Pinero M, Esquivias L (2009) Larnite powders and larnite/silica aerogel composites as effective agents for CO_2 sequestration by carbonation. J Hazard Mater 168(2–3):1397–1403. doi:10.1016/j.jhazmat.2009.03.026
42. Black L, Breen C, Yarwood J, Garbev K, Stemmermann P, Gasharova B (2007) Structural features of C-S-H(I) and its carbonation in air? A raman spectroscopic study. Part II: carbonated phases. J Am Ceram Soc 90(3):908–917. doi:10.1111/j.1551-2916.2006.01429.x
43. Han YS, Hadiko G, Fuji M, Takahashi M (2006) Factors affecting the phase and morphology of $CaCO_3$ prepared by a bubbling method. J Eur Ceram Soc 26(4–5):843–847. doi:10.1016/j.jeurceramsoc.2005.07.050
44. Maciejewski M, Oswald H-R, Reller A (1994) Thermal transformations of vaterite and calcite. Thermochim Acta 234:315–328
45. Justin S, John P (2007) Carbon dioxide in chemical processes. In: Letcher TM (ed) Development and applications in solubility. The Royal Society of Chemistry, UK, pp 337–349
46. Brady PV (1996) Physics and chemistry of mineral surfaces. In: Chemistry and physics of surfaces and minerals. CRC Press LLC, Florida
47. Fernandez Bertos M, Simons SJ, Hills CD, Carey PJ (2004) A review of accelerated carbonation technology in the treatment of cement-based materials and sequestration of CO_2. J Hazard Mater 112(3):193–205. doi:10.1016/j.jhazmat.2004.04.019
48. Pan S-Y, Chiang A, Chang E-E, Lin Y-P, Kim H, Chiang P-C (2015) An innovative approach to integrated carbon mineralization and waste utilization: a review. Aerosol Air Qual Res 15:1072–1091. doi:10.4209/aaqr.2014.10.02
49. Wu HZ, Chang J, Pan ZZ, Cheng X (2009) Carbonate steelmaking slag to manufacture building materials. Adv Mater Res 79–82:1943–1946. doi:10.4028/www.scientific.net/AMR.79-82.1943
50. Fernandez Bertos M, Li X, Simons SJR, Hills CD, Carey PJ (2004) Investigation of accelerated carbonation for the stabilisation of MSW incinerator ashes and the sequestration of CO_2. Green Chem 6(8):428. doi:10.1039/b401872a
51. Johnson DC, MacLeod CL, Carey PJ, Hills CD (2003) Solidification of stainless steel slag by accelerated carbonation. Environ Technol 24(6):671–678
52. Chiang Y-W, Santos RM, Elsen J, Meesschaert B, Martens JA, Van Gerven T (2013) Two-way valorization of blast furnace slag into precipitated calcium carbonate and sorbent materials. Paper presented at the accelerated carbonation for environmental and material engineering, KU Leuven, Belgium
53. Wang Q, Yan P (2010) Hydration properties of basic oxygen furnace steel slag. Constr Build Mater 24:101–109
54. Chang EE, Wang Y-C, Pan S-Y, Chen Y-H, Chiang P-C (2012) CO_2 capture by using blended hydraulic slag cement via a slurry reactor. Aerosol Air Qual Res 12:1433–1443. doi:10.4209/aaqr.2012.08.0210
55. Imppola O (2000) Precipitated calcium carbonate-PCC. In: Pigment Coating and Surface Sizing of Paper. Gummerus Printing, Finland

56. Ropp RC (2013) Encyclopedia of the alkaline earth compounds. Elsevier
57. Yoshioka S, Kitano Y (1985) Transformation of aragonite to calcite through heating. Geochem J 19:24–249
58. Kakizawa M, Yamasaki A, Yanagisawa Y (2001) A new CO_2 disposal process via artificial weathering of calcium silicate accelerated by acetic acid. Energy 26:341–354
59. Teir S, Eloneva S, Zevenhoven R (2005) Co-utilization of CO_2 and calcium silicate-rich slags for precipitated calcium carbonate production (part I). In: Proceedings of the 18th international conference on efficiency; cost; optimization; simulation and environmental impact of energy systems (ECOS 2005). Trondheim, Norway, 20–22 June 2005

Chapter 6
Analytical Methods for Carbonation Material

Abstract Interest in the accelerated carbonation of alkaline solid wastes has sharply escalated because accelerated carbonation of alkaline solid waste is an attractive method for carbon capture and utilization, due to their potential to fix gaseous CO_2 from industry into solid precipitation. The physico-chemical properties of solid wastes can be improved by carbonation, thereby increasing the potential to be used as construction materials such as supplementary cementitious materials and/or aggregates in civil engineering. In this chapter, several advanced analytical methods for material characterization are introduced. For example, an integrated thermal analysis (i.e., modified TG-DTG interpretation) is illustrated to determine carbonation parameters in alkaline solid wastes.

6.1 Integrated Thermal Analysis

Accurate determination of the contents of $CaCO_3$ formation in the course of accelerated carbonation is crucial for assessing the potential for CO_2 capture by alkaline wastes, as well as providing scientific data for interpretation of the carbonation reaction kinetics. To determine the performance of accelerated carbonation via various types of approach and process, thermoanalytical techniques have been utilized to quantify carbonation products, such as calcium carbonate ($CaCO_3$) in solid wastes. The commonly used thermoanalytical techniques include

- Thermogravimetric (TG) analysis,
- Derivative thermogravimetric (DTG),
- Differential thermal analysis (DTA), and
- Differential scanning calorimetry (DSC).

TG analysis determines the weight of sample at different temperatures under an assigned heating programs. The weight loss can be attributed to moisture evaporation and chemical decomposition of compounds into gaseous components.

P.-C. Chiang and S.-Y. Pan, *Carbon Dioxide Mineralization and Utilization*, DOI 10.1007/978-981-10-3268-4_6

Therefore, individual compounds can be quantified because the thermal decomposition temperatures vary among compounds. On the other hand, by taking numerical derivation of the TG curve, a DTG plot can be obtained to provide information on the temperature at the maximum peak and other important peak parameters.

In DSC, a sample cell and a reference were heated equally according to a preset temperature regime. When transformation of the sample occurs, the temperature difference between the sample and the reference cell is measured. The device will increase or reduce the heat input to the sample cell to maintain a zero temperature difference between the sample and the reference cell, establishing a "null balance" [1]. The quantity of the electrical energy supplied to a sample cell is usually expressed in terms of energy per unit time (e.g., watts). Therefore, the amount of energy absorbed or released by the sample can be measured with DSC.

6.1.1 Conventional Thermogravimetric (TG) Analysis

TG analysis has been considered a rapid and accurate method for determining the content of crystalline $CaCO_3$ in samples with simple composition (highly pure) [2]. In the case of cementitious materials such as cement and fly ash [3], however, it was difficult to obtain accurate quantitative amounts of $CaCO_3$ using only TG data. This is attributed to

- The way to interpret TG curves for $CaCO_3$ decomposition in a material was varied among researchers.
- The temperature ranges of thermal decomposition of $CaCO_3$ overlap the calcareous and hydrated components in these materials (as shown in Fig. 6.1).

As shown in Fig. 6.2, two of the most commonly used methods for determining the weight loss of a material by interpreting TG plot are as follows:

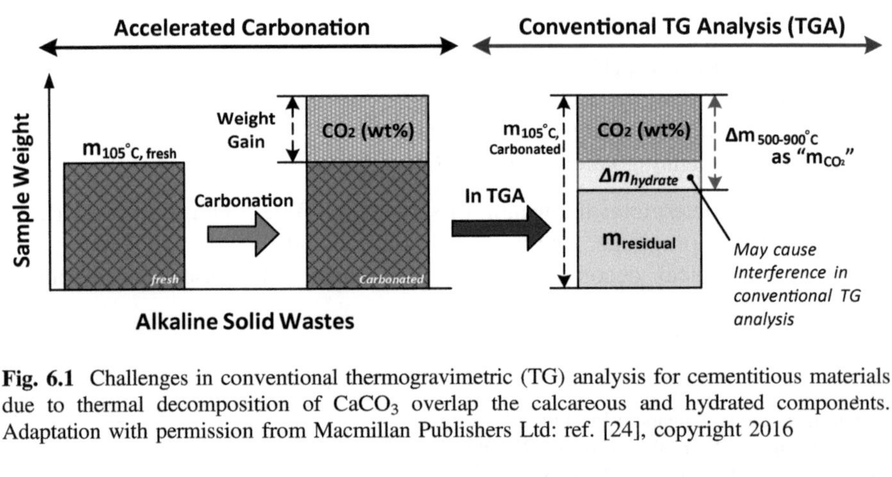

Fig. 6.1 Challenges in conventional thermogravimetric (TG) analysis for cementitious materials due to thermal decomposition of $CaCO_3$ overlap the calcareous and hydrated components. Adaptation with permission from Macmillan Publishers Ltd: ref. [24], copyright 2016

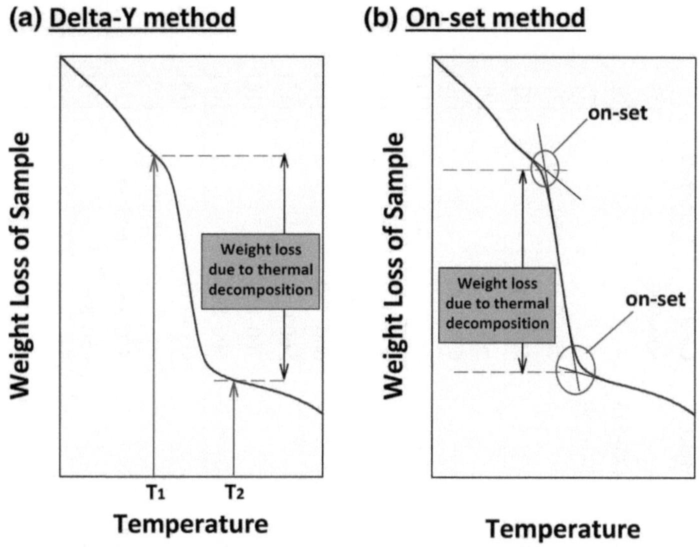

Fig. 6.2 Conventional methods on TG interpretation: **a** delta-Y and **b** on-set methods. Adaptation with permission from Macmillan Publishers Ltd: ref. [24], copyright 2016

- Delta-Y method (Fig. 6.2a): to determine the difference of sample weight directly between two specific temperatures (e.g., T_1 and T_2).
- On-set method (Fig. 6.2b): to extend the straight-line portions of the baseline and the linear portion of the upward/downward slope, mark their intersection, and determine the weight difference between these two intersections.

However, between these two conventional methods, there is a significant difference in the determined weight loss for the same TG plot. For instance, in the case of steel slag, the dehydration of calcium silicate hydrates, calcium aluminate hydrates, and other minor hydrates occurs between 105 and 1000 °C. This would result in a continuous and steady weight loss at 105–1000 °C and especially pronounce at temperatures less than 500 °C [4]. Therefore, consideration must be given to weight loss due to dehydration of the above materials when analyzing, for example, the amounts of $CaCO_3$ (weight loss typically occurs above 500 °C). Otherwise, the $CaCO_3$ contents in solid samples will be overestimated by the conventional delta-Y or on-set methods.

Table 6.1 summarizes the analytical conditions of TG, such as temperature ranges of the thermal decomposition of $Ca(OH)_2$, $MgCO_3$, and $CaCO_3$, for different solid wastes in the literature. As noted in Table 6.1, the evaluation criteria of carbonate products by TG analysis are quite different among the literature because of the wide variance in determining the temperature ranges of product decomposition. This is also largely attributed to the various ways to interpret TG plot, thereby resulting in different bases on the performance evaluation of CO_2 fixation capacity by accelerated carbonation.

Table 6.1 Analytical conditions of TG and the used decomposition temperature for $Ca(OH)_2$, $MgCO_3$, and $CaCO_3$ in different types of raw materials in the literature. Adaptation with permission from Macmillan Publishers Ltd: ref. [24], copyright 2016

Type of materials[a]	Heating rate (°C/min)	Atmosphere	Sample weight (mg)	Temperature of decomposition (°C)				Reference
				$Ca(OH)_2$	$MgCO_3$	$CaCO_3$	Carbonate	
Wollastonite	10	N_2	–	–	–	600–900	–	[5]
Serpentine	10	N_2	–	330–473	473–573	–	–	[6]
Serpentine	10	N_2			300–550			[7]
Concrete	10	–	225	430–530	–	780–900	650–950	[3]
Concrete	20	–	–	425–550	–	550–950	–	[8]
Mortar	10	Air	200	430–520		~750	–	[9]
Mortar/GBFS	10	N_2	–	>380	–	650–790	–	[10]
Steel slag	40	O_2	10–20	–	105–500	–	500–1000	[11]
Steel slag	15	N_2	~100	340–430	–	600–800	–	[12]
BOFS	10	N_2	10–20	–	–	500–780	–	[13]
BOFS	–	–	–	–	–	600–780	–	[14]
EAFS	–	Ar	–	~600	–	>600	–	[15]
CKD/sludge	–	–	–	450–550	–	700–850	–	[16]
CKD	20	N_2	~22	300–500	–	500–800	–	[17]
MSWI-BA	10	Ar	20	–	–	600–750	–	[18]
MSWI-FA	10	–	–	–	–	–	450–900	[19]
APC residue	15	Air	500	400–500	–	750–850	600–850	[20]
APC residue	10	Air	–	400–450	–	650–800	–	[21]
Cockle shell	20	N_2	10–20	–	–	700–900	–	[22]
Synthesis carbonates	50	N_2	5–10	–	515–640	620–780	–	[23]

[a] *GBFS* granulate blast furnace slag; *BOFS* basic oxygen furnace slag; *EAFS* electric arc furnace slag; *CKD* cement kiln dust; *MSWI* municipal solid waste incinerator; *FA* fly ash; *BA* bottom ash; and *APC* air pollution control

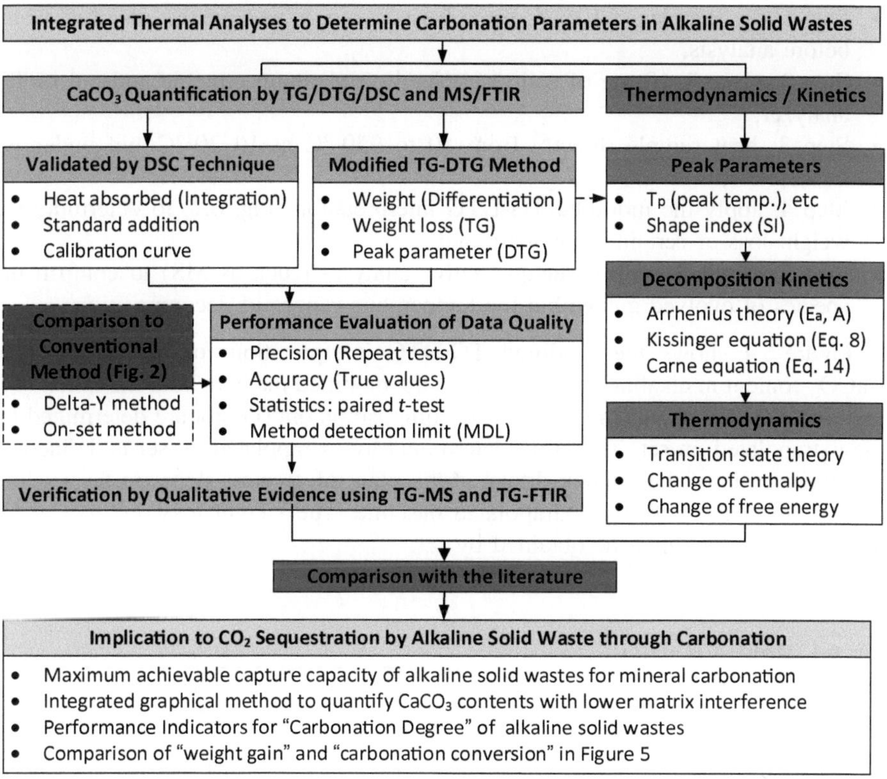

Fig. 6.3 Determination of carbonation conversion in alkaline solid wastes via integrated thermal analysis, including modified TG-DTG method and qualitative analysis via MS and FTIR

To overcome the aforementioned barriers, integrated thermal analyses (including modified TG-DTG method and qualitative analysis via MS and FTIR) are proposed as presented in Fig. 6.3. A modified TG-DTG interpretation is developed to accurately and precisely determine the carbonate content in alkaline solid wastes and validated with DSC analysis, as outlined in Sect. 6.1.2. In addition, the kinetic and thermodynamic parameters of $CaCO_3$ thermal decomposition in solid waste can be determined via various equations, as outlined in Sect. 6.1.4.

6.1.2 Modified TG-DTG Interpretation

In this section, a validated thermal analysis method (i.e., modified TG-DTG interpretation) is illustrated for accurately quantifying the $CaCO_3$ content in solid wastes. The modified TG-DTG interpretation should be generally applicable to various types of alkaline solid wastes. The standard operation procedure is provided as follows:

- Step 1: dry samples at 105 °C for at least 30 min to remove the adsorbed water before analysis,
- Step 2: place 3–10 mg of solid samples in a platinum crucible and put in TG analyzer,
- Step 3: heat sample directly from 50 to 950 °C at 10–20 °C/min under N_2 atmosphere,
- Step 4: apply the modified TG-DTG interpretation (Fig. 6.2) to determine the weight loss of certain compound, and
- Step 5: combine with other qualitative analyses (such as MS) to confirm the species of evolved gas within the temperature ranges of decomposition.

Figure 6.4 shows the modified TG-DTG interpretation for determining the $CaCO_3$ content in alkaline solid wastes. In the modified method, both the initial (T_1) and final (T_2) temperatures of $CaCO_3$ decomposition on TG plot are determined by the extrapolated on-set. The point determined by extrapolated on-set is defined as the intersection of the tangent drawn at the point of greatest slope on the leading edge of the peak with the extrapolated baseline. The weight loss due to $CaCO_3$ decomposition can then be obtained by

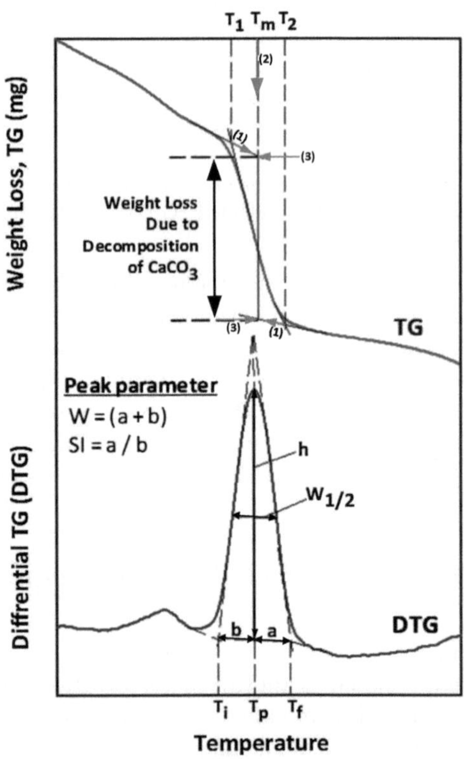

Fig. 6.4 Modified TG-DTG interpretation to determine $CaCO_3$ content in alkaline solid wastes by both TG (*upper curve*) and peak parameters from DTG (*lower curve*). Adaptation with permission from Macmillan Publishers Ltd: ref. [24], copyright 2016

- Step 1: extending the two straight-line portions of the baseline before T_1 and after T_2,
- Step 2: making a vertical line pass through the midpoint (T_m) between T_1 and T_2, and
- Step 3: determining their intersections to the baselines and vertical line. The weight loss between these two intersections was attributed to the $CaCO_3$ decomposition within the solid samples.

For the DTG curve, it can be characterized by the temperature of extrapolated on-set drawn by the beginning (T_i) and final (T_f) of the peak. The center of temperature peak (T_p), half width ($W_{1/2}$), and peak width (W) can be determined accordingly. As shown in Fig. 6.4, the shape index (SI) of the DTG peak is defined as the absolute ratio of the slope of the tangents to the DTG peak at the inflection points. Therefore, the above shape parameters of DTG can be graphically determined.

By the modified TG-DTG interpretation, a positive correlation between $CaCO_3$ content in alkaline wastes and its reactivity with CO_2 through mineral carbonation can be observed. Except for the complex and hydrated compounds in alkaline solid wastes, the weight loss versus decomposition temperature for a material can be separated into

- 50–105 °C: expulsion of surface water,
- 200–300 °C: removal of pore water,
- 400–500 °C: dehydration of crystal water (e.g., $Ca(OH)_2$),
- 500–630 °C: $MgCO_3$ decomposition, and
- 600–850 °C: $CaCO_3$ decomposition.

The matrix interference due to Ca–Al–Si hydrates presented in alkaline solid wastes can be reduced to a level of 10^{-3}. The method detection limit of the modified TG-DTG interpretation is about 0.70%, with a high precision (0.40 ± 0.25%) and accuracy (1.34 ± 0.20%) in the case of BOFS [24].

6.1.3 Key Carbonation Parameters in Solid Wastes

As shown in Fig. 6.2, the CO_2 content (denoted as "CO_2") in the sample can be determined using the modified TG-DTG method by Eq. (6.1):

$$CO_2(\text{wt\%}) = \frac{\Delta m_{CO_2}}{m_{105\,^\circ\text{C}}} \times 100 \tag{6.1}$$

where Δm_{CO_2} (mg) is the weight loss due to the decomposition of $CaCO_3$. $m_{105\,^\circ\text{C}}$ (mg) is the dry weight of the sample.

By doing so in Eq. (6.1), two key carbonation indicators can be calculated, i.e., the weight gain (%) and carbonation conversion. The weight gain (%) of the dry solid waste then can be realized using the value of CO_2 content, according to Eq. (6.2):

$$\text{Weight gain}\,(\%) = \frac{CO_2(\text{wt}\%)}{100 - CO_2(\text{wt}\%)} \times 100 \tag{6.2}$$

The carbonation conversion of solid waste (δ_{CaO}, %) can be calculated by Eq. (6.3), assuming the calcium-bearing compositions are the main reaction species:

$$\delta_{CaO} = \frac{\frac{CO_2(\%)}{100-CO_2(\%)} \times \frac{1}{MW_{CO_2}}}{Ca_{total}/MW_{Ca}} = \frac{\frac{CO_2(\%)}{100-CO_2(\%)} \times \frac{1}{MW_{CO_2}}}{CaO_{total}/MW_{CaO}} \tag{6.3}$$

where MW_{CO_2}, $MW_{Ca,}$ and MW_{CaO} are the molecular weight of CO_2 (44 g/mol), Ca (40 g/mol), and CaO (56 g/mol), respectively. Ca_{total} and CaO_{total} are the percent weight fraction of Ca (normally determined by XRF or by ICP after total digestion) and CaO (normally determined by XRF) in the fresh solid sample, respectively.

Both the weight gain (Eq. 6.2) and carbonation conversion (Eq. 6.3) determined by TGA are frequently expressed as the degree of carbonation for one target material. Figure 6.5 shows the relationship between weight gain and carbonation conversion of alkaline solid wastes. The plot of carbonation conversion versus weight gain is a straight line, and the slope of the straight line is related to the CaO

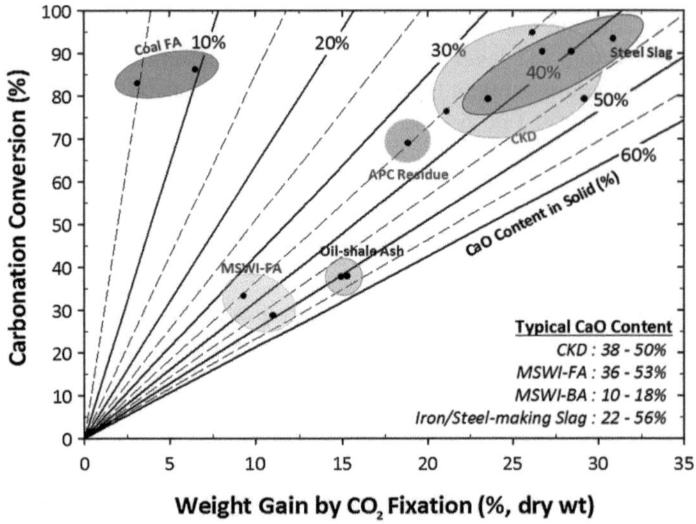

Fig. 6.5 Profile of carbonation conversion and weight gain per dry weight for different solid wastes, including cement kiln dust (CKD), municipal solid waste incinerator (MSWI), fly ash (FA), and bottom ash (BA). Adaptation with permission from Macmillan Publishers Ltd: ref. [24], copyright 2016

content in the solid. It is noted that alkaline solid wastes, such as steel slag and fly ash, have been recognized as effective materials for CO_2 sequestration by mineral carbonation. For example, the cement kiln dust (CKD) and steel slag exhibit the relatively higher CaO content (i.e., 30–50%) and CO_2 capture capacity by the maximum achievable conversion technologies as reported in the literature [6, 11, 14, 25]. Although a higher carbonation conversion of coal fly ash (FA) can be achieved (i.e., 80–90%), the CaO content of coal fly ash is low (i.e., 5–10%), thereby resulting in a relatively low CO_2 capture capacity.

For determining CO_2 fixation capacity via the carbonation reaction, the available methods for the cases of mortars/concrete and cement kiln dust according to their physicochemical properties have been reported by Steinour [26] and Huntzinger et al. [27], which is as follows:

$$\mathrm{ThCO_2}(\%) = 0.785\,(\mathrm{CaO} - 0.56\,\mathrm{CaCO_3} - 0.7\,\mathrm{SO_3}) + 1.091\,\mathrm{MgO} + 1.420\,\mathrm{Na_2O} + 0.935\,\mathrm{K_2O} \quad (6.4)$$

Similarly, for the fresh alkaline solid waste, if it is assumed that all of the CaO content in the solid waste, except that originally bound in $CaSO_4$ and $CaCO_3$ phases, will convert to $CaCO_3$, the theoretical CO_2 capture capacity ($ThCO_2$, as a percentage of dry mass) can be estimated via Eq. (6.5):

$$\mathrm{ThCO_2}(\%) = 0.785\,(\mathrm{CaO} - 0.56\,\mathrm{CaCO_3} - 0.7\,\mathrm{SO_3}) \quad (6.5)$$

where $ThCO_2$ (kg CO_2/kg solid waste) is the theoretical CO_2 capture capacity, CaO (kg CaO/kg solid waste) and SO_3 (kg SO_3/kg solid waste) are the weight fraction of CaO and SO_3 in solid waste measured by XRF, respectively, and $CaCO_3$ (kg $CaCO_3$/kg solid waste) is the weight fraction of $CaCO_3$ analyzed by TGA.

Based on Eq. (6.5), the theoretical amount of CO_2 capture for different solid wastes is fall in the following ranges:

- Blast furnace slag (BFS): 0.252 − 0.322 kg-CO_2/kg-BFS,
- Basic oxygen furnace slag (BOFS): 0.309 − 0.374 kg CO_2/kg BOFS,
- Electric arc furnace oxidizing slag (EAFOS): 0.177 − 0.229 kg-CO_2/kg-EAFOS,
- Electric arc furnace reducing slag (EAFRS): 0.313 − 0.391 kg-CO_2/kg-EAFRS,
- Argon oxygen decarburization slag (AODS): 0.428 − 0.476 kg-CO_2/kg-AODS,
- Ladle furnace slag (LFS): 0.396 − 0.451 kg-CO_2/kg-LFS,
- Coal fly ash (FA): ~ 0.070 kg-CO_2/kg-FA,
- MSWI fly ash (FA): 0.323 − 0.388 kg-CO_2/kg-FA, and
- MSWI bottom ash (BA): 0.124 − 0.158 kg-CO_2/kg-BA.

In some cases, the probability of $MgCO_3$ formation is low due to the relatively low content of MgO in alkaline solid waste. Also due to the relatively low pressure of CO_2 and the short reaction time, limited $MgCO_3$ formation is expected. Typical process conditions for the formation of $MgCO_3$ via aqueous carbonation are (1) p_{CO_2} greater than 100 bar, (2) temperature greater than 144 °C, and (3) a reaction time of hours [28, 29]. The other metal oxide components, such as SiO_2 and P_2O_5, in the fresh solid waste are considered not to contribute to CO_2 fixation.

6.1.4 *Kinetics and Thermodynamics of Thermal Decomposition*

The kinetic (i.e., apparent activation energy, kinetic exponent, and pre-exponential factor) and thermodynamic parameters (i.e., the changes of entropy, enthalpy, and Gibbs free energy for the formation of the activated complex) for the thermal decomposition of a certain compound in a material can be determined by the Kissinger equation and Arrhenius equation, and transition state theory.

6.1.4.1 Kinetic Equations

The Kissinger equation has been extensively applied to evaluate the kinetics of thermal decomposition of a solid material, and the relevant activation energy and reaction order [30–34]. First, the reaction rate of a solid-state reaction can be expressed by means of the general mass action law with the Arrhenius law, as shown in Eq. (6.6):

$$\frac{\mathrm{d}\alpha}{\mathrm{d}t} = k(T)f(\alpha) = A\exp\left(-\frac{E_\mathrm{a}}{RT}\right)f(\alpha) \tag{6.6}$$

where k is the rate constant, T is the absolute temperature (K), α(-) is the reacted fraction, $f(\alpha)$ is an algebraic function depending on the reaction mechanism, A is the pre-exponential factor (1/min), E_a is the apparent activation energy (kJ/mol), and R is the universal gas constant (8.314 J/K mol).

Then, by differentiating Eq. (6.6), if the temperature (T) rises at a constant heating rate ($\beta = \mathrm{d}T/\mathrm{d}t$), Eq. (6.7) can be obtained for a non-isothermal kinetics [34]:

$$\frac{\mathrm{d}^2\alpha}{\mathrm{d}t^2} = \left[\beta\frac{E_\mathrm{a}}{RT^2} + A\exp\left(-\frac{E_\mathrm{d}}{RT}\right)f'(\alpha)\right]\frac{\mathrm{d}\alpha}{\mathrm{d}t} \tag{6.7}$$

where β is the heating rate (K/min), and T_p is the absolute temperature of peak (K).

After that, assuming the maximum rate occurs at a temperature T_p, i.e., $\left[\mathrm{d}(\mathrm{d}\alpha/\mathrm{d}t)_{T_\mathrm{p}}/\mathrm{d}t\right] = 0$, the general form of Kissinger equation for a non-isothermal kinetics can be expressed as Eq. (6.8), for the determination of the activation energy.

$$\ln\left(\frac{\beta}{T_p^2}\right) = -\frac{E_a}{R}\frac{1}{T_p} + \ln\left(-\frac{ARf'(\alpha)}{E_a}\right) = C_1\frac{1}{T_p} + C_2 \tag{6.8}$$

According to Eq. (6.8), the slope of the plot of ln (β/T_p^2) versus $1/T$ gives the apparent activation energy (E_a). The constant term (C_2), i.e., the intercept with the y-axis, is related to both A and $f'(\alpha)$, as shown in Eq. (6.9):

$$C_2 = \ln\left(\frac{ARf'(\alpha)}{E_a}\right) \tag{6.9}$$

Also, we can assume the temperature independence of the pre-exponential factor based on the Arrhenius theory. The rate of a thermally induced solid reaction can be expressed as Eq. (6.10):

$$\frac{d\alpha}{dT} = \frac{Af(\alpha)}{\beta}\exp\left(-\frac{E_a}{RT_p}\right) \tag{6.10}$$

As applied frequently for a description of heterogeneous processes with the surface reaction controlled [31], the $f(\alpha)$ can be expressed by the reaction order kinetic model:

$$f(\alpha) = (1-\alpha)^n \tag{6.11}$$

where n is the kinetic exponent of thermal decomposition reaction. By substitution of n from Eqs. (6.9) and (6.11) into Eq. (6.10) and rearranging, the value of the kinetic exponent (n) can be estimated directly from Eq. (6.12):

$$n = \frac{(1-\alpha_{max})}{d\alpha_{max}/dT}\frac{E_a}{\beta R}\exp\left(C_2\right)\exp\left(-E_a/RT_p\right) \tag{6.12}$$

Furthermore, according to Kissinger [35], the n value for reaction order processes can be estimated by the shape index (SI) if the peak shape is independent of heating rate (Eq. 6.13). The SI value of the DTG curve is an important parameter of thermodynamic analysis. The definition of SI value can refer to Fig. 6.4.

$$n = 1.26\,\mathrm{SI}^{1/2} \tag{6.13}$$

Carne et al. [36] also proposed possibilities to evaluate the n value from the slope of the plot of ln β versus $1/T_p$ as in Eq. (6.14):

$$\frac{d\,ln\beta}{d\left(1/T_p\right)} = -\frac{E_a}{nR} \tag{6.14}$$

6.1.4.2 Thermodynamic Equations

For decomposition thermodynamics of a certain compound in alkaline solid wastes, the transition state theory (so-called activated complex theory) can be applied. It can be described by the general form of the Eyring equation [30] in Eq. (6.15):

$$A = \frac{e \cdot \chi \cdot k_{\mathrm{B}} \cdot T_{\mathrm{p}}}{h} \exp\left(\frac{\Delta S}{R}\right) \tag{6.15}$$

where e is the Neper number (i.e., 2.7183), χ is the transition factor (i.e., 1 for monomolecular reactions), k_{B} is the Boltzmann constant (i.e., 1.381×10^{-23} J K^{-1}), and h is the Plank constant (i.e., 6.626×10^{-34} J s). The change of entropy (ΔS) term can be calculated based on the peak temperature (T_{p}) which characterizes the highest rate for thermal decomposition in the DTG plot. Therefore, the ΔS can be determined by rearranging Eq. (6.15), which is as follows:

$$\Delta S = R \ln\left(\frac{Ah}{e\chi k_{\mathrm{B}} T_{\mathrm{p}}}\right) \tag{6.16}$$

For the activated complex formation, the changes of the enthalpy (ΔH) and Gibbs free energy (ΔG) can be calculated using Eqs. (6.17) and (6.18), respectively:

$$\Delta H = E_{\mathrm{a}} - RT_{\mathrm{p}} \tag{6.17}$$

$$\Delta G = \Delta H^{\neq} - T_{\mathrm{p}} \Delta S \tag{6.18}$$

6.1.5 Case Study: Basic Oxygen Furnace Slag

6.1.5.1 Modified TG-DTG Interpretation

Figure 6.6a, b shows the TG-DTG plots of the fresh and carbonated basic oxygen furnace slag (BOFS), respectively. The results show a continuous weight loss at 200–900 °C in both the fresh and carbonated BOFS, which is attributed to the decomposition of various hydrates, such as α-dicalcium silicate hydrate. Therefore, the thermal decomposition of various hydrates should be considered when determining the carbonate contents in BOFS. Other major minerals in BOFS, including brownmillerite, wollastonite, and larnite, are relatively stable compounds under the temperature range of TG analysis. In the fresh BOFS, the thermal decomposition of portlandite ($Ca(OH)_2$) occurs at temperature 400–500 °C. However, after carbonation, there was a lack of a peak for $Ca(OH)_2$ decomposition in BOFS, indicating

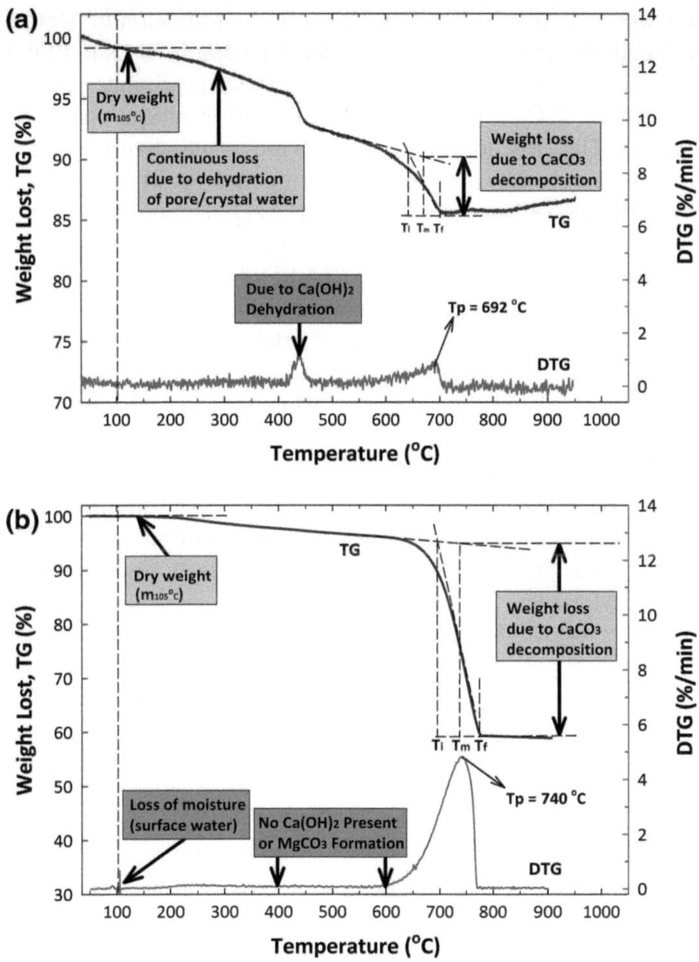

Fig. 6.6 TG-DTG plots for **a** fresh and **b** carbonated BOFS using the modified TG-DTG interpretation. Adaptation with permission from Macmillan Publishers Ltd: ref. [24], copyright 2016

that the $Ca(OH)_2$ was reacted with CO_2 during carbonation. Instead, a great weight loss at 600–800 °C was found in the carbonated BOFS, revealing the formation of $CaCO_3$ after carbonation.

On the other hand, no $MgCO_3$ formation was observed in the carbonated BOFS due to no peak at 500–630 °C in DTG, which was in good agreement with the XRD results [37, 38]. As the aforementioned, the typical conditions for the formation of $MgCO_3$ precipitates by aqueous carbonation are at a temperature over 144 °C [28] for hours [39]. Under the mild condition, with a ratio of Mg^{2+} to Ca^{2+} concentrations higher than 0.5, a metastable (amorphous) hydrated magnesium carbonate phase might be formed [23]. However, since the leaching concentrations of Mg ions from BOFS were low (e.g., 1.7–3.0 mg/L) [40], the formation of $MgCO_3$ crystal is negligible. In other words, the calcium-containing compositions in BOFS should be the major species reacting with CO_2 to form $CaCO_3$ precipitate.

To confirm the $CaCO_3$ content in BOFS, the DSC technique can be coupled with the TG analysis. The DSC technique can provide quantitative measurement on the heat released or absorbed by the specimen during heating. Theoretically, the $CaCO_3$ crystal will start to decompose into $CaO_{(s)}$ and $CO_{2(g)}$ at temperatures above 600 °C, as shown in Eq. (6.19). It is noted that the reaction heat for decomposing one mole of $CaCO_3$ particles at 1000 K is about 170.4 kJ [41]. Since the amounts of heat absorbed can be converted into the weight of $CaCO_3$ decomposed, the correlation between DSC and TG measurements can be established.

$$CaCO_{3(s)} \underset{\Delta}{\longrightarrow} CaO_{(s)} + CO_{2(g)}, \quad \Delta H^{o}_{1000\,K} = 170.4\,kJ/mol \tag{6.19}$$

Figure 6.7 shows the correlation between the amounts of $CaCO_3$ decomposition from modified TG-DTG interpretation (i.e., abscissa) and the heat absorbed from DSC technique (i.e., ordinate). The values of relative percent difference between the weights of $CaCO_3$ determined by thermography and those calculated from DSC are $1.34 \pm 0.20\%$. Moreover, the results of paired-samples t tests indicated that the calculated t-value of 1.595 was less than the tabulated t-value of 2.201, thereby accepting the null hypothesis. In other words, no difference was found in $CaCO_3$ contents calculated from the modified TG-DTG interpretation and DSC method

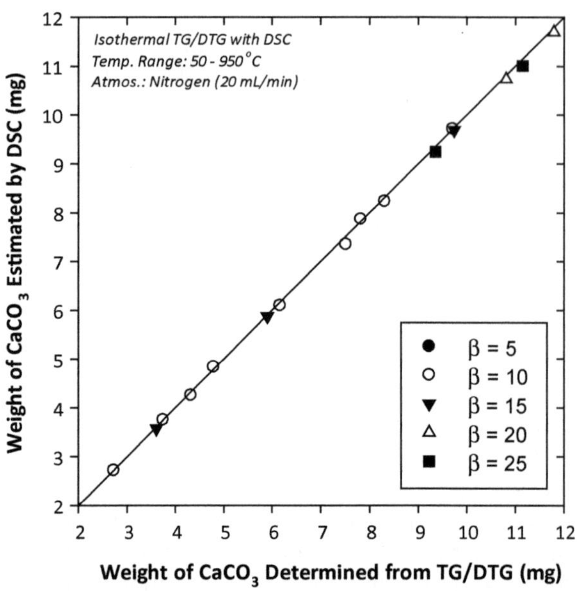

Fig. 6.7 Correlation of TG-DTG and DSC results for reference pure $CaCO_3$ powder under thermal decomposition. Adaptation with permission from Macmillan Publishers Ltd: ref. [24], copyright 2016

($p = 0.139$), with a Pearson's correlation coefficient of 0.9997. This suggests that the modified TG-DTG interpretation should be applicable to provide a precise and accurate analysis of $CaCO_3$ contents in BOFS.

6.1.5.2 Qualitative Analysis by TG-MS

The weight loss between 500 and 900 °C would be simultaneously attributed to the decomposition of carbonates (release CO_2) and hydrates (release H_2O). To identify the types of volatiles and/or gases released during TG analysis, MS and/or FTIR can be used for the evolved gas analysis. For instance, Fig. 6.8a, b show the

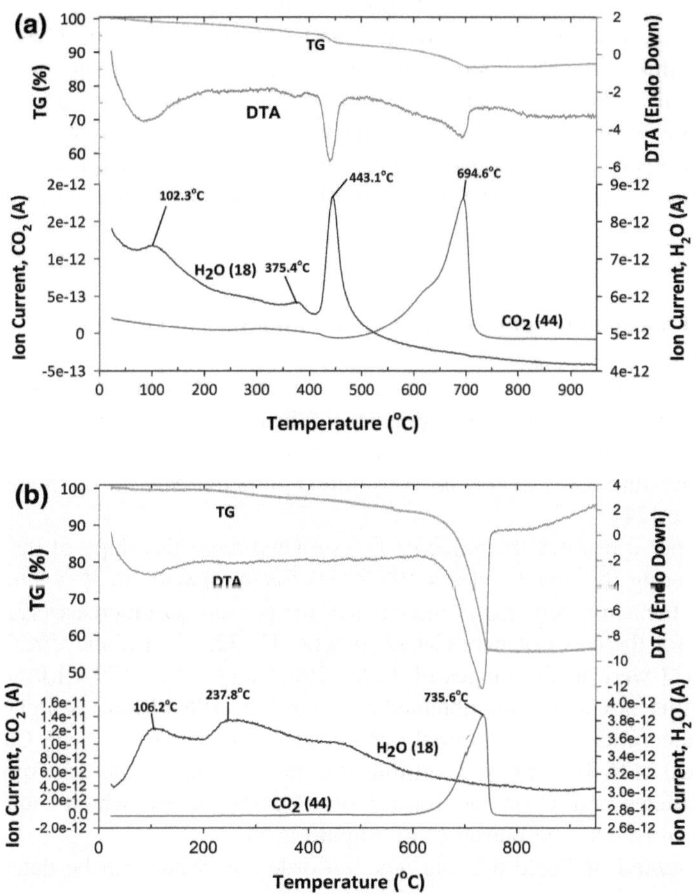

Fig. 6.8 Plots of weight loss (TG Analysis) and mass spectroscopy (ion current for H_2O mass number 18 and CO_2 mass number 44) for **a** fresh and **b** carbonated steel slag. Adaptation with permission from Macmillan Publishers Ltd: ref. [24], copyright 2016

TG-MS plots of both fresh and carbonated BOFS, respectively. The dehydration of different hydrates was found to occur continuously between 50 and 800 °C, especially pronounced before 700 °C. The dissociation of the phases containing H_2O in fresh BOFS consists of three peaks at 102, 375, and 443 °C, which could be attributed to (1) evaporation of surface water, (2) evaporation of pore water, and (3) dehydration of crystal water (i.e., $Ca(OH)_2$), respectively. Similarly, two peaks for the H_2O signal in the carbonated BOFS were observed at 106 and 238 °C, revealing subsequent removal of surface water and pore water, respectively. The Ca $(OH)_2$ content was eliminated after carbonation because no H_2O signal was observed at 400–500 °C. It suggests that the series of H_2O signals in the evolved gas analysis results provides the evidence to the observations in TG-DTG plot.

In addition, in fresh BOFS, CO_2 evolved gas coming from the decomposition of $CaCO_3$ was observed at 695 °C. Similar results were observed in the carbonated BOFS that the loss of mass occurring at 650–750 °C is related to CO_2 emissions, corresponding to the decomposition of $CaCO_3$, with a high temperature peak of 736 °C. This confirmed that the carbonate product is a crystallized $CaCO_3$. Furthermore, no peak of the CO_2 signal was observed at 500–600 °C after carbonation, revealing that the decomposition of $MgCO_3$ crystal was not detected. This provided the rationale that the formation of $MgCO_3$ was negligible in the case of BOFS.

6.1.5.3 Thermal Decomposition Kinetics and Thermodynamics

To determine the kinetic parameters for the thermal decomposition of $CaCO_3$ in BOFS, Table 6.2 presents the influence of heating rate (β) on important peak parameters, including T_i, T_e, W, $W_{1/2}$, H, and SI value. The results indicate that the peak temperature of $CaCO_3$ decomposition (T_p) increases with the increase of the heating rate [24].

Figure 6.9 illustrates the value of E_a estimated from the slope of the Kissinger plot, indicating that the E_a yields 197.7 ± 5.5 kJ/mol with an R^2 value of 0.995. Typically, the activation energy increases as the particle size increases [22, 42]. The E_a values for the cases of pure $CaCO_3$ powder [2, 32, 33, 42] and $CaCO_3$ mixture [22, 30, 43] were in the ranges of 139.0–190.4 and 119.7–179.4 kJ/mol, respectively. In this case study, the obtained E_a value (i.e., 198 kJ/mol) is higher than the value of theoretical E_a for thermal decomposition of isolated calcite $CaCO_3$ (i.e., 175 kJ/mol) [43]. This might be attributed to the fact that calcite was formed inside BOFS particles and/or on the surface of BOFS particles, which required extra energy to overcome the barrier of an impure layer.

As presented in Table 6.3, the reaction order (n) value can be determined by several approaches, including the reaction order model, the SI of peak, and the

Table 6.2 Effect of heating rate (β) on the peak parameters of DTG curve under N_2 atmosphere. Reprinted by permission from Macmillan Publishers Ltd: ref. [24], copyright 2016

β (K min^{-1})	T_i (K)	T_p (K)	T_f (K)	W (K)	$W_{1/2}$ (K)	H (% K^{-1})	T_2/T_1 ratio	SI
1	869.0 ± 1.1	947.4 ± 3.7	958.6 ± 3.2	89.6 ± 2.1	51.8 ± 1.7	0.64 ± 0.01	1.08 ± 0.00	0.14 ± 0.02
5	906.3 ± 3.4	1011.4 ± 4.3	1035.9 ± 5.5	124.1 ± 4.1	70.1 ± 2.2	2.50 ± 0.06	1.09 ± 0.01	0.19 ± 0.02
10	931.3 ± 5.1	1034.0 ± 4.2	1060.3 ± 8.3	129.0 ± 3.2	75.6 ± 4.3	4.45 ± 0.13	1.10 ± 0.01	0.26 ± 0.04
15	943.5 ± 1.8	1048.9 ± 5.7	1079.5 ± 7.3	136.0 ± 5.6	82.5 ± 2.1	6.40 ± 0.23	1.10 ± 0.02	0.26 ± 0.01
20	952.8 ± 0.9	1061.8 ± 2.4	1095.4 ± 0.8	142.5 ± 1.1	85.8 ± 1.5	8.26 ± 0.07	1.10 ± 0.01	0.30 ± 0.01
25	958.3 ± 0.7	1077.7 ± 3.0	1123.0 ± 1.6	164.8 ± 1.4	99.0 ± 2.0	8.86 ± 0.12	1.12 ± 0.00	0.33 ± 0.05
30	963.8 ± 3.0	1087.0 ± 5.2	1136.7 ± 8.5	172.9 ± 7.8	105.3 ± 5.1	10.1 ± 0.4	1.12 ± 0.01	0.39 ± 0.06
35	965.9 ± 0.8	1087.9 ± 3.5	1137.9 ± 6.8	172.0 ± 6.6	103.7 ± 3.0	12.0 ± 0.4	1.12 ± 0.01	0.41 ± 0.01

Note All experiments were carried out with triple duplicates ($N = 3$)

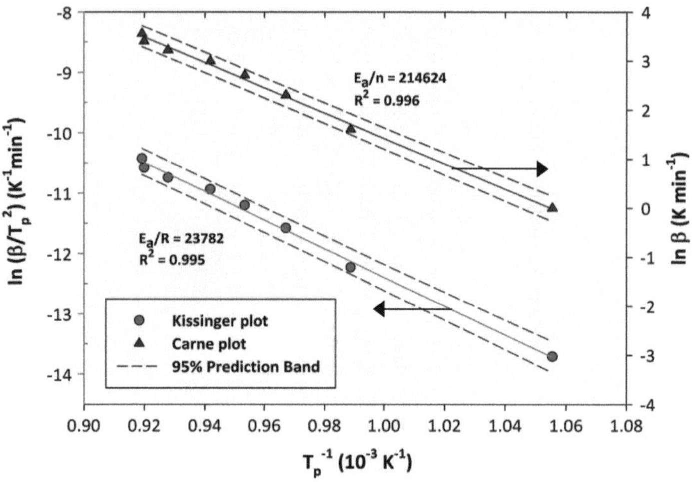

Fig. 6.9 Evaluation of apparent activation energy (via Kissinger plot) and reaction order (via Carne plot) of $CaCO_3$ thermal decomposition in steel slag. Adaptation with permission from Macmillan Publishers Ltd: ref. [24], copyright 2016

Carne equation. The n values determined from the reaction order model and the SI of peaks are 0.11–2.58 and 0.47–0.81, respectively. This indicates that the n value is sensitive to heating conditions. The methods applied for calculation of the n value provide considerably different results. Based on the Carne plot, the n value is estimated to be 0.92 ± 0.03, with the highest R^2 value of 0.996 among the three applied methods. As a result, the $CaCO_3$ decomposition in BOFS can be considered as a first-order reaction, implying interface-controlled growth with grain boundary nucleation after saturation [44]. Similar results ($n \cong 1$) were also observed in the literature [22, 30, 32, 33, 42, 43]. To determine the pre-exponential factor (A), the n value from the Carne method can be used. Substitution of E_a and n values into Eq. (6.7) provides an average value of A of $(2.20 \pm 0.01) \times 10^9$ min^{-1}. It is noted that the A value regularly increased with the heating rate.

On the other hand, the thermodynamic parameters, including ΔS, ΔH, and ΔG, can be determined using general equations. As presented in Table 6.3, the average value of ΔS was estimated to be about −118.82 J/mol K. This indicated that the formation of the activated complex exhibited a more organized structure than the initial substance. In addition, the average values of ΔH and ΔG were 189.04 and 313.16 kJ/mol, respectively [24]. For $CaCO_3$ decomposition, the ΔH values are close to the E_a values. However, significant differences between the values of ΔH and ΔS are observed.

Table 6.3 Kinetic and thermodynamic parameters determined by Kissinger calculation procedure. Adaptation with permission from Macmillan Publishers Ltd: ref. [24], copyright 2016

β (K/min)	$\times 10^{-3}$ 1/ T_p (1/K)	$(1-\alpha_{max})$[a]	$\times 10^{-2}$ $d\alpha_{max}/dT$ (1/K)[a]	n value			A ($\times 10^9$ 1/min)[b]	ΔS (J/mol K)	ΔH (kJ/mol)	ΔG (kJ/mol)
				Kissinger (Eq. 6.12)	Kissinger (Eq. 6.13)	Carne (Eq. 6.14)				
1	1.0555	0.646 ± 0.008	0.657 ± 0.005	2.58	0.47	0.92	2.194	−118.05	189.84	301.69
5	0.9887	0.676 ± 0.018	2.542 ± 0.065	0.68	0.55	0.92	2.202	−118.57	189.31	309.23
10	0.9671	0.671 ± 0.007	4.572 ± 0.115	0.31	0.64	0.92	2.201	−118.75	189.12	311.91
15	0.9534	0.694 ± 0.002	6.507 ± 0.250	0.21	0.65	0.92	2.207	−118.85	189.00	313.66
20	0.9418	0.693 ± 0.004	8.401 ± 0.069	0.16	0.69	0.92	2.207	−118.95	188.89	315.20
25	0.9279	0.694 ± 0.005	9.081 ± 0.131	0.17	0.73	0.92	2.207	−119.08	188.76	317.09
30	0.9199	0.701 ± 0.005	10.35 ± 0.38	0.15	0.79	0.92	2.209	−119.14	188.68	318.19
35	0.9192	0.702 ± 0.007	12.28 ± 0.34	0.11	0.81	0.92	2.209	−119.15	188.67	318.29

[a]Data with 95% confidence interval
[b]The pre-exponential factor was determined by Eqs. (6.7) and (6.14) (n = 0.92)

6.2 Quantitative X-ray Diffraction (QXRD)

Considerable research has been carried out on solid wastes and/or industrial by-products in various domains: accelerated carbonation [37, 45], utilization performance [46, 47], and environmental impact [48, 49]. Although these research studies were performed for a variety of purposes, a common point is that the properties of the material should be characterized in advance. The characterization of a material is usually divided into three parts:

- Physical properties: morphology, fitness, density, solubility, etc.
- Chemical properties: oxide contents, heavy metal leaching amounts, hazardous components, pozzolanic and cementitious properties, etc.
- Mineralogical properties: crystalline, phase fraction, grain size, defect, etc.

The physical properties would greatly influence their reactivity with the environment. The chemical properties would give a first direction of the wastes' hazard potential. The mineralogical analysis of the wastes would provide further understanding on their behavior according to their nature and proportion of various mineral phases. The quantitative mineralogical characterization of waste is an essential step. Nevertheless, accurate analysis of the content of calcium carbonate in alkaline solid wastes is a difficult task due to their complex composition.

The X-ray quantitative phase analysis method can be applied to analyze the content of calcium carbonate in BOFS before and after carbonation. Two types of analytical techniques using X-ray diffraction, such as reference intensity ratio method and Rietveld refinement, can provide precise and accurate information on the fraction of crystal phases in solid wastes, although they are generally time-consuming in sample preparation and data processing.

6.2.1 *Reference Intensity Ratio (RIR)*

Reference Intensity Ratio (RIR) method, based on the measurement of the diffraction intensities (areas) of the characteristic peaks of the minerals, has been used to quantify the mineral phase in solids, e.g., quartz [50] and carbonates [23]. Three samples are needed: (1) the raw material, (2) the pure mineral, and (3) a mixture containing a known mass of pure mineral per gram of raw material. The amount of the certain phase can be calculated as follows [50]:

$$x = a \cdot \frac{I}{I_0} \cdot \frac{I_0 - I'}{I' - I} \tag{6.20}$$

where x is the amount of the certain phase (in g/g of sample), and I, I_0, and I' are the intensities of the characteristic peak of the certain phase in the raw material, in the pure phase, and in the mixture containing a grams of the certain phase per gram of raw material, respectively. The RIR method can be also performed through the use of an internal standard, such as high-purity corundum α-Al_2O_3 crystal powder (>99.5%) [23].

6.2.2 *Rietveld Refinement*

Rietveld method has been considered a powerful tool for crystal structure refinements and quantitatively analysis [51] based on X-raydiffraction pattern. The Rietveld method considers overlapping peaks in the XRD pattern and the effect of preferred orientation [52]. Therefore, this method has been widely used on various materials, of which the mineralogy is well known, for instance, Portland cement [52]. Moreover, it has been applied for the quantification of various mineral crystals in solid wastes [50, 53–55].

Unlike chemical methods and thermal analysis, the Rietveld method does not change the state of samples and avoids side reactions in ambient conditions [52]. Additional advantages of using the Rietveld method for phase quantification included

- Time-consuming calibration measurements can be avoided.
- The phase abundance could be determined if all phases are identified and these crystal structure parameters and chemical composition are known.
- The relative weight fractions of crystalline phases in a multiphase sample could be calculated directly from scale factors of the respective calculated intensities.

Figure 6.10 illustrates the standard operation procedure of QXRD using Rietveld method for quantifying phase fraction in a material:

- Step 1: acquisition of XRD pattern of a material,
- Step 2: identification of the major crystal phases for the XRD pattern, and
- Step 3: application of Rietveld refinement technique to quantify the phase fraction and evaluate the crystal structure.

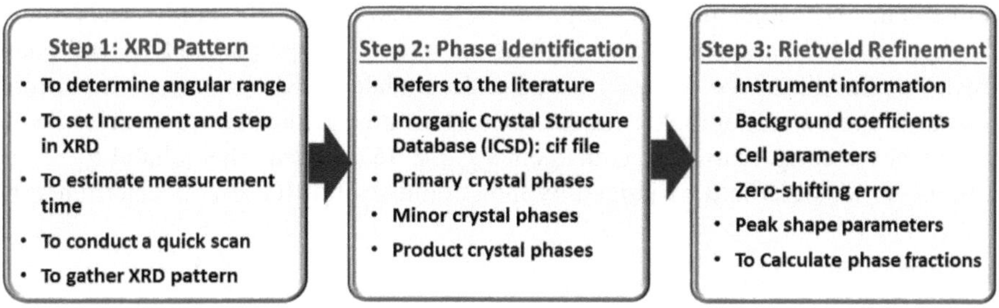

Fig. 6.10 QXRD procedure using Rietveld method for quantifying phase fraction in a material

6.2.2.1 Principles

Several factors, which may be responsible for the discrepancy of the quantitative results, can be obtained from the Rietveld method [51]:

- Sample preparation: It is impossible to get correct results in the case of an inhomogeneous sample.
- Structure model: Large R values occur when incorrect structure input or unknown phase presents.
- Data collection tragedy: Both the step size and scanning speed can cause error in the determination of X-ray diffraction peak positions, intensity, and full width at half maximum (FWHM). Intensity error caused by counting statistic will rise with increasing scanning rate and decreasing count time.

The Rietveld method is a full-pattern analysis that the atom parameters in the unit cell are calculated fitting the entire pattern by the least-squares method so that minimizing the difference (M) between the experimental ($y_{(\text{obs})}$) and calculated ($y_{(\text{calc})}$) XRD diffractogram [51, 56]:

$$M = \sum_{i=1}^{n} w_i \left[y_{i(\text{obs})} - y_{i(\text{calc})} \right]^2 = \text{minimum} \tag{6.21}$$

where w_i is the weight of each observation point, and n is the number of observation points. The sum i is over all data points. Therefore, the missing major phases in the Rietveld method would inevitably involve significant differences between experimental and calculated patterns.

The standard uncertainty of $Y_{O,i}$ (i.e., $\sigma[Y_{O,i}]$) can be determined by measuring the $Y_{O,i}$ intensity for an infinite number of times. To evaluate the goodness of the developed model, a statistical method such as chi-squared (χ^2) test should be introduced as follows:

$$\chi^2 = \frac{1}{n} \sum_{i=1}^{n} \frac{\left[Y_{O,i} - Y_{C,i} \right]^2}{\sigma^2 \left[Y_{O,i} \right]} \tag{6.22}$$

Typically, the χ^2 value would gradually converge to one during the Rietveld refinement. By adjusting the key structure parameters, including background coefficients, cell parameters, zero-shifting error, peak shape parameters, and phase fractions, the difference between actual and simulated XRD patterns can be minimized. If the crystallographic model is correct and chemically reasonable, the χ^2 would never drop below or equivalent to one [57]. Also, the reliability of the refinement result should be judged by the goodness of fit (GOF), as determined by Eq. (6.23):

$$\mathrm{GOF} = R_{wp}/R_{exp} \tag{6.23}$$

where R_p and R_{wp} are the pattern R factor and the weighted pattern R factor, respectively.

6.2.2.2 Available Software for Pattern/Structure Refinement

The refinement of XRD patterns using Rietveld method can be executed by many efficient programs such as GSAS [53, 54], FULLPROF 2000 [52], SIROQUANT [23, 53, 58, 59], X'Pert HighScore Plus [60], DBWS9411 [51], and Maud [50]. In the Rietveld refinement, various corrections can be introduced, such as texture, absorption contrast, sample transparency, displacement, and microabsorption effect.

For instance, the Rietveld method can be performed by the General Crystal Structure Analysis System (GSAS) software with the EXPGUI program. GSAS was created by Larson and Von Dreele [61] of Los Alamos National Laboratory for fitting atomic structural models to single crystal and powder diffraction data, even both simultaneously. In GSAS, the atom parameters including scale factors, background coefficients, zero-shifting error, lattice parameters, profile shape parameters, atomic site occupancies, and phase fractions in the unit cell were refined simultaneously.

The crystal structure parameters used to interpret the XRD patterns can be taken from the ICSD (Inorganic Crystal Structure Database). Typically, the major components in alkaline solid wastes, with the collection codes for each structure, include

- α-dicalcium silicate hydrate ($Ca_2(HSiO_4)(OH)$, abbreviated as C_2–S–H, Code 75277),
- β-larnite (Ca_2SiO_4, abbreviated as C_2S, Code 245080),
- brownmillerite ($Ca_2Fe_{1.014}Al_{0.986}O_5$, Code 98836),
- calcite ($CaCO_3$, Code 169933),
- portlandite ($Ca(OH)_2$, Code 73468),
- wollastonite ($CaSiO_3$, abbreviated as C_1S, Code 240469), and
- wustite (FeO, Code 633038).

6.2.3 *Case Study: Alkaline Solid Wastes*

The Rietveld refinement has been considered to be chemically plausible by viewing the observed and calculated patterns graphically. Kuusik et al. [62] suggested that the compositions of oil shale ash have shown a good correlation between the chemical and quantitative XRD analyses, where the latter can be used for preliminary and rapid analysis. Mahieux et al. [50] determined the mineral composition of

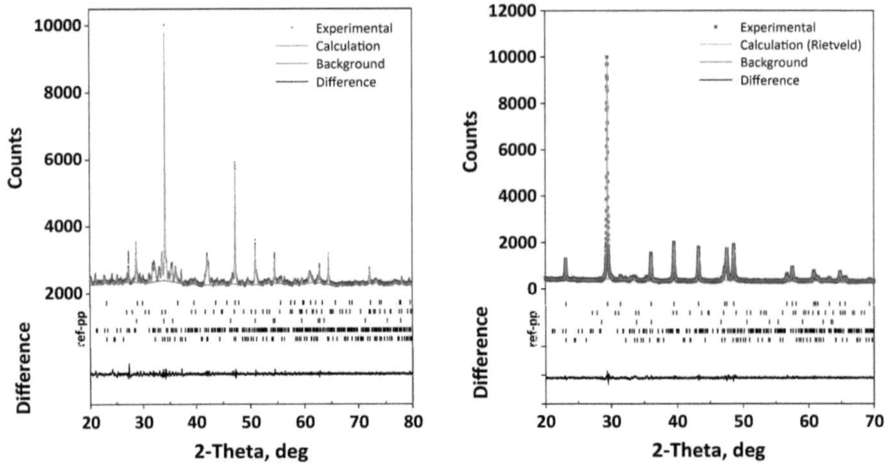

Fig. 6.11 Experimental and calculated XRD diffractogram by Rietveld method in GSAS program for (*left*) fresh and (*right*) carbonated BOFS. Reprinted with the permission from Ref. [40]. Copyright 2013 American Chemical Society

a sewage sludge ash and a municipal solid waste incineration fly ash by both physicochemical analysis and Rietveld method. The results obtained were coherent, suggesting that it is possible to quantify the mineral composition of complex mineral waste with Rietveld method.

For instance, in the case of basic oxygen furnace slag (BOFS), Fig. 6.11 shows the experimental and calculated diffractogram by the Rietveld method for the fresh and carbonated BOFS. The refinement results indicated that only a slight difference was observed in the intensity of major peaks between the experimental and calculated patterns by the Rietveld method. The obtained χ^2 values were 1.87 and 1.94 for fresh and carbonated BOFS, respectively, which was statistically acceptable [57]. Before carbonation, the principle components in BOFS were FeO (23%), $Ca_2Fe_{1.014}Al_{0.986}O_5$ (22%), $Ca(OH)_2$ (19%), C_2–S–H (15%), C_1S (11%), and $CaCO_3$ (10%). In the carbonated BOFS, the content of $CaCO_3$ increased significantly, while the contents of $Ca(OH)_2$, C_1S, C_2–S–H, and $Ca_2Fe_{1.04}Al_{0.986}O_5$ decreased [40]. Therefore, the mineral phases of $Ca(OH)_2$, C_1S, C_2–S–H, and $Ca_2Fe_{1.04}Al_{0.986}O_5$ in BOFS are regarded as the major species reacting with CO_2 to form $CaCO_3$ precipitation.

6.3 Scanning Electronic Microscopy

Scanning electron microscopy (SEM) equipped with X-ray energy-dispersive spectrometer (XEDS) is a useful tool for observing the surface structure of the sample and the elemental analysis of the solid surface. The SEM/XEDS technique involves analysis of thousands of pixels in a short time. The XEDS can record quantum energies between 1 and 40 keV, or higher, simultaneously by means of a multichannel analyzer [63]. Typically, the specimen before and after carbonation is mounted with double-sided carbon tape on an aluminum stub. For better conductivity and reduction of electron charge, the sample is usually coated with a thin layer of platinum.

6.3.1 Types of Techniques in SEM

6.3.1.1 Focused Ion Beam (FIB)

Focused ion beam (FIB) is a technique used in the SEM, where a FIB uses a focused beam of ions to image samples in the chamber, while the SEM uses a focused beam of electrons instead. Most widespread sources of ion beam are liquid metal ion source (LMIS), especially gallium (Ga) ion sources. The melting point of Ga metal is about 30 °C. In a Ga LMIS, gallium metal is placed in contact with a tungsten needle, where the radius of needle tip is typically $\sim$2 nm. The electric field at this tip is greater than 1×10^8 V/cm, causing ionization and field emission of the Ga atoms. Then, source ions are accelerated to an energy between 1 and 50 keV. Unlike an SEM, FIB is inherently destructive to the specimen since the high-energy Ga ions will sputter atoms from the surface. Therefore, the FIB can be used as a micro- and nanomachining tool to modify materials at the micro- and nanoscale.

6.3.1.2 Mapping

Elemental mapping technique uses X-ray counts from thousands of points on a particle surface. The data is collected in less than 1 h and analyzed to provide frequency distribution curves and relative elemental abundance [63]. In general, mapping of Ca, Mg, Fe, Si, C, and O is carried out for alkaline solid wastes to evaluate the distribution of these elements on the sample particles.

6.3.1.3 Cross-Sectional Images

Superficial and cross-sectional observations can be performed on the specimens. For cross-sectional analyses, samples usually are cut and mounted with epoxy resin in a plastic holder. The surface also can be polished with SiC paper and/or alumina. Nital (i.e., a solution of alcohol and nitric acid) etching can be performed for a better characterization of the samples. Similarly, fine gold films can be sputtered on the, otherwise insulating, samples.

6.3.2 Case Study: Carbonation of Steel Slag

Figure 6.12 shows the cross-sectional images and elemental mapping of fresh steel slag. Before carbonation, the entire steel slag is rich in calcium–ferrous–silicate (Ca–Fe–Si oxide) and/or calcium–magnesium–silicate (Ca–Mg–Si oxide) but without carbon (C) element. The distribution of the chemical elements was observed to be inhomogeneous from particle to particle. Meanwhile, the distribution of the calcium is quite concentrated in both of the fresh and carbonated steel slag. Generally, the distribution percentage of the carbon on the surface of the carbonated steel slag is found to be higher than that on the fresh steel slag, which indicates that the CO_2 can be captured successfully by carbonation reaction.

Figure 6.13 shows the cross-sectional images and elemental mapping of carbonated steel slag. After carbonation, the steel slag exhibits rhombohedral crystals, with a size of 1–3 μm, formed uniformly on the surface of the slag, exhibiting an outside $CaCO_3$ product layer (reacted) and an inside metal-rich core (unreacted).

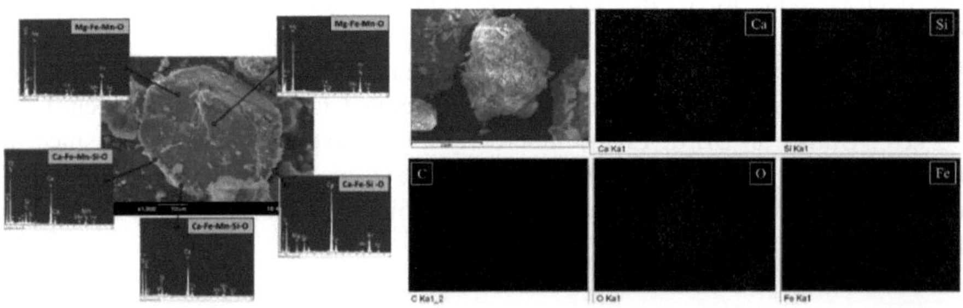

Fig. 6.12 (*Left*) cross-sectional observations and (*right*) elemental mapping of fresh steel slag by SEM/XEDS. Ca, Mg, Fe, Si, C, and O are recorded during the scanning of samples. Reprinted with the permission from ref. [40]. Copyright 2013 American Chemical Society

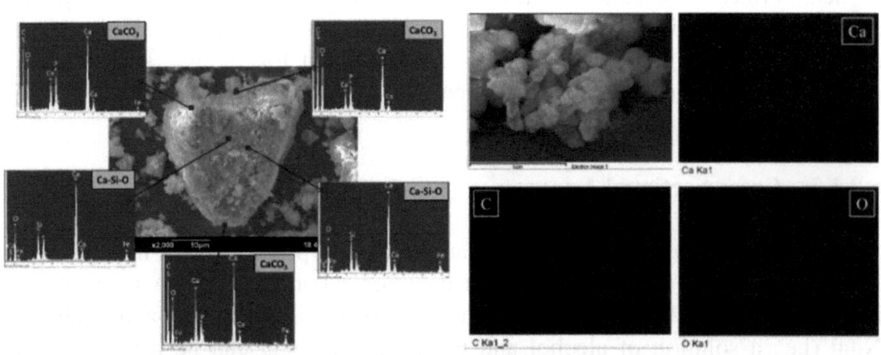

Fig. 6.13 (*Left*) Cross-sectional observations and (*right*) elemental mapping of carbonated steel slag by SEM/XEDS. Ca, Mg, Fe, Si, C, and O are recorded during the scanning of samples. Reprinted with the permission from Ref. [40]. Copyright 2013 American Chemical Society

Moreover, the cubic-shaped crystals coating the surface of carbonated steel slag are composed of calcium, carbon, and oxygen elements, indicating the formation of $CaCO_3$ (calcite). Similarly, the observations of SEM/XEDS and mappings are in good agreement with the results of QXRD using the Rietveld method [40].

References

1. Speyer RF (1994) Thermal analysis of materials. Marcel Dekker Inc., New York
2. Escardino A, García-Ten J, Feliu C, Saburit A, Cantavella V (2013) Kinetic study of the thermal decomposition process of calcite particles in air and CO_2 atmosphere. J Ind Eng Chem 19(3):886–897. doi:10.1016/j.jiec.2012.11.004
3. Villain G, Thiery M, Platret G (2007) Measurement methods of carbonation profiles in concrete: thermogravimetry, chemical analysis and gammadensimetry. Cem Concr Res 37 (8):1182–1192. doi:10.1016/j.cemconres.2007.04.015
4. Marsh BK, Day RL (1988) Pozzolanic and cementitious reactions of fly ash in blended cement pastes. Cem Concr Res 18:301–310
5. Tai CY, Chen WR, Shih S-M (2006) Factors affecting wollastonite carbonation under CO_2 supercritical conditions. AIChE J 52(1):292–299. doi:10.1002/aic.10572
6. Li W, Li B, Bai Z (2009) Electrolysis and heat pretreatment methods to promote CO_2 sequestration by mineral carbonation. Chem Eng Res Des 87(2):210–215. doi:10.1016/j.cherd.2008.08.001
7. Teir S, Eloneva S, Fogelholm C, Zevenhoven R (2009) Fixation of carbon dioxide by producing hydromagnesite from serpentinite. Appl Energy 86(2):214–218. doi:10.1016/j.apenergy.2008.03.013
8. Chang C, Chen J (2006) The experimental investigation of concrete carbonation depth. Cem Concr Res 36(9):1760–1767. doi:10.1016/j.cemconres.2004.07.025
9. Thiery M, Villain G, Dangla P, Platret G (2007) Investigation of the carbonation front shape on cementitious materials: effects of the chemical kinetics. Cem Concr Res 37(7):1047–1058. doi:10.1016/j.cemconres.2007.04.002
10. Bernal SA, de Gutierrez RM, Provis JL, Rose V (2010) Effect of silicate modulus and metakaolin incorporation on the carbonation of alkali silicate-activated slags. Cem Concr Res 40(6):898–907. doi:10.1016/j.cemconres.2010.02.003
11. Huijgen WJJ, Witkamp GJ, Comans RNJ (2005) Mineral CO_2 sequestration by steel slag carbonation. Environ Sci Technol 39(24):9676–9682
12. Santos RM, Van Bouwel J, Vandevelde E, Mertens G, Elsen J, Van Gerven T (2013) Accelerated mineral carbonation of stainless steel slags for CO_2 storage and waste valorization: effect of process parameters on geochemical properties. Int J Greenh Gas Control 17:32–45. doi:10.1016/j.ijggc.2013.04.004
13. Chang EE, Pan SY, Chen YH, Tan CS, Chiang PC (2012) Accelerated carbonation of steelmaking slags in a high-gravity rotating packed bed. J Hazard Mater 227–228:97–106. doi:10.1016/j.jhazmat.2012.05.021
14. Chen Y-T (2008) Effects of process variables on the conversion of BOF slag to carbonate. National Taiwan University, Taipei, Taiwan
15. Lekakh SN, Rawlins CH, Robertson DGC, Richards VL, Peaslee KD (2008) Kinetics of aqueous leaching and carbonization of steelmaking slag. Metall Mater Trans B 39(1): 125–134. doi:10.1007/s11663-007-9112-8
16. Chen Q, Zhang L, Ke Y, Hills C, Kang Y (2009) Influence of carbonation on the acid neutralization capacity of cements and cement-solidified/stabilized electroplating sludge. Chemosphere 74(6):758–764. doi:10.1016/j.chemosphere.2008.10.044

17. Huntzinger DN, Gierke JS, Kawatra SK, Eisele TC, Sutter LL (2009) Carbon dioxide sequestration in cement kiln dust through mineral carbonation. Environ Sci Technol 43(6): 1986–1992
18. Rendek E, Ducom G, Germain P (2006) Carbon dioxide sequestration in municipal solid waste incinerator (MSWI) bottom ash. J Hazard Mater 128(1):73–79. doi:10.1016/j.jhazmat.2005.07.033
19. Li X, Bertos MF, Hills CD, Carey PJ, Simon S (2007) Accelerated carbonation of municipal solid waste incineration fly ashes. Waste Manag 27(9):1200–1206. doi:10.1016/j.wasman.2006.06.011
20. Cappai G, Cara S, Muntoni A, Piredda M (2012) Application of accelerated carbonation on MSW combustion APC residues for metal immobilization and CO_2 sequestration. J Hazard Mater 207–208:159–164. doi:10.1016/j.jhazmat.2011.04.013
21. Baciocchi R, Costa G, Di Bartolomeo E, Polettini A, Pomi R (2009) The effects of accelerated carbonation on $CO(_2)$ uptake and metal release from incineration APC residues. Waste Manag 29(12):2994–3003. doi:10.1016/j.wasman.2009.07.012
22. Hohamed M, Yusup S, Maitra S (2012) Decomposition study of calcium carbonate in cockle shell. J Eng Sci Technol 7(1):1–10
23. Hosseini T, Selomulya C, Haque N, Zhang L (2015) Investigating the effect of the Mg^{2+}/Ca^{2+} molar ratio on the carbonate speciation during the mild mineral carbonation process at atmospheric pressure. Energy Fuel 29(11):7483–7496. doi:10.1021/acs.energyfuels.5b01609
24. Pan SY, Chang EE, Kim H, Chen YH, Chiang PC (2016) Validating carbonation parameters of alkaline solid wastes via integrated thermal analyses: principles and applications. J Hazard Mater 307:253–262. doi:10.1016/j.jhazmat.2015.12.065
25. Eloneva S, Teir S, Salminen J, Fogelholm CJ, Zevenhoven R (2008) Steel converter slag as a raw material for precipitation of pure calcium carbonate. Ind Eng Chem Res 47(18):7104–7111
26. Steinour HH (1959) Some effects of carbon dioxide on mortars and concrete—discussion. J Am Concr Inst 30:905–907
27. Huntzinger DN, Gierke JS, Sutter LL, Kawatra SK, Eisele TC (2009) Mineral carbonation for carbon sequestration in cement kiln dust from waste piles. J Hazard Mater 168(1):31–37. doi:10.1016/j.jhazmat.2009.01.122
28. Teir S (2008) Fixation of carbon dioxide by producing carbonates from minerals and steelmaking slags. Helsinki University of Technology
29. Huijgen WJJ, Comans RNJ (2006) Carbonation of steel slag for CO_2 sequestration: leaching of products and reaction mechanisms. Environ Sci Technol 40(8):2790–2796
30. Georgieva V, Vlaev L, Gyurova K (2013) Non-isothermal degradation kinetics of $CaCO_3$ from different origin. J Chem 2013:1–12. doi:10.1155/2013/872981
31. Ptáček P, Šoukal F, Opravil T, Havlica J, Brandštetr J (2011) The kinetic analysis of the thermal decomposition of kaolinite by DTG technique. Powder Technol 208(1):20–25. doi:10.1016/j.powtec.2010.11.035
32. Li X-G, Lv Y, Ma B-G, Wang W-Q, Jian S-W (2013) Decomposition kinetic characteristics of calcium carbonate containing organic acids by TGA. Arab J Chem. doi:10.1016/j.arabjc.2013.09.026
33. Halikia I, Zoumpoulakis L, Christodoulou E, Prattis D (2001) Kinetic study of the thermal decomposition of calcium carbonate by isothermal methods of analysis. Eur J Min Process Environ Prot 1(2):89–102
34. Chen D, Gao X, Dollimore D (1993) A generalized form of the Kissinger equation. Thermochim Acta 215:109–117
35. Kissinger HE (1957) Reaction kinetics in differential thermal analysis. Anal Chem 29(11): 1702–1706
36. Carne LW, Dynes PJ, Kaelble DH (1973) Analysis of curing kinetics in polymer composites. J Polym Sci Polym Lett Ed 11(8):533–540

37. Pan SY, Chen YH, Chen CD, Shen AL, Lin M, Chiang PC (2015) High-gravity carbonation process for enhancing CO_2 fixation and utilization exemplified by the steelmaking industry. Environ Sci Technol 49(20):12380–12387. doi:10.1021/acs.est.5b02210
38. Pan SY, Chiang PC, Chen YH, Tan CS, Chang EE (2013) Ex situ CO_2 capture by carbonation of steelmaking slag coupled with metalworking wastewater in a rotating packed bed. Environ Sci Technol 47(7):3308–3315. doi:10.1021/es304975y
39. Chang EE, Chiu A-C, Pan S-Y, Chen Y-H, Tan C-S, Chiang P-C (2013) Carbonation of basic oxygen furnace slag with metalworking wastewater in a slurry reactor. Int J Greenh Gas Control 12:382–389. doi:10.1016/j.ijggc.2012.11.026
40. Pan SY, Chiang PC, Chen YH, Chen CD, Lin HY, Chang EE (2013) Systematic approach to determination of maximum achievable capture capacity via leaching and carbonation processes for alkaline steelmaking wastes in a rotating packed bed. Environ Sci Technol 47 (23):13677–13685. doi:10.1021/es403323x
41. Tian S, Jiang J, Hosseini D, Kierzkowska AM, Imtiaz Q, Broda M, Miller CR (2015) Development of a steel-slag-based, iron-functionalized sorbent for an autothermal carbon dioxide capture process. ChemSusChem 8:3839–3846. doi:10.1002/cssc.201501048
42. Criado JM, Ortega A (1992) A study of the influence of particle size on the thermal decomposition of $CaCO_3$ by means of constant rate thermal analysis. Thermochim Acta 195:163–167
43. Escardino A, García-Ten J, Feliu C, Moreno A (2010) Calcium carbonate thermal decomposition in white-body wall tile during firing. I. Kinetic study. J Eur Ceram Soc 30 (10):1989–2001. doi:10.1016/j.jeurceramsoc.2010.04.014
44. Málek J (1995) The applicability of Johnson-Mehl-Avrami model in the thermal analysis of the crystallization kinetics of glasses. Thermochim Acta 267:61–73
45. Santos RM, Chiang YW, Elsen J, Van Gerven T (2014) Distinguishing between carbonate and non-carbonate precipitates from the carbonation of calcium-containing organic acid leachates. Hydrometallurgy 147–148:90–94. doi:10.1016/j.hydromet.2014.05.001
46. Chen KW, Pan SY, Chen CT, Chen YH, Chiang PC (2016) High-gravity carbonation of basic oxygen furnace slag for CO_2 fixation and utilization in blended cement. J Clean Prod 124:350–360. doi:10.1016/j.jclepro.2016.02.072
47. Muhmood L, Vitta S, Venkateswaran D (2009) Cementitious and pozzolanic behavior of electric arc furnace steel slags. Cem Concr Res 39(2):102–109. doi:10.1016/j.cemconres.2008.11.002
48. Xiao L-S, Wang R, Chiang P-C, Pan S-Y, Guo Q-H, Chang EE (2014) Comparative life cycle assessment (LCA) of accelerated carbonation processes using steelmaking slag for CO_2 fixation. Aerosol Air Qual Res 14(3):892–904. doi:10.4209/aaqr.2013.04.012
49. Pan S-Y, Lorente Lafuente AM, Chiang P-C (2016) Engineering, environmental and economic performance evaluation of high-gravity carbonation process for carbon capture and utilization. Appl Energy 170:269–277. doi:10.1016/j.apenergy.2016.02.103
50. Mahieux PY, Aubert JE, Cyr M, Coutand M, Husson B (2010) Quantitative mineralogical composition of complex mineral wastes—contribution of the Rietveld method. Waste Manag 30(3):378–388. doi:10.1016/j.wasman.2009.10.023
51. Liu H, Kuo C (1996) Quantitative multiphase determination using the Rietveld method with high accuracy. Mater Lett 26:171–175
52. Zhou Y, Song W, Zeng X, Xie C (2012) Quantitative X-ray Rietveld analysis of metallic aluminum content in nano-aluminum powders. Mater Lett 67(1):177–179. doi:10.1016/j.matlet.2011.09.051
53. Dermatas D, Dadachov MS (2003) Rietveld quantification of montmorillonites in lead-contaminated soils. Appl Clay Sci 23(1–4):245–255. doi:10.1016/s0169-1317(03)00109-1
54. Fernández-Jiménez A, de la Torre AG, Palomo A, López-Olmo G, Alonso MM, Aranda MAG (2006) Quantitative determination of phases in the alkaline activation of fly ash. Part II: degree of reaction. Fuel 85(14–15):1960–1969. doi:10.1016/j.fuel.2006.04.006

55. Bodénan F, Deniard P (2003) Characterization of flue gas cleaning residues from European solid waste incinerators: assessment of various Ca-based sorbent processes. Chemosphere 51 (5):335–347. doi:10.1016/s0045-6535(02)00838-x
56. Rietveld HM (1969) A profile refinement method for nuclear and magnetic structures. J Appl Crystallogr 2:65–71
57. Toby BH (2012) R factors in Rietveld analysis: how good is good enough? Powder Diffr 21 (01):67–70. doi:10.1154/1.2179804
58. Liu Q, Maroto-Valer MM (2012) Studies of pH buffer systems to promote carbonate formation for CO_2 sequestration in brines. Fuel Process Technol 98:6–13. doi:10.1016/j.fuproc.2012.01.023
59. Uibu M, Kuusik R, Andreas L, Kirsimäe K (2011) The CO_2-binding by Ca-Mg-silicates in direct aqueous carbonation of oil shale ash and steel slag. Energy Procedia 4:925–932. doi:10.1016/j.egypro.2011.01.138
60. Saito T, Sakai E, Morioka M, Otsuki N (2010) Carbonation of r-Ca_2SiO_4 and the mechanism of vaterite formation. J Adv Concr Technol 8(3):273–280
61. Larson AC, Von Dreele RB (1986) Generalized crystal structure analysis system. LAUR 86–748
62. Kuusik R, Uibu M, Kirsimäe K (2005) Characterization of oil shale ashes formed at industrial-scale CFBC boilers. Oil Shale 22:407–420
63. Alsmadi BM, Fox P (2001) Semi-quantitative analysis of changes in soil coatings by scanning electron microscope and energy dispersive X-ray mapping. Colloids Surf A 194:249–261

Chapter 7
Carbonation Mechanisms and Modelling

Abstract From the view point of reaction kinetics, many researchers have attributed the rate-determined step of aqueous carbonation to the reacted (product) layer diffusion. Therefore, this suggests that a well-designed reactor to enhance the mass transfer between the gas, liquid, and solid phases is needed to facilitate the carbonation reaction and increase the carbonation conversion. This chapter provides the principles and mechanisms of carbonation reaction using alkaline solid wastes. The kinetics of the major three steps of carbonation, i.e., metal ion leaching, CO_2 dissolution, and carbonate precipitation, are illustrated. Moreover, several classical heterogeneous kinetic models, e.g., shrinking core model and surface coverage model, are presented. Furthermore, the modeling of mass transfer for carbonation in various reactors and/or processes is summarized and discussed.

7.1 Carbonation Mechanisms

7.1.1 Principles

As discussed in Chap. 5, the process chemistry of accelerated carbonation using alkaline solid wastes can be briefly divided into three steps:

- Step 1: Contemporary dissolution of CO_2 into a liquid phase and conversion of carbonic acid to carbonate/bicarbonate ions
- Step 2: Dissolution of CO_2-reactive species from a solid matrix (irreversible hydration)
- Step 3: Consequent nucleation and precipitation of carbonates

At first, the leaching of Ca-bearing compounds in alkaline solid waste would directly generate the Ca^{2+} in the solution. Secondly, gaseous CO_2 can rapidly dissolve into the alkaline solution, where the predominant carbonate ions (CO_3^{2-}) could reduce the pH of the solution. Since the CO_2 continuously dissolved into the solution during the carbonation, the pH value would decrease gradually to 6.3,

P.-C. Chiang and S.-Y. Pan, *Carbon Dioxide Mineralization and Utilization*, DOI 10.1007/978-981-10-3268-4_7

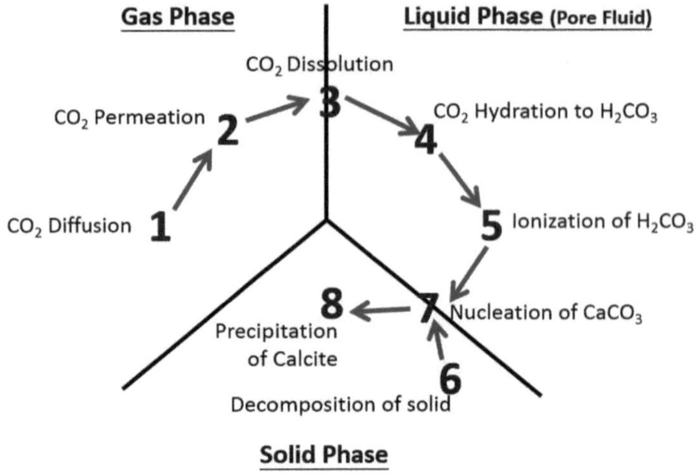

Fig. 7.1 Schematic of accelerated carbonation mechanism

where the bicarbonate ions (HCO_3^-) were found to be dominated. Finally, $CaCO_3$ would be formed by reacting the calcium ions with the carbonate ions (CO_3^{2-}) and bicarbonate ions (HCO_3^-) in the solution under a high pH and circum-neutral condition, respectively. The above statements can be also divided into eight steps, as shown in Fig. 7.1.

It is widely accepted that the carbonation is controlled by the dissolution of CO_2, and therefore the reaction occurs on the liquid side near the interface between gas and liquid phases [1]. It suggests that the irreversible hydration of CaO proceeds spontaneously with carbonation of $Ca(OH)_2$ suspension [2, 3]. Moreover, not only $Ca(OH)_2$ but also other hydrate compounds, such as calcium–silicate–hydrates (C–S–H), can react with CO_2 to produce $CaCO_3$ and a silica gel [4]. The balance between dissolution and precipitation is dependent on the kinetics and solubility of the present feedstock and possible products, where carbonates are probably the most important products [5].

For γ–C_2S phase (dominate component in steel slag), it does not react with water at normal temperature because the leaching of Ca^{2+} and O^{2-} (or OH^-) is very low [4]. Fernández-Bertos et al. [6] suggested that $CaCO_3$ can be directly formed in the carbonation of γ–C_2S without the formation of intermediate products such as Ca $(OH)_2$, due to the catalytic action of capillary water. The breakdown of C_2S and the formation of calcite can be observed by XRD, as indicated in Eq. (7.1):

$$\begin{aligned} &3CaO\bullet SiO_2 + y\,H_2O + (3\text{-}x)\,CO_2 \rightarrow x\,CaO\bullet SiO_2\bullet yH_2O + (3\text{-}x)\,CaCO_3 \\ &\quad\llcorner\!\longrightarrow z\,CO_2 \rightarrow (x\text{-}z)\,CaO\bullet SiO_2\bullet yH_2O + z\,CaCO_3 \end{aligned} \tag{7.1}$$

7.1.2 Key Factors and Operating Parameters

For process design and system optimization, it is crucial to identify the effect of key operating factors on carbonation performance. Carbonation of alkaline solid wastes has been proved to be an effective way to capture CO_2. Several factors can affect the rate and extent of carbonation [7, 8]:

- Transportation-controlled mechanisms such as CO_2 and Ca^{2+} ions diffusion to/from reaction sites
- Boundary layer effects (diffusion across precipitate coatings on particles)
- Dissolution of $Ca(OH)_2$ at the particle surface
- Pore blockage
- Precipitate coating

Table 7.1 summarizes the key factors required for effective accelerated carbonation. These operating factors must be clearly understood, since they determine the economic viability of the technology as well as help to identify the conditions that are the most favorable to the carbonation reaction.

7.1.2.1 Particle Size and Surface Area

Both particle size and specific surface area are the most important factors affecting the dissolution kinetics of materials [9, 10]. It was noted that over 90% and less than 40% of calcium species can be dissolved in 24 h from 45–75 to 150–250-μm-sized ladle slag, respectively [11]. Since solid waste grinding is expected to be a fairly energy-intensive process, it is important to find out the optimal particle size for accelerated carbonation. For the steel slag, average particle sizes of less than 100–150 μm are suggested to be in the optimum range for

Table 7.1 Key operating factors required for effective carbonation

Phases	Physical properties	Chemical properties
Solid phase	• Particle size • Mineralogy • Specific surface area • Porosity/Permeability • Surface activities • Microstructure	• Compositions (e.g., Ca and f-CaO content, Ca/Si ratio, Ferrite/C_3A ratio) • Heavy metals (e.g., Pb, Cd, Ni, Cr) • Free water content • Permeability
Liquid phase	• Temperature • Liquid-to-solid ratio	• Organics/Inorganic • Anions/Cations • pH (or alkalinity) • Permeability
Gas phase	• Partial pressure • Flow rate • Relative humidity • Temperature	• CO_2 concentration • Organics/Inorganic • Particulate matter contents • Other air pollutants (e.g., SO_2)

efficient carbonation [12]. Similarly, Eloneva et al. [13] suggested that a particle size preferably of 100 μm, or no more than 500 μm, should be optimal, based on the kinetics studies of calcium leaching.

7.1.2.2 Reaction Temperature

Reaction temperature is an important operating factor in the carbonation reaction. Temperature exhibits a significant effect on dissolution and carbonation performance, especially during the first 3 h of the dissolution reaction [11]. Each carbonation reaction mechanism exhibits a different sensitivity to the reaction temperature.

- Leaching of metal oxide from solid matrix: The dissolution kinetics of the calcium species could be enhanced by increasing the temperature.
- CO_2 dissolution into solution: The amounts of CO_2 dissolution into the solution decreased at higher temperature, which is detrimental to the carbonation reaction.
- Carbonate precipitate nucleation: The nucleation and growth of $CaCO_3$ are retarded at higher temperatures due to the decreased solubility of CO_2. On the other hand, the carbonation rate significantly increased with increasing reaction temperature.

Typically, a reaction temperature higher than 80 °C is unfavorable for the process design of direct carbonation [14], and 60 °C should be high enough for efficient dissolution and carbonation [15, 16].

7.1.2.3 Liquid-to-Solid (L/S) Ratio

Liquid-to-solid (L/S) ratio determines the leaching capacity of Ca ion (and other ions) from the solid wastes to the solution. In general, an incomplete hydration of calcium-bearing compounds from solid wastes is always observed. For example, the measured leaching concentration of calcium was about 700−900 ppm, equivalent to a fraction of calcium leaching from the steel slag of 0.5−4.0% [17]. If the solubility of pure CaO (i.e., lime) at 20 °C was assumed to be 1.25 g/L [18], the theoretical concentration of calcium ions in water should be 892.9 mg/L. This indicates that the content of lime might be essentially controlling the solubility of CaO-related species in steel slag. In other words, the calcium leaching should be a solubility-controlled step, as a result of equilibrium with specified mineral phases, thereby resulting in an almost constant concentration in solution irrespective of the L/S ratio [17].

For municipal solid waste incineration fly ashes, Li et al. [19] found that the optimum L/S ratio for accelerated carbonation was 0.3 by weight of ash at ambient temperature. Similarly, the optimum L/S ratio was 0.2–0.3 for air pollution control residues and 0.3–0.4 for bottom ash residues [20]. For iron and steel slags, Chang et al. [21, 22] suggested that the optimum L/S ratio for carbonation of steelmaking slag is 10–20 (mL g^{-1}).

7.1.2.4 Solution Compositions (CO_2 Concentration and pH)

The performance of carbonation is highly related to the compositions of the solution. From the CO_2 concentration point of view, the reaction time should be inversely proportional to the CO_2 content of the ingoing gas flow; the lower the gas CO_2 content is, the longer the duration of the carbonation is. However, the CO_2 content of the incoming gas did not seem to have a significant effect on the degree of $CaCO_3$ precipitation.

The pH value of solution is another important factor related to the rate of calcium leaching, the rate of CO_2 dissolution, and the rate of $CaCO_3$ nucleation. The pH decreases continuously as carbonation proceeds due to CO_2 dissolution into solution. There is no change in pH after the carbonation is completed. On the other hand, it has been reported that sodium and potassium ions should cause pH fluctuations and also affect the carbonation rate [23]. It is noted that the dissolution of calcium components in solid wastes is favored at the low pH and high temperature, which is, however, not favored for the precipitation of $CaCO_3$. It also affects the specific species that can be precipitated during the carbonation [24, 25]. It suggests that the optimum pH for aqueous carbonation is around 10, while the dissolution of steel slag occurs under low pH conditions [26].

7.2 Reaction Kinetics

According to the mechanism of carbonation reaction as mentioned before, the overall reaction kinetics for the accelerated carbonation can be divided into three parts: (1) metal ion leaching from solid matrix; (2) CO_2 dissolution into solution; and (3) carbonate precipitation. Several theoretical kinetic models regarding the above three parts are illustrated as follows.

7.2.1 Metal Ion Leaching from Solid Matrix

The release of metal ions to the water solution, so-called leaching, depends on their solution speciation and their affinity to bind to reactive surface in the solid matrix and pore water. To evaluate the leaching performance of a certain process, “mass conservation model” could provide valuable insights into the speciation of metal ions in the solution phase. Moreover, evaluation of the pH dependence of leaching for various metal ions is a good approach to assessing the solid wastes, both for research purposes as well as in the context of the development of regulation and standard operation procedure for leaching tests [27].

7.2.1.1 Mass Conservation Model

The leaching concentrations of various metal ions are generally observed to increase rapidly in the beginning, and then gradually approach a maximum concentration. Therefore, the leaching kinetics of various metal ions (i) from the solid matrix can be evaluated by the mass loss-based method, as shown in Eq. (7.2):

$$r_i = \frac{\mathrm{d}C_i}{\mathrm{d}t} = k_i[C_{\max,i} - C_i]^{n_i} \tag{7.2}$$

where r_i is the leaching rate of various metal ions, k_i is the rate constant of leaching, $C_{\max,i}$ (mg/L) is the maximum leaching concentration for various metal ions in solution, C_i (mg/L) is the leaching concentration of various metal ions, where the background concentration of metal ions originally in the solution was subtracted, and n_i is the order of leaching reaction of various metal ions. Equation (7.3) can be integrated as follows:

$$C_i = C_{\max,i}\left[1 - \mathrm{e}^{-k_i t}\right], \quad \text{for } n = 1 \tag{7.3a}$$

$$C_i = C_{\max,i} - \left[C_{\max,i}^{1-n} - (1-n)k_i t\right]^{1/(1-n)}, \quad \text{for } n \neq 1 \tag{7.3b}$$

In general, both the values of $C_{\max}$ and k are observed to increase as the particle size of solid waste decreased. In the case of steel slag, the leaching behavior of Ca, Na, K, and Fe ions was found to be more sensitive to the concentration driving force than that of others because the obtained n values were greater than one.

7.2.1.2 Leaching Kinetics for Alkaline Solid Wastes

Calcium (Ca), sodium (Na), and potassium (K) metal ions are found to be the major ions leaching out from in the case of steel slag [10]. Steel slag normally contained a great amount of reactive Ca-bearing phases including lime (CaO), portlandite (Ca $(OH)_2$), and larnite (Ca_2SiO_4). This might account for most of the calcium leaching concentration from the steel slag.

On the other hand, both Pb and Zn are considered minor elements that might not affect the carbonation reaction. Similarly, Fe, Al, and Mg are considered slightly released cations because the measured concentrations in the solution typically remain low over the leaching time ($\sim$90 min). It might be attributed to the low mobility of Mg–Fe–Al–Si oxides, where Mg and Al are commonly associated with Fe-bearing phases in steel slag. In most curriculums, Mg release was found to be quite low compared to Ca release, which implies that the formation of $MgCO_3$ should be negligible [10, 16].

Also, the releases of Ni and Cr ions are quite low, and the concentration of Ni and Cr remained almost constant. For example, in the case of steel slag, the concentration of Ni and Cr was approximately 7.1–10.4 and 2.0–3.3 mg/L, respectively, during the entire leaching time of 90 min [10]. According to De Windt et al. [28], the maximum Cr leaching concentration in deionized (DI) water was about 0.015 ppm at a liquid-to-solid (L/S) ratio of 10.

7.2.1.3 Factors Affecting Leaching Kinetics

The rate and extent of calcium leaching are inversely related to particle size and pH, and increased with increasing temperature, pressure, and surface area [28, 29]. The effects of increased reaction temperature from 25 to 50 °C provided a 70% improvement on the dissolution of the primary phase [30]. However, a higher reaction temperature would reduce the dissolution of CO_2 in water, which is disadvantageous to calcite precipitation [21, 31].

Other factors may hinder the leaching of metal ions, thereby leading to a lowering of mass-transfer rate:

- Lack of porosity in solid wastes
- Formation of protective layer on the surface of particles
- Over saturation of metal ions in the solution
- Insufficient stirring and mixing between solid and liquid phases
- Severe agglomeration of (fine) particles in the solution

The leaching of Ca^{2+} ions should result in a withdrawing Ca–silicate core surrounded by a Ca-depleted SiO_2 phase [32, 33]. This SiO_2 rim apparently hindered the diffusion of Ca^{2+} ions from the interior of the particle, resulting in a declining reaction rate. Hence, Ca^{2+} ions' leaching rate, probably determined by diffusion through the Ca-depleted silicate rim, rather than by the boundary layer at the solid–liquid interface, seems to be the main rate-limiting step in aqueous carbonation.

7.2.2 *CO_2 Dissolution into Solution*

The kinetics of dissolution of CO_2 and dehydration of H_2CO_3 has been studied intensively [23, 34]. These two reactions should occur simultaneously, as shown in Eq. (7.4):

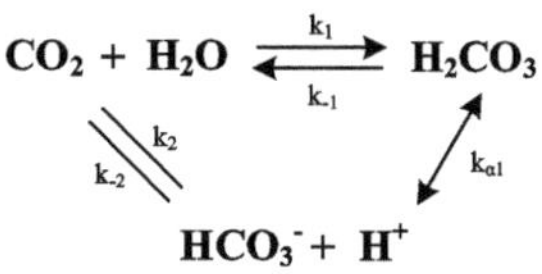

(7.4)

The rate expression of the dissolution–dehydration reaction is:

$$-\frac{d[CO_2]}{dt} = (k_1 + k_2)[CO_2] - k_{-1}[H_2CO_3] - k_{-2}\left[HCO_3^-\right]\left[H^+\right] \quad (7.5)$$

By substituting Eq. (7.4) into Eq. (7.5), we obtain

$$\begin{aligned} -\frac{d[CO_2]}{dt} &= (k_1 + k_2)[CO_2] - (k_{-1} + k_{-2}K_{\alpha 1})[H_2CO_3] \\ &= k_{CO_2}[CO_2] - k_{H_2CO_3}[H_2CO_3] \end{aligned} \quad (7.6)$$

where the overall rate constants in Eq. (7.5) were simplified to be k_{CO_2} and $k_{H_2CO_3}$. The values of k_{CO_2} and $k_{H_2CO_3}$ at 25 °C were 0.032 and 26.6 s^{-1}, respectively [35]. However, at higher pH (e.g., pH higher than 9), an alternative reaction pathway would be expressed as

$$CO_2 + OH^- \overset{k_4}{\leftrightarrow} HCO_3^- \quad (7.7)$$

where k_4 (i.e., 8500 M^{-1} s^{-1} at 25 °C) and k_{-4} (i.e., 0.0002 s^{-1} at 25 °C) are the rate constants [36].

7.2.3 *Carbonate Precipitation*

Carbonation reaction is regulated by solution equilibrium, and the reaction of calcium ions combining with carbonate ions is very fast. Several models, such as simplified first-order model, saturation state model, and heterogeneous dissolution–precipitation model, have been extensively applied for determining the kinetic parameters of carbonation reaction.

7.2.3.1 Simplified First-Order Model

It is suggested that the accelerated carbonation is a first-order reaction, with respect to the concentrations of Ca^{2+} and $CO_3{}^{2-}$ [23]. In other words, the precipitation rate is related to the CO_3^{2-} concentration in the liquid phase, but not to the concentration of other species containing carbonate. Therefore, the rate of carbonation can be described by the following Eq. (7.8):

$$Q_{CO_2} = \frac{d(C_{CaCO_3})}{dt} = -\frac{d(CO_3^{2-})}{dt} = k\left[Ca^{2+}\right]\left[CO_3^{2-}\right] \quad (7.8)$$

where C_{CaCO_3} is the concentration of calcium carbonate (mole L^{-1}), and k is the reaction rate constant ($mol^{-1}\ s^{-1}$). The rate constant (k) is dependent on the reaction temperature according to Arrhenius's law. The Boltzmann distribution law can also be applied to determine the ratio of the molecule having a kinetic energy above the activation energy of the chemical reaction. The k value reported by Ishida and Maekawa [23] was 2.08 ($L\ mol^{-1}\ s^{-1}$) at 25 °C from several sensitivity analyses.

7.2.3.2 Saturation State Model

The kinetic rate laws for weathering of rock-forming minerals can be in principle applied to the leaching of solid wastes, since these materials consist of well-crystallized solid phases. A simple kinetic formulation without any pH dependency can be used for the dissolution of a primary mineral phase (M_i) [28]:

$$r_i = \frac{d[M_i]}{dt} = k_i A_i \left(\frac{Q_i}{K_i^{-1}} - 1 \right) \tag{7.9}$$

where the k_i is the intrinsic rate constant far from equilibrium; A_i is the mineral surface area; Q_i stands for the ion activity product; and K_i is the thermodynamic formation constant. It is noted that the term $(Q_i/K_i^{-1} - 1)$ is the saturation state. At equilibrium, this term becomes zero, and likewise so does the kinetic rate [28].

7.2.3.3 Heterogeneous Dissolution–Precipitation Model

Reaction with CO_2 will lower the pH of slurry; therefore, the minerals will keep dissolving in the course of the carbonation. In other words, as the calcium ions are converted to $CaCO_3$ precipitates, more calcium hydroxide dissolves to equalize the concentration of metal ions. This reveals that precipitation kinetics is not equal to dissolution kinetics. Assuming the heterogeneous dissolution–precipitation reactions occur at the surface of a carbonate, the governing equation of carbonation rate can be expressed as follows [37]:

$$\text{rate} = k_0 \times A_s \times \exp\left(-\frac{E_a}{RT} \right) \times a_{H^+}^{n_H} \times \prod_i a_i^{n_i} \times g(I) \times f(A) \tag{7.10}$$

where k_0 is the rate constant; A_s is the reactive surface are of the mineral; E_a is the activate energy; T is the activation energy (K); the term a_{H+}^{nH} is the pH dependence of the rate of the dissolution–precipitation reaction; the term $\prod_i a_i^{n_i}$ comprises possible catalytic or inhibitory effects linked to other solution; and the terms $g(I)$ and $f(A)$ account for the dependence of the rate on the ionic strength I of the solution, and the ionic strength I of the aqueous solution, respectively. The α is the fractional conversion, which can be expressed by

$$\frac{d\alpha}{dt} = A \times \exp\left(-\frac{E_A}{RT}\right) \times f(\alpha) \tag{7.11}$$

$$g(\alpha) = \int \frac{d\alpha}{f(\alpha)} = \int A \times \exp\left(-\frac{E_A}{RT}\right) dt \tag{7.12}$$

$$f(\alpha) = \frac{-1}{\ln(1-\alpha)} \tag{7.13}$$

Although the Ca^{2+} ion dissolution kinetics can be improved with the increase in temperature, carbonation precipitation is retarded at higher temperatures due to reduced CO_2 solubility [31, 38]. In some circumstance using steel slag under ambient conditions, limited $MgCO_3$ formation from carbonation is expected due to the relatively low magnesium oxide content in the slag, low pressure of CO_2, and short reaction times. Via aqueous carbonation, typical process conditions for the formation of magnesium carbonation are a pressure of CO_2 (p_{CO_2}) higher than 100 bar for hours [39]. However, $MgCO_3$ formation can be observed when natural ores (e.g., serpentine and olivine) are used for carbonation [40, 41], as discussed in Chap. 10.

7.3 Classical Heterogeneous Kinetic Models

Several classical (theoretical) models, such as random pore model [42], overlapping grain model [43], shrinking core model [11, 16], and surface coverage model [44], are available for simulating the performance of accelerated carbonation using a CaO-based material. For the sake of simplicity, different assumptions should be made by each model to avoid complicated calculations.

7.3.1 Shrinking Core Model

Shrinking core model (SCM), developed by Sohn and Szekely [45], has been applied for kinetic analysis of heterogeneous reactions, such as the carbonation reactions of solid particles [11, 22], because of its conceptual and mathematical simplicity. A primary assumption under the SCM is that the reaction occurs first at the outer layer of the particle and then proceeds into the inside of the particle, leaving behind the completely reacted product and the reactive-species-depleted rims referred to as the "ash" and/or "product" layer [11, 22, 46]. Therefore, at any time, an unreacted core of material exists, which shrinks in size during the reaction.

7.3.1.1 Classic Governing Equations

Experimental data can be utilized to determine the kinetics and rate-determining step of a reaction based on the SCM. In the SCM, the governing equations of possible rate-determining steps, as shown in Fig. 7.2, include the following:

- Chemical reaction at the unreacted core surface (C-mechanism)
- Ash-layer diffusion (A-mechanism)
- Fluid-film diffusion (F-mechanism)

Let us consider a fluid–solid reaction of the following general expression:

$$CO_{2(\text{fluid})} + bCaO_{(\text{solid})} \rightarrow CaCO_{3(s)} \tag{7.14}$$

where b (–) is stoichiometric coefficient of a fluid–solid reaction, which can be assigned a value of one for carbonation reaction.

The overall radius of the solid particle (R) is assumed to remain constant, which is a primary assumption of SCM. Therefore, the time required for complete conversion of particle to product (i.e., carbonation conversion $\delta_{CaO} = 100\%$) for the mechanisms of chemical reaction, ash-layer diffusion, and film diffusion (i.e., τ_C, τ_A, and τ_F, respectively) can be obtained via various governing equation, as follows:

1. Chemical reaction at the un-reacted core surface (C-mechanism)

When the chemical reaction between reactants is the rate-limiting step, Eq. (7.15) shows the relationship between the carbonation conversion of alkaline solid waste (δ_{CaO}) and reaction time (t, sec):

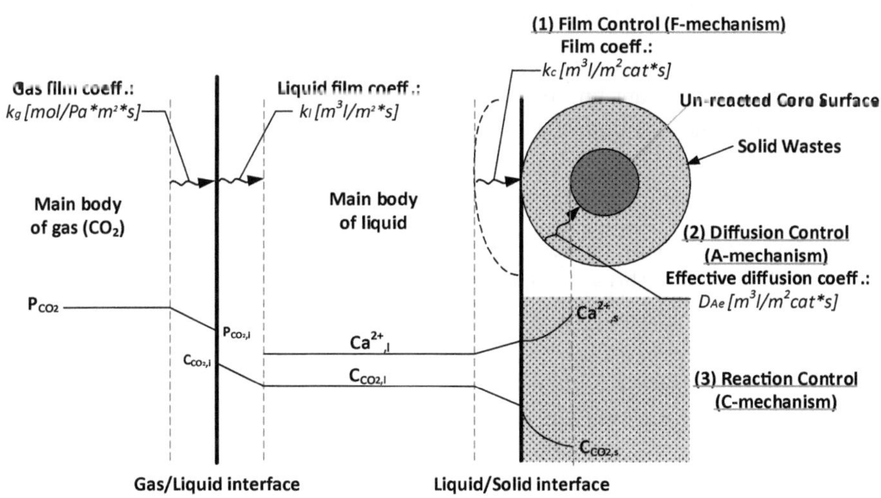

Fig. 7.2 Diagram of phase transition among gas, liquid, and solid phases for shrinking core model

$$t = \tau_{\mathrm{C}} \times \left[1 - (1 - \delta_{\mathrm{CaO}})^{1/3}\right] \quad (7.15)$$

where τ (s) is the time for complete conversion of a reactant particle to product.

For the C-mechanism, the relationship of τ_{C} and the rate constant for the surface reaction (k'', s^{-1}) is shown as Eq. (7.16):

$$\tau_{\mathrm{C}} = \frac{\rho_{\mathrm{B}} R}{b k'' C_{\mathrm{Ag}}} \quad (7.16)$$

where k'' is the first-order rate constant for the surface reaction, the term R (cm) is the radius of the particles, ρ_{B} (mol cm^{-3}) is the molar density of particle, C_{Ag} (mol cm^{-3}) is the concentration of CO_2 in the solution.

2. Ash-layer diffusion (A-mechanism)

When the diffusion of carbonate ions (i.e., reactant) through the ash layer is the rate-limiting step, the relationship between the carbonation conversion and reaction time can be illustrated by Eq. (7.17):

$$t = \tau_{\mathrm{A}} \times \left[1 - 3(1 - \delta_{\mathrm{CaO}})^{2/3} + 2(1 - \delta_{\mathrm{CaO}})\right] \quad (7.17)$$

For the A-mechanism, the relationship of τ_{A} and the effective diffusivity of the reactant through the ash layer (D_{e}, $cm^2\ s^{-1}$) is shown as Eq. (7.18):

$$\tau_{\mathrm{A}} = \frac{\rho_{\mathrm{B}} R^2}{6 b D_{\mathrm{e}} C_{\mathrm{Ag}}} \quad (7.18)$$

3. Fluid-film diffusion (F-mechanism)

When the mass transfer of the reactants through the boundary layer at the liquid–solid interface is the rate-limiting step, Eq. (7.19) shows the relationship between the conversion and reaction time:

$$t = \tau_{\mathrm{F}} \times \delta_{\mathrm{CaO}} \quad (7.19)$$

For the F-mechanism, the relationship of τ_{F} and the mass transfer coefficient (k_{e}, mol $m^{-2}\ Pa^{-1}\ s^{-1}$) is shown as Eq. (7.20):

$$\tau_{\mathrm{F}} = \frac{\rho_{\mathrm{B}} R}{3 b k_{\mathrm{e}} C_{\mathrm{Ag}}} \quad (7.20)$$

7.3.1.2 Modified SCM for Considering Particle Size Changes with Reaction

Due to its conceptual and mathematical simplicity, classical SCM has been used to determine the rate-limiting step in a heterogeneous reaction. However, in the SCM, there is an inherent assumption that complete carbonation conversion of solid particle should be eventually reached. From the experimental data in the literature, the full carbonation conversion of particles will never occur, regardless of time allowed. This implies that other factors, such as mineralogy of particles, particle size changes, and pore blockage, should be considered in addition to the diffusion limit of reactants into the particles [17]. The mineralogical compositions of particles are complex [12], where some of mineral phases do not react during carbonation. In actuality, the thickness of the ash layer would change with the reaction, which might affect the diffusivity of the gaseous reactant in the ash layer. To include the effect of particle size distribution on classical SCM, both "Z factor" and a "λ factor" [47] could be introduced to describe the relationship between the modified particle conversion (δ_{CaO}) and reaction time (t).

1. *Z* factor

Sohn (2004) [48] included a *Z* factor and modified the governing equation for pore diffusion control in SCM, as shown in Eq. (7.21), in the case where the volume of the solid product is different from that of the solid reactant, i.e., *R* changes with conversion:

$$t = \tau_A \times \left[\frac{Z - (Z - (Z - 1)(1 - \delta_{CaO}))^{2/3}}{(Z - 1)/3} - 3(1 - \delta_{CaO})^{2/3} \right] \tag{7.21}$$

where *Z* is the volume of product solid formed from a unit volume of reactant solid (both volumes include those of pores), and the term *r* is the particle size. If *Z* is assigned a value of one (i.e., the overall volume of the solid product is the same as that of the solid reactant), Eq. (7.21) can be simplified to Eq. (7.17) for *Z* = 1 by applying L′ Hospital's rule.

2. λ factor

A "λ factor" [47] can be introduced to modify the carbonation conversion of solid particle with the changes in particle size with reaction, as shown in Eq. (7.22):

$$\delta_{CaO} = 1 - \int_{\left(\frac{t}{-\lambda}\right)^{\frac{1}{\eta - D}}}^{r_{0,\max}} \frac{\left(\frac{t}{\lambda} + r^{\eta - D}\right)^{\frac{\eta}{\eta - D}}}{r^{\eta}} f(r) \mathrm{d}r \tag{7.22}$$

where *D* is the fractal dimension of the particle external surface (i.e., $D \in [2, 3]$), and η is a function of fractal dimension.

If the particle size remained unchanged in the course of reaction, then these modified models should be identical to the traditional SCM. Based on the SEM images, the size of precipitated calcium carbonate on the surface of the steel slag was found to range from 1 to 3 μm, which was significantly smaller than the steel slag [16]. In addition, in reality, the diffusion coefficient of CO_2 in the solution should be time-dependent, since the diffusion rate is a function of concentration gradient. The buffering capacity of the slurry will change with the different L/S ratios and reaction times, thereby resulting in different reactant concentrations (C_{Ag}) in solution because the C_{Ag} also changes as a function of pH [17].

7.3.1.3 Mechanisms and Diffusivity

To scale up and optimize the process, the key parameters for reactor design, such as mass-transfer coefficient and diffusivity of the reactant, can be obtained by the SCM. Information on the kinetic model of carbonation for determining the fluid-film diffusion-controlled and chemical-reaction-controlled conversions is provided in the previous sections. Lekakh et al. [11] evaluated the calcium leaching data of steelmaking slag to determine the rate-limiting step based on shrinking core model and found that during the initial stage (to 30% conversion), the rate-limiting step was chemical reaction controlled, and the later stage was controlled by diffusion. Similar observations were also found in other research that the dissolution step was generally assumed to be the rate-limiting step with respect to the overall carbonation reaction [32, 49].

Compared to the natural weathering process, aqueous carbonation is a rapid reaction because of rapid leaching of calcium ions from the particle surface of solid wastes [12]. For steel slag, it is suggested that the carbonation in a slurry reactor should be controlled by ash diffusion, which is verified through the SEM images of the steelmaking slag surface structure taken during the carbonation [6, 17]. The carbonation reaction would lead the reacted solid with a lower porosity, tortuosity, and pore area due to the formation of calcite. Since the reaction product $CaCO_3$ coated the particle surface, the pores of carbonated particles were blocked by the products. In addition, the calcite precipitate was formed as a protective layer around the reacting particles. Therefore, the diffusion paths were also blocked, resulting in a dynamic equilibrium. This indicates the rationale for applying the SCM for carbonation of alkaline solid wastes.

The carbonation reaction of alkaline solid wastes should be ash-diffusion controlled. The diffusivity (D_e) of reactant is also quite different among the feedstock, types of reactors, and operating conditions. For instance,

- Steelmaking slag with deionized water in a slurry reactor: from 2.88×10^{-7} to 7.28×10^{-7} $cm^2\ s^{-1}$ [22].
- Steelmaking slag with deionized water in high-gravity carbonation: from 5.47×10^{-7} to 1.49×10^{-6} $cm^2\ s^{-1}$ [16].

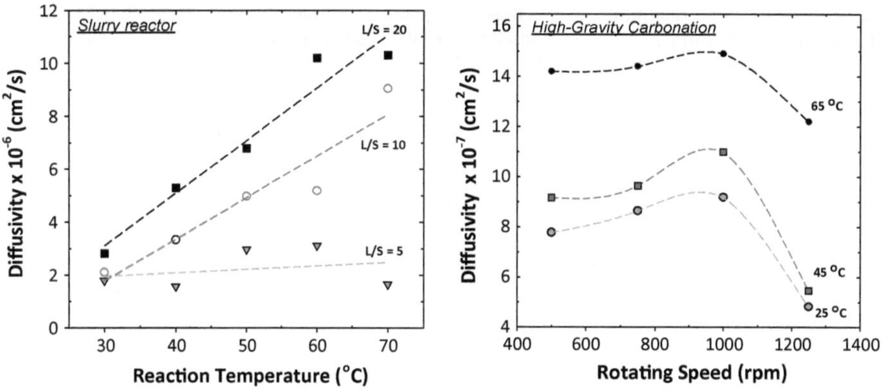

Fig. 7.3 Variation in effective diffusivity with (*left*) reaction temperature in a slurry reactor (P_{CO_2} = 1 bar; CO_2 flow rate = 0.1 L min^{-1}) and (*right*) rotating speed in high-gravity carbonation (P_{CO_2} = 1 bar; slurry flow rate = 1.2 L min^{-1}; D_p ~ 62 μm; L/S = 20 mL g^{-1})

- Steelmaking slag with deionized water in a slurry reactor: from 2.33×10^{-8} to 4.67×10^{-7} cm^2 s^{-1} at temperatures of 30–70 °C [17]
- Bottom ash with cold-rolling wastewater in a slurry reactor: from 1.00×10^{-6} to 2.90×10^{-5} cm^2 s^{-1} [50].

It was noted that the effective diffusivities measured in a slurry reactor and in a rotating packed bed exhibited the same magnitude of 10^{-6} cm^2 s^{-1} [16]. Several examples for the effect of various operating factors on diffusivity are provided, as shown in Fig. 7.3. The D_e value increases as the reaction temperature increases.

7.3.2 Surface Coverage Model

The surface coverage model was originally developed by considering the carbonation and desulfurization of hydrated lime at low temperatures by Shih et al. [51]. Their previous studies demonstrated that the surface coverage model was the most suitable model to describe the reaction between $Ca(OH)_2$ and SO_2 or CO_2. Those reaction behaviors are similar to the carbonation of steelmaking slag. Therefore, a surface coverage model could be utilized to determine the kinetics of the carbonation reaction. Figure 7.4 illustrates the mechanism of accelerated carbonation reaction of alkaline solid wastes, expressed by the surface coverage model.

It is noted that the kinetics of the carbonation reaction of alkaline solid wastes and the reaction rate constant could be described and estimated, respectively, by the surface coverage model [21, 44, 46]. The inherent assumptions of the surface coverage model include the following:

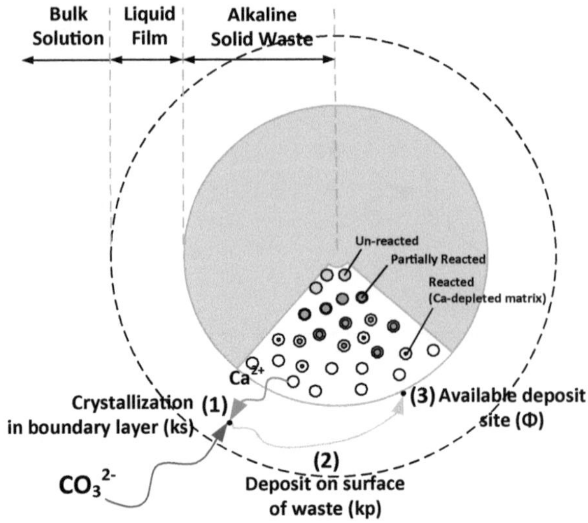

Fig. 7.4 Conceptual diagram of assumptions of surface coverage model for accelerated carbonation reaction using alkaline solid wastes

- Reaction occurs only at un-reacted surface sites without being covered by the reaction product.
- The reaction product is deposited on the surface of the solid reactant.
- The fraction of the active surface sites (φ) that is still not covered by the reaction product changes with reaction times depending on the reaction rate.
- The solid reactant will reach a maximum conversion of reaction ($\delta_{\max}$).

In the surface coverage model, the rate of carbonation conversion (δ_{CaO}) can be expressed by Eq. (7.23) [20, 22]:

$$\frac{\mathrm{d}\delta_{\mathrm{CaO}}}{\mathrm{d}t} = S_{\mathrm{g}}M \cdot r_{\mathrm{s}} = S_{\mathrm{g}}M \cdot k_{\mathrm{s}}\Phi \tag{7.23}$$

where S_g (m^2/g) is the initial specific surface area of alkaline solid wastes; M (g/mole) is the weight of alkaline solid wastes per mole of the reactive species (i.e., CaO); r_s (mole/min/m^2) is the carbonation reaction rate per initial surface area of alkaline solid wastes; and k_s (mole/min/m^2) is the overall rate constant. The fraction of the active surface sites (Φ) that changes with reaction times can be expressed by Eq. (7.24):

$$-\frac{\mathrm{d}\Phi}{\mathrm{d}t} = k_{\mathrm{p}}\Phi^{n-1} \cdot r_{\mathrm{s}} = k_{\mathrm{p}} \cdot k_{\mathrm{s}}\Phi^{n} \tag{7.24}$$

where k_p (m^2/mole) is a proportional constant reflecting the fraction of the active surface that is not covered by the reaction product. The principles and operating details of thermal analysis can be found in Chap. 6. Since calcium-bearing phases were assumed to be the major components participating in the carbonation reaction, the carbonation conversion (δ_{CaO}) of alkaline solid wastes was calculated with a

TGA by measuring the weight loss caused by the thermal decomposition of $CaCO_3$ in alkaline solid wastes.

Moreover, the above kinetic rate constants can be simplified by assuming k_1 (min^{-1}) and k_2 (dimensionless), as shown in Eqs. (7.25) and (7.26), respectively:

$$k_1 = k_s S_g M \tag{7.25}$$

$$k_2 = k_p/(S_g M) \tag{7.26}$$

For the carbonation of hydrated lime, the standard deviations for $n = 1.0$, 1.7, and 2.0 were found to be 0.015, 0.012, and 0.013, respectively, where $n = 1.7$ is the best fitting value [24]. Since these standard deviation values were within the range of experimental error [24], it was difficult to determine the most appropriate rate expression (n value). Consequently, it suggests that the n value in Eq. (7.24) could be assigned to be one for simplicity. In other words, the integration of Eq. (7.23) can be used to describe the relationship between the carbonation conversion and reaction time, in terms of k_1 and k_2, by Eq. (7.27):

$$\delta_{CaO} = [1 - \exp(-k_1 k_2 t)]/k_2, \quad \text{for } n = 1 \tag{7.27}$$

The two terms, k_1 and k_2, in Eq. (7.27) can be obtained accordingly by least squares fitting to the experimental data of carbonation conversion and reaction time. Based on the surface coverage model, it is noted that the carbonation conversion of alkaline solid wastes would reach a maximum value under a specific condition. If the reaction time extends to the saturation time, the exponential term (i.e., exp ($-k_1k_2t$)) in Eq. (7.27) would approach to zero. The carbonation conversion of alkaline solid wastes can be then expressed as a constant value of the reciprocal of k_2. Therefore, Eq. (7.27) can be expressed, in terms of maximum carbonation conversion for alkaline solid wastes (δ_{max}), as Eq. (7.28):

$$\delta_{CaO} = \delta_{max}\left[1 - \exp\left(-k_S k_p t\right)\right] \tag{7.28}$$

It is noted that the overall rate constant (k_s) is related to the following factors:

- The rate of gaseous CO_2 dissolution into solution
- The rate of calcium ions leaching from the solid matrix into solution,
- The rate of calcium carbonate precipitation

The $CaCO_3$ product deposited on the surface of reactant particle would hinder further leaching of the calcium species in the solid matrix into the solution. A higher k_s might result in a greater amount of product formation on the surface of alkaline solid wastes at the beginning of the reaction, which would decrease the dissolution rate of unreacted calcium species from the inside alkaline solid wastes into solution afterward [44].

On the other hand, the active surface site of alkaline solid wastes would be gradually covered by the carbonated product during the reaction. Once the product layer formed around the surface of alkaline solid wastes, the diffusion of reactants through the product layer would be the rate-limiting step for carbonation, hindering the alkaline solid wastes from further carbonation, thereby reaching a maximum carbonation conversion. In the course of carbonation, the surface structure of reacted alkaline solid wastes was found to change, where the small $CaCO_3$ particles were formed on the surface of alkaline solid wastes. Therefore, the carbonation reaction of alkaline solid wastes took place at the surface of alkaline solid wastes and formed a protective layer around the reacting particles. The rate of product deposition on the surface of the alkaline solid wastes ($k_s \cdot k_p$) was found to be higher at lower temperature [44]. This implies that carbonation would reach steady state much faster at lower temperatures than at higher temperatures. The leaching of calcium ions from alkaline solid wastes into solution can be accelerated as the reaction temperature increased. However, at higher temperatures, the precipitation of calcium carbonate was retarded due to the decrease in CO_2 solubility.

Table 7.2 compares the kinetic parameters of carbonation reaction for iron and steel slags via different processes, including high-gravity carbonation, slurry reactor, and autoclave reactor, by the surface coverage model.

The carbonation reaction rate per initial surface area (k_S) in different reactors is summarized as follows:

- Autoclave reactor: 0.8–2.6 mmol/min/m^2
- Slurry reactor: 0.3–0.5 mmol/min/m^2
- High-gravity carbonation process: 0.2–0.3 mmol/min/m^2

This might be attributed to the higher operating pressure of CO_2 (1400 psig) and reaction temperature (160 °C) performed in the autoclave reactor. However, operations with high pressure and temperature maintained in an autoclave reactor over a long period will require more energy and generate extra CO_2 emissions, which is not beneficial to the CO_2 capture process. In addition, in the slurry reactor, carbonation with alkaline wastewater (i.e., CRW) exhibited higher k_s values (i.e., 0.7 mmol/min/m^2) than that using effluent water (i.e., 0.3 mmol/min/m^2) and DI water (i.e., 0.5 mmol/min/m^2). It is evident that the carbonation conversion in the wastewater/solid residues system could be higher than that in the pure water/solid residues system [34, 35]. On the other hand, regardless the initial surface area of particles, the overall rate constant (k_1) for DW in different reactors is summarized as follows:

- High-gravity carbonation process: 0.299 min^{-1}
- Slurry reactor: 0.227 min^{-1}
- Autoclave reactor: 0.033 min^{-1}

It was noted that carbonation of steel slag in the high-gravity carbonation process exhibits superior performance for CO_2 capture due to the highest carbonation conversion of solid waste with the greatest reaction kinetics and a lower reaction time. Generally, the carbonation conversion (δ_{CaO}) of steel slag in the high-gravity carbonation process was found to be greater than that in the other two cases, i.e., the

Table 7.2 Comparison of kinetic parameters for accelerated carbonation of steel slags via various types of processes

Items		Units	Results (surface coverage model)							
Reactor type		–	Autoclave reactor[a] [21]			Slurry reactor[b] [46]			High-gravity carbonation [44]	
Liquid agents		–	DW	DW	DW	DW	EW	CRW	DW	DW
Solid residues		–	BHC	FAS	UFS	BOFS	BOFS	BOFS	BOFS	BOFS
Experiment	δ_{CaO}	–	0.68	0.35	0.38	0.80	0.78	0.89	0.94[c]	0.92[d]
Model parameters	M	g/mole	106.0	144.3	132.0	136.1	136.1	136.1	117.8	117.8
	S_g	m^2/g	0.12	0.24	0.15	3.2	3.2	3.2	9.36	9.36
	k_s	$mole/min/m^2$	0.0026	0.0008	0.0010	0.0005	0.0003	0.0007	0.0003	0.0002
	k_p	$m^2/mole$	21.3	108.1	57.4	634.6	575.5	567.8	1183.0	1163.9
	k_1	1/min	0.033	0.027	0.019	0.227	0.123	0.303	0.299	0.220
	k_2	–	1.67	3.12	2.90	1.457	1.322	1.304	1.073	1.055
	$k_s \cdot k_p$	1/min	0.055	0.084	0.055	0.331	0.163	0.395	0.323	0.234
	r^2	–	0.81	0.85	0.93	0.98	0.98	0.93	0.99	0.99

[a]Operated at an L/S = 10 mL g^{-1}, a partial pressure of CO_2 = 700 psig, 160 °C, and 12 h
[b]Operated for 120 min at 25 °C, CO_2 flow rate = 1 L/min, slurry volume = 350 mL, and L/S ratio = 20:1
[c]Operated at 65 °C and 750 rpm
[d]Operated at 65 °C and 1000 rpm

autoclave reactor and the slurry reactor. It suggests that the carbonation of steel slag should be integrated with co-utilizing alkaline wastewater (i.e., CRW) in the high-gravity carbonation process.

Similarly, the k_p values for high-gravity carbonation is greater than that for slurry reactor and autoclave reactor, as follows:

- High-gravity carbonation process: 1164–1183 m^2/mol
- Slurry reactor: 567–634 m^2/mol
- Autoclave reactor: 21–108 m^2/mol

This indicates that the reaction product covers the surface of solid wastes more uniformly in the cases of the high-gravity carbonation and the slurry reactor, whereas the product builds up more cluster-like and covers less of the surface of steel slag in the autoclave reactor. In all cases, the fresh steelmaking slag possesses a smooth surface before carbonation but then exhibits characteristics such as rhombohedral, cubic, or needle-like $CaCO_3$ crystals on the surface of reacted slags after carbonation [44].

7.4 Mass Transfer Models

Since the accelerated carbonation has been regarded as a diffusion-controlled reaction (i.e., mass transfer limited) [5, 11, 15], mass transfer among phases is a key to effective carbonation for CO_2 capture. The mass transfer model for the carbonation process can be developed based on several classical models, e.g., two-film theory.

7.4.1 General Concepts and Key Parameters

Mass transfer involves the movement of an element or molecular (mass) from one phase into another phase. It occurs in many processes; for example, dissolution of gaseous CO_2 in solution is the mass transfer phenomena. In classical, the mass transfer rate of a component mass can be described by two-film theory, where the rate of mass transfer is proportional to a mass transfer coefficient. The important mass transfer characteristics, especially for CO_2 absorption and/or dissolution, include overall gas-phase mass transfer coefficient (K_Ga) and height of a transfer unit (HTU).

The solid–liquid mass transfer coefficient can be occasionally correlated as itself, where such correlations are specific to the system under consideration and are not generally applicable [52]. The Sherwood number (*Sh*) and the Colburn J_d factor relate the physical properties of the system to the mass transfer coefficient and are more often used. To determine the gas-phase mass transfer coefficient in a heterogeneous system (containing gas, liquid, and solid phases), several assumptions can be made:

- The effect of an inclined gas–liquid interface is neglected.
- The concentration of the liquid at the particle surface is equal to the saturation concentration of the solution (the mass transfer between the liquid and the solid is neglected).
- The changes in size and surface area of solid particles are neglected.
- The solid distribution throughout the bed is uniform.
- The Grashof (*Gr*) number is determined by the mean radius of the packed bed.

As described in the two-film theory, the $K_G a$ (s^{-1}) in a packed bed can be determined by Eq. (7.29):

$$\frac{1}{K_G a_e} = \frac{1}{k_G a_e} + \frac{H}{I(k_L a_e)} \tag{7.29}$$

where k_G ($m\ s^{-1}$) is gas-side mass transfer coefficient; a_e ($m^2\ m^{-3}$) is effective surface area per unit volume of packed bed (gas–liquid interfacial); I (–) is the enhancement factor; and H is the Henry's law constant, expressed as the ratio of the partial pressure in the gas phase to the mass concentration in the liquid phase.

On the other hand, the relationship between mass transfer coefficient and diffusivity can be expressed as Eq. (7.30):

$$k_{CO_2} = \frac{D_{CO_2}}{\delta} \tag{7.30}$$

where D_{CO_2} (m^2/s) is the diffusivity; δ is the film thickness (m); and k_{CO_2} is the mass transfer coefficient (m/s). Table 7.3 presents several key thermodynamic state variables of CO_2 and water for model development. More thermodynamic state parameters for CO_2 and water can be also referred to Chap. 3.

The diffusivity of CO_2 in water (m^2/s) can be estimated by various methods, such as Eq. (7.31) [53] or Eq. (7.32) [54]:

$$D_{CO_2\_H_2O} = 2.35 \times 10^{-6} \times \exp(-2119/T) \tag{7.31}$$

$$D_{CO_2\_H_2O} = 1.81 \times 10^{-6} \times \exp(-16900/RT) \tag{7.32}$$

7.4.2 Incorporation with Reaction Kinetics

Direct carbonation is a heterogeneous reaction, i.e., containing the gas, liquid, and solid phases. It consists of mass transfer and chemical reaction. Therefore, a kinetic model can be developed for determining the reaction rate constant of carbonation (k_c) under different experimental conditions based on the shell mass balance.

Table 7.3 Key thermodynamic state variables of CO_2 and water for mass transfer model development at 1 atm and 30 °C

Abbreviation	Properties	Units	Values
ρ_G	Density of gas (CO_2)	mole cm^{-3}	4.02×10^{-5}
		kg/m^3	1.778
ρ_L	Density of water	kg/m^3	995.7
μ_G	Viscosity of gas (CO_2)	kg/m/s	1.53×10^{-5}
		centipoise (cp)	1.53×10^{-2}
		lb s/ft^2	3.19×10^{-7}
μ_L	Dynamic viscosity of water	kg/m/s	7.97×10^{-4}
ν_G	Kinematic viscosity of gas	m^2/s	8.53×10^{-6}
ν_L	Kinematic viscosity of liquid (water)	m^2/s	8.00×10^{-7}
D_{G_air}	Diffusivity of CO_2 gas in air	m^2/s	1.60×10^{-5}
D_{G_water}	Diffusivity of CO_2 gas in water	m^2/s	1.92×10^{-9}
σ_L	Surface tension of water	kg/s^2	7.12×10^{-2}
σ_c	Surface tension of packing[a]	kg/s^2	7.50×10^{-2}
H_{CO_2}	Henry's constant of CO_2 in water	mole/cm^3/atm	2.98×10^{-5}

[a]For metal packing

The gas phase can be characterized by convection behaviors between the gas inlet and outlet, indicating that the mass transfer phenomenon of CO_2 between the gas and liquid was significant. In addition, the chemical reaction between total inorganic carbon (TIC) concentration in the liquid and the calcium species from the solid should be considered. The following assumptions could be made to simplify the mathematical model:

- The reaction remains at a temperature of 30 °C, where calcium carbonate ($CaCO_3$) is the major product.
- The mass transfer between the liquid and the solid is neglected.
- Total inorganic carbon concentration in liquid phase (C_{TIC}) is a constant.
- The carbonation conversion (δ_{Ca}) in solid phase is a function of reaction time.

Thus, the apparent kinetic model of direct carbonation can be described by the mass balance over a thin shell of gas fluid with the reactor. The CO_2 concentration in exhaust gas (C_{CO_2}, dimensionless) is expressed in Eq. (7.33):

$$\rho_g V_g \frac{dC_{CO_2}}{dt} = \rho_g (Q_{g,i} C_{CO_2,i} - Q_{g,o} C_{CO_2}) - k_L a_p V_l (P_{CO_2} H_{CO_2} - C_{TIC}) \quad (7.33)$$

where ρ_g (mole cm^{-3}) is the CO_2 gas density at the operating condition; V_g (cm^3) and V_l (cm^3) are the volumes of gas and liquid; t (s) is the reaction time; k_L (cm s^{-1}) is the liquid-phase mass transfer coefficient; a_p (cm^2 cm^{-3}) is the specific surface area of packing materials; H_{CO_2} (mole cm^{-3} atm^{-1}) is the Henry's constant of CO_2 in water; and C_{TIC} (mole cm^{-3}) is TIC concentration in the liquid phase. The consumption rate of TIC in the liquid phase can be expressed by Eq. (7.34):

$$V_l \frac{dC_{TIC}}{dt} = k_L a_p V_l (P_{CO_2} H_{CO_2} - C_{TIC}) + k_c C_{TIC} V_l (C_S - C_{S,0}) \tag{7.34}$$

where k_c (cm^3 mol^{-1} s^{-1}) is the rate constant of the carbonation reaction; and C_s (mol cm^{-3}) is the CaO concentration of alkaline solid waste. It is assumed that the reaction was at steady state because the TIC concentration remained relatively unchanged during the carbonation. Therefore, Eq. (7.35) became:

$$k_L a_p V_l (C_{TIC} - P_{CO_2} H_{CO_2}) = k_c C_{TIC} V_l (C_S - C_{S,0}) \tag{7.35}$$

Substituting Eq. (7.35) into Eq. (7.33), the following kinetic equation can be obtained:

$$\rho_g V_g \frac{dC_{CO_2}}{dt} = \rho_g (Q_{g,i} C_{CO_2,i} - Q_{g,o} C_{CO_2,o}) + k_c C_{TIC} V (C_S - C_{S,0})_l \tag{7.36}$$

On the other hand, the rate of the solid concentration can be described by Eq. (7.37):

$$V_s \frac{dC_s}{dt} = k_c C_{TIC} C_s V_l \tag{7.37}$$

It was assumed that C_s was a function of carbonation conversion and also substituted the liquid volume (V_l) by solid volume (V_s), as expressed in Eqs. (7.38) and (7.39), respectively.

$$C_s = C_{s0}(1 - \delta_{CaO}) \tag{7.38}$$

$$V_l = \rho_s V_s \frac{L}{S} \tag{7.39}$$

where C_{s0} is the initial solid concentration (mole cm^{-3}). Thus, we can rearrange and integrate Eq. (7.37) by substituting Eqs. (7.38) and (7.39) into it:

$$\int \frac{d\delta_{CaO}}{1 - \delta_{CaO}} = \rho_s \frac{L}{S} k_c C_{TIC} \int dt \tag{7.40}$$

or

$$\delta_{CaO} = 1 - \exp^{\left(-\rho_s \frac{L}{S} k_c C_{TIC} t\right)} \tag{7.41}$$

As a result, the developed kinetic model could be validated by fitting Eq. (7.41) with the experimental data. Meanwhile, the reaction rate constant (k_c) can be estimated accordingly. According to the results in the literature [55], the rate constant for reaction using wastewater (i.e., cold-rolling wastewater (CRW)) with a slag particle size of less than 125 μm was the highest (216.9 cm^3 mol^{-1} s^{-1}). In addition, the carbonation reaction in the high-gravity carbonation should be accelerated by using wastewater (with a rate constant of 188–217 $cm^3\ mol^{-1}\ s^{-1}$) instead of DI water (with a rate constant of 160–195 $cm^3\ mol^{-1}\ s^{-1}$).

In addition, the liquid-phase mass transfer coefficient (k_L) can be determined using Eq. (7.42) by substituting Eq. (7.41) into Eq. (7.35).

$$k_c C_{TIC} C_{s0} \left(1 - \exp^{\left(-\rho_s \frac{L}{S} k_c C_{TIC} t\right)}\right) = k_L a_p (P_{CO_2} H_{CO_2} - C_{TIC}) \qquad (7.42)$$

The experimental results of carbonation could be well predicted by the developed kinetic model. For example, the mass transfer coefficient ($k_L a$) for reaction using wastewater with a slag particle size of less than 125 μm was the highest (i.e., $9.23 \times 10^{-4}\ s^{-1}$) [55].

7.4.3 *Modelling for Various Types of Reactors*

7.4.3.1 Slurry Bubble Column (SBC)

Slurry bubble column (SBC) mass transfer correlations are often derived from boundary layer theory where the Kolmogroff's isotropic turbulence theory has been applied [52]. The general correlations can be a variation of

$$Sh = 2 + bSc^m \left[ed_p^4/v^3\right]^n \qquad (7.43)$$

where e is the energy dissipation term and defined as $U_g g$. The term of m is often set equal to 1/3 as predicted by boundary layer theory.

Research has conducted different kinetic models to determine the rate-limiting step of mineral carbonation in a conventional bubble column [11, 22]. In fact, liquid–solid mass transfer is important and considered as the rate limiting factor in many cases of mineral carbonation [52, 56]. This might be due to the fact that minerals in solid matrix dissolve partly, and the passive layers are gradually formed, which increase resistance to mass transfer, and eventually lead to incomplete conversion.

7.4.3.2 Rotating Packed Bed (RPB)

Rotating packed bed (RPB) has been intensively applied and studied in many fields including gas absorption [57–59], stripping [60, 61], distillation [62], ozonation [63], deaeration [61], and biofuel production [64]. An RPB reactor has been introduced to improve the mass transfer rate among phases due to its high centrifugal forces and great micro-mixing ability. It thus can enhance mass transfer between gas and liquid, and even between liquid and solid. The features of utilizing the RPB reactor for process intensification include [1, 58, 62, 65, 66] the following:

- It can provide a mean acceleration of hundreds, and even thousands, of times greater than the force of gravity.
- It can effectively lead to the formation of thin liquid films and micro- or nano-droplets.
- The overall mass transfer coefficient in an RPB, especially the liquid-side mass transfer coefficient, was greater than that in a packed column.
- The volumetric gas–liquid mass transfer coefficients are an order of magnitude higher than those in a conventional packed bed.
- Dramatic reductions in equipment size over that required for equivalent mass transfer in a gravity-flow-packed bed.
- If the tangential gas velocity in the rotor is nearly the same as that in the packing zone, the gas-side mass transfer coefficient is believed to be in the same range as that of the conventional packed columns.

1. Assumptions

Performance of the carbonation reaction using an RPB reactor (so-called high-gravity carbonation) was found to be better than using an autoclave or slurry reactor [10, 16, 67]. Over the past years, several theoretical models have been developed for describing mass transfer phenomena of RPB for gas–liquid absorption [56, 68]. However, a few studies on gas–liquid–solid mass transfer for the high-gravity carbonation of alkaline solid wastes have appeared in the literature. This is attributed to the fact that sophisticated assumptions and complicated, partial differential equation programming are usually required to determine the gas–liquid–solid mass transfer.

The solid–liquid mass transfer coefficient is occasionally correlated as itself; however, such correlations are specific to the system under consideration and are not generally applicable [52]. Precipitation is usually a rapid reaction and the mixing (especially micro-mixing) of the process is very important for particle size distribution [69]. Therefore, in high-gravity carbonation process, the carbonation reaction would be controlled by the intrinsic reaction kinetics due to the excellent micro-mixing and the fine particle size (with a particle size of 44–88 μm) [16, 55, 67]. Before the passive layers around particles are formed, the minerals in solid particle can be rapidly dissolved into solution when the solid particle is moved through the packed bed. This indicates that the mass transfer between liquid and solid phases may not be the rate-limiting factor in the high-gravity carbonation system.

2. Gas-phase mass transfer coefficient

The driving force between the saturated CO_2 concentration in the bulk gas and the CO_2 concentration on the surface of liquid film can be determined by mass balance over a thin shell of fluid with the RPB, as shown in Eq. (7.44):

$$\frac{1}{\rho_{CO_2}}\frac{dM_G}{dV} = K_G a_e (C_G^* - C_G^{'}) \tag{7.44}$$

where dV can be expressed by Eq. (7.45):

$$dV = 2\pi rh \cdot dr \tag{7.45}$$

By substitution of Eq. (7.45) into Eq. (7.44), the overall gas-phase mass transfer coefficient ($K_G a$, s^{-1}) can be obtained as Eq. (7.46):

$$K_G a_e = \frac{M_G}{\rho_G h\pi(r_o^2 - r_i^2)}(\mathrm{NTU_G}) = \frac{Q_G}{h\pi(r_o^2 - r_i^2)}\ln\left(\frac{C_{G,i}}{C_{G,o}}\right) \tag{7.46}$$

where $C_{G,i}$ (%) and $C_{G,o}$ (%) are CO_2 contents in inlet and outlet gas streams, respectively. The terms of r_o (m) and r_i (m) are the outside and inside radii of packed bed, respectively. In general, the $K_G a$ value increases with the gas flow rate, with the liquid flow rate, and mainly with the rotor speed [70]. Also since the $K_G a$ values in an RPB are an order of magnitude higher than those in a conventional packed bed, the reactor size of RPB could be much smaller than that of a conventional reactor such as slurry reactor and bubble column [58, 62].

Beside the $K_G a$ value, the height of transfer unit (HTU) and the area of transfer unit (ATU) are important mass transfer characteristics. The values of HTU and ATU can be calculated using Eqs. (7.47) and (7.48), respectively [71, 72]:

$$\mathrm{HTU} = \frac{r_o - r_i}{\mathrm{NTU}} = \frac{r_o - r_i}{\ln(C_{G,i}/C_{G,o})} \tag{7.47}$$

$$\mathrm{ATU} = \frac{\pi(r_o^2 - r_i^2)}{\mathrm{NTU}} = \frac{\pi(r_o^2 - r_i^2)}{\ln(C_{G,i}/C_{G,o})} \tag{7.48}$$

3. Liquid-phase mass transfer coefficient

In spite of the significant differences between RPB and traditional packed bed column, penetration theory can be capable of describing the liquid-side mass transfer behavior fairly well in RPB [73]. These correlations are most often expressed in terms of dimensionless numbers in the form of a power series. As described in the penetration theory, the liquid-side mass transfer coefficient (k_L) can be expressed in Eq. (7.49):

$$k_L = a\left(\frac{D_L}{d_p}\right) Sc_L^b Re_L^c Gr_L^d \tag{7.49}$$

where the terms of a, b, c, and d are constants. The Grashof number, representing the ratio of gravitational to viscous forces, can be determined by Eq. (7.50):

$$Gr_{\rm L} = g d_{\rm p}^3 \left(\frac{\rho_{\rm L}}{\mu_{\rm L}}\right)^2 \tag{7.50}$$

In Eq. (7.50), the g value (m/s^2) can be replaced by the centrifugal acceleration in the RPB, which can be determined by Eq. (7.51):

$$g = a_{\rm m} = \left(\frac{r_{\rm o}^2 + r_{\rm i}^2}{2}\right)^{1/2} \omega^2 \tag{7.51}$$

where ω is the rotational speed (rad s^{-1}) at the mean radius of packed bed.

4. Empirical mass transfer models

In a conventional packed bed column, the commonly used model for liquid-side ($k_{\rm L}$) and gas-side ($k_{\rm G}$) mass transfer coefficient is originally developed by Onda et al. [74], as shown in Eqs. (7.52) and (7.53), respectively. It was suggested that the constant of 5.23 in Eq. (7.53) should be well correlated by changing the constant into 2.00 for smaller packings (when the diameter of packings is less than 1.5 cm).

$$k_{\rm L} = 0.0051 \left(\frac{a_{\rm t}}{a_{\rm w}}\right)^{2/3} Re_{\rm L}^{2/3} Sc_{\rm L}^{-1/2} \left(\frac{\rho_{\rm L}}{\mu_{\rm L} g}\right)^{-1/3} (a_{\rm p} d_{\rm p})^{0.4} \tag{7.52}$$

and

$$k_{\rm G} = 5.23 (a_{\rm p} D_{\rm G}) Re_{\rm G}^{0.7} Sc_{\rm G}^{1/3} (a_{\rm p} d_{\rm p})^{-2} \tag{7.53}$$

It is difficult to determine mass transfer coefficients separated from volumetric mass transfer coefficients $k_{\rm L}a_{\rm e}$ and $k_{\rm G}a_{\rm e}$, since the effective interfacial area ($a_{\rm e}$) between the liquid phase and vapor phase is usually not known [62]. Several correlations could be used to estimate the wetted surface area ($a_{\rm w}$), as shown in Eq. (7.54) [73, 75]. It suggests that the $a_{\rm w}/a_{\rm t}$ predicted value should be reliable under high-gravity RPB process, if packing materials with small static hold up (i.e., large packing size) the wetted area ($a_{\rm w}$) may equal the interfacial area ($a_{\rm e}$).

$$\frac{a_{\rm w}}{a_{\rm t}} = \frac{a_{\rm e}}{a_{\rm t}} = 1 - \exp\left[-1.45 \left(\frac{\sigma_{\rm c}}{\sigma_{\rm L}}\right)^{0.75} Re_{\rm L}^{0.1} We_{\rm L}^{0.2} Fr_{\rm L}^{-0.05}\right] \tag{7.54}$$

Pan et al. [76] developed a correlation for $K_{\rm G}a$ in the high-gravity carbonation process. In this model, the important operating parameters for high-gravity carbonation include gas flow rate, slurry flow rate, liquid-to-solid ratio, and rotation speed. The relevant models were obtained by the nonlinear regression of experimental data for estimation of both the $K_{\rm G}a$ and HTU values:

$$K_{G}a_{e} = 0.01\left(\frac{a_{t}D_{G}}{d_{p}}\right)Re_{G}^{-1.16}Gr_{G}^{0.33}Re_{L}^{2.12} \tag{7.55}$$

$$\mathrm{HTU} = 0.0003\left(\frac{a_{t}D_{G}}{d_{p}}\right)Re_{G}^{2.16}Gr_{G}^{-0.33}Re_{L}^{-2.12} \tag{7.56}$$

where the ranges of the dimensionless groups in this correlation are as follows:

$$7.8 < Re_{G} < 15.9 \tag{7.57}$$

$$1.3 < Re_{L} < 2.2 \tag{7.58}$$

$$2.3 < Gr_{G} < 26.8 \tag{7.59}$$

Both the $K_{G}a$ and HTU values varied with the centrifugal acceleration to the 0.33 power. The dependence of liquid velocity on $K_{G}a$ value was much higher than that of gas velocity, indicating that the high-gravity carbonation of steel slag could exhibit a mass transfer resistance lay on the liquid side [76]. Figure 7.5 shows the comparison of the estimated $K_{G}a$ value by the models with the experimental data. The result indicates that the $K_{G}a$ values predicted by the developed model were similar to the experimental value, where the experimental $K_{G}a$ values lay within $\pm 20\%$ of the values estimated by Eq. (7.55).

In contrast, the effect of gas velocity on the HTU value was similar to that of liquid velocity. The overall mass transfer coefficient in high-gravity carbonation, especially the liquid-side mass transfer coefficient ($k_{L}a$), was greater than in packed bed columns [76]. It is noted that since the tangential gas velocity in the rotor is

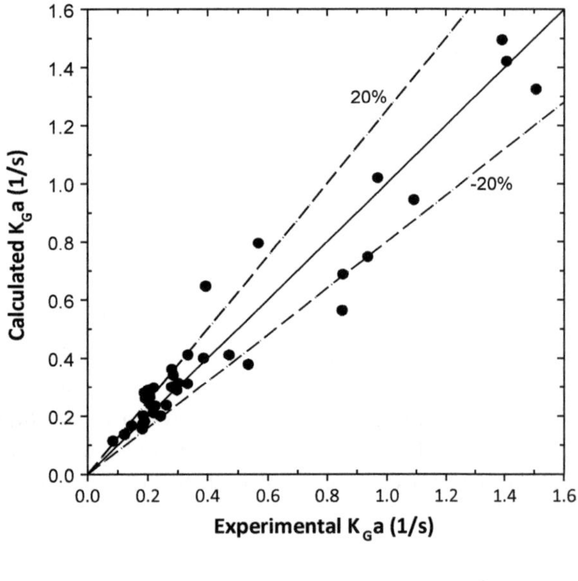

Fig. 7.5 Comparison of experimental and calculated $K_{G}a$ values for high-gravity carbonation process model. Reprinted by permission from Macmillan Publishers Ltd: Ref. [76], copyright 2015

nearly the same as that in the packing zone, the gas-side mass transfer coefficient should be in the same range as that of the conventional packed column [66]. The results in the literature indicate [55] that the $k_{L}a$ value for high-gravity reaction with a steel slag particle size of less than 125 μm was 9.23×10^{-4} s^{-1}, based on the shell mass balance model.

References

1. Wang M (2004) Controlling factors and mechanism of preparing needlelike $CaCO_3$ under high-gravity environment. Powder Technol 142(2–3):166–174. doi:10.1016/j.powtec.2004.05.003
2. Thiery M, Villain G, Dangla P, Platret G (2007) Investigation of the carbonation front shape on cementitious materials: effects of the chemical kinetics. Cem Concr Res 37(7):1047–1058. doi:10.1016/j.cemconres.2007.04.002
3. Montes-Hernandez G, Perez-Lopez R, Renard F, Nieto JM, Charlet L (2009) Mineral sequestration of CO_2 by aqueous carbonation of coal combustion fly-ash. J Hazard Mater 161 (2–3):1347–1354. doi:10.1016/j.jhazmat.2008.04.104
4. Tsuyoshi S, Etsuo S, Minoru M, Nobuaki O (2010) Carbonation of γ-Ca_2SiO_4 and the mechanism of vaterite formation. J Adv Concr Technol 8(3):273–280
5. Pan S-Y, Chang EE, Chiang P-C (2012) CO_2 capture by accelerated carbonation of alkaline wastes: a review on its principles and applications. Aerosol Air Qual Res 12:770–791. doi:10.4209/aaqr.2012.06.0149
6. Fernandez Bertos M, Simons SJ, Hills CD, Carey PJ (2004) A review of accelerated carbonation technology in the treatment of cement-based materials and sequestration of CO_2. J Hazard Mater 112(3):193–205. doi:10.1016/j.jhazmat.2004.04.019
7. Huntzinger DN, Gierke JS, Kawatra SK, Eisele TC, Sutter LL (2009) Carbon dioxide sequestration in cement kiln dust through mineral carbonation. Environ Sci Technol 43 (6):1986–1992
8. Uibu M, Kuusik R (2009) Mineral trapping of CO_2 via oil shale ash aqueous carbonation: controlling mechanism of process rate and development of continuous-flow reactor system. Oil Shale 26(1):40. doi:10.3176/oil.2009.1.06
9. Huijgen WJJ, Ruijg GJ, Comans RNJ, Witkamp GJ (2006) Energy consumption and net CO_2 sequestration of aqueous mineral carbonation. Ind Eng Chem Res 45(26):9184–9194
10. Pan SY, Chiang PC, Chen YH, Chen CD, Lin HY, Chang EE (2013) Systematic approach to determination of maximum achievable capture capacity via leaching and carbonation processes for alkaline steelmaking wastes in a rotating packed bed. Environ Sci Technol 47 (23):13677–13685. doi:10.1021/es403323x
11. Lekakh SN, Rawlins CH, Robertson DGC, Richards VL, Peaslee KD (2008) Kinetics of aqueous leaching and carbonization of steelmaking slag. Metall Mat Trans B 39(1):125–134. doi:10.1007/s11663-007-9112-8
12. Costa G (2009) Accelerated carbonation of minerals and industrial residues for carbon dioxide storage. Università delgi Studi di Roma
13. Eloneva S, Said A, Fogelholm C-J, Zevenhoven R (2012) Preliminary assessment of a method utilizing carbon dioxide and steelmaking slags to produce precipitated calcium carbonate. Appl Energy 90(1):329–334. doi:10.1016/j.apenergy.2011.05.045
14. Kodama S, Nishimoto T, Yamamoto N, Yogo K, Yamada K (2008) Development of a new pH-swing CO_2 mineralization process with a recyclable reaction solution. Energy 33(5):776–784. doi:10.1016/j.energy.2008.01.005

15. Chang EE, Wang Y-C, Pan S-Y, Chen Y-H, Chiang P-C (2012) CO_2 capture by using blended hydraulic slag cement via a slurry reactor. Aerosol Air Qual Res 12:1433–1443. doi:10.4209/aaqr.2012.08.0210
16. Chang EE, Pan SY, Chen YH, Tan CS, Chiang PC (2012) Accelerated carbonation of steelmaking slags in a high-gravity rotating packed bed. J Hazard Mater 227–228:97–106. doi:10.1016/j.jhazmat.2012.05.021
17. Pan SY, Liu HL, Chang EE, Kim H, Chen YH, Chiang PC (2016) Multiple model approach to evaluation of accelerated carbonation for steelmaking slag in a slurry reactor. Chemosphere 154:63–71. doi:10.1016/j.chemosphere.2016.03.093
18. Zumdahl SS (2009) Chemical principles, 6th edn. Houghton Mifflin Company
19. Li X, Bertos MF, Hills CD, Carey PJ, Simon S (2007) Accelerated carbonation of municipal solid waste incineration fly ashes. Waste Manag 27(9):1200–1206. doi:10.1016/j.wasman.2006.06.011
20. Fernandez Bertos M, Li X, Simons SJR, Hills CD, Carey PJ (2004) Investigation of accelerated carbonation for the stabilisation of MSW incinerator ashes and the sequestration of CO_2. Green Chem 6(8):428. doi:10.1039/b401872a
21. Chang EE, Pan S-Y, Chen Y-H, Chu H-W, Wang C-F, Chiang P-C (2011) CO_2 sequestration by carbonation of steelmaking slags in an autoclave reactor. J Hazard Mater 195:107–114. doi:10.1016/j.jhazmat.2011.08.006
22. Chang EE, Chen CH, Chen YH, Pan SY, Chiang PC (2011) Performance evaluation for carbonation of steel-making slags in a slurry reactor. J Hazard Mater 186(1):558–564. doi:10.1016/j.jhazmat.2010.11.038
23. Ishida T, Maekawa K (2000) Modeling of pH profile in pore water based on mass transport and chemical equilibrium theory. In: Proceedings of JSCE, 2000
24. Druckenmiller ML, Maroto-Valer MM (2005) Carbon sequestration using brine of adjusted pH to form mineral carbonates. Fuel Process Technol 86(14–15):1599–1614. doi:10.1016/j.fuproc.2005.01.007
25. Tai CY, Chen WR, Shih S-M (2006) Factors affecting wollastonite carbonation under CO_2 supercritical conditions. AIChE J 52(1):292–299. doi:10.1002/aic.10572
26. Park A, Fan L (2004) Mineral sequestration: physically activated dissolution of serpentine and pH swing process. Chem Eng Sci 59(22–23):5241–5247. doi:10.1016/j.ces.2004.09.008
27. Dijkstra JJ, Meeussen JCL, Comans RNJ (2004) Leaching of heavy metals from contaminated soils: an experimental and modeling study. Environ Sci Technol 38(16):4390–4395. doi:10.1021/es049885v
28. De Windt L, Chaurand P, Rose J (2011) Kinetics of steel slag leaching: batch tests and modeling. Waste Manag 31(2):225–235. doi:10.1016/j.wasman.2010.05.018
29. Iizuka A, Fujii M, Yamasaki A, Yanagisawa Y (2004) Development of a new CO_2 sequestration process utilizing the carbonation of waste cement. Ind Eng Chem Res 43(24):7880–7887. doi:10.1021/ie0496176
30. Alexander G, Maroto-Valer MM, Aksoy P, Schobert H (2005) Development of a CO_2 sequestration module by integrating mineral activation and aqueous carbonation. The Pennsylvania State University
31. Park AHA, Jadhav R, Fan LS (2003) CO_2 mineral sequestration: chemically enhanced aqueous carbonation of serpentine. Can J Chem Eng 81(3–4):885–890
32. O'Connor WK, Dahlin DC, Rush GE, Dahlin CL, Collins WK (2002) Carbon dioxide sequestration by direct mineral carbonation: process mineralogy of feed and products. Miner Metall Proc 19(2):95–101
33. Santos A, Ajbary M, Morales-Florez V, Kherbeche A, Pinero M, Esquivias L (2009) Larnite powders and larnite/silica aerogel composites as effective agents for CO_2 sequestration by carbonation. J Hazard Mater 168(2–3):1397–1403. doi:10.1016/j.jhazmat.2009.03.026
34. Morel FMM, Hering JG (1993) Principles and applications of aquatic chemistry. Wiley
35. Edsall JT, Wyman J (1958) Biophysical chemistry. Academic Press
36. Stumm W, Morgan JJ (1996) Aquatic chemistry. A Wiley-Interscience Publication, Wiley

37. Lagasa AC (1995) Fundamental approaches to describing mineral dissolution and precipitation rates. In: White AF, Brantley SL (eds) Reviews in mineralogy, vol 31. Mineralogical Society of America, Washington, D.C., pp 23–86
38. Gerdemann SJ, O'Connor WK, Dahlin DC, Penner LR, Rush H (2007) Ex situ aqueous mineral carbonation. Environ Sci Technol 41(7):2587–2593
39. Huijgen WJJ, Comans RNJ (2006) Carbonation of steel slag for CO_2 sequestration: leaching of products and reaction mechanisms. Environ Sci Technol 40(8):2790–2796
40. Alexander G, Mercedesmarotovaler M, Gafarovaaksoy P (2007) Evaluation of reaction variables in the dissolution of serpentine for mineral carbonation. Fuel 86(1–2):273–281. doi:10.1016/j.fuel.2006.04.034
41. Krevor SCM, Lackner KS (2011) Enhancing serpentine dissolution kinetics for mineral carbon dioxide sequestration. Int J Greenhouse Gas Control 5(4):1073–1080. doi:10.1016/j.ijggc.2011.01.006
42. Bhatia SK, Perlmutter DD (1983) Effect of the product layer on the kinetics of the CO_2-lime reaction. AIChE J 29(1):79–86
43. Liu W, Dennis JS, Sultan DS, Redfern SAT, Scott SA (2012) An investigation of the kinetics of CO_2 uptake by a synthetic calcium based sorbent. Chem Eng Sci 69:644–658
44. Pan S-Y, Chiang P-C, Chen Y-H, Tan C-S, Chang EE (2014) Kinetics of carbonation reaction of basic oxygen furnace slags in a rotating packed bed using the surface coverage model: maximization of carbonation conversion. Appl Energy 113:267–276. doi:10.1016/j.apenergy.2013.07.035
45. Sohn HY, Szekely J (1973) Reactions between solids through gaseous intermediates—I reactions controlled by chemical kinetics. Chem Eng Sci 28:1789–1801
46. Chang EE, Chiu A-C, Pan S-Y, Chen Y-H, Tan C-S, Chiang P-C (2013) Carbonation of basic oxygen furnace slag with metalworking wastewater in a slurry reactor. Int J Greenhouse Gas Control 12:382–389. doi:10.1016/j.ijggc.2012.11.026
47. Li X, Yang Z, Zhao J, Wang Y, Song R, He Y, Su Z, Lei T (2015) A modified shrinking core model for the reaction between acid and hetero-granular rough mineral particles. Hydrometallurgy 153:114–120. doi:10.1016/j.hydromet.2015.03.001
48. Sohn HY (2004) The effects of reactant starvation and mass transfer in the rate measurement of fluid–solid reactions with small equilibrium constants. Chem Eng Sci 59(20):4361–4368. doi:10.1016/j.ces.2004.06.033
49. Daval D, Martinez I, Corvisier J, Findling N, Goffé B, Guyot F (2009) Carbonation of Ca-bearing silicates, the case of wollastonite: experimental investigations and kinetic modeling. Chem Geol 265(1–2):63–78. doi:10.1016/j.chemgeo.2009.01.022
50. Chang EE, Pan SY, Yang L, Chen YH, Kim H, Chiang PC (2015) Accelerated carbonation using municipal solid waste incinerator bottom ash and cold-rolling wastewater: performance evaluation and reaction kinetics. Waste Manag 43:283–292. doi:10.1016/j.wasman.2015.05.001
51. Shih SM, Ho CS, Song YS, Lin JP (1999) Kinetics of the reaction of Ca(OH)(2) with CO_2 at low temperature. Ind Eng Chem Res 38(4):1316–1322
52. Arters DC, Fan L-S (1986) Experimental methods and correlation of solid liquid mass transfer in fluidized beds. Chem Eng Sci 41(1):107–115
53. Versteeg GF, Swaalj WV (1988) Solubility and diffusivity of acid gases (carbon dioxide, nitrous oxide) in aqueous alkanolamine solutions. J Chem Eng Data 33(1):29–34
54. Frank MJW, Kuipers JAM, van Swaaij WPM (1996) Diffusion coefficients and viscosities of $CO_2 + H_2O$, $CO_2 + CH_3OH$, NH3 + H_2O and $NH_3 + CH_3OH$ liquid mixtures. J Chem Eng Data 41:297–302
55. Chang EE, Chen T-L, Pan S-Y, Chen Y-H, Chiang P-C (2013) Kinetic modeling on CO_2 capture using basic oxygen furnace slag coupled with cold-rolling wastewater in a rotating packed bed. J Hazard Mater 260:937–946. doi:10.1016/j.jhazmat.2013.06.052
56. Munjal S, Dudukovic MP, Ramachandran P (1989) Mass-transfer in rotating packed beds—I. Dev Gas-Liquid Liquid-Solid Mass Transf Correl Chem Eng Sci 44(10):2245–2256

57. Cheng H-H, Tan C-S (2011) Removal of CO_2 from indoor air by alkanolamine in a rotating packed bed. Sep Purif Technol 82:156–166. doi:10.1016/j.seppur.2011.09.004
58. Tan C, Chen J (2006) Absorption of carbon dioxide with piperazine and its mixtures in a rotating packed bed. Sep Purif Technol 49(2):174–180. doi:10.1016/j.seppur.2005.10.001
59. Lin CC, Lin YH, Tan CS (2010) Evaluation of alkanolamine solutions for carbon dioxide removal in cross-flow rotating packed beds. J Hazard Mater 175(1–3):344–351. doi:10.1016/j.jhazmat.2009.10.009
60. Guo F, Zheng C, Guo K, Feng YD, Gardner NC (1997) Hydrodynamics and mass transfer in crossflow rotating packed bed. Chem Eng Sci 52(21–22):3853–3859
61. Chen YS, Lin FY, Lin CC, Tai CYD, Liu HS (2006) Packing characteristics for mass transfer in a rotating packed bed. Ind Eng Chem Res 45(20):6846–6853
62. Kelleher T, Fair JR (1996) Distillation studies in a high-gravity contactor. Ind Eng Chem Res 35:4646–4655
63. Chen YH, Chang CY, Su WL, Chen CC, Chiu CY, Yu YH, Chiang PC, Chiang SIM (2004) Modeling ozone contacting process in a rotating packed bed. Ind Eng Chem Res 43(1): 228–236
64. Chen YH, Huang YH, Lin RH, Shang NC (2010) A continuous-flow biodiesel production process using a rotating packed bed. Bioresour Technol 101(2):668–673. doi:10.1016/j.biortech.2009.08.081
65. Lin C, Chen B (2008) Characteristics of cross-flow rotating packed beds. J Ind Eng Chem 14 (3):322–327. doi:10.1016/j.jiec.2008.01.004
66. Chandra A, Goswami PS, Rao DP (2005) Characteristics of flow in a rotating packed bed (HIGEE) with split packing. Ind Eng Chem Res 44(11):4051–4060
67. Pan SY, Chiang PC, Chen YH, Tan CS, Chang EE (2013) Ex situ CO_2 capture by carbonation of steelmaking slag coupled with metalworking wastewater in a rotating packed bed. Environ Sci Technol 47(7):3308–3315. doi:10.1021/es304975y
68. Zhao B, Su Y, Tao W (2014) Mass transfer performance of CO_2 capture in rotating packed bed: dimensionless modeling and intelligent prediction. Appl Energy 136:132–142. doi:10.1016/j.apenergy.2014.08.108
69. Xiang Y, Wen LX, Chu GW, Shao L, Xiao GT, Chen JF (2010) Modeling of the precipitation process in a rotating packed bed and its experimental validation. Chin J Chem Eng 18(2): 249–257
70. Liu H, Kuo C (1996) Quantitative multiphase determination using the Rietveld method with high accuracy. Mater Lett 26:171–175
71. Sandilya P, Rao DP, Sharma A (2001) Gas-phase mass transfer in a centrifugal contactor. Ind Eng Chem Res 40:384–392
72. Cheng H-H, Shen J-F, Tan C-S (2010) CO_2 capture from hot stove gas in steel making process. Int J greenhouse gas control 4(3):525–531. doi:10.1016/j.ijggc.2009.12.006
73. Tung HH, Mah RSH (1985) Modeling liquid mass transfer in HIGEE separation process. Chem Eng Commun 39:147–153
74. Onda K, Takeuchi H, Okumoto Y (1968) Mass transfer coefficients between gas and liquid phases in packed columns. J Chem Eng Jpn 1(1):56–62
75. Perry RH, Chilton CH (1973) Chemical engineers' handbook. McGraw-Hill, New York. doi:10.1002/aic.690200140
76. Pan SY, Eleazar EG, Chang EE, Lin YP, Kim H, Chiang PC (2015) Systematic approach to determination of optimum gas-phase mass transfer rate for high-gravity carbonation process of steelmaking slags in a rotating packed bed. Appl Energy 148:23–31. doi:10.1016/j.apenergy.2015.03.047

Chapter 8
Applications of Carbonation Technologies

Abstract Reduction of CO_2 emission in industries and/or power plants should be a portfolio option; for example, CO_2 capture and alkaline solid waste treatment can be combined through an integrated approach, i.e., accelerated carbonation. Gaseous CO_2 is fixed as thermodynamically stable solid precipitates, which are rarely released after mineralization. In addition, the proximity between the industrial CO_2 emission and the waste residue sources reduces transportation costs. This chapter presents an integrated approach (i.e., accelerated carbonation technology) to capturing CO_2 while improving the physico-chemical properties of alkaline solid wastes for the utilization in civil engineering. The carbonation of industrial alkaline wastes, such as steel slags and metalworking wastewater, has been proved to be an effective way to capture CO_2.

8.1 Concept of Accelerated Carbonation

The concept behind carbonation reaction is similar to natural weathering processes, where CO_2 reacts with metal oxide-bearing materials, such as CaO and MgO, to form stable and insoluble carbonates (such as $CaCO_3$ and $MgCO_3$). In this section, different types of carbonation technologies are first reviewed and discussed. The concept of the system mass and energy balance is then provided. Several key performance indicators for carbonation reaction are also illustrated.

8.1.1 Types of Carbonation Technologies

Figure 8.1 shows various approaches to integrating solid waste treatment with CO_2 fixation and utilization, including the following:

- Indirect carbonation (using natural ore or alkaline solid wastes)
- Direct carbonation (using natural ore or alkaline solid wastes)
- Carbonation curing (for the production of cement mortar or concrete block)

P.-C. Chiang and S.-Y. Pan, *Carbon Dioxide Mineralization and Utilization*, DOI 10.1007/978-981-10-3268-4_8

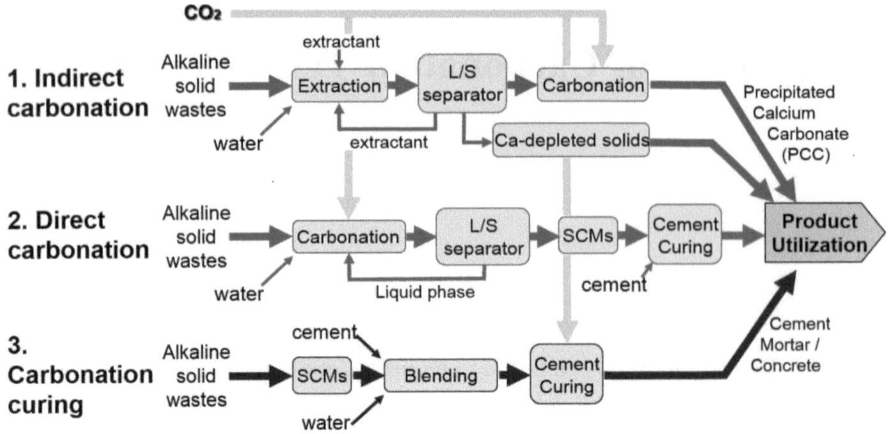

Fig. 8.1 Various approaches to integrating alkaline solid waste treatment with CO_2 fixation and utilization through carbonation process

For the indirect carbonation using alkaline solid wastes, high purity of $CaCO_3$ precipitates (e.g., 99%) can be produced and implemented as high value industrial materials, such as coating pigments and filters [1–3]. In contrast, the direct carbonation can fix CO_2 into carbonate precipitates with alkaline solid wastes, which is superior in providing high storage capacity and long storage time [4–6]. Another approach to integrating CO_2 with alkaline solid wastes treatment is the carbonation curing for blended cement or concrete through the injection of pressurized CO_2 gas (normally using pure CO_2 stream) into a sealed chamber.

8.1.2 System Mass and Energy Balance

The CO_2 mass balance can be carried out in the system by measuring the CO_2 loss in the gas phase as well as the increase of inorganic carbon content in both the liquid (i.e., total inorganic carbon (TIC) concentration) and solid phases (i.e., $CaCO_3$ content). Figure 8.2 shows the systematic diagram among gas, liquid, and solid phases.

Accordingly, the CO_2 mass balance equation can be developed as shown in Eq. (8.1):

$$\rho_{CO_2,i}Q_{g,i}C_{g,i} - \rho_{CO_2,o}Q_{g,o}C_{g,o} = Q_l \Delta C_{TIC} + Q_{sl}C_{s,act}(S/L) \tag{8.1}$$

where ρ_g (i.e., 1.796 g L^{-1}) is the density of CO_2 at standard temperature and pressure (STP); ΔC_{TIC} (g/L) is the concentration difference of the TIC in the liquid agents before and after carbonation; Q_l (L/min) is the liquid flow rate; Q_s (L/min) is the solid input rate; $C_{s,act}$ (g) is the mass of CO_2 captured in the solid waste; and L/S (mL/g) is the liquid-to-solid ratio. The molar volume of CO_2 is generally assumed

Fig. 8.2 Schematic of mass balance among gas, liquid, and solid phases (Q_g: Gas flow rate (L/min); C_g: CO_2 Concentration (%); C_{TIC}: TIC concentration in liquid phase (g/L); Q_l: Liquid flow rate (L/min); $C_{s,act}$: CO_2 capture on solid (g); L/S: Liquid-to-solid ratio (mL/g))

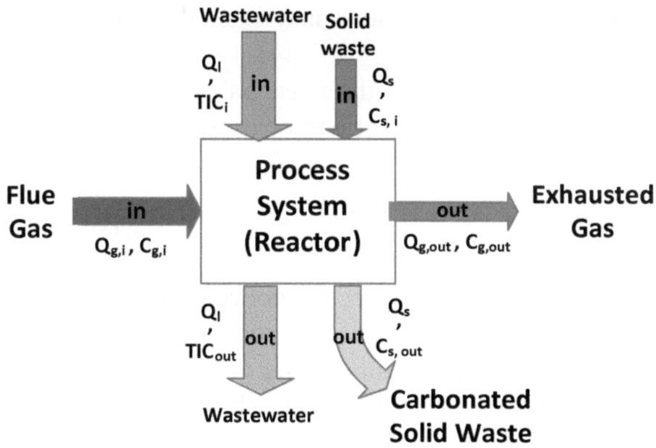

to be 24.5 L mol^{-1} at standard temperature and pressure (STP). The STP is defined as the temperature of 0 °C (or 32 °F) and the pressure of 760 mmHg (or 14.7 psi).

In the gas phase, CO_2 concentrations in the inlet and exhausted gas streams could be measured using a non-dispersive infrared (NDIR) CO_2 analyzer. The CO_2 removal efficiency (η, %) of the carbonation process can be calculated by Eq. (8.2):

$$\eta = \frac{\left(\rho_{CO_2,i} Q_{g,i} C_{g,i} - \rho_{CO_2,o} Q_{g,o} C_{g,o}\right)}{\rho_{CO_2,i} Q_{g,i} C_{g,i}} \times \% \tag{8.2}$$

where $\rho_{CO_{2,i}}$ and $\rho_{CO_{2,0}}$ (g/L) are the CO_2 mass density at the temperature of the inflow and exhaust gas streams, respectively; $Q_{g,i}$ (m^3/min) and $Q_{g,o}$ (m^3/min) are the volumetric flow rate of the inlet gas and exhaust gas, respectively; and $C_{g,i}$ (%) and $C_{g,o}$ (%) are the CO_2 concentration in the inlet gas and exhaust gas, respectively, which can be measured by a portable gas analyzer. If pure CO_2 gas is utilized for carbonation, since the CO_2 concentration in the input and output streams should be 99.9%, the amounts of CO_2 captured by the slurry reactor can be determined by the difference in the volumes of the gas flows measured at the inlet and outlet of the slurry reactor.

In the solid phase, the carbonation conversion (also referred to as carbonation degree) of alkaline solid wastes was determined by thermal analysis (see details in Chap. 6). Generally, the weight loss between 500 and 850 °C is caused mainly by the decomposition of $CaCO_3$ based on Eq. (8.3):

$$CaCO_{3(s)} \xrightarrow{500\sim850\,^\circ C} CaO_{(s)} + CO_{2(g)} \tag{8.3}$$

For direct carbonation, it is observed that most of the gaseous CO_2 was fixed as $CaCO_3$ precipitate (i.e., solid phase). Less than 0.1% of the gaseous CO_2 was dissolved as HCO_3^- ions and existed in the liquid phase after 120-min reaction time.

On the other hand, the mass balance of calcium species between the liquid and solid phases for indirect carbonation was determined by Eq. (8.4):

$$C_{Ca,i}V_l = C_{Ca,o}V_l + \frac{m_{CaCO_3}}{MW_{CaCO_3}} \times MW_{Ca} \quad (8.4)$$

where $C_{Ca,i}$ and $C_{Ca,o}$ are the concentrations of calcium ions in the solution before and after carbonation, respectively.

In addition, the errors of the mass balance for Eqs. (8.3) and (8.4) were evaluated by the recovery ratio (R), which is defined by Eq. (8.5):

$$R(\%) = \frac{\text{Production}}{\text{Input} - \text{Output}} \times \% \quad (8.5)$$

8.1.3 Novel Concept and Process Intensification

Mineralization of CO_2 by accelerated carbonation of alkaline wastes has the potential not only to sequester CO_2 but also to upgrade the physicochemical properties of waste streams. One of the challenges of accelerated carbonation using alkaline waste is to accelerate the reaction and exploit the heat of reaction to minimize energy and material losses. Also, pretreatments and addition of chemicals tend to consume additional energy and materials. More challenges of accelerated carbonation can be referred to Chap. 5.

To overcome the aforementioned challenges, various novel concepts and process intensification should be considered. For example, the rate of carbonation reaction can be enhanced by

- Adding acids or caustic alkali metal hydroxide
- Utilizing a bicarbonate/salt ($NaHCO_3$/NaCl) mixture
- Employing various pretreatment techniques (i.e., including physical and chemical pretreatment, electrolysis and heat pretreatment, mechanical activation methods)
- Using wastewater/brine as liquid agents
- Process intensification concepts such as
  - Thermal heat activation [7, 8]
  - Chemical activation using solvents, acids, and bases [8, 9]
  - Co-utilization with wastewater (or brine solution) [10–12]
  - Biological enhancement using enzymes and microorganisms
  - Centrifugal force [13, 14]
  - Ultrasonic vibration [9, 15, 16]
  - Electrolysis treatment [7]

In the following section, indirect carbonation and direct carbonation using alkaline solid wastes, as well as carbonation curing processes, are discussed.

8.2 Indirect Carbonation

Several effective extractants, such as acetic acid and ammonium salts (e.g., CH_3COONH_4, NH_4NO_3, NH_4SO_4, and NH_4HSO_4), are commonly used in the extraction stage [17–20]. Table 8.1 presents the types of acid and ammonium salt extractants for metal ion extraction. For ammonium salts, the acid type of the solvent is a more important factor affecting the efficiency of calcium extraction rather than the acidity of the solvent [18].

8.2.1 Acidic Extraction Process

Research on the performance of enhancing dissolution using acetic acid solutions and other chemicals has also been investigated in the literature [22, 23]. In the presence of acetic acid, the solubility of metal ions in slag was found to decrease with an increase in temperature [24]. Better chemical extractions of calcium occur at low temperatures (30 °C) at long durations ($\geq$2 h). In addition, the concentration of acetic acid within the range of 0–10 wt% in the solution exhibited a dramatic

Table 8.1 Types of acid and ammonium salt extractants for metal ion extraction

Types	Extractants	Name	PH	Acidity (pK_a)	Tested concentration
Acid	HCl	Hydrogen chloride	3.72	−3.0	–
	H_2SO_4	Sulphuric acid	–	−3; 1.99	–
	HNO_3	Nitric acid	–	−1.44	–
	H_3PO_4	Phosphoric acid	–	2.148; 7.198; 12.319	–
	$ClCH_2COOH$	Haloacetic acid	–	2.86	–
	CH_3COOH	Acetic acid	5.32	4.76	–
Base (salt)	NH_4NO_3	Ammonium nitrate	9.8–10.5	Solid form	0.5–1 M [18]
	CH_3COONH_4	Ammonium acetate	9.5	Solid form	0.5–1 M [18]
	NH_4Cl	Ammonium chloride	9.0–9.6	Solid form	2 M at 60 °C [21]
	$(NH_4)_2SO_4$	Ammonium sulfate	9.6	Solid form	0.5–1 M [18]

effect upon the extraction of calcium [24]. Compared to only 30 mg/L of calcium after 2 h without acetic acid addition, 4000 mg/L of calcium can be extracted in 5 min using an aqueous solution of 10 wt% acetic acid. However, a further increase of acetic acid concentration to 33.3 wt% did not result in a significant difference from that of 10 wt% acetic acid. Therefore, approximately 6–7 mL glacial acetic acid per gram of slag is suggested for effective calcium extraction [24]. After extraction, a maximum carbonation conversion of 86% can be achieved under a CO_2 concentration of 10% at 30 °C, with a $CaCO_3$ purity of 99.8%. Roughly 4.7 tons of steel slag are consumed for one ton CO_2 capture, with 2.3 tons of end $CaCO_3$ products and 3.4 tons of residual slag [23].

8.2.2 *Base Extraction Process*

In base extraction process, ammonium chloride (NH_4Cl) is commonly used due to its relative ease of recycling. The ammonia being stripped out during extraction step can be recycled and utilized for pH adjustment in order for the leachate to minimize the consumption of ammonia during the operation [21]. When ammonium chloride is used, the extraction can be improved as the particle size of slag decreased and the reaction temperature increased, with a maximum extraction yield of 60% [25]. As the extraction reaction proceeded, the deposition of an inert layer (such as SiO_2) formed on the reactive surface, thereby limiting further diffusion of extractant into slag particles [26].

The overall carbonation reactions of leachate can be expressed as shown in Eq. (8.6):

$$2\,NH_4OH + (Ca, Mg)Cl_2 + CO_2 \rightarrow (Ca, Mg)CO_3 + 2\,NH_4Cl + H_2O \tag{8.6}$$

In fact, at the beginning, the CO_2 will react with NH_4OH, as described in Eqs. (8.7) and (8.8):

$$NH_4OH_{(aq)} + CO_{2\,(aq)} \rightarrow NH_4HCO_{3\,(aq)} \tag{8.7}$$

$$2\,NH^+_{4\,(aq)} + HCO^-_{3\,(aq)} + OH^-_{(aq)} \rightarrow (NH_4)_2CO_{3(aq)} + H_2O_{(aq)} \tag{8.8}$$

As carbonation proceeds, the dissolved CO_2 will continuously consume OH^- in solution, thereby inhabiting reaction (8.8). In this case, the produced NH_4HCO_3 (Eq. 8.7) will begin to react with calcium/magnesium ions to form calcium/magnesium bicarbonate, which is described in Eq. (8.9)

$$(Ca, Mg)^{2+}_{(aq)} + 2\,HCO^-_{3(aq)} \rightarrow (Ca, Mg)(HCO_3)_{2(aq)} \tag{8.9}$$

The efficiency of calcium extraction was found to be higher in the order of $(NH_4)_2SO_4 < NH_4Cl < CH_3COONH_4 < NH_4NO_3$, regardless of their solution concentrations [18]. Similarly, the efficiency of various extractants on the carbonation behavior was found to be in the same sequence [18]. It was noted that the maximum CO_2 capture capacity can reach around 210 kg CO_2 per ton of slag, with the consideration of the contribution of $Mg(HCO_3)_2$ in capturing CO_2 [21]. In this case, the precipitates obtained under optimized carbonation conditions were rich in $CaCO_3$ with a purity of 96 ± 2%.

On the basis of experimental analysis, the energy consumption of the proposed process using ammonium chloride was estimated as 300 kWh/ton-CO_2 [25]. This value is smaller than other CO_2 sequestration processes; for instance, 470–640 kWh/t-CO_2 for amine absorption and geological sequestration [27].

8.2.3 *Multistage Indirect Carbonation*

Multistage indirect carbonation, in the presence of additives, can reach high carbonation efficiency under mild operating conditions within a short residence time [28, 29]. It was noted that high-purity carbonate precipitates (e.g., spherical vaterite) can be generated by multistage indirect carbonation [18]. For example, two-stage (or so-called pH swing) indirect carbonation processes are one type of the multistage indirect carbonation. They have been evaluated to optimize the efficiency of the dissolution and carbonation processes, as well as to achieve a pure precipitate product for specific industrial applications. The pH-swing using a weak base—strong acid solution, with a fine particle size (e.g., less than 63 μm) at 80 °C, in which the maximum calcium extraction ratio and carbonation conversion achieved, was 60 and 70%, respectively [25].

The requirement of an energy-intensive process for chemical regeneration might be a limiting factor for large-scale deployment. It was noted that energy and chemical costs could be reduced by carrying out the reaction between hydroxide and CO_2 at high pressure (i.e., 2.5 MPa) and temperature (i.e., 450 °C) [30]. This potentially makes the hydroxide route technically achievable on an industrial scale.

8.3 Direct Carbonation

Recently, the ex situ direct aqueous carbonation has aggressively developed for carbon fixation and utilization. There is more than enough natural ore on Earth to sequester the CO_2 emissions from all fossil fuels [6]. However, the existing methods for mineral carbonation are still expensive due to the energy-intensive mining process. To effectively lower the carbonation process, alkaline solid wastes are thus used in the carbonation process because they are relatively cheaper feedstocks than natural ores. Accelerated carbonation involves the following three steps (see details in Chap. 5).

- Calcium leaching from alkaline solid matrix into solution
- Gaseous CO_2 dissolution in solution
- Calcium carbonate precipitation

For various types of processes, the mass transfer steps, i.e., CO_2 dissolution into solution and the diffusion of reactants in solid matrix, are typically considered to be the rate-limiting step. Consequently, it is essential to improve the mass transfer phenomena to achieve a rapid reaction.

Accelerated carbonation has been focused on assessing and maximizing the performance of CO_2 capture capacity by optimizing the operating conditions, such as pressure, temperature, liquid-to-solid (L/S) ratio, gas humidity, gas flow rate, liquid flow rate, particle size, and solid pretreatment in the literature [1, 28, 31–34]. In the following part, several important processes, such as slurry reactor, autoclave reactor, high-gravity carbonation, and ultrasonic technology, are introduced.

8.3.1 Autoclave Reactor

A schematic diagram demonstrating the carbonation of alkaline solid wastes in an autoclave reactor is shown in Fig. 8.3. CO_2 was injected continuously into the reactor at a designated pressure and a constant flow rate. The key operational factors, including the reaction time (t), liquid-to-solid ratio (L/S), reaction temperature (T), CO_2 pressure (P), and initial pH, are varied systematically with the various feedstocks to minimize energy and chemical consumption.

In the case of operation at 160 °C and 700 psig, the carbonation of the ultra-fine slag (UFS), fly ash slag (FAS), and blended hydraulic cement slag (BHCS) are 38.1, 34.7, and 68.3%, respectively [35]. Accordingly, the actual CO_2 capture capacities per kg of dry solid are approximately 130, 110, and 280 kg CO_2 for the UFS, FAS, and BHCS, respectively. The carbonation reaction exhibited a stationary phase after 60 min because of the formation of a calcium-depleted SiO_2 rim on the surface of particle. This would strongly block the reactive surface sites, thereby inhibiting the further release of calcium ions from the particle.

The CO_2 pressure in the autoclave reactor could vary between two various conditions: a normal condition (700 psig) and a supercritical condition (1300 psig). The conversion of alkaline solid wastes in supercritical CO_2 was found to be slightly less than that in normal CO_2 [35], due to the inhibition of $CaCO_3$ crystal growth under this supercritical pressure. However, in general, a high pressure increases the rate of the carbonation reaction, which precludes the formation of $CaCO_3$ crystals and inhibits the reaction. In the supercritical conditions (e.g.,

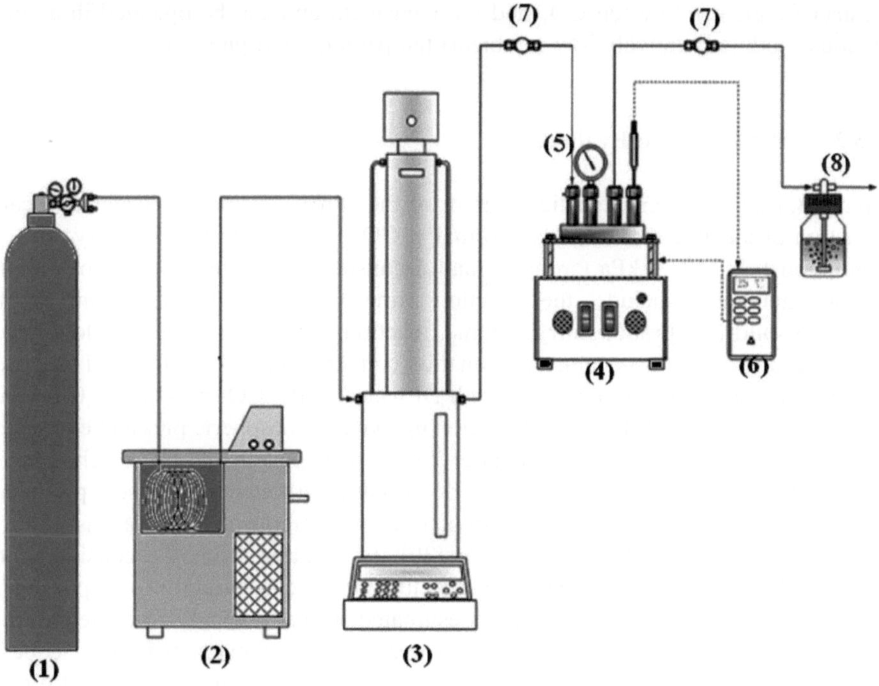

Fig. 8.3 Schematic diagram of the experimental setup of autoclave reactor for the carbonation of alkaline solid wastes (*1*) gas cylinder; (*2*) circulating bath; (*3*) syringe pump; (*4*) magnetic stirrer and heater; (*5*) autoclave reactor; (*6*) thermo couple and pH analyzer; (*7*) needle valve; and (*8*) vent to hood

1300 psig), the CO_2 fluid has a relatively higher solubility of liquid and lower dynamic viscosity of gas. The conversion of BHCS increases due to its superior CO_2 solubility at higher pressure and temperatures. The temperature significantly influenced the carbonation conversion of BHCS, with increasing temperatures resulting in higher conversions.

8.3.2 Slurry Reactor

A slurry reactor contains fine solid particles suspended in a liquid. It is frequently used in the chemical and/or biochemical industries due to its ability to enhance mass transfer [36]. The first application of the two-phase fluidization system was made by Winkler in 1922. The gas–liquid–particle three-phase system was then developed, which seems to be a more efficient tool for a chemical reactor [37]. Compared to that using an autoclave reactor, slurry reactor can provide a higher contact frequency between CO_2 and the feedstock, and can be operated in a continuous mode at relatively low (ambient) temperature and pressure.

8.3.2.1 Performance Evaluation

Figure 8.4 shows the schematic diagram of the slurry reactor for accelerated carbonation of alkaline solid wastes. Normally, CO_2 is injected into the slurry reactor continuously at 101.3 kPa (ambient) and a constant flow rate. The key operation factors generally include the reaction time (or hydraulic retention time), liquid-to-solid (L/S) ratio, slurry volume, reaction temperature, and gas flow rate.

In the slurry reactor system, a carbonation conversion of 53.8% for steel slag can be achieved at an L/S ratio of 20:1 (with DI water) and a CO_2 pressure of 0.1 MPa at 25 °C for 60 min [38]. The carbonation uptake at atmospheric pressure exceeded 80% of the reference 56-day value after 7 days of exposure and had reached 95% within 28 days. Although the atmospheric pressure carbonation treatment provided a slightly lower carbonation conversion, it was accomplished with a gas of low pressure and dilute CO_2 concentration [39]. In addition, with the consideration given to the grinding process and blowers, the energy consumption for capturing 1 ton of CO_2 using the slurry reactor is estimated to be 180.3 kWh (in the case of MSWI bottom ash) [40]. It thus suggests that an economical approach to sequestering CO_2 in alkaline solid wastes with minimum energy consumption and direct use of flue gas could be achieved.

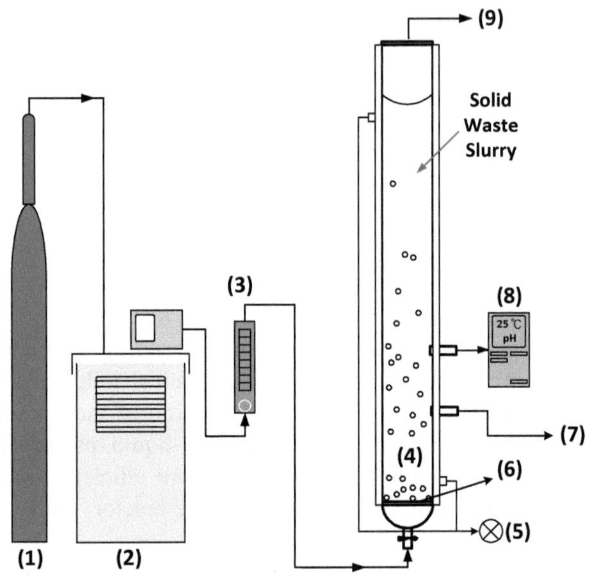

Fig. 8.4 Schematic diagram of the experimental setup of slurry reactor for accelerated carbonation of alkaline solid wastes (*1*) gas cylinder; (*2*) circulating bath; (*3*) rotameter; (*4*) slurry reactor; (*5*) heating jacket; (*6*) gas distributor; (*7*) sampling; (*8*) thermo couple and pH analyzer; and (*9*) vent to hood

8.3.2.2 Key Operating Factors

Since the mass transfer between the liquid and gas phases could be enhanced by increasing the CO_2 flow rate, the conversion of steel slag was observed to improve when the flow rate increased [11]. However, the conversion decreases moderately with further increasing flow rate in the slurry reactor [38]. This is because the channeling effect in the slurry reactor becomes significant at a high flow rate, resulting in a poor gas–liquid mass transfer rate and a decrease of carbonation conversion. Moreover, a higher CO_2 flow rate delivered a greater amount of CO_2 into the reactor. At a constant reaction time, the pH value drops more rapidly with a high flow rate than with a low flow rate, which is unfavorable to carbonation reaction. Therefore, the flow rate should be limited to a certain value that is able to make the slurry reactor in a fluidization mode.

The carbonation conversion of alkaline solid waste largely depends on the reaction temperature since the temperature would affect several system parameters, including the reaction kinetics, equilibrium, CO_2 dissolution, and calcium leaching simultaneously. In general, reaction rate would significantly increase with increasing reaction temperature, since the reaction rate constant exponentially increases with increasing temperature as expressed by the Arrhenius equation. However, because the carbonation is an exothermic reaction, an increase in temperature may also lead to a decrease in the equilibrium constant based on Le Chatelier's principle. Similarly, the rate of the CO_2 dissolution and the calcium ion leaching at higher temperature would decrease and increase, respectively.

It suggests that the maximum carbonation conversion of steel slag should be at a temperature of 60 °C [38]. In this case, the carbonation reaction can be generally categorized into two main regimes:

- Temperatures below 60 °C: The carbonation conversion increased with increasing reaction temperature due to the higher leaching rate of calcium ions. The $CaCO_3$ crystallization reaction was thus accelerated at higher temperature in this regime (30–60 °C).
- Temperatures above 60 °C: Boiling in the slurry reactor was accompanied by the low dissolution of CO_2, resulting in a decrease in the carbonation conversion. In this regime (60–80 °C), the CO_2 solubility is likely to be the key factor affecting the carbonation conversion.

The liquid-to-solid (L/S) ratio represents the weight ratio of water solution to alkaline solid wastes. It suggests that the L/S ratio should be greater than 5 mL/g to achieve a better carbonation of alkaline solid wastes [40]. Otherwise, the significant amounts of foreign ions would dissolve in the slurry, resulting in a competitive reaction and lowering the rate of calcium dissolution. It was noted that a maximum carbonation conversion of steel slag achieved in the slurry reactor was 89% at an L/S ratio of 20. When the L/S ratio was lower than 20, the slurry could not mix well in the slurry reactor, leading to poor mass transfer between liquid and solid phases. However, when the L/S ratio was higher than 20, a lower leaching concentration of calcium ions in solution occurred since the amount of solid waste per unit volume of liquid solution was lower.

8.3.3 High-Gravity Carbonation (HiGCarb)

A rotating packed-bed (RPB) reactor is designed to enhance mass transfer between gas and liquid phases via high centrifugal acceleration. RPB reactor can provide a mean acceleration of up to 1000 times greater than the force of gravity, thereby leading to the formation of thin liquid films and tiny liquid droplets (microscale to nanoscale) [41–47]. With this feature, the mass transfer between gas (CO_2) and aqueous solution in an RPB reactor can be significantly improved, resulting in a high capture efficiency within a short contact period. Moreover, using an RPB reactor can intensify the mixing of microscopic amounts of material by molecular diffusion (so-called micromixing efficiency) due to the formation of thin film flow on the packings. As a result, the excellent micromixing abilities of the RPB reactor can be applied for numerous applications of contact between two different phases (also refer to Chap. 17 for details), such as

- Absorption: for example, CO_2 capture through the use of absorbents (e.g., alkanolamine, piperazine, and NaOH).
- Distillation: for example, petrochemical process.
- Stripping: for example, ammonia stripping.
- Ozonation: for example, water disinfection.
- Biodiesel production: for example, transesterification.

In the case of accelerated carbonation using an RPB reactor, it involves in a three-phase mixing (i.e., gas, liquid, and solid phases) system. Since the aqueous carbonation of alkaline solid wastes was believed to be a diffusion-controlled reaction, an RPB reactor was introduced to improve the mass transfer rate among phases due to its high centrifugal forces and effective micromixing ability. This approach using an RPB reactor for accelerated carbonation of solid wastes is so-called high-gravity carbonation (HiGCarb) process [10, 13, 14, 48, 49], as shown in Fig. 8.5. The reactants are presumed to be intensively mixed within a short time, typically $\sim$1 min, in the packed-bed rotator while the carbonation reaction proceeds quickly.

8.3.3.1 Batch Operation

Figure 8.6 shows the schematic diagram of experimental setup for HiGCarb process. The slurry of alkaline solid waste is constantly stirred in all experiments. In the experiments, CO_2 gas stream flows inward from the outer edge of the RPB under a pressure driving force, while the slurry of alkaline solid waste is pumped

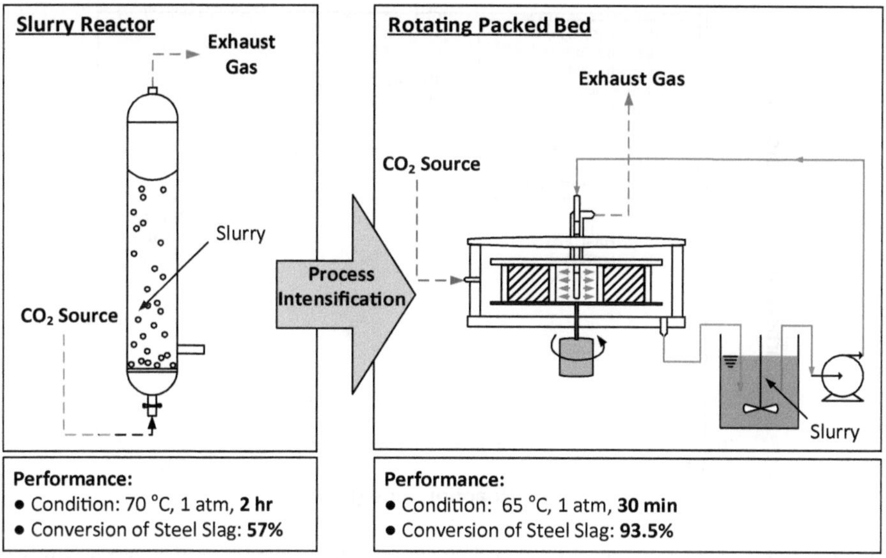

Fig. 8.5 Schematic diagram of process intensification via rotating packed-bed reactor for accelerated carbonation, i.e., high-gravity carbonation (HiGCarb) process

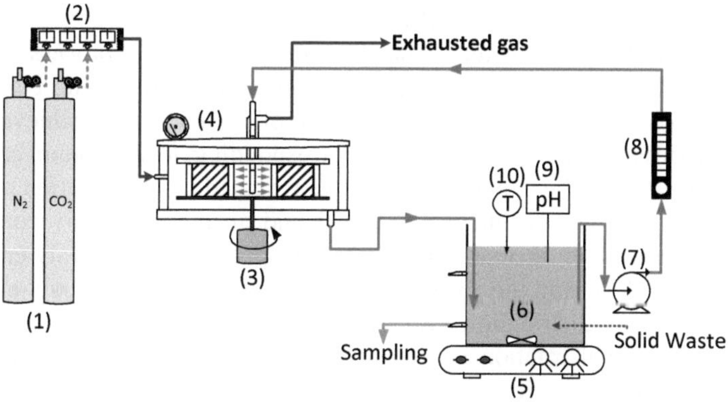

Fig. 8.6 Schematic diagram of the experimental setup for high-gravity carbonation (HiGCarb) process (*1*) gas cylinder; (2) mass controller; (*3*) rotor; (*4*) high-gravity RPB reactor; (*5*) stirring and heating machine; (*6*) slurry storage tank; (*7*) pump; (*8*) rotameter; (*9*) pH meter; and (*10*) thermometer

from the repository to the inner edge of the RPB. The slurry moves outward and leaves from the outer edge of the RPB under centrifugal force. As shown in Fig. 8.5, both CO_2 and slurry are mixed countercurrently within the RPB, in which CO_2 is dissolved in the liquid phase and reacts with the reactive species dissociated

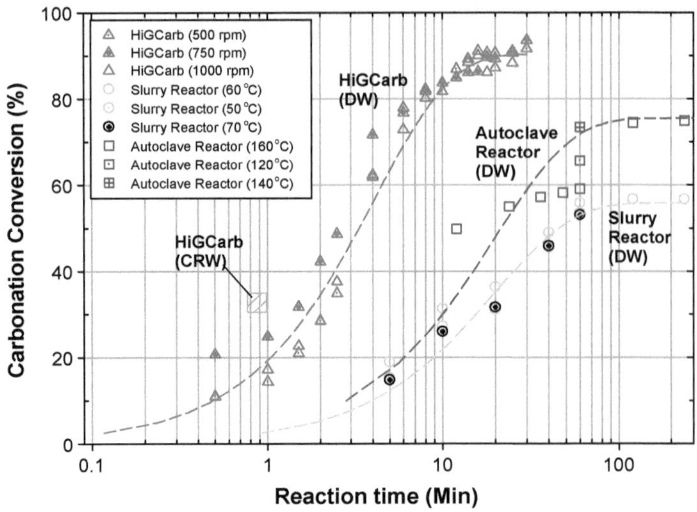

Fig. 8.7 Effect of reaction time on carbonation conversion of steel slag using autoclave reactor [35], slurry reactor [38], and high-gravity carbonation (HiGCarb) with cold-rolling wastewater [13] and DI water [49]

from the slag. The slurry is discharged from the bottom of the RPB, whereas the exhausted CO_2 streams are expelled from the top.

Chang et al. [49] conducted a batch-modulus HiGCarb process under low temperature (i.e., less than 70 °C) at ambient pressure. A maximum basic oxygen furnace slag (BOFS) carbonation conversion of 94% can be achieved by the HiGCarb process at 65 °C for 30 min, corresponding to a capture capacity of 290 kg CO_2/t-BOFS. The experimental results revealed a rapid reaction rate during the first 5 min at a gas-to-slurry ratio of 20 and an approximately constant rate thereafter. In general, an increase in the rotation speed would not only enhance the mass transfer of CO_2 from the gas phase to the liquid phase, but would also reduce the residence time distribution (RTD) of the slurry within the packed bed. Therefore, an optimal rotation speed should be determined to balance the mass transfer and reaction kinetics.

Figure 8.7 shows the comparison of carbonation conversion for steel slag using autoclave reactor, slurry reactor, and HiGCarb process with cold-rolling wastewater (CRW) and DI water. The HiGCarb process evidenced a relatively higher CO_2 removal efficiency and carbonation conversion of alkaline solid waste with a short reaction time than those in an autoclave or a slurry reactor [13, 34, 48, 49]. It was also found that carbonation conversion of steel slags could be significantly enhanced by using wastewater (i.e., CRW) in the HiGCarb process [13].

8.3.3.2 Continuous Field Operation

For continuous operation using the HiGCarb process, CO_2 fixation and waste treatment at a blast furnace plant in steel industry are performed using the BOFS. In this operation tests, the packed bed is in horizontal rotation with a mean diameter and height of 46.5 and 19.9 cm, respectively. The maximal rotation speed of the packed bed is designed as 900 rpm, to provide a centrifugal acceleration of up to 2065 m/s^2 (about 210 g). The CO_2 source with an average concentration of 30 $\pm$ 2% was supplied directly from a hot-stove furnace at the steel industry; no capture or concentrated processes were required in advance.

High CO_2 removal efficiency can be achieved with a retention time of less than 1 min under ambient temperature and pressure conditions [13]. A maximal CO_2 removal of 97% was achieved at a G/L ratio of 40, with a capture capacity of 165 kg CO_2 per day, as shown in Fig. 8.8. The CO_2 removal efficiency moderately increases as the rotation speed increases up to 500 rpm. This is due to the reduction of mass transfer resistance (i.e., reduced liquid film thickness) by increasing the rotation speed within this range, which is favorable to the carbonation reaction [13].

In addition, the increase of slurry flow rate can also improve the liquid-side mass transfer. In the HiGCarb process, a higher slurry flow rate brings a faster initial radial velocity of slurry droplets and causes a relatively thinner liquid film and a higher relative velocity between the slurry and the CO_2, thereby enhancing the mass transfer between the liquid and gas and intensifying the micromixing. However, it

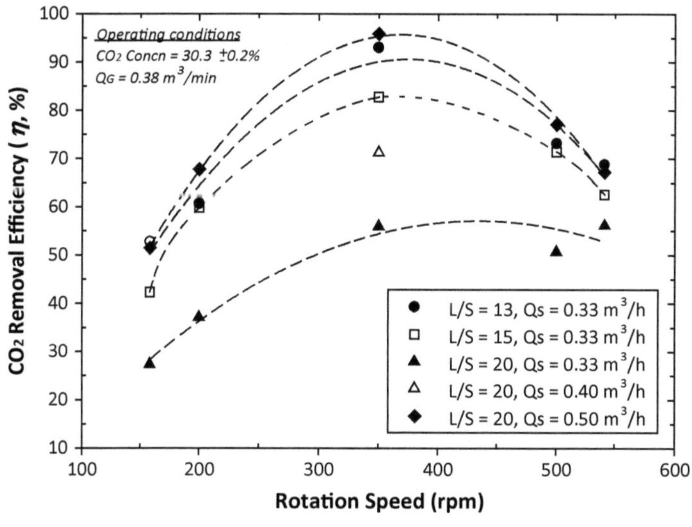

Fig. 8.8 Influence of rotation speed, liquid-to-solid (L/S) ratio, and slurry flow rate on CO_2 removal efficiency in hot-stove gas using high-gravity carbonation process. Reprinted with the permission from ref. [50]. Copyright 2015 American Chemical Society

also results in a lower RTD within the packing zone of the RPB reactor at higher slurry flow rates. This implies that the overall mass transfer resistance of carbonation reaction in the HiGCarb process might be mainly led on the liquid-phase side according to the two-film theory. Furthermore, the amount of calcium left in solid waste decreased as the calcium rapidly dissolved into solution, thereby resulting in a diminished concentration gradient needed to maintain the dissolution rate.

Similar observations can be found by adjusting other operating parameters. At higher gas flow rates, an increase in gas–liquid contact area and a reduction in gas-phase mass transfer resistance occur. However, the CO_2 removal efficiency was observed to decrease at a higher gas flow rate (i.e., a higher G/L ratio). It is confirmed that the high-gravity process is a liquid-side mass transfer controlled reaction. This suggests that both the rotation speed and G/L ratio should be the key factors for scale-up design of the HiGCarb process.

8.3.4 Ultrasonic Carbonation

Since the carbonation reaction is regarded as diffusion controlled (i.e., mass transfer limited), intensification of mass transfer efficiency among phases was essential to improve CO_2 capture capacity and reduce energy consumption and operating costs. A slurry reactor incorporated with ultrasound vibration has been introduced to accelerate the precipitation rate of calcium carbonate via ultrasonic irradiation [9, 15, 16, 51]. The results indicate that the efficiency of physical mixing, particle breakdown, and removal of passivation layers increased with sound waves with frequencies in the range of 16–100 kHz. Therefore, a better conversion can be achieved in a shorter time compared to that without ultrasound; for instance, the carbonation conversion of combustion ashes increased from 27 to 83% with ultrasound for 40 min [51].

8.3.5 Biologically Enhanced Carbonation

8.3.5.1 Microorganism

Microorganisms are the microscopic living organism, such as bacteria, protozoa, and algae. The anaerobic and aerobic bacteria can positively promote carbonation by enhancing the solubility and dissolution of minerals [52]. The absorption of cations by a cell wall (net negative) can increase the cation concentration within the cell, which can be used to facilitate mineral precipitation. The "carbonation ponds" or "basins" with a high alkalinity are designed to use a natural cyanobacteria-dominated consortium for the photoautotrophs to thrive and precipitate carbonates [53]. This process also can serve as silicon sinks by the formation of mineralized cell walls (i.e., frustule) from [$SiO_2 \cdot nH_2O$] in the bioreactor.

Another approach is to utilize algae and cyanobacteria to perform photosynthesis in the presence of sunlight. The sunlight is used as energy by proteins (so-called reaction centers) to convert gaseous CO_2 into biosubstance (i.e., chemical energy) [52, 53]. The reaction centers (proteins) contain green chlorophyll pigments, which can strip electrons from suitable substances such as water to produce water. This part of mechanism is called light reactions. Then, the CO_2 can be used to convert into carbohydrate molecules, such as sugar, where the chemical energy is stored in. This part of mechanism is called light-independent (dark) reactions, or Calvin–Benson cycle, as shown in Eq. (8.10).

$$3\,CO_2 + 9\,ATP + 6\,NADPH + 6\,H^+ \rightarrow C_3H_6O_3 - \text{phosphate} + 9\,ADP + 8\,P_i + 6\,NADP^+ + 3\,H_2O \tag{8.10}$$

Usually, plants can convert light energy into chemical energy with a photosynthetic efficiency of 3–6% [54]. For comparison, solar panels can convert light energy into electric energy at an efficiency of 6–20% for large-scale production and >40% in the laboratory scale.

8.3.5.2 Enzymatic Carbonation

Typically, the supply of CO_2 is limiting the rate of carbonation due to the slow kinetics of CO_2 dissolution. Enzymatic carbonation relying on carbonic anhydrase (CA) could be an effective approach to overcoming the above process hurdle, even in an industrial environment, or when utilized in an open-air environment. The CA family can catalyze the rapid interconversion of gaseous CO_2 and water to form bicarbonate (HCO_3^-) and protons (H^+), as shown in Eq. (8.11). There are at least five distinct CA families (i.e., α-CA, β-CA, γ-CA, δ-CA, and ε-CA), where these families exhibit no significant amino acid sequence similarity. The catalytic rates of various CA enzyme families typically range between 10^4 and 10^6 reactions per second [55].

$$CO_{2(g)} + H_2O_{(aq)} \xrightarrow{\text{Carbonic Anhydrase}} H_2CO_{3(aq)} \rightarrow HCO^-_{3(aq)} + H^+_{(aq)} \tag{8.11}$$

It is noted that the CA enzyme could enhance the carbonation of Ca- and Mg-bearing materials [56, 57]. For example, a CA enzyme-based membrane system was able to remove 90% of the CO_2 supplied [58]. However, the instability and high material costs of CA are still the challenges for practical and large-scale applications [28].

8.4 Integrated Carbonation with Brine (Wastewater) Treatment

Brine or wastewater is a saline-based waste solution (e.g., total dissolved solid is generally more than 50,000 mg/L) produced from industrial procedures, such as oil and natural gas extraction (known as oil-field brines). It can be used as the liquid agents in the carbonation reaction [12, 59–61]. Most wastewater treatments use chemicals as a neutralizing agent to adjust pH of wastewater and enhance metal precipitation. However, the use of chemicals carries with it a high economic and environmental cost because it is a "resource" and not a "residue."

The wastewater or brine solution can be used for accelerated carbonation process to enhance (1) the rate of calcium ion leaching from solid wastes, or (2) gaseous CO_2 dissolution into the liquid phase. The total dissolved solid (TDS) in wastewater is generally higher than that in tap water due to high concentrations of anions and cations. The criteria of wastewater and/or brine property for carbonation are illustrated as follows [12, 61]:

- A pH of the solution over 9.0 is favored to the precipitation of carbonate because CO_3^{2-} dominates under a basic condition.
- The selected solution should contain neither bicarbonate nor carbonate ions.

Under the appropriate conditions, CO_2 would dissolve in the brine solution to initiate a series of reactions that ultimately lead to the bonding of carbonate ions to various metal cations inherent in brine or wastewater to form carbonate precipitates, such as calcite ($CaCO_3$), magnesite ($MgCO_3$), and dolomite ($CaMg(CO_3)_2$).

8.4.1 Enhanced Calcium Leaching for Carbonation

The carbonation reaction can be enhanced by coupling with brine solution and/or wastewater, such as metalworking wastewater. Pan et al. [48] found that the leaching concentration of calcium ions in metalworking wastewater was higher than that in DI water, thereby resulting in a greater carbonation reaction rate and higher CO_2 capture capacity. Figure 8.9 shows the leaching concentration of calcium ions in DI water and metalworking wastewater (e.g., CRW) for different particle sizes of steel slag (i.e., BOFS). The leaching rate and capacity of calcium ions from steel slag in metalworking wastewater were greater than that in tap water. The measured pH values of metalworking wastewater ranged from 11.20 to 11.87. A maximum Ca^{2+} concentration of 2600 mg/L was measured in the alkaline wastewater with a particle size of steel slag less than 125 μm. The leaching of metal ions from the solid waste into solution would be higher with smaller particle size.

The high concentration of Na^+ and Cl^- in wastewater might accelerate the leaching behavior of Ca-bearing phases in solid wastes. The presence of inorganic ionic species in solution, such as sodium and chloride, can promote the dissolution

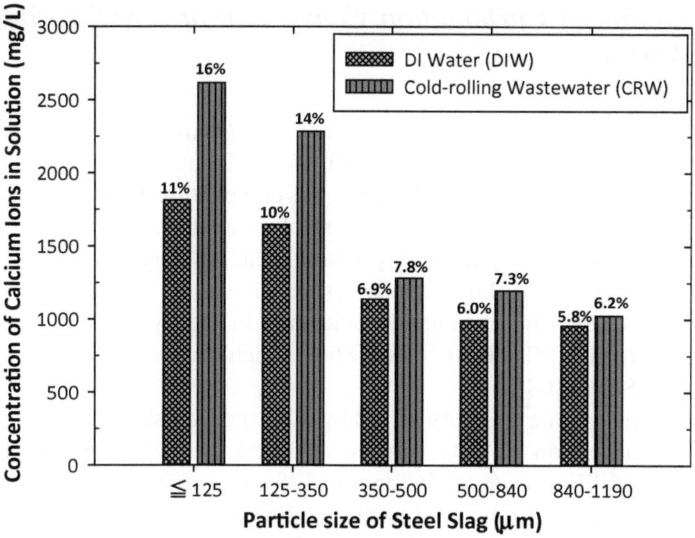

Fig. 8.9 Leaching concentrations of calcium ion in deionized (DI) water and cold-rolling wastewater under different particle sizes of steel slag for 90 min. Percentages shown on each bar represent the fraction of leaching concentration in totally leaching concentration

of silicate-bearing minerals due to the formation of surface complexes, leading to the reductive (and oxidative) dissolution of minerals [62–64]. O'Connor et al. [65] found that the addition of the sodium chloride can improve the extent of the carbonation reaction. Similarly, the use of the caustic alkali metal hydroxide can enhance the carbonation of metal silicates via the following mechanism [66]:

$$3CaSiO_{3(s)} + \underline{\mathbf{Na}}OH_{(aq)} + CO_{2(g)} \rightarrow CaCO_{3(s)} + \underline{\mathbf{Na}}Ca_2Si_3O_8(OH)_{(s)} \quad (8.12)$$

Jo et al. [67] found that Ca leaching from ordinary Portland cement (OPC) in aqueous solution and the hydration of C–S–H into $Ca(OH)_2$ in the OPC were enhanced by the presence of NaCl. In the presence of 0.2–2.0 M NaCl, the carbonation of OPC can thus be enhanced by 7.5–18.4%, in terms of the amount of $CaCO_3$ precipitation, which might be attributed to the increase in C–S–H dissolution by interaction with chloride ions. This suggests that a greater reactivity of the calcium-bearing complex toward chloride than other ions was found, thereby resulting in greater dissolution rates of minerals in the presence of chloride. Likewise, Nyambura et al. [68] suggest that the carbonation conversion of solid wastes is higher in the brine/solid residues system, compared to the water/solid residues.

8.4.2 Enhanced Carbonation Conversion of Alkaline Solid Wastes

In general, the carbonation conversion in the brine/solid residues system is higher than that in the water/solid residues [11, 50, 68]. O'Connor et al. [65] found that the addition of sodium chloride can enhance the carbonation reaction. Similarly, the concentrations of sodium and chloride in alkaline wastewater are generally higher than those in DI water. As a result, Nyambura et al. [68] found that the carbonation conversion in the brine/solid residues system should be higher than that in the water/solid residues. The carbonation efficiency in the fly ash (FA)/brine system under a pressure of 4 MPa at 30 °C was 86%, which was superior to the water/brine system (i.e., 68.2%) [68].

A maximum capture capacity of 283 g CO_2 per kg slag, corresponding to a carbonation conversion of 89%, can be achieved in alkaline wastewater with a reaction time of 120 min at 25 °C and an L/S ratio of 20 [11]. The highest carbonation conversion was observed in the alkaline wastewater system, compared to the neutralized effluent water and DI water. This might be attributed to more CO_2 dissolved in the liquid phase under higher pH, thereby enhancing the carbonation reaction. It thus suggests that the carbonation reaction of steel slag was enhanced more in the alkaline wastewater system than in the DI water system.

8.4.3 Improved Water Quality After Carbonation

Although the concentration of Fe, Sr, and Ba decreases in the solution, iron hydroxide, strontium carbonates, and barium carbonates were not detected by XRD, most likely, due to the small amounts of iron (i.e., 434 ppm) and barium (i.e., 917 ppm) present in the fresh brines [61]. After carbonation, the concentrations of TIC and magnesium in wastewater increased significantly, while the concentrations of sodium and calcium were found to be decreased as well as the pH value. The decrease of the calcium concentration after carbonation might be caused by the formation of the calcium carbonate in the course of carbonation. In addition, the concentrations of potassium, iron, and nitrate remained roughly constant for wastewater. It was noted that the CO_2 was absorbed into the slurry, thus leading to the formation of calcium carbonate precipitations and the increase of the TIC concentration in the slurry [11].

8.5 Carbonation Curing Process

The use of supplementary cementitious materials (SCMs) is a sustainable practice to make environmentally friendly cement and concrete industries. Aside from using CO_2 in direct carbonation of solid wastes for the production of SCMs, the CO_2 can be used in the curing process during the preparation of cement mortar, so-called "carbonation curing" process. Carbonation is a common type of attack in mortars and concretes, where CO_2 penetrates into the mortar and reacts with Ca^{2+} from calcium hydroxide (CH), calcium silicate hydrate (C–S–H), and the hydrated calcium aluminates and ferroaluminates. As a result, various types of calcium carbonate ($CaCO_3$), silica gel, and hydrated Al/Fe oxides can be formed during carbonation curing.

The carbonation curing process can significantly improve the mechanical properties and durability of the cementitious materials. This early-age carbonation in curing is the reaction between calcium silicates or early hydration products with carbon dioxide producing a hybrid binder structure of calcium silicate hydrates and calcium carbonates [69]. With the carbonation curing process, the initial porosity of mortar reduces since the volume of the formed $CaCO_3$ is about 12% greater than the volume of the original CH phase. This would result in a higher compressive strength at early ages of carbonation, thereby preventing further CO_2 diffusion and carbonation attack [70, 71]. Ghouleh et al. [72] suggest that the improvement on the early strength could be attributed to carbonation of the γ-C_2S component in solid waste, as shown in Eq. (8.13).

$$2(2CaO \cdot SiO_2) + CO_2 + 3\,H_2O \rightarrow 3\,CaO \cdot 2SiO_2 \cdot 3H_2O + CaCO_3 \tag{8.13}$$

However, if the porosity of the carbonated pastes is sufficiently high to permit constant CO_2 diffusion, the CH will be depleted and the interlayer calcium from C–S–H will react with CO_2. Polymerization of the silicate chains in C–S–H occurs afterward and causes a volumetric shrinkage, resulting in material cracks and coarse porosity [70].

Several studies have been carried out to evaluate the effect of carbonation curing on the properties of blended cement [73] and concrete masonry [69]. Table 8.2 presents the performance of carbonation curing for cements or concretes with various types of solid wastes. Wu et al. [74] demonstrated that, after carbonation curing, the blended cement with steel slags exhibited higher strength and eligible soundness. With carbonation curing process, early hydration products in cement are converted to a crystalline microstructure, and subsequent hydration transforms amorphous carbonates into more crystalline calcite [69].

In addition, Mo et al. [73] showed that carbonation curing using higher pressure CO_2 could result in faster strength development due to the rapid penetration of CO_2 and carbonation of blended cement. Salman et al. [78] found that a higher strength for cylindrical specimens (d = 55 mm, h = 75 mm) made out of compressed continuous casting slag could be achieved within 6 h by the carbonation curing

Table 8.2 Carbonation curing process for various types of alkaline solid wastes as construction cements in cement or concrete

Waste sources	Carbonation curing condition (pure CO_2)[a]	Feedstock and carbonation process	Substitution ratio (wt%)	Performance and product properties[c]	References
Steel slag and blast furnace slag	0.54 MPa at 88 °C for 2 h, RH 12.6%	Water content of 12–13%, with 2% additives	Steel slag: blast furnace slag = 41:57	f_c = 2.1 MPa (uncarbonated) and 18.4 MPa (carbonated) CO_2 gain = 17.7%	[74]
Steel slag	0.54 MPa at 20 °C for 2 h	Particle size = 5–24 μm; L/S ratio = 0.125; eligible autoclave soundness	n.r.[b]	f_c = 4.65 MPa (uncarbonated); 44.97 MPa (carbonated) Capacity: 107.9 CO_2 g/kg specimen Porosity: 21.8% → 13.3%	[75]
Steel slag	0.32 MPa at 86 °C for 2 h	Particle size < 75 μm; optimal water content = 19%	50	f_c = 4.0–6.3 MPa at 3-d; 18.5–24.6 MPa at 28-d CaO_f = 3.9% → 1.1%	[76]
Steel slag	0.15 MPa for 2 h	Particle size (SS) = 212 μm; w/s ratio = 0.15; compacted by 16 MPa; sealed hydration for 28 d	100	f_c = 0.7 MPa (uncarbonated curing); 109.3 MPa (carbonation curing) CO_2 uptake of SS = 13.94 wt%	[72]
Argon oxygen decarburization slag	0.8 MPa at 80 °C for 5 h	Normal curing (5% CO_2 at 1 atm, 22 °C, RH 80%) followed by carbonation; particle size < 500 μm; w/s ratio = 0.15 (wt)	n.r.[b]	f_c = 40.7 MPa CO_2 uptake of AODS = 6.72 wt%. Porosity: 35.7% (3 d) → 33.9% (3 weeks) Precipitation of calcite, aragonite and vaterite	[77]
Hydrated fly ash	0.1 MPa at 23 °C for 14 d, RH 55-65%	Normal curing (23 °C, RH 98%) followed by carbonation; water to binder = 0.4	Fly ash: MgO: cement = 1:1:3	f_c = 62.9 MPa (uncarbonated curing); 67.2 MPa (3 h carbonation curing); 90.0 MPa (14 d carbonation curing) Pore volumes reduced by 32.1% (3 h)	[73]

[a] *RH* relative humidity
[b] *n.r.* not report
[c] f_c = compressive strength of mortar; f_t = tensile strength of mortar; f_f = flexural strength of mortar

process at higher temperature and pressure (i.e., 60 °C and 1.2 MPa CO_2), compared to 4-week carbonation at atmospheric conditions (22 °C, fixed humidity 80, 5% CO_2 at atmosphere pressure). This would help to reduce the costs of the carbonation curing process.

References

1. Azdarpour A, Asadullah M, Mohammadian E, Hamidi H, Junin R, Karaei MA (2015) A review on carbon dioxide mineral carbonation through pH-swing process. Chem Eng J 279:615–630. doi:10.1016/j.cej.2015.05.064
2. Said A, Mattila HP, Jarvinen M, Zevenhoven R (2013) Production of precipitated calcium carbonate (PCC) from steelmaking slag for fixation of CO_2. Appl Energy 112:765–771. doi:10.1016/j.apenergy.2012.12.042
3. Pérez-Moreno S, Gázquez M, Bolívar J (2015) CO_2 sequestration by indirect carbonation of artificial gypsum generated in the manufacture of titanium dioxide pigments. Chem Eng J 262:737–746
4. Seifritz W (1990) CO_2 disposal by means of silicates. Nature 345(6275):486
5. Bobicki ER, Liu Q, Xu Z, Zeng H (2012) Carbon capture and storage using alkaline industrial wastes. Prog Energy Combust Sci 38(2):302–320. doi:10.1016/j.pecs.2011.11.002
6. Lackner KS (2003) A guide to CO_2 sequestration. Science 300(5626):1677–1678
7. Li W, Li B, Bai Z (2009) Electrolysis and heat pretreatment methods to promote CO_2 sequestration by mineral carbonation. Chem Eng Res Des 87(2):210–215. doi:10.1016/j.cherd.2008.08.001
8. Maroto-Valer MM, Fauth DJ, Kuchta ME, Zhang Y, Andrésen JM (2005) Activation of magnesium rich minerals as carbonation feedstock materials for CO_2 sequestration. Fuel Process Technol 86(14–15):1627–1645. doi:10.1016/j.fuproc.2005.01.017
9. Rao A, Anthony EJ, Manovic V (2008) Sonochemical treatment of FBC ash: a study of the reaction mechanism and performance of synthetic sorbents. Fuel 87(10–11):1927–1933. doi:10.1016/j.fuel.2007.11.007
10. Chang EE, Chen T-L, Pan S-Y, Chen Y-H, Chiang P-C (2013) Kinetic modeling on CO_2 capture using basic oxygen furnace slag coupled with cold-rolling wastewater in a rotating packed bed. J Hazard Mater 260:937–946. doi:10.1016/j.jhazmat.2013.06.052
11. Chang EE, Chiu A-C, Pan S-Y, Chen Y-H, Tan C-S, Chiang P-C (2013) Carbonation of basic oxygen furnace slag with metalworking wastewater in a slurry reactor. Int J Greenhouse Gas Control 12:382–389. doi:10.1016/j.ijggc.2012.11.026
12. Druckenmiller ML, Maroto-Valer MM (2005) Carbon sequestration using brine of adjusted pH to form mineral carbonates. Fuel Process Technol 86(14–15):1599–1614. doi:10.1016/j.fuproc.2005.01.007
13. Pan SY, Chiang PC, Chen YH, Tan CS, Chang EE (2013) Ex Situ CO_2 capture by carbonation of steelmaking slag coupled with metalworking wastewater in a rotating packed bed. Environ Sci Technol 47(7):3308–3315. doi:10.1021/es304975y
14. Pan S-Y, Chiang P-C, Chen Y-H, Tan C-S, Chang EE (2014) Kinetics of carbonation reaction of basic oxygen furnace slags in a rotating packed bed using the surface coverage model: maximization of carbonation conversion. Appl Energy 113:267–276. doi:10.1016/j.apenergy.2013.07.035
15. Santos RM, Ceulemans P, Van Gerven T (2012) Synthesis of pure aragonite by sonochemical mineral carbonation. Chem Eng Res Des 90(6):715–725. doi:10.1016/j.cherd.2011.11.022
16. Santos RM, François D, Mertens G, Elsen J, Van Gerven T (2013) Ultrasound-intensified mineral carbonation. Appl Therm Eng 57(1–2):154–163. doi:10.1016/j.applthermaleng.2012.03.035

17. Santos RM, Chiang YW, Elsen J, Van Gerven T (2014) Distinguishing between carbonate and non-carbonate precipitates from the carbonation of calcium-containing organic acid leachates. Hydrometallurgy 147–148:90–94. doi:10.1016/j.hydromet.2014.05.001
18. Jo H, Park S-H, Jang Y-N, Chae S-C, Lee P-K, Jo HY (2014) Metal extraction and indirect mineral carbonation of waste cement material using ammonium salt solutions. Chem Eng J 254:313–323. doi:10.1016/j.cej.2014.05.129
19. Dri M, Sanna A, Maroto-Valer MM (2013) Dissolution of steel slag and recycled concrete aggregate in ammonium bisulphate for CO_2 mineral carbonation. Fuel Process Technol 113:114–122. doi:10.1016/j.fuproc.2013.03.034
20. Kakizawa M, Yamasaki A, Yanagisawa Y (2001) A new CO_2 disposal process via artificial weathering of calcium silicate accelerated by acetic acid. Energy 26:341–354
21. Sun Y, Yao MS, Zhang JP, Yang G (2011) Indirect CO_2 mineral sequestration by steelmaking slag with NH4Cl as leaching solution. Chem Eng J 173:437–445
22. Eloneva S, Teir S, Salminen J, Fogelholm CJ, Zevenhoven R (2008) Fixation of CO_2 by carbonating calcium derived from blast furnace slag. Energy 33(9):1461–1467
23. Eloneva S, Puheloinen EM, Kanerva J, Ekroos A, Zevenhoven R, Fogelholm CJ (2010) Co-utilisation of CO_2 and steelmaking slags for production of pure $CaCO_3$-legislative issues. J Clean Prod 18:1833–1839
24. Teir S, Eloneva S, Fogelholm C-J, Zevenhoven R (2007) Dissolution of steelmaking slags in acetic acid for precipitated calcium carbonate production. Energy 32(4):528–539. doi:10.1016/j.energy.2006.06.023
25. Kodama S, Nishimoto T, Yamamoto N, Yogo K, Yamada K (2008) Development of a new pH-swing CO_2 mineralization process with a recyclable reaction solution. Energy 33(5): 776–784. doi:10.1016/j.energy.2008.01.005
26. Park A, Fan L (2004) mineral sequestration: physically activated dissolution of serpentine and pH swing process. Chem Eng Sci 59(22–23):5241–5247. doi:10.1016/j.ces.2004.09.008
27. New Energy and Industrial Technology Development Organization (NEDO) (1993) A survey on the current state of research and development for techniques to recover and sequester CO_2 from thermal power plants (II) (Karyoku hatsuden puranto karano CO_2 kaishu sisutemu ni kansuru chosa (II)) The Institute of Applied Energy (IAE). NEDO, Tokyo
28. Sanna A, Uibu M, Caramanna G, Kuusik R, Maroto-Valer MM (2014) A review of mineral carbonation technologies to sequester CO_2. Chem Soc Rev 43(23):8049–8080. doi:10.1039/c4cs00035h
29. Jung S, Wang LP, Dodbiba G, Fujita T (2014) Two-step accelerated mineral carbonation and decomposition analysis for the reduction of CO(2) emission in the eco-industrial parks. J Environ Sci (China) 26(7):1411–1422. doi:10.1016/j.jes.2014.05.006
30. Nduagu E, Bjorklof T, Fagerlund J, Makila E, Salonen J, Geerlings H, Zevenhoven R (2012) Production of magnesium hydroxide from magnesium silicate for the purposes of CO_2 minralization—Part 2: Mg extraction modeling and applcation to different Mg silicate rocks. Miner Eng 30(1):87–94
31. Costa G, Baciocchi R, Polettini A, Pomi R, Hills CD, Carey PJ (2007) Current status and perspectives of accelerated carbonation processes on municipal waste combustion residues. Environ Monit Assess 135(1–3):55–75. doi:10.1007/s10661-007-9704-4
32. Huntzinger DN, Gierke JS, Kawatra SK, Eisele TC, Sutter LL (2009) Carbon dioxide sequestration in cement kiln dust through mineral carbonation. Environ Sci Technol 43(6):1986–1992
33. Haug TA, Kleiv RA, Munz IA (2010) Investigating dissolution of mechanically activated olivine for carbonation purposes. Appl Geochem 25(10):1547–1563. doi:10.1016/j.apgeochem.2010.08.005
34. Pan S-Y, Chang EE, Chiang P-C (2012) CO_2 capture by accelerated carbonation of alkaline wastes: a review on its principles and applications. Aerosol Air Qual Res 12:770–791. doi:10.4209/aaqr.2012.06.0149

35. Chang EE, Pan S-Y, Chen Y-H, Chu H-W, Wang C-F, Chiang P-C (2011) CO_2 sequestration by carbonation of steelmaking slags in an autoclave reactor. J Hazard Mater 195:107–114. doi:10.1016/j.jhazmat.2011.08.006
36. Alper E, Wichtendahl B, Deckwer WD (1980) Gas absorption mechanism in catalytic slurry reactor. Chem Eng Sci 35:217–222
37. Ostergaard K (1968) Gas-liquid-particle operations in chemical reaction engineering. Adv Chem Eng 7:71–137. doi:10.1016/s0065-2377(08)60081-2
38. Chang EE, Chen CH, Chen YH, Pan SY, Chiang PC (2011) Performance evaluation for carbonation of steel-making slags in a slurry reactor. J Hazard Mater 186(1):558–564. doi:10.1016/j.jhazmat.2010.11.038
39. Monkman S, Shao Y, Shi C (2009) Carbonated ladle slag fines for carbon uptake and sand substitute. J Mater Civ Eng 21:657–665. doi:10.1061//asce/0899-1561/2009/21:11/657
40. Chang EE, Pan SY, Yang L, Chen YH, Kim H, Chiang PC (2015) Accelerated carbonation using municipal solid waste incinerator bottom ash and cold-rolling wastewater: performance evaluation and reaction kinetics. Waste Manag 43:283–292. doi:10.1016/j.wasman.2015.05.001
41. Lin C, Chen B (2008) Characteristics of cross-flow rotating packed beds. J Ind Eng Chem 14(3):322–327. doi:10.1016/j.jiec.2008.01.004
42. Wang M (2004) Controlling factors and mechanism of preparing needlelike $CaCO_3$ under high-gravity environment. Powder Technol 142(2–3):166–174. doi:10.1016/j.powtec.2004.05.003
43. Kelleher T, Fair JR (1996) Distillation studies in a high-gravity contactor. Ind Eng Chem Res 35:4646–4655
44. Yu C-H, Huang C-H, Tan C-S (2012) A review of co_2 capture by absorption and adsorption. Aerosol Air Qual Res 12:745–769. doi:10.4209/aaqr.2012.05.0132
45. Chen YH, Huang YH, Lin RH, Shang NC (2010) A continuous-flow biodiesel production process using a rotating packed bed. Bioresour Technol 101(2):668–673. doi:10.1016/j.biortech.2009.08.081
46. Chen YH, Chang CY, Su WL, Chen CC, Chiu CY, Yu YH, Chiang PC, Chiang SIM (2004) Modeling ozone contacting process in a rotating packed bed. Ind Eng Chem Res 43(1):228–236
47. Cheng H-H, Tan C-S (2011) Removal of CO_2 from indoor air by alkanolamine in a rotating packed bed. Sep Purif Technol 82:156–166. doi:10.1016/j.seppur.2011.09.004
48. Pan SY, Chiang PC, Chen YH, Chen CD, Lin HY, Chang EE (2013) Systematic approach to determination of maximum achievable capture capacity via leaching and carbonation processes for alkaline steelmaking wastes in a rotating packed bed. Environ Sci Technol 47(23):13677–13685. doi:10.1021/es403323x
49. Chang EE, Pan SY, Chen YH, Tan CS, Chiang PC (2012) Accelerated carbonation of steelmaking slags in a high-gravity rotating packed bed. J Hazard Mater 227–228:97–106. doi:10.1016/j.jhazmat.2012.05.021
50. Pan SY, Chen YH, Chen CD, Shen AL, Lin M, Chiang PC (2015) High-gravity carbonation process for enhancing CO_2 fixation and utilization exemplified by the steelmaking industry. Environ Sci Technol 49(20):12380–12387. doi:10.1021/acs.est.5b02210
51. Rao A, Anthony EJ, Jia L, Macchi A (2007) Carbonation of FBC ash by sonochemical treatment. Fuel 86(16):2603–2615. doi:10.1016/j.fuel.2007.02.004
52. Huang C-H, Tan C-S (2014) A review: CO_2 utilization. Aerosol Air Qual Res 14:480–499. doi:10.4209/aaqr.2013.10.0326
53. McCutcheon J, Power IM, Harrison AL, Dipple GM, Southam G (2014) A greenhouse-scale photosynthetic microbial bioreactor for carbon sequestration in magnesium carbonate minerals. Environ Sci Technol 48(16):9142–9151. doi:10.1021/es500344s
54. Miyamoto K (2009) Biological energy production. In: Renewable biological systems for alternative sustainable energy production (FAO Agricultural Services Bulletin - 128). Food and Agriculture Organization of the United Nations

55. Lindskog S (1997) Structure and mechanism of carbonic anhydrase. Pharmacol Ther 74(1): 1–20
56. Li W, Chen W-S, Zhou P-P, Zhu S-L, Yu L-J (2013) Influence of initial calcium ion concentration on the precipitation and crystal morphology of calcium carbonate induced by bacterial carbonic anhydrase. Chem Eng J 218:65–72. doi:10.1016/j.cej.2012.12.034
57. Li W, Chen W-S, Zhou P-P, Yu L-J (2013) Influence of enzyme concentration on bio-sequestration of CO_2 in carbonate form using bacterial carbonic anhydrase. Chem Eng J 232:149–156. doi:10.1016/j.cej.2013.07.069
58. Figueroa JD, Fout T, Plasynski S, McIlvried H, Srivastava RD (2008) Advances in CO_2 capture technology—The U.S. Department of Energy's carbon sequestration program. Int J Greenhouse Gas Control 2(1):9–20. doi:10.1016/s1750-5836(07)00094-1
59. Uibu M, Kuusik R (2009) Mineral trapping of CO_2 via oil shale ash aqueous carbonation: controlling mechanism of process rate and development of continuous-flow reactor system. Oil Shale 26(1):40. doi:10.3176/oil.2009.1.06
60. Liu Q, Mercedes Maroto-Valer M (2011) Investigation of the effect of brine composition and pH buffer on CO_2-brine sequestration. Energy Procedia 4:4503–4507. doi:10.1016/j.egypro.2011.02.406
61. Liu Q, Maroto-Valer MM (2012) Studies of pH buffer systems to promote carbonate formation for CO_2 sequestration in brines. Fuel Process Technol 98:6–13. doi:10.1016/j.fuproc.2012.01.023
62. Brady PV (1996) Physics and chemistry of mineral surfces. Chemistry and physics of surfaces and minerals. CRC Press LLC, Florida
63. Hangx SJT (2005) Behaviour of the CO_2-H_2O system and preliminary mineralisation model and experiments. Subfurface mineralisation: rate of CO_2 mineralisation and geomechanical effects on host and seal formations. Shell International Exploration and Production (leader CATO WP 4.1)
64. Krevor SCM, Lackner KS (2011) Enhancing serpentine dissolution kinetics for mineral carbon dioxide sequestration. Int J Greenhouse Gas Control 5(4):1073–1080. doi:10.1016/j.ijggc.2011.01.006
65. O' Connor WK, Dahlin DC, Rush GE, Gerdemann SJ, Penner LR, Nilsen DN (2005) Aqueous mineral carbonation: Mineral availability, pretreatment, reaction parameters, and process studies. Albany Research Center (ARC), U.S.A
66. Beard JS, Blencoe JG, Anovitz LM, Palmer DA (2004) Carbonation of metal silicates for long-term CO_2 sequestration. Canada Patent
67. Jo H, Jang Y-N, Young Jo H (2012) Influence of NaCl on mineral carbonation of CO_2 using cement material in aqueous solutions. Chem Eng Sci 80:232–241. doi:10.1016/j.ces.2012.06.034
68. Nyambura MG, Mugera GW, Felicia PL, Gathura NP (2011) Carbonation of brine impacted fractionated coal fly ash: implications for CO_2 sequestration. J Environ Manage 92(3):655–664. doi:10.1016/j.jenvman.2010.10.008
69. El-Hassan H, Shao Y (2015) Early carbonation curing of concrete masonry units with Portland limestone cement. Cement Concr Compos 62:168–177. doi:10.1016/j.cemconcomp.2015.07.004
70. Borges PHR, Costa JO, Milestone NB, Lynsdale CJ, Streatfield RE (2010) Carbonation of CH and C-S–H in composite cement pastes containing high amounts of BFS. Cem Concr Res 40:284–292. doi:10.1016/j.cemconres.2009.10.020
71. Zhang F, Mo L, Deng M (2015) Mechanical strength and microstructure of mortars prepared with MgO-CaO-Fly ash-Portland cement blends after accelerated carbonation. J Chin Ceram Soc 43(8):1–8. doi:10.14062/j.issn.0454-5648.2015.08.01
72. Ghouleh Z, Guthrie RIL, Shao Y (2015) High-strength KOBM steel slag binder activated by carbonation. Constr Build Mater 99:175–183. doi:10.1016/j.conbuildmat.2015.09.028
73. Mo L, Zhang F, Deng M (2015) Effects of carbonation treatment on the properties of hydrated fly ash-MgO-Portland cement blends. Constr Build Mater 96:147–154. doi:10.1016/j.conbuildmat.2015.07.193

74. Wu HZ, Chang J, Pan ZZ, Cheng X (2009) Carbonate steelmaking slag to manufacture building materials. Adv Mater Res 79–82:1943–1946. doi:10.4028/www.scientific.net/AMR.79-82.1943
75. Wu HZ, Chang J, Pan ZZ, Cheng X (2011) Effects of carbonation on steel slag products. Adv Mater Res 177:485–488. doi:10.4028/www.scientific.net/AMR.177.485
76. Liang XJ, Ye ZM, Chang J (2012) Early hydration activity of composite with carbonated steel slag. J Chin Ceram Soc 40(2):228–233 (in Chinese)
77. Salman M, Cizer Ö, Pontikes Y, Santos RM, Snellings R, Vandewalle L, Blanpain B, Van Balen K (2014) Effect of accelerated carbonation on AOD stainless steel slag for its valorisation as a CO_2-sequestering construction material. Chem Eng J 246:39–52. doi:10.1016/j.cej.2014.02.051
78. Salman M, Cizer Ö, Pontikes Y, vandewalle L, blanpain B, Van Balen K (2013) Carbonation potential of continuous casting stainless steel slag. Paper presented at the Acelerated Carbonation for Environmental and Material Engineering KU Leuven, Belgium

Chapter 9
System Analysis

Abstract Accelerated carbonation may include a large amount of energy requirements and the high costs. The challenges encountered are to accelerate the reaction and utilize the heat of reaction to maximize the overall capture capacity and minimize energy demand, as well as to determine the optimal operating conditions. As a result, significant technological breakthroughs, such as reactor design, waste-to-resource supply chain, and system optimization, are needed before deployment can be considered. In addition, it is important to evaluate greenhouse gas emission reduction by geographic region, engineering performance, environmental benefit, and economic viability for decision making. Therefore, in this chapter, the methodologies and tools pertaining to the geospatial analysis, response surface analysis, life cycle assessment, and cost–benefit analysis are illustrated.

9.1 Geospatial Analysis

Geospatial analysis (or so-called spatial analysis) is an approach to applying statistical analysis and other analytic techniques to evaluate data that contains a geographic or spatial aspect. Geospatial data has great potential to contribute in many societal applications, such as climate change, smart city, disease surveillance and response, and transportation. Such data analysis typically employs computer techniques and software capable of rendering maps processing spatial data and terrestrial or geographic datasets, such as the use of geographic information system (GIS) and geomatics. The spatial statistics primarily result from observation rather than experimentation.

9.1.1 Geographic Information System (GIS)

GIS can provide a variety of capabilities designed to capture, store, manipulate, analyze, manage, and present all types of geographic data, and utilize geospatial analysis in a variety of contexts, operations, and applications. It was originally

P.-C. Chiang and S.-Y. Pan, *Carbon Dioxide Mineralization and Utilization*, DOI 10.1007/978-981-10-3268-4_9

developed for issues in the environmental and life science, particularly ecology, geology ad epidemiology. Nowadays, it has extended to almost all areas including natural resources, defense, intelligence, utilities, social science, medicine, public safety, disaster risk reduction and management (DRRM), and adaptation to climate change.

The advanced GIS-based operations and facilities can augment the core tools employed in spatial analysis throughout the analytical process, such as exploration of data and identification of patterns and relationships. Several functions of geospatial analysis using GIS are briefly illustrated as follows:

- **Vector-based GIS operations**: related to map overlay (combining two or more map layers to the predefined rules), simple buffering (identifying regions of a map within a specified distance of one or more features) and similar basic operations.
- **Raster-based GIS operations**: related to a range of actions applied to the grid cells of one or more maps, involving filtering and/or algebraic operations. It is frequently used in the environmental sciences and remote sensing.
- **Surface analysis**: related to examining the properties of physical surface, such as gradient, aspect, and visibility, involving the analysis of surface-like data "fields."
- **GIS-based network analysis**: related to evaluating the properties of natural and man-made networks to examine the behavior of flows within and around such networks. It may be used to address a wide range of practical issues, such as route selection, facility location, and transportation research.
- **Geovisualization**: related to the creations and manipulations of images, maps, diagrams, charts, 3D views, and their associated tabular datasets.

Figure 9.1 shows an example of GIS-based analysis for the source of alkaline solid wastes in Taiwan. In Taiwan, since different types of alkaline solid wastes may not be generated closely in distance, the centralized accelerated carbonation process may not be feasible due to the institutional, regulatory, technological, and financial barriers. However, it is possible to integrate the carbonation process directly by stacks to avoid additional long-distance piping for CO_2. By utilizing the GIS-based analysis, the best location to obtain the cost-effective solution could be identified throughout the facility location analysis. The best solution also should be determined by considering the aspects of environment (impacts) and economic (costs and benefits), which are illustrated in Sects. 2 and 3, respectively, in this chapter.

9.1.2 Big Data Analysis and Data Visualization

Geospatial data is typically a large data set. Therefore, big data analysis for geospatial data has been received considerable attention to allow users to analyze huge amounts of geospatial data. Big data analysis can be considered as "structured and unstructured datasets with massive volumes that cannot be easily captured,

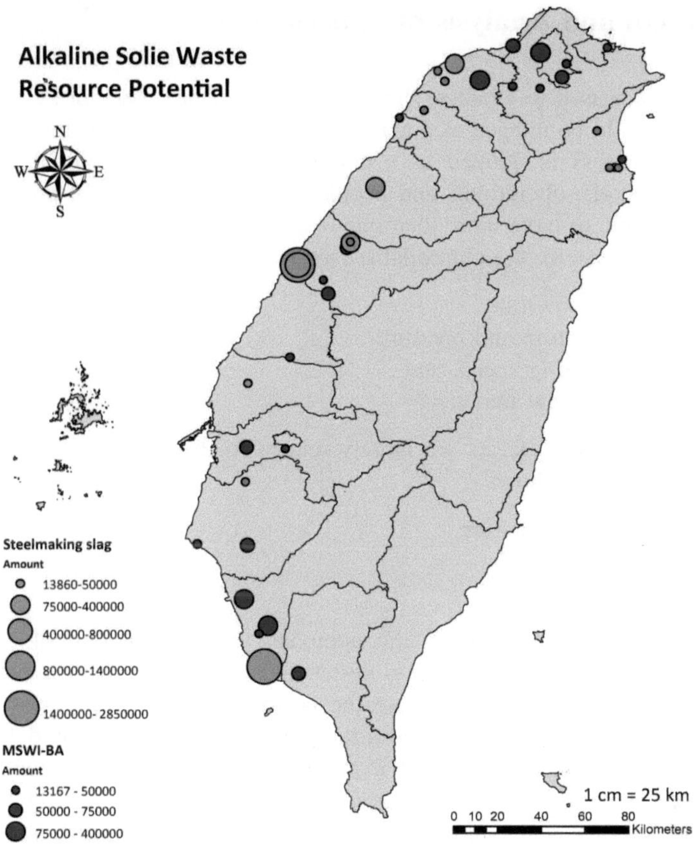

Fig. 9.1 GIS-based analysis for the sources of alkaline solid wastes in Taiwan

stored, manipulated, analyzed, managed, and presented by traditional hardware, software, and database technologies [1].” According to the estimation by United Nations Initiative on Global Geospatial Information Management (UN-GGIM), 2.5 quintillion bytes of data is being generated every day, and a large portion of the data is location-aware [2].

Visualizations are widely acknowledged as a part of analysis process, where we could explore the data and build hypotheses during this process. Interactive and exploratory visualization environments help at the early stages to identify the real information with big data. So far, there are still several challenges and issues for future directions, including the following:

- Efficient representation and modeling for geospatial big data (e.g., real-time data handling).
- Analyzing, mining, and visualizing geospatial data for decision support.
- Quality assurance and control of geospatial big data from new sources.

9.2 Design and Analysis of Experiments

Experiments are used to evaluate the performance of processes or systems. Every experiments could be imaged as a general model of a process system, as shown in Fig. 9.2. A process is defined as a combination of input materials, resources, machines, methods, operations, and people. As a result, a typical process system could be possible to have more than one response or output.

In a process system, the controllable factors may include the following:

- Gas and liquid flow rate,
- Reaction temperature and pressure,
- Rotating and stirring speed, and
- Formulation of input materials.

The uncontrollable factors are largely related to the surrounding environment and people, such as follows:

- Environmental temperature,
- Humidity, and
- Different operators.

Figure 9.3 illustrates a systematic approach to determination of good engineering practice (GEP) for a process system via an Engineering–Environmental–Economic (3E) analysis. To evaluate the performance of a process system, the system boundary and scope of work for the process should be clearly defined in the beginning. Through a well-designed experimental set, both the time and costs for performance evaluation would be minimized. Along the way, the key parameters for reactor design (such as dimensions and operating factors) can also be determined by the prediction model for further scale-up of the process. On the other hand, both the environmental impacts and economic costs of a process should be quantified and estimated via life cycle assessment and technoeconomic assessment, respectively. By collecting and analyzing the above information, the GEP can be

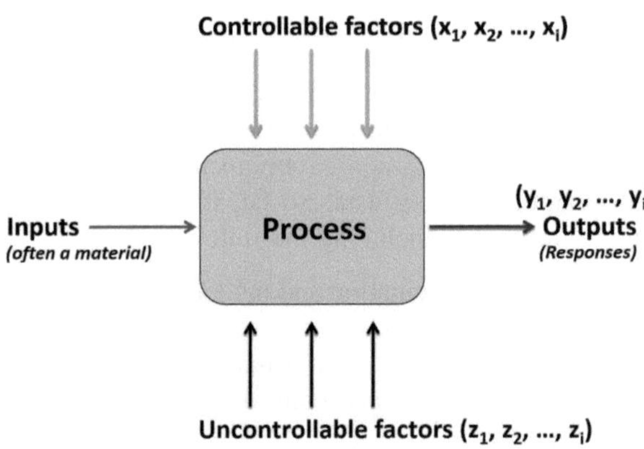

Fig. 9.2 Conceptual diagram of a process system for design and analysis of experiments

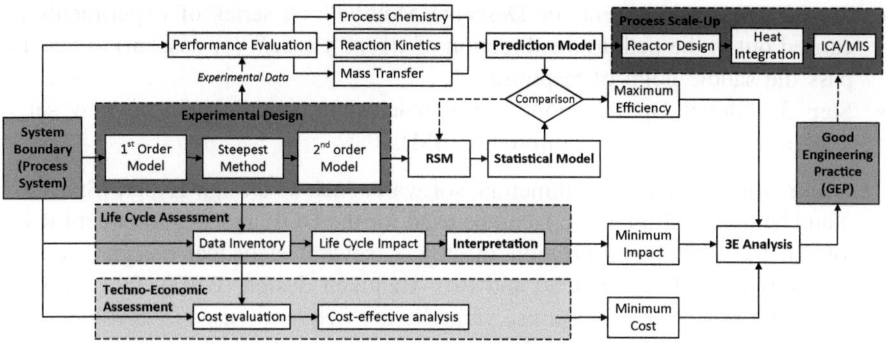

Fig. 9.3 Systematic approach to determination of good engineering practice (GEP) for a process system via an Engineering–Environmental–Economic (3E) analysis

systematically obtained via the 3E analysis to facilitate the process scale-up and demonstration toward commercialization.

In this section, the concept of experimental design and its subsequent analysis of experiments, i.e., via a response surface methodology (RSM), are illustrated. In the field of chemical engineering, "experimental design," "regression analysis," and "operations research" are three major courses to become a qualified process system engineer.

9.2.1 Experimental Design

Design of experiments (DOE) refers to a structured and organized methods used to determine the interactions between factors affecting a process and the output of that process statistically. It is usually carried out through observance of forced changes made methodically as directed by mathematically systematic tables. Four principles of DOE are as follows:

- Control: to know the potential effect of factors that are not being studied.
- Randomization: subjects to treatments to even out effects that we cannot control.
- Replicate: to avoid unrepresentative sample if only a single test was carried out.
- Block: to reduce the effects of identifiable attributes of the subjects, preferably controlled.

For a typical application of experimental design for a chemical process, the procedure of characterizing the key operating factors of a process is as follows:

- Step 1 "Local Optimization": In this stage, experiments with randomly determined ranges for different operating factors (n) are typically conducted by a 2^n factorial design. The analysis of variance (ANOVA) is a powerful tool to determine the most significant variable(s) affecting response(s).

- Step 2 "Steepest Ascent (or Descent) Method": A series of experiments are carried out in the path to the direction of steepest ascent (or descent) to reach or pass the saddle point of response.
- Step 3 "Global Optimization": After reached the saddle point, a new set of experiments are designed through (RSM).

A numerous number of commercial software, such as Design Expert (Stat-Ease, Inc.) and Minitab (Minitab Inc.), can be used for the DOE and its subsequent RSM. To obtain the global optimization in step 3, several statistical designs, such as central composite design (CCD) and Box–Behnken design (BBD), for building a second-order model of the response variables are frequently used in developing the response surface methodology (RSM). These designs should be sufficient to fit a quadratic model, that is, one containing squared terms and products of two factors.

- Central composite design (CCD)
  - A factorial design in various numeric (or nominal) factors, each having two levels, usually coded as −1 and +1.
  - A set of center points is carried out at the medians of the values used in the factorial portion, which is often replicated to improve the precision of the experiment.
  - A set of axial points will take on value both below and above the median of the two factorial levels.
- Box–Behnken design (BBD)
  - Each factor is placed at one of three equally spaced values, usually coded as −1, 0, and +1.
  - The ratio of the number of experimental points to the number of coefficient in the quadratic model should keep in the range of 1.5 to 2.6.
  - The estimation variance should more or less depend only on the distance from the center, especially for the designs with 4 and 7 factors.

9.2.2 Response Surface Methodology

Response surface methodology (RSM) is considered as a collection of statistical design, empirical modeling methodologies, and numerical optimization techniques used to optimize response variables. It is developed based on the experimental design from the statistics point of view to examine the maximal response of system performance. With a proper design of experiments, the RSM can be applied to determine the operating conditions for maximal responses, e.g., CO_2 removal efficiency, from a statistical point of view. Based on the developed RSM, several candidates can be revealed to determine the maximum objective function, such as carbonation conversion of alkaline solid wastes. It thus can be introduced to evaluate the effects of the relating operational parameters, such as reaction time, temperature, and liquid-to-solid ratio on the carbonation conversions of alkaline solid wastes.

The response surface can be analyzed with various commercial software packages, such as Design Expert (Stat-Ease Inc., USA) and Minitab (Minitab Inc., USA). By employing the software, RSM can explore the relationships between several explanatory variables and one (or more) response variables by using a sequence of designed experiments. The experimental variables and responses can be analyzed by regression procedure using a second-order polynomial equation:

$$y = \beta_0 + \sum \beta_i x_i + \sum \beta_{ii} x_i^2 + \sum \beta_{ij} x_i x_j + \varepsilon \tag{9.1}$$

where y is the response variable; x_i and x_j are the independent variables; β_0, β_0, β_{ii}, and β_{ij} are incept, linear, quadratic, and interaction constant coefficients, respectively, and ε is the residual.

In the RSM, the least-squares estimation is used to determine the model parameters in an approximating polynomial equation. This polynomial equation, representing the response surface, can be a quadratic- or cubic-order model. The analysis of the obtained response surface is generally equivalent to the analysis of the actual system if the fitted surface is a satisfactory estimation of the true response function. From the statistical point of view, the significance of each individual coefficient term in the response surface model can be determined from the analysis of variance (ANOVA) using the p value and/or F value with 95% confidence level.

As indicated in Fig. 9.3, a well design of experiments is crucial to determine the optimal objective for a process system. In the real application of RSM, sometimes, the operating factor of the experiments (such as L/S ratio) determined as the optimum conditions might be peripheral. This means that the obtained surface model is not suitably used to predict the values outside the assigned experimental ranges.

9.2.3 *Multi-response Surface Optimization*

Multi-response surface (MRS) optimization is an approach to simultaneously dealing with several responses that are conflicting; for instance, to maximize the performance and efficiency of a process while minimizing the operating costs of the process. The MRS optimization often encounters several challenges, such as follows [3]:

- Correlation among multiple responses,
- Robustness measurement of multivariate process,
- Confliction among multiple goals,
- Prediction performance of the process model, and
- Reliability assessment for optimization results.

In the following part, we provide an example on accelerated carbonation process for CO_2 capture and utilization to illustrate the applications and constrains of MRS. Based on the response surfaces developed by the CCD method, the MRS can be used to balance the CO_2 removal efficiency (η) and energy consumption (ψ) for high-gravity carbonation process. The optimization design objectives are used to maximize the CO_2 removal efficiency while minimizing the energy consumption of the carbonation process, as expressed in Eqs. (9.2) and (9.3):

$$\text{Max (CO}_2\text{ removal efficiency)}: \eta = f(X_1, X_2, X_3, X_4) > 90\% \quad (9.2)$$

$$\text{Min (energy consumption)}: \psi = f(X_1, X_2, X, X_4) < 250\,\text{kWh per t-CO}_2 \quad (9.3)$$

The effects of four different operating factors (e.g., X_1, X_2, X_3, and X_4) on both η and ψ responses can be examined with the analysis of variance (ANOVA) table and visualized by a response surface model from a statistical point of view. The mathematical models of η and ψ responses, as formulated in terms of the operating factors with coded values, can be obtained from the RSM, as shown in Eqs. (9.4) and (9.5), respectively:

$$\eta(\%) = 39.1 + 10.7 * X_1 + 14.7 * X_3 - 15.8 * X_4 - 20.8 * X_1^2 + 136.3 * X_2^2 \quad (9.4)$$

$$\psi(\text{kWh}) = 432.2 - 32.6 * X_1 + 270.1 * X_2 + 21.0 * X_3 + 15.3 * X_4 + 38.6 * X_1X_3 - 35.1 * X_1X_4 - 59.1 * X_3X_4 + 110.6 * X_1^2 \quad (9.5)$$

The ANOVA results will indicate that the developed response models are significant if the p value was less than 0.05; for instance, the CO_2 removal efficiency was presented in Table 9.1.

Calculated by Eqs. (9.4) and (9.5), Fig. 9.4a, b present the 2D contour plots of CO_2 removal efficiency and energy consumption, respectively. These 2D plots can be also converted into 3D response surface plots.

Table 9.1 Analysis of variance (ANOVA) table of RSM for CO_2 removal efficiency

Source	SS	DF	MS	*F* value	*p* value (Prob > *F*)
Model	19597.1	5	3919.4	65.6	<0.0001
X_1: rotation speed	1664.4	1	1664.4	27.9	<0.0001
X_3: slurry flow rate	4099.9	1	4099.9	68.6	<0.0001
X_4: L/S ratio	3173.4	1	3173.4	53.1	<0.0001
X_1^2	2131.2	1	2131.2	35.7	<0.0001
X_2^2	6493.3	1	6493.3	108.7	<0.0001
Residual	2091.0	35	59.7		
Cor total	21688.1	40			

SS Sum of squares; *DF* degree of freedom; *MS* mean of square. RSM was performed using Design Expert statistical software (Stat-Ease Inc., Minneapolis, USA)

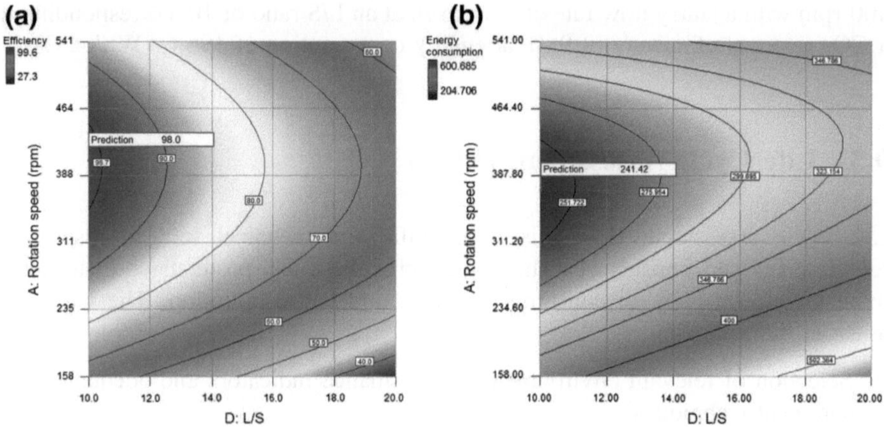

Fig. 9.4 2D contour plots of **a** CO_2 removal efficiency and **b** energy consumption, in terms of rotation speed and liquid-to-solid (L/S) ratio. Reprinted with the permission from Ref. [4]. Copyright 2015 American Chemical Society

Thus, the optimum ranges of the operating factors can be determined graphically by setting optimization objectives for each targeted response (e.g., CO_2 removal efficiency and energy consumption) and then creating an overlay contour that highlights an area of operability. As shown in Fig. 9.5, the optimal rotation speed and gas flow rate should be in the ranges of 259.2–410.2 rpm and 0.34–0.45 m^3/min, respectively. According to the MRS analysis, one of the optimum candidates within the above optimal ranges was estimated to be operated at a rotation speed of

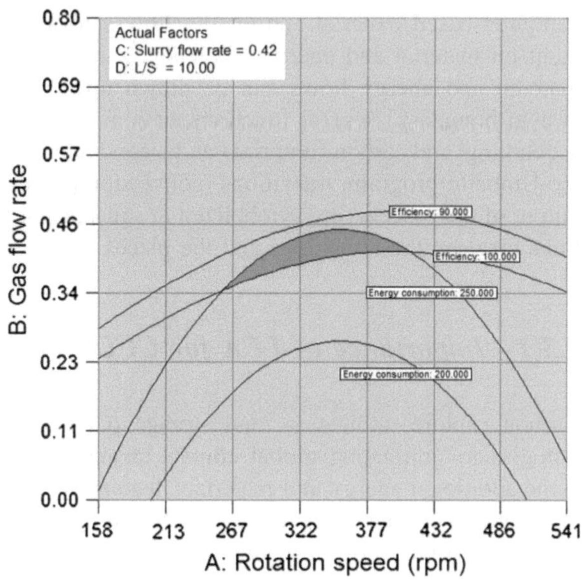

Fig. 9.5 Optimal operating conditions of high-gravity carbonation process in the cases of an L/S ratio of 10 and a slurry flow rate of 0.42 m^3/h. Reprinted with the permission from Ref. [4]. Copyright 2015 American Chemical Society

400 rpm with a slurry flow rate of 0.36 m^3/h at an L/S ratio of 10, corresponding to a CO_2 removal efficiency of 98% at energy consumption of 196.6 kWh/t-CO_2.

9.3 Life Cycle Assessment (LCA)

Life cycle assessment (LCA) is a scientific and technical tool to assess the requirements and impacts of technologies, processes, and products so as to determine their propensity to consume resources and cause pollution. The LCA can assist in the following:

- Selection of relevant environmental performance indicators and adequate measurement techniques.
- Identification of improvement opportunities for a product (service or technology) throughout its life cycle.
- Decision making in industry, and governmental and non-governmental organizations.
- Marketing opportunities for a product (service or technology) by using LCA data for ecolabeling, environmental product declaration, etc.

With the LCA, the environmental impacts of each input and output material and/or energy source for a process or product can be quantified by various software packages. Table 9.2 presents the notable feature of various software packages, such as SimaPro (Pré Consultants B.V., the Netherlands), Umberto (ifu hamburg, Germany), and GaBi (Thinkstep, Germany), associated with their available database.

Taking the Umberto software as an example, it is frequently used for computer-based material and energy flow analysis to achieve successful management on material and energy efficiency, and process optimization. A balance for material and energy flows can be drawn up in Umberto to analyze the defined system boundary. Several functions such as life cycle assessment, integrated cost accounting, and carbon footprint can be conducted in the framework of Umberto. In the Umberto program, transitions (conversion processes) and places (the input and output of the process) are symbolized in squares and circles, respectively. Then, the links between the transitions and the places can be established with arrows.

9.3.1 Importance of LCA for CCUS Technologies

Carbon capture, utilization, and storage (CCUS) are prominent emerging technologies for mitigating global climate change. The CCUS processes would consume additional energy and materials, leading to further emissions of CO_2 into the atmosphere. On the other hand, CCUS may increase other environmental impacts such as eutrophication or acidification due to the increase in the concentrations of

Table 9.2 Comparison of software for life cycle assessment (LCA)

Software	Company	Database	Notable features
SimaPro	Netherlands	Ecoinvent, ELCD, US-LCI, LCAfood, etc.	• Determine the impacts with statistical accuracy • Calculate carbon footprint of products and systems
Umberto	Germany	Ecoinvent, GaBi, etc.	• Support ISO compliant LCAs and Footprints • Flexibility in specifying processes • Update data directly using Excel
GaBi	Germany	GaBi	• Database documentation in accordance with International Life Cycle Database formatting • Collect data across every point of process
Quantis Suite	Switzerland	Predefined	• feature non-expert access capabilities • Result analysis at the inventory level
EarthSmart	USA	Ecoinvent, US-EI, Australian LCI, etc.	• Rapid creation of professional reports that can be updated almost instantaneously • Easy to upload data from SimaPro into the tool
DoITPro	Taiwan (ROC)	DoITPro	• Incorporate with Taiwan localized database • Support carbon footprint analysis

other pollutants. As a result, an LCA of the CCUS process is particularly important, where the effects of CCUS on changes in the environmental impacts should be systematically weighed and compared.

As proved by the results of LCA, CCUS can offer great potential of decreasing greenhouse gas (GHG) emissions. For instance, the GHG emissions from UK fossil fuel plant stations with CO_2 capture and storage can be reduced by 75–84% relative to a subcritical pulverized coal power plant [5]. In pulverized hard coal combustion, the energy penalty ranged from 5 to 18% with applications of various types of processes for CO_2 capture and storage [6].

9.3.2 Methodology

Life cycle assessment (LCA) is a systematic approach to estimating the performance and its associated environmental impacts of a product or a service during the life cycle. In other words, the purpose of LCA is to quantify the environmental impacts and benefits of technologies or services (e.g., GHG emissions and eutrophication) over their lifetimes using a "cradle to grave" approach [7]. The framework and structure of LCA have been proposed by the International Standards Organization (ISO). ISO 14000 is a family of standards related to environmental management, which are generally accepted by all stakeholders and the international community. The ISO standards provide principles, framework, methodological requirement, and guidelines for conducting LCA work, including the following:

- ISO 14040:2006: Environmental management—Life cycle assessment—Principles and framework;
- ISO 14041:1999: Environmental management—Life cycle assessment—Goal and scope definition and inventory analysis;
- ISO 14042:2000: Environmental management—Life cycle assessment—Life cycle impact assessment; and
- ISO 14043:2000: Environmental management—Life cycle assessment—Life cycle interpretation.

As shown in Fig. 9.6, according to ISO 14040, the framework for LCA includes the following:

- Definition of goal and scope,
- Life cycle inventory (LCI) analysis,
- Life cycle impact assessment (LCIA), and
- Interpretation.

In particular, in dealing with the energy issues by LCA, several key elements should be considered in the first stage of LCA, such as

- Energy penalty,
- Scale-up challenges,
- Non-climate environmental impacts,
- Uncertainty management,
- Policy-making needs, and
- Market effects.

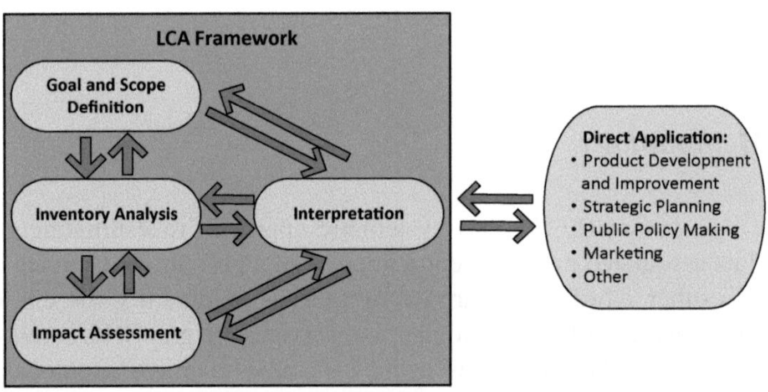

Fig. 9.6 Phases of life cycle assessment (LCA) according to ISO 14040

9.3.2.1 Goal and Scope

LCA can quantify the potential environmental impacts throughout the life cycle (i.e., cradle to grave) consideration of a product and/or process from raw material acquisition through production, use, and disposal. According to ISO 14040, several key items should be precisely defined in this step to ensure the validity of the results:

- Goal and objectives,
- Method,
- The function of the system,
- Functional unit (i.e., comparison basis),
- System boundary,
- Time and system options,
- Allocation procedures,
- Types of environmental impacts,
- Data requirements, and
- Assumptions and limitations.

The scope describes the depth of the study, showing that the purpose can be fulfilled with the actual extent of the limitations.

The functional unit (FU) is a key element of LCA which has to be clearly defined. The FU is a measure of the function of the studied system for providing a reference to which the inputs and outputs can be related, thereby enabling comparison of two essential different systems. For example, the FU for a paint process may be defined as the unit surface protected for 15 years. Therefore, a comparison of the environmental impacts of two different paint systems with the same functional unit is therefore possible. Typically, the FU for accelerated carbonation was set to be “1 kg CO_2 captured by process.”

In the scope phase, the system boundaries define the types of unit processes to be included in the LCA study based on a subjective choice. The features of system boundaries include the following.

- Boundaries between the technological system and nature (or ecosystem): Usually, it begins at the extraction point of raw materials and energy carriers from nature, while ends at the waste generation and/or heat production.
- Space (geographic area): In most LCA studies, geography plays a crucial role, e.g., infrastructures vary from one region to another. Moreover, the sensitivity of ecosystem to environmental impacts differs regionally.
- Time horizon: Basically, the LCA is performed to evaluate the present impacts and predict the future scenarios. Limitations to time horizon are included, such as types of technologies (or product) and pollutants life span.

9.3.2.2 Life Cycle Inventory (LCI)

In this step, a balance for material and energy flows is drawn up for calculating and analyzing of the investigated system. The data collection forms must be properly designed for optimal collection. Inventory analysis compiles an inventory of the relevant inputs and outputs of a product, service, and/or process. Normally, the life cycle inventory (LCI) data of materials and processes within the system boundary can be processed from the database in LCA software. Several guidelines are recommended for the LCI phase:

- If possible, allocation should be avoided. Otherwise, the inputs and outputs should be partitioned between its different functions (or products) in the approach to reflecting the underlying physical relationships between them.
- When the data situation is unclear, an assumption has to be made and its influence has to be investigated via sensitivity analysis in results.

The data collection is the most resource consuming part of the LCA. The inputs to the process system, including energy (e.g., electricity consumption) and materials (e.g., water, alkaline solid wastes, and CO_2 source), should be identified. The outputs comprise pollutant emissions and treated wastes, where the output solid residues generated from the process are generally assumed to be treated by landfill.

9.3.2.3 Life Cycle Impact Assessment (LCIA)

LCA could expand the database on environmental concerns beyond a single issue (global warming) to a wider range of environmental issue (acidification, ecotoxicity, human toxicity, etc.). In the phase of life cycle impact assessment (LCIA), the significance of potential environmental impacts is evaluated using the results from the LCI phase. Usually, three methods, i.e., EPS (Environmental Priority Strategies), ECO (Ecological scarcity), and ET (Environmental Theme), are used for LCIA. In a formal LCA, three steps are required to conduct an LCIA as presented in Fig. 9.7.

According to ISO 14040, several steps should be conducted during this phase:

- Category: selection of impact categories, category indicators, and characterization models (mandatory).
- Classification: Assignment of individual inventory parameters to impact categories (mandatory).
- Characterization: conversion of LCI results to common units within each impact category to aggregate the results into category indicator results (mandatory).
- Normalization: calculation of the magnitude of the category indicator relatively to reference information (optional).

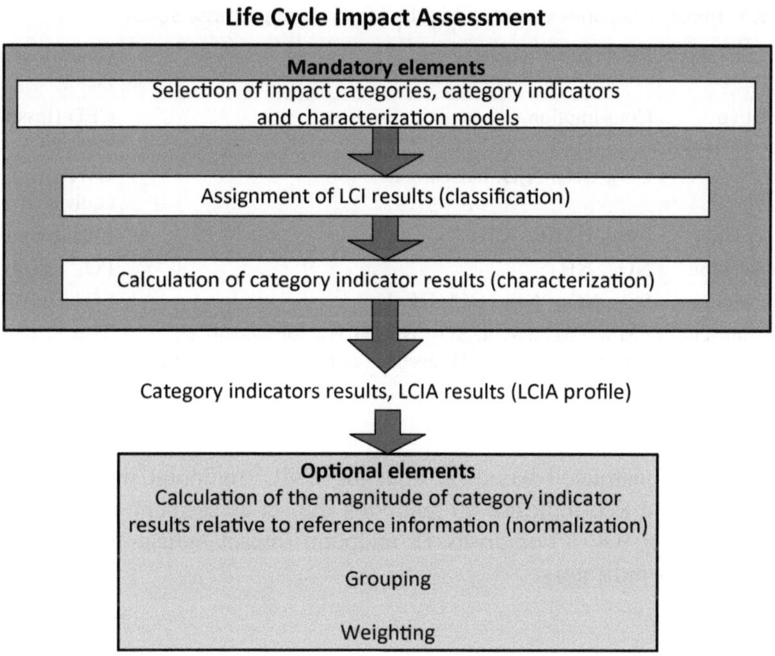

Fig. 9.7 Elements of life cycle impact assessment (LCIA) according to ISO 14040

- Weighting: conversion of different impact categories to a common unit by using factors based on value choices (optional).
- Grouping: assignment of impact categories into one or more groups sorted after geographic relevance, company priorities, etc. (optional).

In the LCA, the environmental impacts of a process can be assessed by (1) midpoint impact indicators or (2) endpoint impact indicators:

- Midpoint impact: A problem-oriented approach describing environment themes.
- Endpoint impact: A damage-oriented approach expressing a consistent and concise view of, typically, ecosystem quality, human health, and resource depletion.

As shown in Table 9.3, the various midpoint impacts can be normalized into endpoint impacts, such as the human health potential in the dimension unit of DALY (disability-adjusted life year). The environmental impacts can be evaluated and quantified using various types of valuation systems, such as ReCiPe, Eco-indicator 99, and IMPACT 2002+.

Table 9.3 Impact categories considered in the life cycle impact assessment

Impact category	Relevant parameters	Characterization factor
Cumulative energy demand	Consumption of energetic resources	CED (fossil and nuclear)
Global warming	CO_2, CH_4, N_2O, halocarbons	GWP100, CO_2-equivalent
Summer smog	NO_X, NMHC, CH_4	Ethene-equivalent
Eutrophication	NO_X, NH_3	PO_4^{3-}-equivalent
Acidification	SO_2, NO_X, NH_3, HCl, HF, H_2S	SO_2-equivalent
Health impacts	PM_{10}, $PM_{2.5}$, soot, SO_2, NO_X, CH_4, formaldehyde, benzene, B(a)P, PAH, arsenic, cadmium, dioxin, furan	Year of life lost (YOLL)

1. ReCiPe
   ReCiPe is constructed based on both the CML (midpoint impact assessment approach) and Eco-indicator 99 (endpoint impact assessment approach) [8]. As shown in Fig. 9.8, it comprises 18 midpoint impact indicators and three endpoint impact indicators.

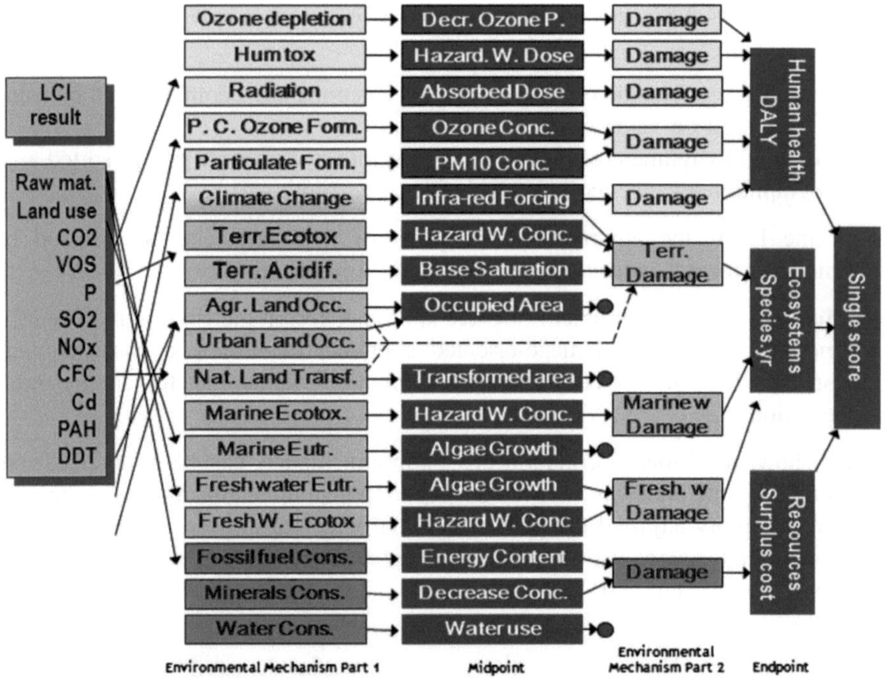

Fig. 9.8 Overall scheme of the ReCiPe framework. Reproduced from Ref. [8] by permission of Mark Goedkoop in PRé Consultants bv

The midpoint indicators in ReCiPe include the following:

- Decreased ozone potential,
- Hazardous waste dose,
- Absorbed dose,
- Ozone concentration,
- PM_{10} concentration,
- Infrared forcing,
- Hazardous waste concentration for terrestrial ecotoxicity,
- Base saturation,
- Occupied area,
- Transformed area,
- Hazardous waste concentration for marine ecotoxicity,
- Algae growth for marine eutrophication,
- Algae growth for fresh water eutrophication,
- Hazardous waste concentration for fresh water ecotoxicity,
- Energy content,
- Decreased concentration, and
- Water use.

Since the large number of midpoint indicators with a very abstract meaning is difficult to elucidate and interpret, the endpoint indicators with a more understandable meaning could be utilized for interpretation. The above midpoint indicators are further weighted and converted to three endpoint indicators, i.e., (1) human health potential, (2) ecosystem quality potential, and (3) ecosystem resources potential.

2. Eco-indicator 99
   Eco-indicator 99 is a damage-oriented method (i.e., endpoint) for life cycle impact assessment. The standard Eco-indicator values can be regarded a dimensionless figure [9]. In the Eco-indicator 99, the environmental impacts are defined with three types of damages:
   - Human health: the number and duration of diseases, and life year lost due to premature death from environment causes. This includes climate change, ozone layer depletion, carcinogenic effects, respiratory effects, and ionizing nuclear radiation.
   - Ecosystem quality: the effect on species diversity, especially for vascular plants and lower organisms. This includes ecotoxicity, acidification, eutrophication, and land-use.
   - Resources: the surplus energy needed in future to extract lower quality mineral and fossil resources. This includes depletion of agricultural and bulk resources as sand and gravel.
3. Impact 2002+
   IMPACT 2002+, adapted by Quantis, is a method of a combined midpoint/damage approach via several midpoint categories to several damage

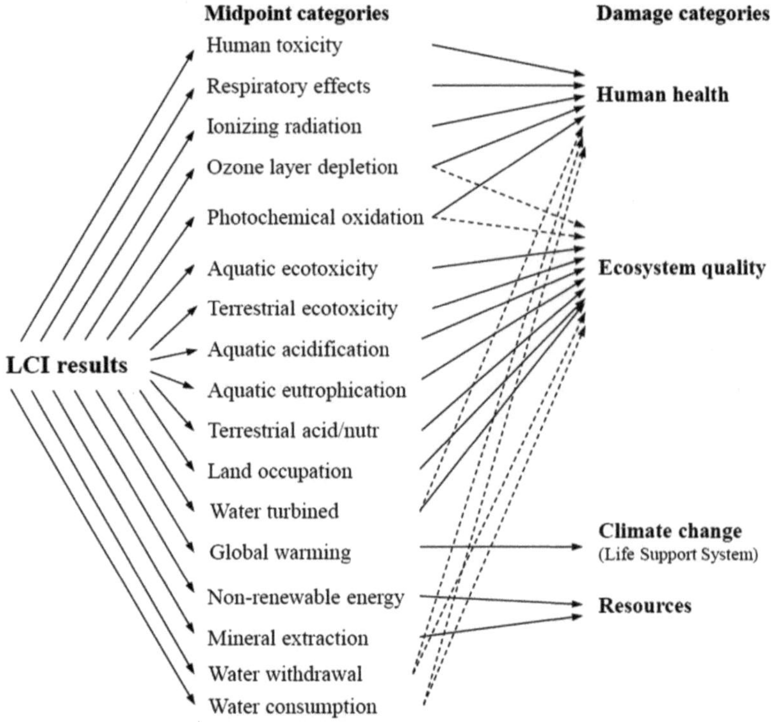

Fig. 9.9 Overall scheme of the IMPACT 2002+ framework. Reprinted from Ref. [10], with kind permission from Springer Science+Business Media

categories. As shown in Fig. 9.9, the midpoint categories comprise human toxicity (e.g., human toxicity carcinogenic effects and human toxicity non-carcinogenic effects), respiratory effects (due to inorganics), ionizing radiation, ozone layer depletion, photochemical oxidation, aquatic ecotoxicity, terrestrial ecotoxicity, aquatic acidification, aquatic eutrophication, terrestrial acidification/nutrification, land occupation, water turbined, global warming, non-renewable energy consumption, mineral extraction, water withdrawal, and water consumption. These midpoint categories can be further converted to four damage categories, i.e., human health, ecosystem quality, climate change, and resources.

9.3.2.4 Interpretation

The interpretation phase aims to analyze the consistency with the goal and scope defined in first step and to reach conclusions and recommendations. According to ISO 14040 and 14044, the two key elements of an LCA are the assessment of the entire life cycle of the investigated system and the assessment of a variety of environmental impacts. In this phase, results from the LCI and LCIA are combined together to provide a complete and unbiased account of the study. Three main elements are comprised in this phase:

- Identification: significant issues should be identified based on the results of the LCI and LCIA phases of a LCA.
- Evaluation: completeness, sensitivity, and consistency checks should be considered.
- Decision making: The strategies and decisions should be made according to the conclusions and recommendations of the LCA studies.

Uncertainties from conceptual errors (assuming importance of irrelevant information and disregarding relevant information) might exist throughout the LCA procedure. Therefore, a sensitivity analysis should be conducted to determine the effects of assessment parameters such as quantifiable data components (generation rate and composition of wastes) on the LCA results.

9.3.3 Case Study: Accelerated Carbonation

Although accelerated carbonation can potentially fix and store CO_2 over a geological time scale, the environmental effects of the process involved need to be considered from the perspective of LCA. The LCA of the accelerated carbonation process is particularly important, since energy is used in transporting, grinding, sieving, pressuring, heating, and operating the reactor. Accelerated carbonation may increase other environmental impacts, such as human toxicity, eutrophication, and acidification. For instance, to achieve an acceptable reaction rate for industrial applications, alkaline solid wastes are usually ground down to fine particles (approximately 100 μm) prior to use. The effect and toxicity of finely ground particles seem to cause significant health issues related to particulate formation, which is often believed to be a leading cause of respiratory disease [11–13]. Therefore, the effects of accelerated carbonation on changes in the environmental impacts should be carefully weighed and compared.

For instance, Fig. 9.10 shows the scope of work for evaluating the environmental impacts of accelerated carbonation process for carbon capture and utilization. The pretreatment of alkaline wastes and carbonation process should be employed within the system boundary. In general, alkaline solid wastes can be characterized by its strongly alkaline nature and significant levels of metal ions, especially calcium. After carbonation, the physico-chemical properties of solid wastes can be improved and suitably used as green materials, such as supplementary cementitious materials (SCMs) for replacing portions of the Portland cement in concrete/cement.

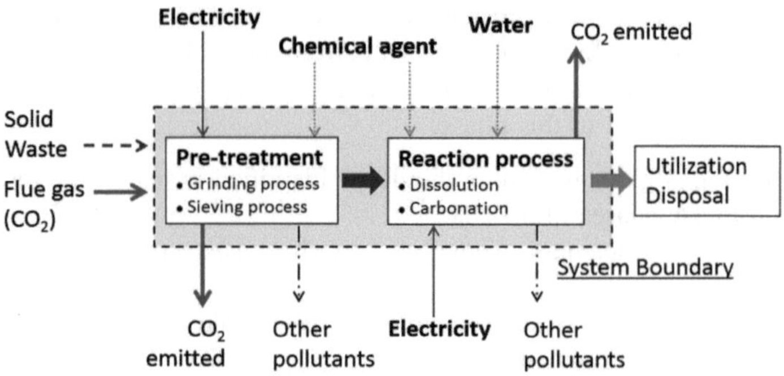

Fig. 9.10 System boundary of accelerated carbonation process in life cycle assessment (LCA)

The carbonated product can be utilized as construction aggregate (large particles) or SCM (fine particles) to avoid the use of aggregate and cement clinker in concrete. Therefore, it can be expected that huge environmental benefits, in terms of global warming potential and natural resource depletions, are achievable with the use of alkaline solid waste in blended cement. According to the LCA results by Chang et al. [14], a net capture capacity of ~100 kg CO_2 per ton of BHC can be achieved in an autoclave reactor. In the study, energy consumption, including grinding, sieving, pressuring, and heating, was taken into account into the LCA. Similarly, the net CO_2 emission of the aqueous carbonation of serpentine by the pH-swing method was estimated to be negative 0.02 kg CO_2 per kg of feedstock input [15]. However, the pH-swing method consumed base and acid solvents, which would cause adverse effects on the environment.

On the other hand, for one ton of cement clinker production, 1.5–1.7 tons of natural resource (e.g., limestone and clay) and 0.11–0.13 ton of coal are used [16]. In addition, concrete made from Portland cement carries a carbon footprint of 0.73–0.99 ton CO_2 per ton cement [17]. To evaluate the worldwide impacts and benefits, the annual world cement production is expected to reach 3.7–4.4 Gt by 2050, compared to ~2.5 Gt in 2006 [17]. In the case of steel slag, blended cements with up to 60% residual steel slag could exhibit similar properties, comparable to those of Portland cement.

9.4 Cost–Benefit Analysis (CBA)

Alkaline solid wastes can be used to fix a great amount of CO_2, especially if the wastes are generated nearby the point source of CO_2, for achieving both the environmental and economic benefits. In addition to determining the environmental benefits by LCA, the economic cost analysis for a developed process can be performed by the cost–benefit analysis (CBA). Since costs and benefits may be tangible or intangible, not all of them are reasonably assigned a monetary value.

A CBA can provide a distinct value that represents the total marketable services costs to the total non-marketable services and impacts. One of the commonly used indicators for CBA is the benefit–cost ratio (BCR), which can be calculated according to Eq. (9.6):

$$\mathrm{BCR} = R't/C't \tag{9.6}$$

where $R't$ is the revenue gained at the tth year, $C't$ is the cost investment at the tth year, and t is the expected operation period. The BCR considers the amount of monetary gain realized by performing a project versus the amount of costs by executing the project. A higher BCR indicates the more economically viable to the invested project. The rule of thumb is that the project is economically feasible (a good investment) if the BCR was greater than one.

Depreciation represents the periodic transfer of the cost of a fixed asset to an expense account. In accountancy, the deprecation refers to (1) the decrease in value of assets (fair value depreciation) and (2) the allocation of the cost of assets to periods where the assets are used (depreciation with the matching principle). Typically, the depreciation rate can be assumed within the range of 5 and 10%.

The investment costs of a process or service include (1) capital cost; (2) operation and maintenance (O&M); (3) land costs; (4) sale tax and surcharges (without value-added tax); and (5) environmental cost due to secondary pollutions. In the case of accelerated carbonation process, the benefits (or profits) are mainly from (1) carbon credit from certified emission reductions (CER), (2) sale profits of stabilized solid wastes as green products, and (3) avoidance of treatment fees for solid wastes and/or wastewater. To estimate the cost and benefit, the impacts of carbonation on steel and cement industries should be assessed, where the evaluation framework can be referred to the literature [18].

A major difficulty in CBA is in placing monetary values on most costs and benefits [19]. In this case, detailed inflow and outflow cash mush be estimated by considering the discounted value by various tools, such as interest rate, price inflation, and opportunity costs. In the following section, the net present value on estimating the discounted cash over the periods of time is illustrated. Moreover, the concepts of cost-effectiveness and cost-optimal beyond CBA are introduced afterward.

9.4.1 Net Present Value (NPV)

Some indices can be used to obtain the discounted values that are required for CBA. Net present value (NPV) method is widely used throughout economics, finance, and accounting, which is intended to determine the difference between the present value of cash inflows and the present value of cash outflows. It is usually used in capital budgeting to analyze the profitability of a project or investment. In finance, the NPV is defined as the sum of the present values (PVs) of incoming (e.g., profits and benefits) and outgoing (e.g., costs) cash flows over a period of operation. The PV of each cash flow, such as benefits and costs, can be discounted back and determined by the following formula:

$$\mathrm{PV} = \frac{R_t}{(1+i)^t} \tag{9.7}$$

where R_t is the net cash flow (e.g., benefits or costs); i is the discount rate. As a result, the NPV is the sum of all terms of cash flow.

$$\mathrm{NPV} = \sum_{t=0}^{N} \frac{R_t}{(1+i)^t} \tag{9.8}$$

where N is the total number of periods.

NPV can measure the excess or shortfall of cash flows, in terms of PV, above the cost of funds. It is a central tool of analysis in discounted cash flow (DCF) and is a standard method for using the "time value of money" to assess long-term projects. The concept of time value of money indicates that cash flows in different periods of time cannot be accurately compared, unless they have been adjusted to reflect their value at the same period of time (in this instance, $t = 0$). The PV of each cash flow must be determined to provide the meaningful comparison between cash flows at different periods of time. There are three assumptions in this type of DCF (i.e., NPV) analysis:

- The investment horizon of all projects considered is equally acceptable to the investor. For instance, a 2-year project is not necessary preferable to a 20-year project.
- Each project is equally speculative. Normally, a 10% discount rate is the appropriate and stable rate.
- The shareholders cannot get above a 10% return on their money if the level of risk for all possibility was equivalent.

In addition, the rate used to discount future cash flows to the present value is a key variable of the NPV analysis. An NPV analysis using variable discount rates may adequately reflect the situation than that from a constant discount rate for the NPV calculation. It suggests that a direct comparison should be carried out between the profitability of the project and the desired rate of return.

Furthermore, NPV does not provide an overall picture of the gain or loss of executing a certain project. Typically, internal rate of return (IRR) or other efficiency measures should be applied as a complement to NPV. The IRR implies during project operation, all cash flows that firmly received will be reinvested by desired interest rate, i.e., minimum attractive rate of return (MARR), than market lending interest rate. Therefore, IRR is an overestimate indicator, which provides multiple solutions of rates of returns that are unreasonable. In contrast, external rate of return (ERR) is more realistic than IRR, which implies all cash flows that firmly received will be reinvested by lending interest rate (lower than MARR). A project with higher IRR or ERR is better for investors [19]. It is noted that the NPV comes from various positive and negative cash flows, discounted to present time together (t = 0) or starting time when the initial capital is invested and/or equipment is installed.

9.4.2 Cost Effectiveness and Cost Optimal

CBA differs from the other commonly used cost-effectiveness analysis since most research in engineering economics and financial feasibility involves cost-effectiveness analysis rather than CBA. To evaluate the energy (or engineering) and economic performance of a certain process or service, a comparative methodology for determining cost optimal (CO) levels of engineering performance requirements, as shown in Fig. 9.11. The framework of the CO methodology can be referred to the delegated regulation 244/2012 [20]. The method defines a reference scenario, representing the local situation or the state-of-the-art process and compares several alternative solutions based on their engineering performance (such as energy efficiency). The methodology assumes that all of the alternatives (shown in dots) can be represented in a form of U-shaped curve. The less expensive solution is thus identified at the bottom of curve.

The payback period is a good indicator of the economic feasibility of an alternative process because it estimates the length of time when all discounted net cash flow has recovered the initial investment for the investor. A short payback period indicates that the project has a good implementation potential. In this case, investors would be a greater economic incentive to invest more money in expanding the

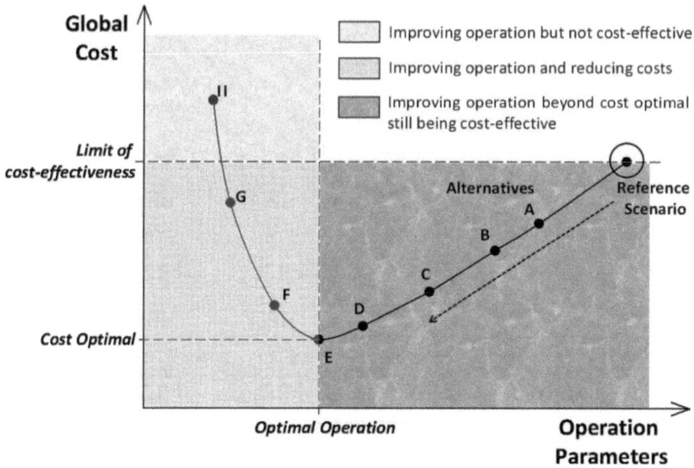

Fig. 9.11 Graphical presentation of conceptual relations between cost-effectiveness and cost optimal

capacity of the plant or building up a new plant. Otherwise, it would not be preferable if its initial costs are higher than the amount of the investor is able to invest [21]. It is noted that the CO method follows the above concepts to determine the best payback period.

9.5 Engineering, Environmental, and Economic (3E) Triangle Model

To accelerate technology development and implementation, a balanced performance in the 3Es (Engineering, Environmental, and Economic) is quite important. Any large-scale implementation of carbon capture, utilization, and storage (CCUS) technologies must be less energy and material intensive as well as economically viable.

9.5.1 Principles

A triangle model has been extensively utilized for assessments of new energy technologies or sustainability trends [22, 23]. As shown in Fig. 9.12, the 3E triangle

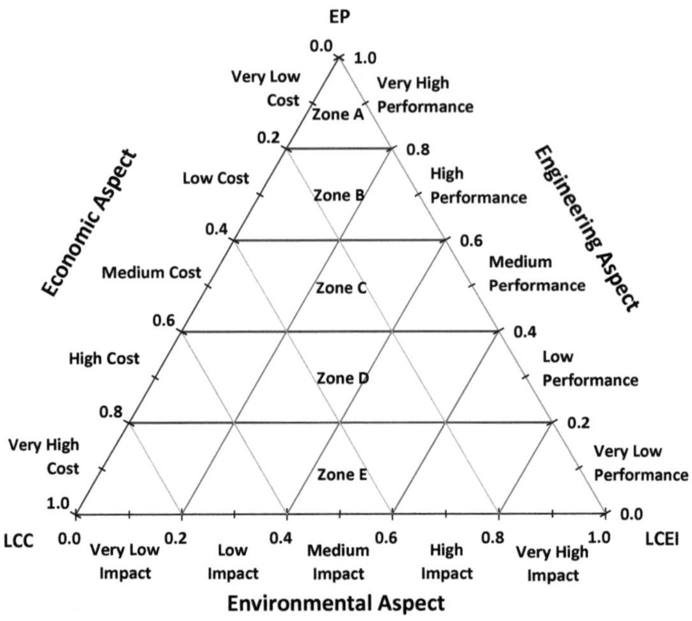

Fig. 9.12 Engineering, Environmental, and Economic (3E) triangle model for determining good engineering practice of a process. LCEI (*x*-axis): life cycle environmental impact; EP (*y*-axis): engineering performance; LCC (*z*-axis): life cycle cost

model considers the aspects of life cycle environmental impact (LCEI) on the x-axis, engineering performance (EP) on the y-axis, and life cycle cost (LCC) on the z-axis. In general, each of the 3E perspectives exhibit equal importance, i.e., each weighting one-third, for performance evaluation. Therefore, the 3E performance of a process can be conducted in a triangle model to provide a holistic assessment.

Each axis, ranging from 0 to 1 in the counterclockwise direction, can be divided into five levels: very low (0–0.2), low (0.2–0.4), medium (0.4–0.6), high (0.6–0.8), and very high (0.8–1.0). Therefore, the performance of a process can be subdivided into five zones within the 3E triangle model:

- Zone A: excellent scenario,
- Zone B: good scenario,
- Zone C: fair scenario,
- Zone D: bad scenario, and
- Zone E: poor scenario.

For example, a point located in "zone A" represents an operation that offers very high engineering performance with a very low environmental impact at a very low cost. In comparison, an operation located in "zone E" is considered as poor performance with a very high environmental impact at a very high cost.

9.5.2 Key Performance Indicators (KPIs)

9.5.2.1 Engineering Aspect

Engineering aspect considers the flexibility, reliability, efficiency, productivity, and redundancy of a process. For instance, in a mineral carbonation process, the engineering performance (EP) could include the capacities for CO_2 removal, wastewater neutralization, and carbonated product. More details on the RSM and for evaluating the engineering performance can be referred to this chapter and Chap. 19, respectively.

For process energy consumption, feedstock grinding is a commonly used and typically energy-intensive process for accelerated carbonation. The energy consumption of material grinding (E_G) can be estimated by Bond's equation [24], as shown in Eq. (9.9). It is noted that the Bond's equation should give the most accurate estimation of grinding energy requirement within the conventional grinding range of 25,000 to 20 μm [25].

$$E_G = w_i\left(\frac{10}{\sqrt{D_{P80}}} - \frac{10}{\sqrt{D_{F80}}}\right) \quad (9.9)$$

where D_{F80} (μm) and D_{P80} (μm) are the 80% passing size of feed and product, respectively. The w_i (kWh/ton) is the work index of ground material, which expresses the resistance of the material to crushing and grinding. The work index is subject to variations because of variations in the inherent properties of materials, variations in the grinding environment, and variations in the mechanism of energy transfer from the grinding equipment to its charge [26]. The work index for various types of materials can be found as follows:

- Natural ores:
  - Bauxite: 25.66 kWh/ton [27].
  - Dolomite: 9.82–12.19 kWh/ton for feed particle sizes of 3300 and 830 μm, respectively [28].
  - Feldspar from 11.14 [26] to 12.80 kWh/ton [29].
  - Hematite: 12.68 kWh/ton [29].
  - Kaolin: 10.52 kWh/ton [26].
  - Limestone: 22.67 kWh/ton [27].
  - Quartz: from 11.38 [26] to 12.77 kWh/ton [29].
- Alkaline solid wastes:
  - Iron slag: 30.40 kWh/ton [30]

Any variation in the work index in tests at different product sizes indicates that the material is non-homogeneous to size reduction. As a result, laboratory evaluation should preferably be performed at the product size required in commercial grinding.

9.5.2.2 Environmental Aspect

For the environmental aspect, the life cycle assessment (LCA) approach is considered as the most suitable tool for environmental assessment of technologies and/or processes along their entire life cycles [31–34]. Various types of environmental indicators, expected for their relevance in the process, should be selected to determine the life cycle environmental impact (LCEI) in the 3E triangle model. The remaining impact categories could be excluded for the 3E analysis, once they did not exhibit significant difference among the scenarios. More details on environmental impact assessment via life cycle assessment (LCA) can be referred to Chap. 9.

9.5.2.3 Economic Aspect

Economic aspect can be determined using various types of economic indicators, such as costs of operation and maintenance, operation profits, and payback period. In an accelerated carbonation process, the profits from carbon price (tax) can be expected since the CO_2 emission from industry is reduced. Moreover, the carbonated product from carbonation process could be sold as SCMs in blended

Table 9.4 Price of commonly used materials in construction engineering. Information is available in the literature [35]

Materials	Category	Unit	Locations	
			Mainland	Taiwan[a]
Steel slag (BOFS)	Amount	Mt/y	60	1.2
	Price	USD/t	6.0	
Fly ash (APC)	Amount	t/y		
	Price	USD/t		6.6–10.0
Cement (CEM I/42.5)	Amount	Mt/y	–	13.2 (in 2016)
	Price	USD/t	49.8 (in 2008)	73.0 (in 2016)
Fine aggregate	Price	USD/m^3	–	21.7 (in 2016)
Coarse aggregate	Price	USD/m^3	–	19.7 (in 2016)
Fe_2O_3 (>50%)	Price	USD/t	75.5 (in 2008)	–

[a]1 USD = 30 NTD

cement or aggregates in concrete block. Table 9.4 presents the price of commonly used materials in construction engineering. More details on the CBA can be referred to this chapter.

9.5.3 Data Analyses and Interpretation

Since the complex relationships among 3E aspects can be easily visualized on a ternary plot among the different scenarios, the triangle graphical presentation can be used for evaluating key factors that are related but also complementary [36, 37].

9.5.3.1 Data Evaluation

Depending on evaluation goals and objectives, various key performance indicators (KPIs) should be selected for evaluating the entire process using a triangle model. From the 3E perspectives, the KPIs can be categorized into engineering performance (EP), life cycle environmental impact (LCEI), and economic cost (EC). The determined KPIs can be thus arranged as a data matrix for LCEI, EP, and EC indicators, as shown in Eqs. (9.10) to (9.12):

$$\mathrm{LCEI}_{yi} = (e_{yi}) = \begin{bmatrix} e_{11} & \cdots & e_{1m} \\ \vdots & \ddots & \vdots \\ e_{s1} & \cdots & e_{sm} \end{bmatrix} \tag{9.10}$$

$$\mathrm{EP}_{yj} = (p_{yj}) = \begin{bmatrix} p_{11} & \cdots & p_{1m} \\ \vdots & \ddots & \vdots \\ p_{s1} & \cdots & p_{sm} \end{bmatrix} \tag{9.11}$$

$$\mathrm{EC}_{yk} = (c_{yk}) = \begin{bmatrix} c_{11} & \cdots & c_{1m} \\ \vdots & \ddots & \vdots \\ c_{s1} & \cdots & c_{sm} \end{bmatrix} \tag{9.12}$$

where LCEI_{yi}, EP_{yj}, and EC_{yk} are the original data matrix for LCEI, EP, and EC indicators, respectively. "y" is the yth studied object, which is different scenarios with various evaluation targets (e.g., various levels of CO_2 capture performance); i, j, and k are the ith selected LCEI indicator, the jth selected EP indicator, and the kth selected EC indicator, respectively.

9.5.3.2 Data Normalization and Standardization

Because the KPI usually include multiple dimensions and the value range of KPI data varies widely, feature rescaling methods should be adopted to make the features independent of each other and to scale the range between 0 and 1. Various types of feature rescaling methods, such as direct and inverse rescaling, are available to be applied for different types of situations.

1. Direct Rescaling
   Equation (9.13) can be used for a "direct rescaling," where the maximal value in a certain row represents the highest level of dimensionless feature, i.e., "1". For instance, a high treatment scale of a process represents a high performance on engineering aspect.

$$\mathrm{X}'_{yj} = \left(x'_{yj}\right) = \left(\frac{x_{yj} - \min\limits_{y=1}\left(x_{yj}\right)}{\max\limits_{y=1}\left(x_{yj}\right) - \min\limits_{y=1}\left(x_{yj}\right)}\right) = \begin{bmatrix} x'_{11} & \cdots & x'_{1m} \\ \vdots & \ddots & \vdots \\ x'_{s1} & \cdots & x'_{sm} \end{bmatrix} \tag{9.13}$$

$$(y = 1, 2, \ldots, s; j = 1, 2, \ldots, n)$$

   where $\mathrm{X'}_{yi}$ is the standardized data matrix for direct rescaling indicators.
2. Inverse Rescaling
   Equation (9.14) can be used for an "inverse rescaling," where the maximal value possesses the lowest level of dimensionless feature, i.e., "0". For instance, a high product sale profit represents a low economic impact.

$$T'_{yi} = \left(t'_{yi}\right) = \left(\frac{\max\limits_{y=1}\left(t_{yi}\right) - t_{yi}}{\max\limits_{y=1}\left(t_{yi}\right) - \min\limits_{y=1}\left(t_{yi}\right)}\right) = \begin{bmatrix} t'_{11} & \cdots & t'_{1m} \\ \vdots & \ddots & \vdots \\ t'_{s1} & \cdots & t'_{sm} \end{bmatrix} \tag{9.14}$$

$$(y = 1, 2, \ldots, s; \quad i = 1, 2, \ldots, m)$$

where T'_{yj} is the standardized data matrix for inverse rescaling indicators.

9.5.3.3 Weighting Factors

The synthetic KPI indexes for LCEI, EP, and LCC can be calculated by Eq. (9.15):

$$\mathrm{KPI}_y = \sum_{i=1}^{m} \left(\mathrm{KPI}'_{yi} \cdot W_{yi}\right) \tag{9.15}$$

where the W_i is the weighting factor of each KPI. The weighting factors (W_i) of each KPI can be determined by an ad hoc committee, so-called expert consulting, via the Delphi method [38]. The Delphi method is a widely used methodology for gathering data from respondents within their domain of expertise [39, 40]. It originally developed as a systematic and interactive forecasting method, which relies on a panel of experts. Typically, the experts answer questionnaires in at least two rounds until getting a consensus.

References

1. Li S, Dragicevic S, Castro FA, Sester M, Winter S, Coltekin A, Pettit C, Jiang B, Haworth J, Stein A, Cheng T (2016) Geospatial big data handling theory and methods: a review and research challenges. ISPRS J Photogramm Remote Sens 115:119–133. doi:10.1016/j.isprsjprs.2015.10.012
2. Lee J-G, Kang M (2015) Geospatial big data: challenges and opportunities. Big Data Res 2 (2):74–81. doi:10.1016/j.bdr.2015.01.003
3. Wang J, Ma Y, Ouyang L, Tu Y (2016) A new Bayesian approach to multi-response surface optimization integrating loss function with posterior probability. Eur J Oper Res 249(1):231–237. doi:10.1016/j.ejor.2015.08.033
4. Pan SY, Chen YH, Chen CD, Shen AL, Lin M, Chiang PC (2015) High-gravity carbonation process for enhancing CO_2 fixation and utilization exemplified by the steelmaking industry. Environ Sci Technol 49(20):12380–12387. doi:10.1021/acs.est.5b02210
5. Odeh NA, Cockerill TT (2008) Life cycle GHG assessment of fossil fuel power plants with carbon capture and storage. Energy Policy 36(1):367–380. doi:10.1016/j.enpol.2007.09.026
6. Marx J, Schreiber A, Zapp P, Haines M, Hake JF, Gale J (2011) Environmental evaluation of CCS using life cycle assessment—a synthesis report. Energy Procedia 4:2448–2456
7. de Haes HAU, Heijungs R (2007) Life-cycle assessment for energy analysis and management. Appl Energy 84(7–8):817–827

8. Goedkoop M, Heijungs R, Huijbregts M, De Schryver A, Struijs J, van Zelm R (2013) ReCiPe 2008: a life cycle impact assessment method which comprises harmonised category indicators at the midpoint and the endpoint level. Ruimte en Milieu, The Netherlands
9. MHSPE (2000) Eco-indicator 99 manual for designers. Ministry of Housing, Spatial Planning and the Environment, The Netherlands
10. Jolliet O, Margni M, Charles R, Humbert S, Payet J, Rebitzer G, Rosenbaum R (2003) IMPACT 2002+: a new life cycle impact assessment methodology. Int J Life Cycle Assess 8 (6):324–330. doi:10.1007/bf02978505
11. Koornneed J, Nieuwlaar E (2009) Environmental life cycle assessment of CO_2 sequestration through enhanced weathering of olivine. Working paper
12. IEA (2013) Mineralisation—carbonation and enhanced weathering. International Energy Agency
13. Giannoulakis S, Volkart K, Bauer C (2014) Life cycle and cost assessment of mineral carbonation for carbon capture and storage in European power generation. Int J Greenhouse Gas Control 21:140–157. doi:10.1016/j.ijggc.2013.12.002
14. Chang EE, Pan S-Y, Chen Y-H, Chu H-W, Wang C-F, Chiang P-C (2011) CO_2 sequestration by carbonation of steelmaking slags in an autoclave reactor. J Hazard Mater 195:107–114. doi:10.1016/j.jhazmat.2011.08.006
15. Park A, Fan L (2004) Mineral sequestration: physically activated dissolution of serpentine and pH swing process. Chem Eng Sci 59(22–23):5241–5247. doi:10.1016/j.ces.2004.09.008
16. Kumar S, Kumar R, Bandopadhyay A (2006) Innovative methodologies for the utilisation of wastes from metallurgical and allied industries. Resour Conserv Recycl 48(4):301–314. doi:10.1016/j.resconrec.2006.03.003
17. Hasanbeigi A, Price L, Lin E (2012) Emerging energy-efficiency and CO_2 emission-reduction technologies for cement and concrete production: a technical review. Renew Sustain Energy Rev 16(8):6220–6238. doi:10.1016/j.rser.2012.07.019
18. Tomás RAF, Ramôa Ribeiro F, Santos VMS, Gomes JFP, Bordado JCM (2010) Assessment of the impact of the European CO_2 emissions trading scheme on the Portuguese chemical industry. Energy Policy 38(1):626–632. doi:10.1016/j.enpol.2009.06.066
19. Lee D-H (2016) Cost-benefit analysis, LCOE and evaluation of financial feasibility of full commercialization of biohydrogen. Int J Hydrogen Energy 41(7):4347–4357. doi:10.1016/j.ijhydene.2015.09.071
20. EU Commission (2012) Delegated Regulation (EU) no. 244/2012. Offic J Eur Commun
21. Araújo C, Almeida M, Bragança L, Barbosa JA (2016) Cost-benefit analysis method for building solutions. Appl Energy 173:124–133. doi:10.1016/j.apenergy.2016.04.005
22. Xu F-L, Zhao S-S, Dawson RW, Hao J-Y, Zhang Y, Tao S (2006) A triangle model for evaluating the sustainability status and trends of economic development. Ecol Model 195(3–4):327–337. doi:10.1016/j.ecolmodel.2005.11.023
23. Zhang J, Yang G, Pu L, Peng B (2014) Trends and spatial distribution characteristics of sustainability in Eastern Anhui Province, China. Sustainability 6(12):8398–8414. doi:10.3390/su6128398
24. Bond FC (1961) Crushing and grinding calculations. Part 1. Br Chem Eng 6:378–385
25. Hukki RT (1961) Proposal for a solomonic settlement between the theories of von Rittinger, Kick, and Bond. Translation. Soc Mining Eng AIME 220:403–408
26. Ipek H, Ucbas Y, Hosten C (2005) The bond work index of mixtures of ceramic raw materials. Miner Eng 18(9):981–983. doi:10.1016/j.mineng.2004.12.014
27. Ozkahraman HT (2005) A meaningful expression between bond work index, grindability index and friability value. Miner Eng 18(10):1057–1059. doi:10.1016/j.mineng.2004.12.016
28. Magdalinovic N, Trumic M, Trumic G, Magdalinovic S, Trumic M (2012) Determination of the Bond work index on samples of non-standard size. Int J Miner Process 114–117:48–50. doi:10.1016/j.minpro.2012.10.002
29. Gent M, Menendez M, Toraño J, Torno S (2012) A correlation between Vickers hardness indentation values and the bond work index for the grinding of brittle minerals. Powder Technol 224:217–222. doi:10.1016/j.powtec.2012.02.056

30. Kodama S, Nishimoto T, Yamamoto N, Yogo K, Yamada K (2008) Development of a new pH-swing CO_2 mineralization process with a recyclable reaction solution. Energy 33(5):776–784. doi:10.1016/j.energy.2008.01.005
31. Khoo HH, Tan RBH (2006) Life cycle investigation of CO_2 recovery and sequestration. Environ Sci Technol 40(12):4016–4024
32. von der Assen N, Jung J, Bardow A (2013) Life-cycle assessment of carbon dioxide capture and utilization: avoiding the pitfalls. Energy Environ Sci 6(9):2721. doi:10.1039/c3ee41151f
33. von der Assen N, Voll P, Peters M, Bardow A (2014) Life cycle assessment of CO_2 capture and utilization: a tutorial review. Chem Soc Rev 43(23):7982–7994. doi:10.1039/c3cs60373c
34. Von der Assen N, Lorente Lafuente AM, Peters M, Bardow A (2015) Environmental assessment of CO_2 capture and utilisation. In: Styring P, Quadrelli EA, Armstrong K (eds) Carbon dioxide utilisation. Elsevier, New York
35. Zhang T, Yu Q, Wei J, Li J, Zhang P (2011) Preparation of high performance blended cements and reclamation of iron concentrate from basic oxygen furnace steel slag. Resour Conserv Recycl 56(1):48–55. doi:10.1016/j.resconrec.2011.09.003
36. Yi Q, Feng J, Wu Y, Li W (2014) 3E (energy, environmental, and economy) evaluation and assessment to an innovative dual-gas polygeneration system. Energy 66:285–294. doi:10.1016/j.energy.2014.01.053
37. Silvestre JD, de Brito J, Pinheiro MD (2013) From the new European Standards to an environmental, energy and economic assessment of building assemblies from cradle-to-cradle (3E-C2C). Energy Build 64:199–208. doi:10.1016/j.enbuild.2013.05.001
38. Dalkey N, Helmer O (1963) An experimental application of the DELPHI method to the use of experts. Manage Sci 9(3):458–467. doi:10.1287/mnsc.9.3.458
39. Galo JJM, Macedo MNQ, Almeida LAL, Lima ACC (2014) Criteria for smart grid deployment in Brazil by applying the Delphi method. Energy 70:605–611. doi:10.1016/j.energy.2014.04.033
40. Al-Saleh YM, Vidican G, Natarajan L, Theeyattuparampil VV (2012) Carbon capture, utilisation and storage scenarios for the Gulf Cooperation Council region: a Delphi-based foresight study. Futures 44(1):105–115. doi:10.1016/j.futures.2011.09.002

Part III
Types of Feedstock for CO_2 Mineralization

Chapter 10
Natural Silicate and Carbonate Minerals (Ores)

Abstract Natural silicate-type (e.g., wollastonite) and carbonate-type (e.g., limestone) ores containing alkaline earth metal are suitable for mineral carbonation reaction due to their excellent theoretical sequestration potential. Over geologic time, the natural weathering of carbonate-type ores captures and sequesters atmospheric CO_2. On the other hand, vast quantities of silicate-type minerals are required to sequester a significant fraction of emitted CO_2. In this chapter, the physico-chemical properties of various types of natural ores and/or minerals are illustrated. Two promising processes utilizing natural ores to capture CO_2 from the flue gas, (1) accelerated mineral carbonation and (2) accelerated carbonate weathering, are discussed in detail in terms of theoretical process chemistry and practical applications, including challenges and barriers.

10.1 Types of Natural Minerals/Ores

Natural silicate minerals are suitable feedstock for natural carbonation (i.e., weathering) because they are rich in calcium or magnesium content. They are abundant but generally difficult to access. Based on their chemical properties, they can be categorized into two types: (1) silicate ores and (2) carbonate ores. The majority of previous researches have focused on the aqueous carbonation of naturally existing silicate minerals, for example, wollastonite [1–3], serpentine [4–6], and olivine [7, 8]. According to Lackner [9], although there are natural ores on earth to sequester CO_2 emissions from all fossil fuels, cost-effective methods for accelerating carbonation (for silicate-type ores or alkaline solid wastes) or accelerating weathering (for carbonate-type ores) should be developed.

Different types of natural ores are involved with different reaction with CO_2. Table 10.1 presents the physico-chemical properties of these natural ores. Natural silicate ores, suitable materials subject to accelerated mineral carbonation (see details in Sect. 10.2 in this chapter), include wollastonite ($CaSiO_3$), serpentine

P.-C. Chiang and S.-Y. Pan, *Carbon Dioxide Mineralization and Utilization*, DOI 10.1007/978-981-10-3268-4_10

Table 10.1 Physico-chemical properties of natural silicate-type and carbonate-type ores

Properties	Items	Unit	Silicate-type				Carbonate-type	
			Serpentine	Wollastonite	Olivine	Talc	Dolomite	Limestone
Physical	Density	g/cm_3	2.5–2.6	2.5–2.6	3.3–4.3	2.5–2.8	2.8–2.9	2.71
	Hardness	–	3–5	5–5.5	6.5–7.0	1.0	3.5–4.0	3.0
Chemical	CaO	%	–	45.53		0–18	30.41	56.0
	MgO	%	43.0	0.03		9–32	21.87	–
	SiO_2	%	44.1	52.91		42–61	–	–
	Al_2O_3	%	–	0.07		0–11	–	–
	FeO	%	–	0.38		3–20	–	–
	MnO	%	–	0.56		–	–	–
	CO_2	%	–	–		–	47.72	44.0
	H_2O	%	12.9	–		–	–	–

($Mg_3Si_2O_5(OH)_4$), olivine ($(Mg,Fe)_2SiO_4$), talcum ($Mg_3Si_4O_{10}(OH)_2$), pyroxene ($(Ca,Na,Fe,Mg)_2(Si,Al)_2O_6$), and amphibole. On the other hand, natural carbonate ores, such as limestone ($CaCO_3$), marble ($CaCO_3$), and dolomite ($CaMg(CO_3)_2$), are suitable for accelerated carbonate weathering (see details in Sect. 10.3 in this chapter).

10.1.1 Silicate Ores

1. Wollastonite ($CaSiO_3$)

Wollastonite is a mineral rich in calcium inosilicate ($CaSiO_3$), containing small amounts of iron, magnesium, and manganese substituting for calcium. It has a specific gravity of 2.87–3.09 and is known as hard material with the Mohs hardness scale of 4.5–5.0. Moreover, it exhibits brightness and whiteness with low moisture and oil absorption. Therefore, it is used primarily in ceramics, paint filler, friction products, and plastics. In 2014, world reserves of wollastonite were estimated to exceed 90 million tons, where the major producers were China (~300,000 ton/y), India (~160,000 ton/y), USA (~67,000 ton/y), Mexico (~55,000 ton/y), and Finland (~11,500 ton/y) [10].

2. Serpentine

Serpentine is not a single mineral and can be classified into two main groups: (1) Antigorite represents more solid form; and (2) chrysotile represents the fibrous form, especially asbestos. These kaolinite–serpentine groups are greenish and brownish and used as a source of magnesium and asbestos. It has a specific gravity of 2.2–2.9 and is known as soft material with the Mohs hardness scale of 2.5–4.0. The serpentine group is composed of several related minerals, where a generic formula could be presented as follows:

$$X_{2-3}Si_2O_5(OH)_4 \quad (10.1)$$

where X could be Mg, Fe^{2+}, Fe^{3+}, Ni, Al, Zn, or Mn. For example, the antigorite can be expressed by $(Mg,Fe)_3Si_2O_5(OH)_4$, while the chrysotile is $Mg_3Si_2O_5(OH)_4$. Chrysotile is a group of polymorphous minerals (different crystal lattice) with the same chemical composition.

3. Olivine (Mg_2SiO_4)

Olivine is a mineral rich in hydrated magnesium silicate, expressed as the chemical formula $(Mg,Fe)_2SiO_4$. It is a common mineral in the earth's subsurface. It has a specific gravity of 3.2–4.5 and is known as a hard material with the Mohs hardness scale of 6.5–7.0. Olivine gives its name to the group of minerals with a related structure (the olivine group), which includes tephroite (Mn_2SiO_4), monticellite ($CaMgSiO_4$), and kirschsteinite ($CaFeSiO_4$).

4. Talc ($Mg_3Si_4O_{10}(OH)_2$)

Talc is a clay mineral rich in hydrated magnesium silicate expressed as the chemical formula $H_2Mg_3(SiO_3)_4$ or $Mg_3Si_4O_{10}(OH)_2$. It is commonly used as a substance for baby powder (or talcum powder). It has a specific gravity of 2.5–2.8 and is known as the softest material with the Mohs hardness scale of 1. Since 1994, the production and apparent consumption of talc decreased by 44 and 34%, respectively [10]. In 2014, the major talc producers were China ($\sim$2200 t/y), India ($\sim$660 t/y), Korea ($\sim$540 t/y), USA ($\sim$535 t/y), and Brazil ($\sim$500 t/y) [10].

10.1.2 Carbonate Ores

1. Limestone

Limestone, a naturally sedimentary rock with 92–98% $CaCO_3$ [11], composes largely of the minerals of calcite and/or aragonite, which are different crystal forms of calcium carbonate ($CaCO_3$). In other words, limestone is an abundant and inexpensive natural ores as $CaCO_3$ sources. Limestone also is the major ingredient in cement industry. Consequently, clinker manufacturing plants are colocated with limestone quarries to economically transport limestone from the supply quarry to the plant [12].

Limestone has numerous applications in building and construction such as aggregate, white pigment, and filler. Currently, more than 20% of limestone production and processing in the USA results in waste limestone fines (<10 mm), which could be further utilized or being accumulated at limestone mining and processing sites [13].

2. Marble

Marble is a non-foliated metamorphic rock, which is primarily composed of recrystalized carbonates, such as calcite ($CaCO_3$) and dolomite ($CaMg(CO_3)_2$), with trace amounts of clay, micas, quartz, pyrite, iron oxides, and graphite. It usually forms when limestone is subjected to the heat and pressure of metamorphism. Moreover, the Mohs hardness scale of marble is around 3, indicating that it is easy to carve. Therefore, it is commonly used for sculpture or served as a material for building.

10.2 Accelerated Mineral Carbonation: Silicate Ores

10.2.1 Natural Mineral Carbonation

Carbonation can occur naturally. Natural carbonation is also well known as "weathering," which eventually removes CO_2 from the atmosphere by neutralizing the acid with mineral alkalinity. Natural weathering occurs by the reaction between natural alkaline silicates and atmospheric CO_2, as shown in Eqs. (10.2) and (10.3). It is noted that the gaseous CO_2 in the atmosphere is fixed and stored in the form of bicarbonate ions via natural carbonation.

$$\begin{aligned} &CaSiO_{3(s)} + 2CO_{2(aq)} + H_2O_{(l)} \rightarrow Ca^{2+}_{(aq)} + 2HCO^{-}_{3\ (aq)} + SiO_{2(s)} \\ &\Delta H = -63\ \mathrm{kJ/mol}\ CO_2 \end{aligned} \tag{10.2}$$

$$\begin{aligned} &Mg_2SiO_{4(s)} + 4\ CO_{2(aq)} + 2\ H_2O_{(l)} \rightarrow 2\ Mg^{2+}_{(aq)} +\ 4\ HCO^{-}_{3(aq)} +\ SiO_{2(s)} \\ &\Delta H = -280\ \mathrm{kJ/mol}\ CO_2 \end{aligned} \tag{10.3}$$

Another pathway of CO_2 fixation in nature is through the formation of carbonate precipitates. Atmospheric CO_2 dissolves in rainwater, producing weak carbonic acid (${CO_3}^{2-}$), and becomes slightly acidic by nature. Once contact with the rainwater, calcium and magnesium silicates are leached from the mineral matrix. Rainwater thus carries the leached calcium and magnesium ions to rivers and subsequently to the ocean, where solid carbonates (M^{2+} represents alkaline earth metal element), such as calcium and magnesium precipitates, are formed as shown in Eq. (10.4):

$$M^{2+}_{(aq)} + CO^{2-}_{3(aq)} \rightarrow M^{2+}CO^{-}_{3(aq)} \rightarrow M^{2+}CO^{-}_{3(s)} \tag{10.4}$$

Chemical weathering of natural mineral ores is one of the main mechanisms contributing to the global geochemical carbon cycle. Nature's carbon cycle sequesters CO_2 as mineral carbonates by precipitation from the ocean in a geological timescale [14]. However, the kinetics of natural carbonation is extremely slow due to the relatively low CO_2 concentration, about 0.03–0.06% [15].

According to an investigation by Haug et al. [7], the weathering rate of olivine is estimated to be $10^{-8.5}$ mol/(m^{2}s) using the average ground temperature in Norway of 6 °C and a pH of 5.6, which can be considered as the acidity of rainwater.

10.2.2 Process Chemistry and Reaction Mechanisms

Accelerated mineral carbonation (or accelerated weathering) utilizes natural ores, such as wollastonite ($CaSiO_3$), larnite (Ca_2SiO_4), olivine (Mg_2SiO_4), serpentine ($Mg_3Si_2O_5(OH)_4$), and fayalite (Fe_2SiO_4) to react with high-purity CO_2, as shown in Eqs. (10.5)–(10.10). These processes are based on acid–base reactions in which carbonic acid is neutralized by a base from alkaline minerals. It has been extensively studied in the literature [14, 16, 17]. It was found that the reaction between minerals and CO_2 would form at least one type of carbonate product [18]. Compared to the other proposed technologies (see Chap. 2), the main advantage of mineral carbonation is that it represents the form of permanent CO_2 storage with environmentally friendly product, i.e., various types of carbonate precipitates.

$$\text{Wollastonite}: CaSiO_{3(s)} + CO_{2(g)} + 2\ H_2O_{(l)} \rightarrow CaCO_{3(s)} + H_4SiO_{4(aq)} \quad (10.5)$$

$$\text{Larnite}: Ca_2SiO_{4(s)} + 2\ CO_{2(g)} \rightarrow 2\ CaCO_{3(s)} + SiO_{2(s)} \quad (10.6)$$

$$\text{Olivine}: Mg_2SiO_{4(s)} + 2\ CO_{2(g)} + 2\ H_2O_{(l)} \rightarrow 2MgCO_{3(s)} + H_4SiO_{4(aq)} \quad (10.7)$$

$$\text{Olivine}: 2\ Mg_2SiO_{4(s)} + CO_{2(g)} + 2\ H_2O_{(l)} \rightarrow MgCO_{3(s)} + Mg_3Si_2O_5(OH)_{4(s)} \quad (10.8)$$

$$\begin{aligned}\text{Serpentine}: &\ Mg_3Si_2O_5(OH)_{4(s)} + 3\ CO_{2(g)} + 2\ H_2O_{(l)} \\ &\rightarrow 3MgCO_{3(s)} + 2\ H_4SiO_{4(aq)} + 2\ H_2O_{(l)}\end{aligned} \quad (10.9)$$

$$\text{Fayalite}: Fe_2SiO_{4(s)} + 2\ CO_{2(g)} + 2\ H_2O_{(l)} \rightarrow 2FeCO_{3(s)} + H_4SiO_{4(aq)} \quad (10.10)$$

Typically, the natural minerals containing alkaline earth metals are suitable for mineral carbonation reaction due to their excellent theoretical sequestration potential. For instance [19]:

- wollastonite: 0.36 g CO_2/g dry sample,
- olivine (forsterite): 0.55 g CO_2/g dry sample, and
- serpentine (lizardite): 0.40 g CO_2/g dry sample.

In reality, Park and Fan [20] showed that the net CO_2 emission of the aqueous carbonation of serpentine by the pH-swing method was −0.02 kg CO_2/kg feedstock. However, prior investigations using the pH-swing method consumed a great amount of base and acid solvents, which would cause adverse effects on the environment.

10.3 Accelerated Carbonate Weathering: Carbonate Ores

10.3.1 Process Chemistry and Reaction Mechanisms

Accelerated carbonate weathering imitates the natural carbonate weathering reaction according to Eq. (10.11). Natural limestone composes largely of the minerals of calcite and/or aragonite, which are different crystal forms of calcium carbonate ($CaCO_3$). Since limestone is much abundant and inexpensive source of calcium carbonate, extensive studies have been carried out on the accelerated weathering of limestone (AWL) [11, 21]. In principle, according to the theoretical consideration, capturing one ton of CO_2 will require 2.3 tons of $CaCO_3$ ores and 0.3 tons of water, thereby generating 2.8 tons of bicarbonate ions.

$$CO_{2(g)} + H_2O_{(l)} + CaCO_{3(s)} \rightarrow Ca^{2+}_{(aq)} + 2\ HCO^{-}_{3(aq)} \tag{10.11}$$

The above reaction can be expressed via several steps. First, gaseous CO_2 should be dissolved into solution (liquid phase) to form carbonic acid, as shown in Eqs. (10.12) and (10.13):

$$CO_{2(g)} \rightarrow CO_{2(aq)} \tag{10.12}$$

$$CO_{2(aq)} + H_2O \rightarrow H_2CO_{3(aq)} \tag{10.13}$$

After that, the carbonic acid is reacted with calcium carbonate in natural ores, as presented in Eq. (10.14):

$$H_2CO_{3(aq)} + CaCO_{3(s)} \rightarrow Ca^{2+}_{(aq)} + 2\ HCO^{-}_{3(aq)} \tag{10.14}$$

Therefore, the dissolved calcium and bicarbonate ions can be directly released and diluted in the ocean. The overall process is geochemically equivalent to continental and marine carbonate weathering, which will otherwise naturally consume anthropogenic CO_2 over many millennia [11].

10.3.2 Seawater as The Liquid Phase for Reaction

In Eq. (10.11), the reaction can be spontaneously driven to the right under elevated CO_2 concentrations (or high pressure). In the case of using seawater with 0.15 atm partial pressure of CO_2 in flue gas, a carbonic acid solution will be produced, with a pH of 5.7 at equilibrium condition [13].

Using seawater for the reaction is less effective in dissolving $CaCO_3$ since the seawater already contains significant quantities of HCO_3^- and CO_3^{2-} that act as a buffer for pH and $CaCO_3$ saturation state depression. Nevertheless, calcium carbonate undersaturation and $CaCO_3$ dissolution could be easily achieved when seawater is equilibrated with CO_2 in concentrations typical of flue gas (e.g., 5–15 V%) [22]. It was noted that the dissolution rates of calcium carbonate would be on the order of 2×10^{-6} mmol s^{-1} per cm^2 of reactive surface area of carbonate minerals [23].

10.3.3 CO_2 Emulsion with Limestone for Storage in Ocean

Concentrated CO_2 storage in deep ocean encounters several challenges: (1) It must be injected at sufficient depth ($\sim$500 m) in the ocean since liquid CO_2 is less dense than water and poorly miscible with water; and (2) it might form carbonic acid and lower the pH of seawater, which might pose adverse impacts on marine biota. To overcome the above issues, several researches [24] have been conducted to inject an emulsion of CO_2 in water stabilized by limestone ($CaCO_3$) particles into the ocean. Under specific conditions, liquid CO2 will form an emulsion in water in the presence of limestone powders. Since the CO_2 emulsion (globule) is denser than seawater, it could sink deeper from the injection point rather than buoying upward. In addition, the $CaCO_3$-coated CO_2 droplets will not acidify the seawater.

Through proper mixing in a high-pressure batch reactor, a stable emulsion could be formed with globules consisting of an inner core of liquid CO_2 coated with a sheath $CaCO_3$ (limestone) particles dispersed in solution, as shown in Fig. 10.1

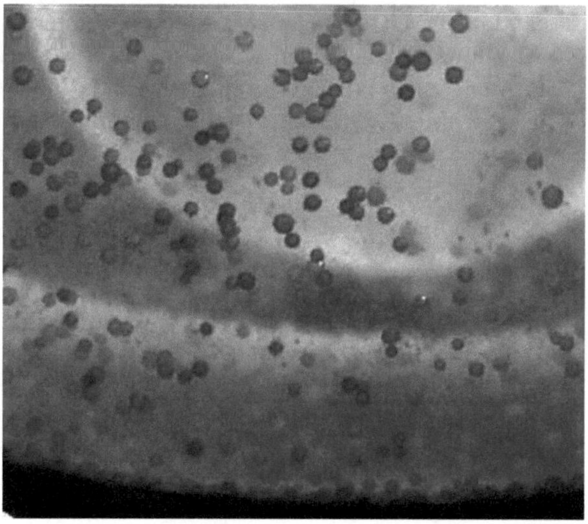

Fig. 10.1 CO_2 globules coated with a sheath of limestone particles and settling out of suspension. Reproduced from Ref. [24] with the permission from USDOE, National Energy Technology Laboratory

[24]. It is estimated that approximately 0.50–0.75 tons of pulverized limestone should be required to fix one ton of CO_2 into stable emulsions. It was noted that globules of 100–200 μm diameter were observed by using limestone particles with a size range of 6–13 μm, at a pH value of solution at 7–10 [24]. Furthermore, the seawater could be used instead of deionized water to form a stable emulsion.

10.4 Feedstock Processing and Activation

In mineral carbonation, the most important factor is the surface activities of the feedstock. Various studies have addressed various methods to accelerate the kinetics of direct aqueous carbonation, including physico-chemical pretreatment [25, 26], electrolysis [27], thermal heat or steam pretreatment [16, 27], and mechanical activation methods [4, 7, 18, 20]. It is possible to enhance the reactivity of the minerals considerably, while these methods require additional chemicals and energy.

10.4.1 Physico-chemical Pretreatment

In general, the chemical activations were found to be more effective than physical activation at increasing the surface area. Chemical activation utilizes the solvent, acid or base agents, to weaken the Ca or Mg bonds in a silicate structure. This would improve the dissolution kinetics of minerals, thereby increasing the efficiency of substantial carbonation. Many chemical solutions, such as ammonia, acetone, and hydrogen chloride, have been evaluated in the literature [26, 28, 29].

The "pH-swing" process is also a similar approach to enhancing the mineral dissolution and CO_2 conversion, which is commonly used in indirect carbonation [30]. It allows silicate minerals to be dissolved at a relatively low pH (<4.0) by acid treatment and then precipitates carbonates at a relatively high pH (>10.0) with base addition. This would result in more control of the carbonation process (see discussion in Chap. 8). However, it is much more expensive than traditional carbonation due to the cost of huge usage of chemicals.

10.4.2 Thermal Heat or Steam Pretreatment

As a pretreatment for natural minerals, thermal (heat) activation is commonly used to remove chemically bound water in the ores. In particular, for serpentines containing up to 13% of chemically bound water, the crystalline features are changed to amorphous ones following the decomposition of hydroxyl groups after heating to 600–650 °C [20, 27]. For heat pretreatment, O'Connor et al. [16] found that the maximum achievable conversion of serpentine could increase up to 78% using a supercritical CO_2 (185 atm) in 30 min at 155 °C and 15% solid contents.

After thermal heat treatment, the porosity and surface area of materials increase, creating the structural instability. This allows to promote the rate of carbonation afterward [20, 27]. However, the challenges in this process are that huge amounts of energy are required to maintain high temperatures, making it impractical for use as a large-scale treatment. Thus, another new approach is utilizing the exothermic reaction from mineral carbonation. It has been proved to be self-sufficient in terms of energy [31].

10.5 Challenges and Barriers in Mineral Carbonation

For mineral carbonation, extensive research studies have focused on the identification of reaction pathways and kinetics by bench-scale experiments and characterization of the mineralogy of the reactants and products. Although the CO_2 storage capacity of the natural Ca–Mg–silicate minerals is sufficient to fix the CO_2 emitted from the combustion of fossil fuels, mineral carbonation of natural ores is suffered from (1) slow kinetics and (2) energy demanding. These factors become the major challenges that hinder mineral carbonation from the large-scale use of silicate minerals. Thus, it suggests that research and implementation of these technologies require new collaborative efforts among the crushed stone and cement industries, electric utilities, and the science and engineering communities [12].

10.5.1 Slow Mineral Dissolution and Reaction Kinetics

Pure calcium or magnesium oxides are rarely found in nature minerals due to their high reactivity. Instead, they are present in a variation of oxides in a silicate matrix, such as $CaSiO_3$. Conversion of the silicate to carbonate does not occur in the solid state, but appears to require mineral dissolution in the aqueous phase [16]. In general, the dissolution of the mineral is considered to be the rate-limiting step in aqueous mineral carbonation process [20]. In other words, the accelerated mineral carbonation (or accelerated weathering) reaction is limited by the mass transfer phenomena, thereby resulting in a slow reaction kinetics and a huge reactor size. It is estimated that a reactor volume of roughly equivalent to a 60-m cube could achieve a 20% reduction of the CO_2 emissions from a typical 500-MW coal-fired power plant [32].

Furthermore, the types of reaction product are directly related to the required operating conditions. Typically, under ambient condition, calcium carbonate ($CaCO_3$) could be easily formed, compared to magnesium carbonate ($MgCO_3$). It is noted that mineral carbonation of olivine (rich in Mg contents) at 115 bar and 185 °C is limited by $MgCO_3$ (magnesite) precipitation, not the dissolution rate of olivine [18].

10.5.2 High Energy Consumption

The principle drawback of mineral carbonation using natural materials is the necessity of mining. Beside the energy-intensive mining process, with the present technology, there is always a net demand for high-grade energy to drive the mineral carbonation process [14, 33]. For example,

- preparation of the solid minerals, including mining, transport, grinding, and activation when necessary,
- processing, including the energy associated with recycling and possible losses of additives or catalysts,
- separation of valuable metals (e.g., iron oxide) from feedstock prior to accelerated carbonation or accelerated weathering, and
- disposal of carbonates and by-products.

The use of alkaline solid wastes has received a great attention as an alternative to natural feedstock since no additional mining process was required. Usually, the proximity between the CO_2 emission and the residue sources reduces transportation costs [34]. In addition, in an oxidizing atmosphere, the presence of magnetic material, especially oxides of iron, will negatively affect mineral carbonation by forming a passive layer of hematite [35]. Therefore, the separation of magnetic particles (e.g., iron oxide) from the mineral feedstock prior to mineral carbonation is recommended, where gravity and magnetic separation are the commonly used techniques [35].

The relevant details regarding accelerated carbonation using alkaline solid wastes, such as iron/steel slags and air pollution control ashes, are illustrated in the following chapters.

References

1. Tai CY, Chen WR, Shih S-M (2006) Factors affecting wollastonite carbonation under CO_2 supercritical conditions. AIChE J 52(1):292–299. doi:10.1002/aic.10572
2. Daval D, Martinez I, Corvisier J, Findling N, Goffé B, Guyot F (2009) Carbonation of Ca-bearing silicates, the case of wollastonite: experimental investigations and kinetic modeling. Chem Geol 265(1–2):63–78. doi:10.1016/j.chemgeo.2009.01.022
3. Kakizawa M, Yamasaki A, Yanagisawa Y (2001) A new CO_2 disposal process via artificial weathering of calcium silicate accelerated by acetic acid. Energy 26:341–354
4. Mckelvy MJ, Chizmeshya AVG, Diefenbacher J, Bearat H, Wolf G (2004) Exploration of the role of heat activation in enhancing serpentine carbon sequestration reactions. Environ Sci Technol 38(24):6897–6903
5. Krevor SCM, Lackner KS (2011) Enhancing serpentine dissolution kinetics for mineral carbon dioxide sequestration. Int J Greenhouse Gas Control 5(4):1073–1080. doi:10.1016/j.ijggc.2011.01.006
6. Alexander G, Mercedesmarotovaler M, Gafarovaaksoy P (2007) Evaluation of reaction variables in the dissolution of serpentine for mineral carbonation. Fuel 86(1–2):273–281. doi:10.1016/j.fuel.2006.04.034

7. Haug TA, Kleiv RA, Munz IA (2010) Investigating dissolution of mechanically activated olivine for carbonation purposes. Appl Geochem 25(10):1547–1563. doi:10.1016/japgeochem.2010.08.005
8. Teir S, Bacher J, Kentta E, Satlin J (2013) Silica produced from olivine for inkjet paper coating. In: Paper presented at the accelerated carbonation for environmental and material engineering, KU Leuven, Belgium
9. Lackner KS (2003) A guide to CO_2 sequestration. Science 300(5626):1677–1678
10. USGS (2015) Mineral Commodity Summaries 2015. U.S. Geological Survey
11. Caldeira KG, Knauss KG, Rau GH (2004) Accelerated carbonate dissolution as a CO_2 separation and sequestration strategy. Lawrence Livermore National Laboratory
12. Langer WH, San Juan CA, Rau GH, Caldeira K (2009) Accelerated weathering of limestone for CO_2 mitigation opportunities for the stone and cement industries. Min Eng 61(2):27–32
13. McClellan G, Eades J, Fountain K, Kirk P, Rothfuf C (2002) Research and technoeconomic evaluation: uses of limestone byproducts. University of Florida
14. Gerdemann SJ, O'Connor WK, Dahlin DC, Penner LR, Rush H (2007) Ex situ aqueous mineral carbonation. Environ Sci Technol 41(7):2587–2593
15. Lackner KS (2002) Carbonate chemistry for sequestering fossil carbon. Annu Rev Energy Environ 27(1):193–232. doi:10.1146/annurev.energy.27.122001.083433
16. O'Connor WK, Dahlin DC, Rush GE, Dahlin CL, Collins WK (2002) Carbon dioxide sequestration by direct mineral carbonation: process mineralogy of feed and products. Miner Metall Proc 19(2):95–101
17. Costa G, Baciocchi R, Polettini A, Pomi R, Hills CD, Carey PJ (2007) Current status and perspectives of accelerated carbonation processes on municipal waste combustion residues. Environ Monit Assess 135(1–3):55–75. doi:10.1007/s10661-007-9704-4
18. Haug TA, Munz IA, Kleiv RA (2011) Importance of dissolution and precipitation kinetics for mineral carbonation. Energy Procedia 4:5029–5036. doi:10.1016/j.egypro.2011.02.475
19. O' Connor WK, Dahlin DC, Rush GE, Gerdemann SJ, Penner LR, Nilsen DN (2005) Aqueous mineral carbonation: mineral availability, pretreatment, reaction parameters, and process studies. Albany Research Center (ARC), U.S.A
20. Park A, Fan L (2004) mineral sequestration: physically activated dissolution of serpentine and pH swing process. Chem Eng Sci 59(22–23):5241–5247. doi:10.1016/j.ces.2004.09.008
21. Symonds RT, Lu DY, Macchi A, Hughes RW, Anthony EJ (2009) CO_2 capture from syngas via cyclic carbonation/calcination for a naturally occurring limestone: modelling and bench-scale testing. Chem Eng Sci 64(15):3536–3543. doi:10.1016/j.ces.2009.04.043
22. Rau GH (2011) CO_2 mitigation via capture and chemical conversion in seawater. Environ Sci Technol 45:1088–1092
23. Arakaki T, Mucci A (1995) A continuous and mechanistic representation of calcite reaction-controlled kinetics in dilute solution at 25 & #xB0;C and 1 atm total pressure. Aquat Geochem 1:105–130
24. Plasynski S, Beckert H, Golumb DS (2008) Laboratory investigations in support of carbon dioxidide-limestone sequestration in the ocean. National Energy Technology Laboratory
25. Maroto-Valer MM, Tang Z, Zhang Y (2005) CO_2 capture by activated and impregnated anthracites. Fuel Process Technol 86(14–15):1487–1502. doi:10.1016/j.fuproc.2005.01.003
26. Maroto-Valer MM, Fauth DJ, Kuchta ME, Zhang Y, Andrésen JM (2005) Activation of magnesium rich minerals as carbonation feedstock materials for CO_2 sequestration. Fuel Process Technol 86(14–15):1627–1645. doi:10.1016/j.fuproc.2005.01.017
27. Li W, Li B, Bai Z (2009) Electrolysis and heat pretreatment methods to promote CO_2 sequestration by mineral carbonation. Chem Eng Res Des 87(2):210–215. doi:10.1016/j.cherd.2008.08.001
28. Power IM, Harrison AL, Dipple GM, Southam G (2013) Carbon sequestration via carbonic anhydrase facilitated magnesium carbonate precipitation. Int J Greenhouse Gas Control 16:145–155. doi:10.1016/j.ijggc.2013.03.011

29. Jung S, Dodbiba G, Fujita T (2014) Mineral carbonation by blowing incineration gas containing CO_2 into the solution of fly ash and ammonia for ex situ carbon capture and storage. Geosyst Eng 17(2):125–135. doi:10.1080/12269328.2014.930358
30. Park A-HA, Fan L-S (2004) Mineral sequestration: physically activated dissolution of serpentine and pH swing process. Chem Eng Sci 59(22–23):5241–5247. doi:10.1016/j.ces.2004.09.008
31. Moazzem S, Rasul MG, Khan MMK (2013) Energy recovery opportunities from mineral carbonation process in coal fired power plant. Appl Thermal Eng 51(1–2):281–291. doi:10.1016/j.applthermaleng.2012.09.021
32. Rau GH, Knauss KG, Langer WH, Caldeira K (2007) Reducing energy-related CO_2 emissions using accelerated weathering of limestone. Energy 32(8):1471–1477. doi:10.1016/j.energy.2006.10.011
33. IPCC (2005) IPCC Special Report on Carbon dioxide Capture and Storage. Intergovernmental Panel on Climate Change, Cambridge. ISBN-13 978-0-521-86643-9
34. Ben Ghacham A, Cecchi E, Pasquier LC, Blais JF, Mercier G (2015) CO_2 sequestration using waste concrete and anorthosite tailings by direct mineral carbonation in gas-solid-liquid and gas-solid routes. J Environ Manage 163:70–77. doi:10.1016/j.jenvman.2015.08.005
35. Veetil SP, Mercier G, Blais J-F, Cecchi E, Kentish S (2015) Magnetic separation of serpentinite mining residue as a precursor to mineral carbonation. Int J Miner Process 140:19–25. doi:10.1016/j.minpro.2015.04.024

Chapter 11
Iron and Steel Slags

Abstract Nowadays, iron and steel industries are moving toward environmental sustainability through careful control of greenhouse gas emissions and appropriate management of iron/steel manufacturing residues generated. Application of fresh iron and steel slags as an alternative to standard materials has been known for a number of years around the world. It was frequently used in asphalt mixtures, other layers of pavement structure, unbound base courses, and embankments. However, several barriers, including volume expansion of blended materials and concerns about environmental impacts and social acceptance, have been encountered. In this chapter, the types of iron and steel industries are first illustrated. The physico-chemical properties of four different types of iron and steel slags, including blast furnace slag, basic oxygen furnace slag, electric arc furnace slag, and ladle furnace slag, are then illustrated. In addition, the challenges in direct use of slag in civil engineering are summarized.

11.1 Iron and Steel Industries

Iron and steel mills are energy- and material-intensive industries because they have brought down the time for transformation from iron ore to steel within a day. Typically, the manufacturing process of iron and steel can be divided into two major groups: (1) integrated mill using natural ore to produce iron and steel and (2) electric arc furnace using scrap to remake iron and steel.

11.1.1 Integrated Mill Process

Figure 11.1 shows the schematic diagram of an integrated mill manufacturing process. It comprises of all the functions for primary steel production, including iron making (convert ore to liquid iron), steelmaking (convert pig iron to liquid steel), casting (solidification of the liquid steel), roughing and/or billet rolling

P.-C. Chiang and S.-Y. Pan, *Carbon Dioxide Mineralization and Utilization*, DOI 10.1007/978-981-10-3268-4_11

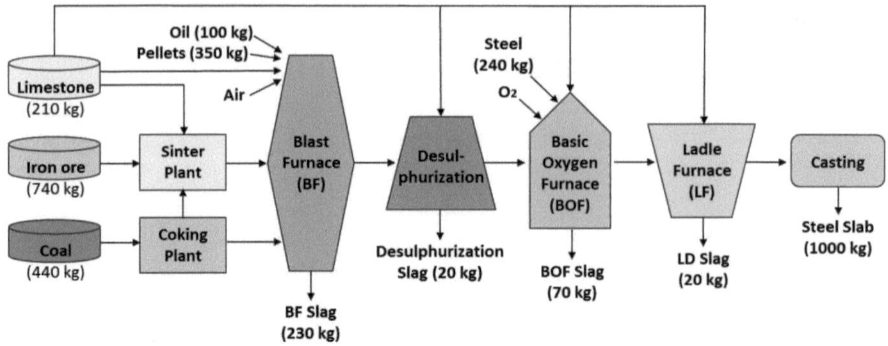

Fig. 11.1 Schematic diagram of the integrated steel manufacturing process. Rough values of some of the raw material and by-product streams are presented in parentheses

(reduce size of blocks), and product rolling (shaping). The principal raw materials for an integrated mill are coal (or coke), iron ore, and limestone, which are charged in batches into a blast furnace. In the blast furnace, the iron compounds in the ore react with excess oxygen and become liquid iron. Then, the liquid iron will either case into pig iron product, or direct to other vessel for further steelmaking (e.g., basic oxygen furnace). Typically, the final products from an integrated mill plant are large structural sections (such as heavy plate and railway rails) and long products (such as bars and pipes).

11.1.2 Electric Arc Furnace (EAF)

An electric arc furnace (EAF) is a furnace that heats a charged material by means of an electric arc. Figure 11.2 shows a typical EAF process for stainless steel manufacturing using scrap material. The EAF normally comprises of three graphite electrodes, molten bath, tapping spout, refractory movable brick roof, and brick shell. The EAF is built on a tilting platform so that the liquid steel can be poured into another vessel for transport. After the EAF process, the molten liquid steel is sent to ladle refining furnaces, such as argon oxygen decarburization (AOD) process and desulfurization process, to produce high-quality steel prior to casting.

In a typical EAF process, the energy sources include electrical energy and chemical energy resulting from oxidation reactions, while the energy sinks comprise the off-gas extraction and furnace cooling systems. Industrial EAF temperatures can be up to 1800 °C (3272 °F). To melt a tonne of scrap steel, the theoretical minimum amount of energy required is 300 kWh (melting point at 1520 °C or 2768 °F). In Germany, the electrical energy requirements of EAF plants are approximately 500 kWh/t_{steel} [1]. The energy supplied to the EAF process, including electrical energy and chemical energy, is released from the combustion of

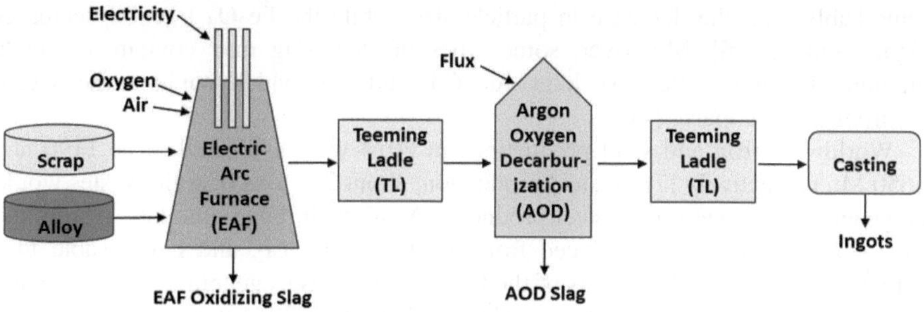

Fig. 11.2 Stainless steel manufacturing process from scrap material

nature gas, liquefied petroleum gas or oil, and the oxidation of elements in the melt during refining (e.g., C, Si, Al, Fe, Cr, and Mn) [1].

It is noted that the total energy consumption and CO_2 emissions levels in the EAF process are significantly lower than that for oxygen steelmaking due to the high proportion of recycled scrap (i.e., 1389–4250 kWh per ton steel production) and the use of electrical energy (i.e., 0.15–1.08 t-CO_2 per ton steel production) [2]. On the other hand, the total primary energy consumption of steelmaking defined in a global life cycle inventory study is 5960–8810 kWh/t with a total CO_2 emission between 1.61 and 2.60 t-CO_2 per ton steel production [2].

11.2 Types of Iron and Steel Slags

Steel slag is an inevitable solid waste of the steel manufacturing industry characterized by its strongly alkaline nature and significant levels of metal ions, especially calcium. Appropriate stabilization and utilization of iron and steel slags are challenging because these materials are highly active chemically. In the USA, Japan, German, and France, the fraction of slag utilization is close to 100%, where a half of the produced slag has been directly used as road base, with the remainder used for sintering and iron-making recycling [3]. The utilization of slags can not only provide a partial solution to environmental pollution caused by improper management of slag but also modify the microstructure and significantly improve the durability of cement/concrete. There are five main types of iron and steelmaking slags, including electric arc furnace (EAF), blast furnace (BF), desulphurization (DS), basic oxygen furnace (BOF), and ladle furnace (LF) slags, each named for the processes where they are produced.

Iron and steel slags can be generally categorized as iron slag, carbon steel slag, and stainless steel slag [3]. The chemical composition of iron and steel slags, primarily in the form of Ca and Mg silicates, is similar to natural sand, gravel, and crushed stones. The composition of slags varies widely depending on the source and particle sizes [4]. For instance, the CaO and SiO_2 contents in BOF slag increase remarkably with the decrease in particle size, while the Fe_2O_3 fraction decreases significantly [5, 6]. Moreover, some types of steel slag may contain a notable amount of heavy metals, such as As, Cd, and Hg, which could cause severe environmental impacts [4].

Worldwide iron and steel productions in 2014 were estimated to be 1190 and 1650 Mt, respectively [7]. In steel production, about 2–4 tons of solid wastes would be generated for one ton of steel produced. As a result, huge amounts of iron and steel slags are annually produced from the BF, BOF, LF, and EAF. Table 11.1 summarizes the typical amounts of the typical solid wastes generated from iron and steelmaking plants. Worldwide iron slag production in 2014 was estimated on the order of 310–370 Mt, while steel slag was about 170–250 Mt [7]. In the USA, the steel produced via either the integrated manufacturing process (e.g., BOF) or the EAF process is approximately 37 and 63% of total steel production, respectively [7].

Figure 11.3 shows a normalized phase diagram of $CaO(MgO)–SiO_2(Na_2O, K_2O)–Al_2O_3(Fe_2O_3)$ for various types of iron and steel slags. It is noted that the mineralogical compositions of the steel slags are generally more complex than those of other types of residues, such as air pollution control ashes [15]. They are a consolidated mix of many compounds, e.g., primarily calcium, iron, silicon, aluminum, magnesium, and manganese oxides, presented in different phases [16]. If the slag is used in cement and concrete, the CaO and SiO_2 contents are primarily related to the hydraulic and pozzolanic properties, respectively. The CaO and MgO contents also determine the CO_2 fixation capacity for accelerated carbonation or CO_2 curing for cement. On the other hand, the Fe_2O_3 content contributes to hardness and grindability of a slag; for instance, the BOFS, DSS, and EAFOS are relatively hard materials due to their high Fe_2O_3 content. Consequently, the mineralogical investigation is crucial to identify the possible CO_2-reacting phases and the main reaction products contributed by the carbonation process.

Table 11.1 Types of solid wastes generated from iron and steelmaking plants

Source of generation	Solid wastes	Hot metal (kg/t)	References
Blast furnace	Blast furnace slag (BFS)	250–420	[4, 8]
	Blast furnace dust	5–30	[4, 9]
Basic oxygen furnace	Basic oxygen furnace slag (BOFS)	50–200	[10–12]
Steel melting shop	Ladle furnace slag (LFS)	150–200	[4]
	Ladle furnace sludge	15–16	[4]
	Refractory, bricks	5–12	[4, 9]
	Fly ash	7–15	[9]
Electric arc furnace	Electric arc furnace slag (EAFS)	200	[13]
	Electric arc furnace dust	10–20	[14]
Rolling mills	Mill sludge	12	[4]
	Mill scale	22–33	[4, 12]

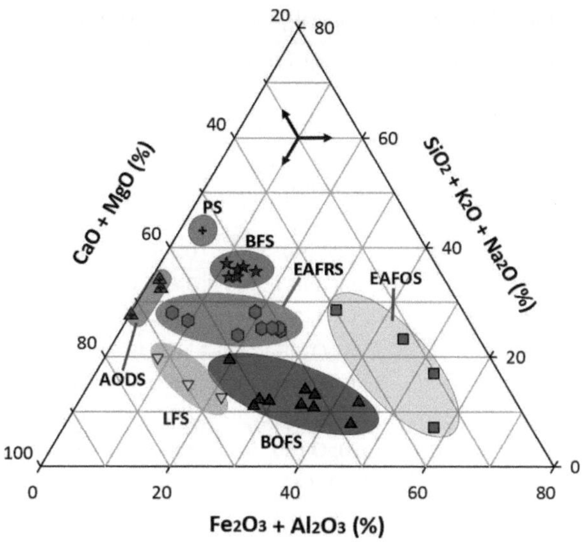

Fig. 11.3 Normalized CaO (MgO)–SiO_2(Na_2O,K_2O)–Al_2O_3(Fe_2O_3) phase diagram for various types of iron and steel slags. Acronyms: blast furnace slag (*BFS*); basic oxygen furnace slag (*BOFS*); electric arc furnace reducing slag (*EAFRS*); electric arc furnace oxidizing slag (*EAFOS*); ladle furnace slag (*LFS*); phosphorus slag (*PS*); argon oxygen decarburization slag (*AODS*)

In the following sections, the physicochemical properties of blast furnace slag (BFS), basic oxygen furnace slag (BOFS), electric arc furnace slag (EAFS), and ladle furnace slag (LFS) are illustrated.

11.2.1 Blast Furnace Slag (BFS): Physico-chemical Properties

Blast furnace slag (BFS) is a coproduct of the pig iron production and has been widely used in various applications for many decades. Two types of BFS are generated from steel plants: (1) air cooled BFS and (2) granulated (crystallized) BFS. A portion of the BFS is drenched in water, thereby resulting in a glassy granulated material with latent hydration properties [17]. Table 11.2 presents the physico-chemical properties of BFS used in the literature. The specific gravity of BFS is about 2.90 with bulk density ranging from 1.0 to 1.3 g/cm^3. In addition, BFS contains numerous inorganic constituents, including SiO_2 (30–36%), CaO (32–45%), Al_2O_3 (10–16%), MgO (6–9%), and Fe_2O_3 (0.2–1.5%), where their crystal structure is mainly amorphous [18]. It is also noted that the BFS is rich in silicon–aluminum components, which is suitable as the raw materials for geopolymer preparation.

BFS is whitish in color with low iron content and high calcium content, being similar to the compositions of clinker. Therefore, air-cooled BFS is frequently used as a substitute for stones, aggregates, and chips for road construction [4]. Similarly, granulated BFS can be used as raw components for cement production or as a standardized cement component for the granulated state or as aggregate for the

Table 11.2 Physico-chemical properties of blast furnace slags (BFS) used in the literature

Items	Components	BFS used in the literature						
		[4, 19]	[20]	[21]	[22]	[23]	[24]	[17]
Physico-chemical properties	Density (g/cm^3)	1.2–1.3	–	–	–	–	–	1.0–1.3
	Surface area (cm^2/g)	–	–	4000	–	–	–	–
	Basicity (-)	–	–	–	1.34	–	–	–
	LOI (%)	0.4	0.0	–	0.01	–	4.57	–
XRF (solid phase)	SiO_2 (%)	34.6–36.0	33.2	32.3	30.2	32.3	31.0	41.2
	CaO (%)	39.0–41.0	38.2	39.4	40.4	45.2	37.2	32.7
	Al_2O_3 (%)	11.0–12.5	16.3	10.5	10.8	10.6	14.3	14.3
	Fe_2O_3 (%)	0.5	0.3	0.5	0.6	0.7	1.5	0.8
	P_2O_5 (%)	–	0.10	–	–	0.1	–	–
	MgO (%)	8.4–8.7	8.1	8.7	–	6.4	9.1	7.3
	SO_3 (%)	–	N.D.[a]	3.2	3.2	2.0	0.1	0.8
	MnO (%)	0.3	–	0.5	–	0.3	–	0.8
	TiO_2 (%)	–	–	–	6.0	0.5	–	1.1
	K_2O (%)	–	–	0.8	0.6	0.4	0.6	0.9

[a]N.D. Not detected (under detection limit)

crystallized state in construction [4, 25]. The granulated BFS can be safely used in the cement manufacturer; e.g., up to ~50% of clinker could be replaced by granulated BFS as additives in the production of cement [4]. Since the excellent hydration properties of BFS, the proven benefits in cement quality by partially replacing Portland cement with BFS include the following [4]:

- low heat of hydration,
- good long-term strength,
- controlled alkali–silica reaction,
- resistance to acid,
- improved durability by the addition of granulated slag, and
- cost-saving measure to the cement industries.

11.2.2 *Basic Oxygen Furnace Slag (BOFS): Physico-chemical Properties*

Basic oxygen furnace slag (so-called steel converter slag) was produced simultaneously with the steel as the lime reacts with the impurities in the basic oxygen furnace. In the BOF, the oxygen is blown into the furnace to remove the carbon

Table 11.3 Chemical properties of basic oxygen furnace slags (BOFS) used in the literature and its associated CO_2 capture capacity

Items	Components	BOFS-1	BOFS-2	BOFS-3 [20]	BOFS-4 [26]
XRF (solid phase)	SiO_2 (%)	8.6–13.1	7.8–12.1	19.5	11.1
	CaO (%)	40.1–45.0	38.4–39.3	47.6	46.4
	Al_2O_3 (%)	1.7–2.1	1.0–2.7	2.67	1.2
	Fe_2O_3 (%)	28.3–32.0	22.5–38.1	16.8	23.2
	P_2O_5 (%)	1.4–2.4	–	1.76	2.1
	MgO (%)	4.5–7.5	8.6–9.0	7.58	8.4
	Cr_2O_3 (%)	–	–	–	–
	TiO_2 (%)	0.5–0.9	0.9	–	0.4
	SO_3 (%)	0.4–1.2	–	N.D.[a]	–
	ZnO (%)	–	0.03	–	–
	MnO (%)	2.0–4.1	3.6–4.3	–	2.7
Chemical analysis	f-CaO (%)	2.0–8.1	–	–	–
	f-MgO (%)	–	–	–	–
	$Ca(OH)_2$ (%)	1.1–7.3	–	–	–
Thermal analysis	$CaCO_3$ (%)	1.9–4.0	–	–	–
	LOI (%)	0–8.3	9.0	0	–

[a]N.D. Not detected (under detection limit)

from the molten iron, thus producing low-carbon steel [25]. Depending on the grade of steel produced, approximately 100–200 kg of BOFS can be generated for each ton of steel production [10]. Table 11.3 presents the chemical properties of basic oxygen furnace slag (BOFS). The major oxides in BOFS are CaO (38–48%), Fe_2O_3 (17–38%), SiO_2 (9–20%), and MgO (6–9%). A portion of CaO in BOFS presents in terms of $Ca(OH)_2$ (1–7%) and free-CaO (2–8%). Since CaO is a lipophilic molecule, the BOFS can be closely integrated with the asphalt, reducing the phenomenon of stripping and enhancing the resistance to water [26].

Table 11.4 presents the mineralogical composition of BOFS in the literature. β-C_2S (38–52%) is the main form of calcium silicate species in BOFS, and it is the active polymorph presented in clinker [27, 28]. Other common mineralogy of BOFS includes C_2F (20–30%), C_3S, C_4AF, RO phase (CaO-FeO-MgO solid solution), olivine, and merwinite [29, 30]. It is noted that the C_3S phase is present in basic BOFS if the CaO/SiO_2 ratio is higher than 2.7 [31]. Moreover, the minor phases, such as calcite ($CaCO_3$), periclase (MgO), and quartz (SiO_2), have been identified in particular samples [25, 32]. The contents of SiO_2 and Al_2O_3 in BOFS are significantly lower than those of granulated BFS, while the Fe_2O_3 content in BOFS is much higher than that of granulated BFS.

Table 11.4 Mineralogical compositions of basic oxygen furnace slags (BOFS) in the literature

Mineral names	Chemical formula	References		
		[33]	[34]	[25]
Srebrodolskite	$Ca_2Fe_2O_5$ (C_2F)	32.3	37.2	30.3
β-C_2S	Ca_2SiO_4	30.8	21.5	41.8
γ-C_2S	Ca_2SiO_4	–	7.3	
Lime	CaO	8.8	10.6	2.0
Iron	Fe	–	0.4	–
Wustite	FeO	6.7	1.1	11.9
Hematite	Fe_2O_3	–	4.1	–
Magnetite	Fe_3O_4	–	0.8	–
Ferrosilite	$FeSiO_3$	–	2.8	–
Fayalite	Fe_2SiO_4	3.8	5.5	–
Portlandite	$Ca(OH)_2$	3.1	–	6.3
Wollastonite	$CaSiO_3$	2.9	1.1	–
Brucite	$Mg(OH)_2$	2.4	–	–
Ferrosilite	$(Fe(II),Mg)_2Si_2O_6$	2.2	–	–
Clinoenstatite	$Mg_2Si_2O_6$	1.9	–	–
Enstatite	$MgSiO_3$	–	3.8	–
Periclase	MgO	1.1	0.7	–
Calcite	$CaCO_3$	–	–	2.9
Quartz	SiO_2	–	2.8	4.7
Other (<1 wt%)	–	4.0	–	–

With similar chemical compositions to Portland cement, the exploitation of BOFS in some hydraulic binders could be considered. However, due to the high Fe_2O_3 content in BOFS, the absolute density is as high as 3.7 g/cm^3, thereby requiring a greater grinding work than clinker. Moreover, since iron is relatively inert with water contact, the high iron content would minimize the hydraulic activity of blended cement [25]. As a result, a reclamation process for iron components in BOFS prior to carbonation or utilization should be considered.

Furthermore, due to its low content of C_3S as well, blended cement and concrete with BOFS exhibit poor hydraulic activity at early ages of activation [35]. This can be enhanced by the addition of $CaCl_2$ to improve the mechanical properties because it can form a complex with $Ca(OH)_2$ [25, 36]. Among several available admixtures ($CaCl_2$, CaO, NaCl, and Na_2SiO_3), only $CaCl_2$ is effective in improving the mechanical properties of blended cement at early ages [25].

Table 11.5 Chemical properties of electrical arc furnace slag (EAFS) used in the literature

Items	Components	EAFOS-1	EAFOS-2	EAFOS-3 [38]	EAFRS
XRF (solid phase)	SiO_2 (%)	6.0–15.4	13.1–23.3	14.9–42.2	26.6–28.2
	CaO (%)	24.4–29.1	23.9–33.0	5.6–39.6	40.8–49.4
	Al_2O_3 (%)	12.2–14.1	4.1–7.4	1.8–12.3	8.4–17.6
	Fe_2O_3 (%)	27.4–34.4	24.1–36.8	0.9–48.3	1.1–1.6
	Na_2O (%)	0.1–0.2	–	–	–
	P_2O_5 (%)	1.19–1.24	–	–	–
	MgO (%)	2.9–3.4	5.1–12.0	1.9–17.6	6.2–9.8
	Cr_2O_3 (%)	0.7–1.0	0.8	–	–
	TiO_2 (%)	–	0.6–0.9	–	–
	SO_3 (%)	–	0.03–0.14	0.01–0.08	0.04–0.38
	K_2O (%)	1.1–1.5	–	–	–
	PbO (%)	0.22–0.32	–	–	–
	MnO (%)	5.6–15.6	4.2	–	–

11.2.3 Electric Arc Furnace Slag (EAFS): Physico-chemical Properties

Electric arc furnace slag (EAFS) can be categorized into two types:

- EAF oxidizing slag (EAFOS), with a lower CaO content and
- EAF reducing slag (EAFRS), with a higher CaO content.

Table 11.5 presents the chemical properties of different kinds of EAFS. Compared to EAFRS or other steel slags, the EAFOS exhibits great hardness and grinding resistance due to its high iron oxide content (24–37%), commonly displaying resistance to erosion and fragmentation [37]. Moreover, the concentrations of free-lime and total sulfate (expressed as SO_3) present in EAFS are usually below 0.1 and 0.6%, respectively.

Rapid cooling of EAFS is necessary to avoid the defects of disintegration in slag by converting the unstable polymorph C_2S phase to the low-temperature stable form (such as γ-C_2S), resulting in an increase of 10% in volume [37]. In general, both oxidizing and reductive EAFS can be used as aggregates and cementitious materials in concrete, respectively [39]. The porous nature of EAFS could provide a strong interlocking effect for concrete or cement development [38]. According to X-ray diffraction analysis, the main crystalline phases in EAFS include the following [32, 37]:

- Iron oxides
  - wustite [FeO],
  - magnesioferrite [$MgFe_2O_4$],
  - magnetite [Fe_3O_4], and
  - Hematite [Fe_2O_3].

- Silicates
  - larnite [Ca_2SiO_4],
  - bredigite/merwinite [$Ca_{14}Mg_2(SiO_4)_8$], and
  - gehlenite [$Ca_2Al(Al,Si)_2O_7$].
- Manganese oxides
  - birnessite [$(Na_{0.3}Ca_{0.1}K_{0.1})(Mn^{4+},Mn^{3+})_2O_4 \cdot 1.5\ H_2O$],
  - hausmannite [Mn_3O_4],
  - rutile/hollandite [MnO_2], and
  - groutellite.

Due to its low pozzolanic activity and the possibility of long-term expansion reaction, fresh EAFS might not be a suitable feedstock as supplementary cementitious materials and/or active additions. According to the result in the literature [32], cement mortar blended with fresh EAFS did not exhibit pozzolanic activity within 90 days. Regarding the long-term volume instability, the presence of the periclase crystal (MgO) in fresh EAFS should be the main challenge. It was noted that the free-MgO content in EAFS was typically below 1% for the various particle sizes [32]. The non-combined MgO contents in EAFS should be a vitreous or amorphous phase, which is responsible for the long-term expansion reaction [32].

11.2.4 Ladle Furnace Slag (LFS): Physico-chemical Properties

The ladle refining process is regarded as the secondary metallurgy process, whose purpose is to convert the molten pig iron and/or steel scraps into high-quality steel. Ladle furnace slag (LFS), so-called refining slag, comes from the refining process of pig iron or molten steel via various types of ladle furnace converters [4, 40]. For example, the desulphurization slag (DS) would be generated from the desulphurization process for molten liquid steel. Table 11.6 presents the chemical properties of different kinds of LFS, including DS, argon oxygen decarburization slag (AODS), and phosphorus slag (PS). It is noted that calcium, silicon, magnesium, aluminum, and iron oxides represent more than 95% of the total mass in LFS. The extractable CaO content (e.g., free-CaO) is typically higher than 6% in the as-received LFS with small particle (less than 75 μm), while lower than 2% for LFS with large particles (75–600 μm) [41].

Calcium silicates are the major components in the fresh LFS, where the typical mineral phases include diopside [$MgCaSi_2O_6$], merwinite [$Ca_3Mg(SiO4)_2$], rankite [$Ca_3Si_2O_7$], wollastonite [$CaSiO_3$], larnite [β-Ca_2SiO_4], bredigite [$Ca_7Mg(SiO_4)_4$], ingesonite [(γ-Ca_2SiO_4)], and calcium–olivine. During the cooling process of LFS at temperatures of 450–500 °C, the high-temperature stable form of dicalcium silicate, i.e., larnite (β-C_2S), could be gradually converted into the low-temperature stable form of belite (γ-C_2S). This is responsible for the initial cracking of the LFS

Table 11.6 Physico-chemical properties of ladle refining furnace slags used in the literature

Items[a]	Components	LFS(unspecified)	AODS	PS	DS [42]
Physical properties	Specific surface (cm^2/g)	–	–	–	3829
	Specific gravity	–	–	–	2.38
XRF (solid phase)	SiO_2 (%)	12.6–19.8	32.5–34.1	43.1	15.7
	CaO (%)	50.5–57.5	54.5–54.8	46.7	66.2
	Al_2O_3 (%)	4.3–18.6	1.1–1.4	2.6	3.5
	Fe_2O_3 (%)	1.6–3.3	0.2–0.3	0.8	7.4
	Na_2O (%)	0.03–0.07	–	–	–
	P_2O_5 (%)	~0.01	–	–	0.6
	MgO (%)	7.5–11.9	8.0–9.0	1.2	2.3
	Cr_2O_3 (%)	0.01–0.1	~0.8	–	–
	TiO_2 (%)	0.18–0.89	~0.4	–	–
	SO_3 (%)	0.9–1.5 (as S)	~0.2	–	3.4
	K_2O (%)	0.01–0.02	–	–	–
	MnO (%)	0.36–0.52	~0.4	0.1	0.5
Chemical analysis	f-CaO (%)	3.5–19.0	–	–	–
	f-MgO (%)	3.0–8.0	–	–	–
	$Ca(OH)_2$ (%)	–	–	–	–
Thermal analysis	$CaCO_3$ (%)	–	–	–	–
	LOI (%)	1.2–5.5	0.1	0.4	–

[a]Argon oxygen decarburization slag (*AODS*); phosphorus slag (*PS*); desulfurization slag (*DS*)

monolithic mass and the resultant by-product in the form of a gravel. Other major crystal phases in fresh LFS include portlandite [$Ca(OH)_2$], hematite [Fe_2O_3], wustite [FeO], magnetite [Fe_3O_4], and periclase [MgO], with minor compounds such as jasmundite, fluorite, brucite, and aluminum [9, 40].

A fine LFS with high fineness and specific surface area can be used to construction or industrial processes [40]. However, practical application of the LFS differs from BFS, BOFS, and EAFS due to its greater variety in composition and structure [40] than EAFS [43] and BOFS [44, 45]. The use of LFS in each particular engineering application should be carefully analyzed in accordance with the composition and structure. It suggests that the observable prevalence of aggregate particle sizes should be about 50–60 μm [40].

11.3 Challenges in Treatment, Disposal, and/or Utilization

Huge amounts of iron and steel slags are generated annually worldwide from integrated manufacturing processes and/or the electric arc furnace. Conventionally, the treatment approach for BOFS is dumping and partial reuse as an aggregate for civil engineering due to its good technical properties. Due to the excellent characteristics, the iron and steel slags could be utilized in various fields and applications:

- artificial reef [46];
- aggregate in concrete [47, 48] and in asphalt mixture [49];
- supplementary cementitious materials [50, 51];
- catalyst in humification [52] or Fenton reaction [53];
- enhanced chemical oxidation for groundwater remediation [54];
- adsorbent for hazardous substances from wastewater, such as ionic copper and lead [55] and phosphates [56];
- geopolymer for immobilizing heavy metals, such as Cu^{2+}, Zn^{2+}, Pb^{2+}, and Cd^{2+} [22], or structural geopolymer concrete [57]; and
- ceramic tiling [17].

For instance, the Fe oxides and/or oxyhydroxides (Fe-oxys) components in BOFS can provide a high adsorption capacity for phosphates. Since the adsorption mechanism involves an inner-sphere ligand exchange, the affinity of phosphates to Fe-oxys depends on the complexing capacity of anions by the ligand exchange, attractive (or repulsive) electrostatic interactions with charged surface [56]. In this case, the pH of the solution plays a crucial role in the phosphate removal efficiency.

In the USA, steel slag is currently sold and used for various applications, predominantly for asphalt and concrete aggregate and road base [58]. Research on the utilization of iron and steel slags has been carried out in the following fields: (a) aggregate in concrete or asphalt, (b) cement industry, (c) road construction, (d) glassmaking, (e) ceramic tiles, (f) soil conditioner for agriculture, (g) aquaculture, and (h) land amendments [59], as shown in Table 11.7. However, conventional uses of untreated iron and steel slags in civil engineering have encountered several technological barriers, such as fatal volume expansion, heavy metal leaching, and low cementitious property of slag.

Table 11.7 Utilization of iron and steel slags around the world

Types of slags	Production techniques	Typical applications
Blast furnace slag (BFS)	Air cooled	Base, subbase, concrete aggregate, filter aggregate, construction fill, scour protection, rockwool, etc.
Granulated blast furnace slag (GBFS)	Molten slag quenched with high-pressure water sprays	Subbase, construction fill, construction sand, stabilizing binder, grit blasting, glass, cement manufacture, etc.
Basic oxygen furnace slag (BOFS)	Air cooled and watered	Base, subbase, asphalt aggregate, sealing aggregate, construction fill, subsoil drains, grit blasting, adsorbent for phosphate, etc.
Electric arc furnace slag (EAFS)	Air cooled and watered	Base, subbase, subsoil drains, grit blasting, asphalt aggregate, etc.

In the following parts, conventional pathways of direct utilization for fresh iron and steel slags in civil engineering, such as (1) road and pavement engineering, (2) hydraulic engineering, (3) fine aggregates in concrete, and (4) supplementary cementitious materials in blended cement are illustrated.

11.3.1 Coarse Aggregates in Road and Pavement Engineering

Road consists of a number of layers, including asphalt surface layer, asphalt binder course, asphalt base course, unbound base course, frost protection layer, foundation, and subgrade. In modern road, steel slag can be used as a coarse aggregate in surface layers of the pavement or in unbound bases and subbases, especially in asphaltic surface layers due to its high level of strength, high binder adhesion, and high frictional and abrasion resistance [4, 52, 60]. In Japan and Europe, approximately 60% of steel slag has been used for road engineering; in the UK, 98% of steel slag is utilized as aggregates of cement and bituminous pavement [3].

Excellent mechanical properties can be exhibited when fresh steel slag was used as coarse aggregates in bituminous mixes asphalt [61], cold-mix recycling asphalt pavement [62], hot-mix asphalt [63], and stone mastic asphalt (SMA) pavement [49]. According to Asi [64], the asphalt concrete containing 30% of fresh steel slag exhibited the highest skid number, compared to the Superpave, conventional SMA, and Marshall mixes. For high-tracked road layers, especially for asphaltic surface layers, several unique features have qualified them as coarse aggregates in concrete [60, 62], such as the following:

- high polished stone value,
- binder adhesion >90%,
- enhanced Marshall stability, resilient modulus, and tensile strength, and
- improved resistance to moisture damage.

In the case of SMA containing fresh steel slag, the performance as a surface friction course was comparable with that of conventional asphalt pavement [49] and even better in some aspects [65]. Although the substitution of steel slag for basalt increases the optimal bitumen content slightly, the performances of volume stability for SMA mixture can meet the related requirements of specifications [49].

11.3.2 Coarse Aggregates in Hydraulic Engineering

Steel slag can be used for dams and dikes, stabilization of river bottoms, refilling of erosion areas on river bottoms, and stabilization of riverbanks. Due to its high apparent density, strength, rough texture, and durability, steel slag can be processed as high-quality aggregates comparable with natural aggregates. For application in hydraulic engineering, steel slag with bulk density and particle sizes higher than 3.2 g/cm^3 and 10 mm, respectively, is typically required to offer a high level of strength, abrasion, and a rough texture. Therefore, steel slag is qualified as construction materials for hydraulic engineering purposes to prevent the erosion of fine particles [3, 60]. These features of aggregates can ensure long-term resistance to dynamic forces coming from waves and river flow [60]. In Germany, approximately 400,000 tons of steel slag are annually used as aggregates for the stabilization of riverbanks and riverbeds against erosion [60].

11.3.3 Fine Aggregates in Concrete Blocks

Iron and steel slags can be used as fine aggregates for high-strength and refractory concrete [3]. The compressive strength of concrete using steel slags as fine aggregates is 1.1–1.3 times higher than that of conventional concrete [66], becoming a high-strength concrete of greater than 70 MPa [67] and even reaching 100 MPa [68]. Moreover, other important properties of concrete can be improved, such as fire resistance up to 400 °C [69], excellent water-reducing effect [70], and excellent performance of anti-chloride ion penetration [68]. Several methods have been proposed to improve the performance of steel slag as fine aggregates. For example, Ducman and Mladenovic [71] developed a process for reheating slag up to temperatures of ~1000 °C prior to its use for refractory concrete. The results indicated that the final products blended with the pretreated EAFS exhibited comparable mechanical properties to the conventional bauxite aggregates.

11.3.4 Supplementary Cementitious Materials (SCM) in Cement

Iron and steel slags with a fine particle size (typically less than 125 μm) can be used as supplementary cementitious materials (SCMs) in cement and concrete. With the use of slag as SCM, formation of the hydrated products of calcium aluminates could provide small or moderate development of mechanical strength [40]. The cementitious properties of slag normally increase with its basicity, such as the presence of C_3S, C_2S, and C_4AF. For instance, anhydrous calcium silicates and aluminates of slag could be potentially reactive with water at room temperature [40, 45].

BFS (iron slag) is particularly suitable for high-performance concrete because the hardening (hydration) of concrete with BFS is generally slower than that with Portland cements [72]. The ground BFS powder can be used along with clinker and gypsum for the production of Portland cement [73] and belite cement [74], due to the compositions similar to those of clinker (high CaO to FeO ratio), or used as

SCMs in cement and concrete [75]. It was noted that the compressive strength of a mixture of phosphogypsum, granulated BFS, steel slag, and limestone exceeded 40 MPa at 28-d [76]. The hydration products were found to be mainly ettringite and C–S–H gel.

Similarly, a certain amount of steel slag addition as SCMs in cement could improve the porosity and pore distribution, and increase the consistency of cement [77]. It was noted that blended cement with 30% steel slag fine powder addition could meet the Turkish standard requirements for Portland cement [78]. However, fresh BOFS (one type of steel slag) typically contains 3–10% free-CaO and 1–5% free-MgO [6], which would lead to fatal expansion of hardened cement-BOFS paste. This has limited its application as SCMs in cements or aggregates in concrete. Furthermore, because of its high iron oxide content, fresh BOFS may not be a suitable material as SCMs. In some circumstances, BOFS needs to be subjected to metal recovery processes before its application and utilization [4].

11.4 Incorporated CO_2 Emission Reduction with Slag Stabilization

Iron and steel mills accounts for 22% of total industrial energy use in 2011 [79] and 6–7% of total CO_2 emissions worldwide [80]. Although CO_2 emission from iron and steel industrial plants is relatively lower than that from power generation plants, effective control of greenhouse gas emission in the steelmaking industry is important in moving toward environmental sustainability. Beside the CO_2 emission, large amounts of iron and steel slags are generated from the iron and steel productions. During steel production, 2–4 tons of solid wastes are generated per ton of steel produced [4]. Appropriate stabilization and utilization of iron and steel slags are challenging because these materials are highly active chemically.

Substitution of iron and steel slags in cement and concrete can not only conserve dwindling resources but also reduce the environmental impacts of quarrying and exploitation of the raw materials for cement manufacturing. Aside from iron and steel slags, different types of solid wastes, such as fly ash, silica fume, by-product gypsum, rice husk ash, red mud, and lignin-based materials, are suitable to partially substitute clinker or Portland cement in mortar. It is noted that during cement production, a large amount of air pollutants and CO_2 will be generated. The carbon footprint of cement industry is approximately 0.8–1.3 ton CO_2 per ton of cement production in the dry process [81], associated with a large amount of SO_2 emission depending upon the type of fuel used.

As a result, an integrated approach to capturing CO_2 while also improving the environmental and mechanical properties of iron/steel slags appears to be highly desirable. CO_2 capture by carbonating the iron and steel slags could be an interesting option for reducing CO_2 emissions from steel plant. Recently, accelerated carbonation of iron and steel slags has been progressively evaluated to determine the CO_2

fixation potential for various types of slags [51, 82, 83]. Prior to going through the attractive process, in this section, the iron and steel manufacturing processes are briefly illustrated. While global theoretical CO_2 emission reduction potential of steel slag carbonation is only 170 Mt/year (estimation of the world's annual slag production vary from 220 and 420 Mt, with a CaO content of 34 $\sim$ 52%), the reduction could be quite significant for an individual steel mill [84].

References

1. Kirschen M, Risonarta V, Pfeifer H (2009) Energy efficiency and the influence of gas burners to the energy related carbon dioxide emissions of electric arc furnaces in steel industry. Energy 34(9):1065–1072. doi:10.1016/j.energy.2009.04.015
2. Sandberg H, Lagneborg R, Lindblad B, Axelsson H, Bentell L (2001) CO_2 emissions of the Swedish steel industry. Scand J Metall 30:420–425
3. Yi H, Xu G, Cheng H, Wang J, Wan Y, Chen H (2012) An overview of utilization of steel slag. Procedia Environ Sci 16:791–801. doi:10.1016/j.proenv.2012.10.108
4. Das B, Prakash S, Reddy PSR, Misra VN (2007) An overview of utilization of slag and sludge from steel industries. Resour Conserv Recycl 50(1):40–57. doi:10.1016/j.resconrec.2006.05.008
5. Pan SY, Chiang PC, Chen YH, Chen CD, Lin HY, Chang EE (2013) Systematic approach to determination of maximum achievable capture capacity via leaching and carbonation processes for alkaline steelmaking wastes in a rotating packed bed. Environ Sci Technol 47(23):13677–13685. doi:10.1021/es403323x
6. Zhang T, Yu Q, Wei J, Li J, Zhang P (2011) Preparation of high performance blended cements and reclamation of iron concentrate from basic oxygen furnace steel slag. Resour Conserv Recycl 56(1):48–55. doi:10.1016/j.resconrec.2011.09.003
7. USGS (2015) Mineral Commodity Summaries 2015. U.S. Geological Survey
8. Eloneva S, Teir S, Salminen J, Fogelholm CJ, Zevenhoven R (2008) Steel converter slag as a raw material for precipitation of pure calcium carbonate. Ind Eng Chem Res 47(18):7104–7111
9. Nayak NP (2008) Characterization and utilization of solid wastes generated from Bhilai steel plant. National Institute of Technology, Rourkela
10. Mahieux PY, Aubert JE, Escadeillas G (2008) Utilization of weathered basic oxygen furnace slag in the production of hydraulic road binders. Constr Build Mater 23:742–747
11. Dippenaar R (2004) Industrial uses of slag: the use and re-use of iron and steelmaking slags. In: VII International conference on molten slags fluxes and salts, 2004. The South African Institute of Mining and Metallurgy, pp 57–70
12. World Bank Group (1998) Project guidelines: iron and steel manufacturing. Pollution prevention and abatement handbook. WORLD BANK GROUP, USA
13. Kishore K (2015) Sand for concrete from steel mills induction furnace waste slag. engineeringcivil.com. http://www.engineeringcivil.com/sand-for-concrete-from-steel-mills-induction-furnace-waste-slag.html. Accessed 10 Dec 2015
14. Pickles CA (2009) Thermodynamic analysis of the selective chlorination of electric arc furnace dust. J Hazard Mater 166(2–3):1030–1042. doi:10.1016/j.jhazmat.2008.11.110
15. Costa G (2009) Accelerated carbonation of minerals and industrial residues for carbon dioxide storage. Università delgi Studi di Roma
16. Bonenfant D, Kharoune L, Sauve S, Hausler R, Niquette P, Mimeault M, Kharoune M (2008) CO_2 sequestration potential of steel slags at ambient pressure and temperature. Ind Eng Chem Res 47(20):7610–7616

17. Bayer Ozturk Z, Eren Gultekin E (2015) Preparation of ceramic wall tiling derived from blast furnace slag. Ceram Int 41(9):12020–12026. doi:10.1016/j.ceramint.2015.06.014
18. Teir S, Eloneva S, Fogelholm C-J, Zevenhoven R (2007) Dissolution of steelmaking slags in acetic acid for precipitated calcium carbonate production. Energy 32(4):528–539. doi:10.1016/j.energy.2006.06.023
19. Li P, Pan SY, Pei S, Lin YPJ, Chiang PC (2016) Challenges and perspectives on carbon fixation and utilization technologies: an overview. Aerosol Air Qual Res 16(6):1327–1344. doi:10.4209/aaqr.2015.12.0698
20. Han GX, An XH, Jin F, Chen CJ, Zhou H (2016) Self-compacting concrete prepared with solid waste. In: Progress in civil, architectural and hydraulic engineering IV edn. Taylor & Francis Group, London, UK
21. Martinez-Lopez R, Escalante-Garcia JI (2016) Alkali activated composite binders of waste silica soda lime glass and blast furnace slag: strength as a function of the composition. Constr Build Mater 119:119–129. doi:10.1016/j.conbuildmat.2016.05.064
22. Huang X, Huang T, Li S, Muhammad F, Xu G, Zhao Z, Yu L, Yan Y, Li D, Jiao B (2016) Immobilization of chromite ore processing residue with alkali-activated blast furnace slag-based geopolymer. Ceram Int 42(8):9538–9549. doi:10.1016/j.ceramint.2016.03.033
23. Vilaplana JL, Baeza FJ, Galao O, Alcocel EG, Zornoza E, Garcés P (2016) Mechanical properties of alkali activated blast furnace slag pastes reinforced with carbon fibers. Constr Build Mater 116:63–71. doi:10.1016/j.conbuildmat.2016.04.066
24. Zhao J, Wang D, Yan P, Zhao S, Zhang D (2016) Particle characteristics and hydration activity of ground granulated blast furnace slag powder containing industrial crude glycerol-based grinding aids. Constr Build Mater 104:134–141. doi:10.1016/j.conbuildmat.2015.12.043
25. Belhadj E, Diliberto C, Lecomte A (2012) Characterization and activation of basic oxygen furnace slag. Cement Concr Compos 34(1):34–40. doi:10.1016/j.cemconcomp.2011.08.012
26. Li Q, Ding H, Rahman A, He D (2016) Evaluation of Basic Oxygen Furnace (BOF) material into slag-based asphalt concrete to be used in railway substructure. Constr Build Mater 115:593–601. doi:10.1016/j.conbuildmat.2016.04.085
27. Wu XQ, Zhu H, Hou XK, Li HS (1999) Study on steel slag and fly ash composite Portland cement. Cem Concr Res 29:1103–1106
28. Huang YH, Liu CJ (2008) Analysis on comprehensive utilization of electric furnace slag. Ind Heat 37(5):4–6 (in Chinese)
29. Shi CJ, Qian JS (2000) High performance cementing materials from industrial slags—a review. Resour Conserv Recy 29(3):195–207
30. Birat J-P (2009) Steel and CO_2—the ULCOS Program, CCS and mineral carbonation using steelmaking slag. In: 1st International slag valorisation symposium, Leuven, 2009
31. Shi C, Qian J (2000) High performance cementing materials from industrial slags—a review. Resour Conserv Recy 29(3):195–207
32. Frías Rojas M, Sánchez de Rojas MI (2004) Chemical assessment of the electric arc furnace slag as construction material: Expansive compounds. Cem Concr Res 34 (10):1881–1888. doi:10.1016/j.cemconres.2004.01.029
33. Bodor M, Santos RM, Cristea G, Salman M, Cizer Ö, Iacobescu RI, Chiang YW, van Balen K, Vlad M, van Gerven T (2016) Laboratory investigation of carbonated BOF slag used as partial replacement of natural aggregate in cement mortars. Cem Concr Compos 65:55–66. doi:10.1016/j.cemconcomp.2015.10.002
34. Santos RM, Ling D, Sarvaramini A, Guo M, Elsen J, Larachi F, Beaudoin G, Blanpain B, Van Gerven T (2012) Stabilization of basic oxygen furnace slag by hot-stage carbonation treatment. Chem Eng J 203:239–250. doi:10.1016/j.cej.2012.06.155
35. Murphy JN, Meadowcroft TR, Barr PV (1997) Enhancement of the cementitious properties of steelmaking slag. Can Metall Quart 36:315–331
36. Ramachandran V (1977) Calcium chloride in concrete. Mag Concr Res 29:1–216
37. Luxán MP, Sotolongo R, Dorrego F, Herrero E (2000) Characteristics of the slags produced in the fusion of scrap steel by electric arc furnace. Cement Concr Compos 30:517–519

38. Chang JJ, Yeih W, Chung TJ, Huang R (2016) Properties of pervious concrete made with electric arc furnace slag and alkali-activated slag cement. Constr Build Mater 109:34–40. doi:10.1016/j.conbuildmat.2016.01.049
39. Kuo W-T, Shu C-Y, Han Y-W (2014) Electric arc furnace oxidizing slag mortar with volume stability for rapid detection. Constr Build Mater 53:635–641. doi:10.1016/j.conbuildmat.2013.12.023
40. Setién J, Hernández D, González JJ (2009) Characterization of ladle furnace basic slag for use as a construction material. Constr Build Mater 23:1788–1794. doi:10.1016/j.conbuildmat.2008.10.003
41. Monkman S, Shao Y, Shi C (2009) Carbonated ladle slag fines for carbon uptake and sand substitute. J Mater Civ Eng 21:657–665. doi:10.1061//asce/0899-1561/2009/21:11/657
42. Kuo W-T, Wang H-Y, Shu C-Y (2014) Engineering properties of cementless concrete produced from GGBFS and recycled desulfurization slag. Constr Build Mater 63:189–196. doi:10.1016/j.conbuildmat.2014.04.017
43. Frías RM, Sánchez dRM (2004) Chemical assessment of the electric arc furnace slag as construction material: expansive compounds. Cem Concr Res 34(10):1881–1888
44. Poh HY, Ghataora GS, Gharizeh N (2006) Soil stabilization using basic oxygen slag fines. J Mater Civ Eng (ASCE) 18(2):229–240
45. Shi C, Quian J (2000) High performance cementing materials from industrial slags—a review. Resour Conserv Recy 29(3):195–207
46. Huang X, Wang Z, Liu Y, Hu W, Ni W (2016) On the use of blast furnace slag and steel slag in the preparation of green artificial reef concrete. Constr Build Mater 112:241–246. doi:10.1016/j.conbuildmat.2016.02.088
47. Geiseler J (1996) Use of steelworks slag in Europe. Waste Manag 16(1–3):59–63
48. De Windt L, Chaurand P, Rose J (2011) Kinetics of steel slag leaching: batch tests and modeling. Waste Manag 31(2):225–235. doi:10.1016/j.wasman.2010.05.018
49. Wu SP, Xue YJ, Ye QS, Chen YC (2007) Utilization of steel slag as aggregates for stone mastic asphalt (SMA) mixtures. Build Environ 42(7):2580–2585. doi:10.1016/j.buildenv.2006.06.008
50. Pan SY, Chen YH, Chen CD, Shen AL, Lin M, Chiang PC (2015) High-gravity carbonation process for enhancing CO_2 fixation and utilization exemplified by the steelmaking industry. Environ Sci Technol 49(20):12380–12387. doi:10.1021/acs.est.5b02210
51. Chen KW, Pan SY, Chen CT, Chen YH, Chiang PC (2016) High-gravity carbonation of basic oxygen furnace slag for CO_2 fixation and utilization in blended cement. J Clean Prod 124:350–360. doi:10.1016/j.jclepro.2016.02.072
52. Qi G, Yue D, Fukushima M, Fukuchi S, Nie Y (2012) Enhanced humification by carbonated basic oxygen furnace steel slag–I. Characterization of humic-like acids produced from humic precursors. Bioresour Technol 104:497–502. doi:10.1016/j.biortech.2011.11.021
53. Chiou CS, Chang CF, Chang CT, Shie JL, Chen YH (2006) Mineralization of reactive black 5 in aqueous solution by basic oxygen furnace slag in the presence of hydrogen peroxide. Chemosphere 62(5):788–795. doi:10.1016/j.chemosphere.2005.04.072
54. Tsai TT, Kao CM, Wang JY (2011) Remediation of TCE-contaminated groundwater using acid/BOF slag enhanced chemical oxidation. Chemosphere 83(5):687–692. doi:10.1016/j.chemosphere.2011.02.023
55. Kang HJ, An KG, Kim DS (2004) Utilization of steel slag as an adsorbent of ionic lead in wastewater. J Environ Sci Health A Tox Hazard Subst Environ Eng 39(11–12):3015–3028
56. Han C, Wang Z, Yang W, Wu Q, Yang H, Xue X (2016) Effects of pH on phosphorus removal capacities of basic oxygen furnace slag. Ecol Eng 89:1–6. doi:10.1016/j.ecoleng.2016.01.004
57. Islam A, Alengaram UJ, Jumaat MZ, Bashar II, Kabir SMA (2015) Engineering properties and carbon footprint of ground granulated blast-furnace slag-palm oil fuel ash-based structural geopolymer concrete. Constr Build Mater 101:503–521. doi:10.1016/j.conbuildmat.2015.10.026

58. Stolaroff J, Lowry G, Keith D (2005) Using CaO- and MgO-rich industrial waste streams for carbon sequestration. Energy Convers Manag 46(5):687–699. doi:10.1016/j.enconman.2004.05.009
59. CSC Group (2003) Slag Utilization. 4 edn., Kaohsiung
60. Motz H, Geiseler J (2001) Products of steel slags an opportunity to save natural resources. Waste Manag 21:285–293
61. San Jose′ JT, Urı′a A (2001) Escorias de horno de arco ele′ctrico en mezclas bituminosas. Arte Cem 1905:122–125
62. Ameri M, Behnood A (2012) Laboratory studies to investigate the properties of CIR mixes containing steel slag as a substitute for virgin aggregates. Constr Build Mater 26:475–480
63. Ahmedzadea P, Sengoz B (2009) Evaluation of steel slag coarse aggregate in hot mix asphalt concrete. J Hazard Mater 165:300–305
64. Asi IM (2007) Evaluating skid resistance of different asphalt concrete mixes. Build Environ 42:325–329
65. Xue YJ, Wu SP, Hou HB, Zha J (2006) Experimental investigation of basic oxygen furnace slag used as aggregate in asphalt mixture. J Hazard Mater B 138:261–268
66. Qasrawi H, Shalabi F, Asi I (2009) Use of low CaO unprocessed steel slag in concrete as construction and building materials 23:1118–1125
67. Papayianni I, Anastasiou E (2010) Production of high-strength concrete using high volume of industrial by-products. Constr Build Mater 24:1412–1417
68. Wen XL, Ouyang D, Pan P (2011) Research of high anti-chloride ion permeability of C100 concrete mixed with steel slag. Concrete 6:73–75 (in Chinese)
69. Netinger I, Kesegic I, Guljas I (2011) The effect of high temperatures on the mechanical properties of concrete made with different types of aggregates. Fire Saf J 16:425–430
70. Chen DY, Tan KF (2006) Study on mineral admixture of concrete prepared with electric furnace slag. Bull Chin Ceram Soc 25(6):73–75 (in Chinese)
71. Ducman V, Mladenovic A (2011) The potential use of steel slag in refractory concrete. Mater Ch 62:716–723
72. Slag: A Sound Choice in Favour of Ecology. Germany
73. Tsakiridis PE, Papadimitriou GD, Tsivilis S, Koroneos C (2008) Utilization of steel slag for Portland cement clinker production. J Hazard Mater 152:805–811
74. Iacobescua RI, Koumpouri D, Pontikesc Y, Sabana R, Angelopoulos GN (2011) Valorisation of electric arc furnace steel slag as raw material for low energy belite cements. J Hazard Mater 196:287–294
75. Tiifekqi M, Demirbas A, Genc H (1997) Evaluation of steel furnace slags as cement additives. Cem Concr Res 27(11):1713–1717
76. Huang Y, Liu Z (2010) Investigation on phosphogypsum–steel slag–granulated blast-furnace slag-limestone cement. Constr Build Mater 24:1296–1301
77. Feng CH, Dou Y, Li DX (2011) Steel slag used as admixture in composite cement. J Nanjing Univ Technol (Nat Sci Ed) 33 (1):74–79 (in Chinese)
78. Altun IA, Yilmaz I (2002) Study on steel furnace slags with high MgO as additive in Portland cement. Cem Concr Res 32(8):1247–1249
79. IEA (2014) Tracking clean energy progress 2014—energy technology perspectives 2014 excerpt iea input to the clean energy ministerial. International Energy Agency, France
80. Doucet FJ (2010) Effective CO_2-specific sequestration capacity of steel slags and variability in their leaching behaviour in view of industrial mineral carbonation. Miner Eng 23(3):262–269. doi:10.1016/j.mineng.2009.09.006
81. Chandra S (1997) Waste materials used in concrete manufacturing. Noyes Publications, New Jersey, USA
82. Pan S-Y, Chiang A, Chang E-E, Lin Y-P, Kim H, Chiang P-C (2015) An innovative approach to integrated carbon mineralization and waste utilization: a review. Aerosol Air Qual Res 15:1072–1091. doi:10.4209/aaqr.2014.10.02

83. Huijgen WJJ, Comans RNJ (2006) Carbonation of steel slag for CO_2 sequestration: leaching of products and reaction mechanisms. Environ Sci Technol 40(8):2790–2796
84. Eloneva S, Teir S, Revitzer H, Salminen J, Said A, Fogelholm CJ, Zevenhoven R (2009) Reduction of CO_2 emissions from steel plants by using steelmaking slags for production of marketable calcium carbonate. Steel Res Int 80(6):415–421

Chapter 12
Fly Ash, Bottom Ash, and Dust

Abstract Fly ashes, suspended in the exhaust gases, can be collected by electrostatic precipitators or baghouse filter. On the other hand, the portion of the non-combustible residues from combustion in an incinerator or furnace that fall by themselves to the bottom hopper of a furnace or incinerator is referred as bottom ash. Both the fly ash and bottom ash can be used in various applications, such as concrete production, embankments, cement clinkers, and road subbase construction. In this chapter, the physico-chemical properties of fly ash and bottom ash are illustrated. In addition, the challenges in direct utilization of ashes, such as heavy metal leaching, are discussed.

12.1 Introduction

Solid ashes, such as air pollution control (APC) residues, bottom ash, and dust, are typical by-products from industrial or combustion processes. Due to the wide varieties of input feedstocks in physico-chemical properties, the produced ashes may contain hazardous compounds which might be harmful to human health and ecosystem quality. APC residues are the solid output of the flue gas treatment equipment after incineration or combustion processes. Typically, they comprise the fly ash from incineration (fine particles) together with the reagents (such as lime and activated carbon) used in the treatment equipment. On the other hand, bottom ash will generate in the incineration or combustion (furnace), which is considered part of the non-combustible residue. It usually refers to coal combustion ash in an industrial context, or bottom ash in a municipal solid waste incinerator (MSWI).

In the following content, the physico-chemical properties of different types of ashes are illustrated. The challenges in direct utilization of the ashes are also addressed.

P.-C. Chiang and S.-Y. Pan, *Carbon Dioxide Mineralization and Utilization*, DOI 10.1007/978-981-10-3268-4_12

12.2 Fly Ash

Fly ash (FA), called APC residues from combustion process, are classified as hazardous wastes according to the Commission Decision 2000/532/EC due to the character of fine particle size and the potential release of persistent organic pollutants (POPs), metals, and soluble salts [1, 2]. Thus, they may contain the following:

- Volatile contaminants: chlorides, metals, etc.,
- Compounds created in the incineration process: dioxins, etc., and
- Other materials from flue gas treatment process: sulfates, alkalinity, etc.

In general, the FA is captured by electrostatic precipitators or other particle filtration equipment such as baghouse filter. In the United Kingdom, FA is also referred as pulverized fuel ash, which is mainly the product of coal combustion. In the USA, coal FA can be divided into two categories: (1) class C or (2) class F in accordance with the ASTM C618. The primary difference between the two US classifications of coal FA is the total content of silica, alumina, and ferrite in the material. Typically, the class C FA (FA-C) has much higher calcium content than class F FA (FA-F) [3]. This excess content of calcium is responsible for the self-cementing nature of FA-C as the principal reactive phases of FA-C are anhydrite ($CaSO_4$) and lime (CaOH).

12.2.1 Physico-chemical Properties

Table 12.1 presents the chemical properties of various types of FA used in the literature and its associated CO_2 capture capacity. FA is a heterogeneous material, with a spherical shape in size from 0.5 to 300 μm. According to the XRF results, the chemical composition of FA varies widely depending on the source and on particle sizes. For example, the chemical composition of fresh by-product ash consists mainly of CaO ($\sim$62.8%) and SO_3 ($\sim$31.0%). The mineral composition of FA includes merwinite [$(Ca_3Mg)(SiO_2)_4$], periclase (MgO), anhydrite [$Ca(SO)_4$], stishovite (SiO_2), and calcite ($CaCO_3$), where $Ca(SO)_4$ is the main phase. Therefore, the SO_3 composition in the fresh FA is mainly related to $CaSO_4 \cdot 2H_2O$ (gypsum), which has a profound effect on the initial cement chemical as well as mechanical strength.

In the fresh FA, Ca–Mg–Si oxide is typically present in significant quantities [4]. The contents of MgO, free-CaO, and $Ca(OH)_2$ in the fresh FA are generally related to a poor durability of cement mortar (e.g., autoclave expansion).

12.2.2 Challenges in Direct Utilization

During combustion in the boiler, a significant amount of limestone ($CaCO_3$) is introduced to suppress sulfur-species pollution from petroleum coke. Thus, the

Table 12.1 Chemical properties of various types of fly ash (FA) used in the literature and its associated CO_2 capture capacity

Items[a]	Components (%)	Coal FA (Victorian)	Coal FA	MSWI-FA	By-product lime ash [1]
XRF (solid phase)	SiO_2	58.6–60.0	51.2	9.2–13.0	3.08
	CaO	5.9–7.5	9.2	24.8–29.7	62.8
	Al_2O_3	19.1–19.7	26.0	2.1–2.5	1.01
	Fe_2O_3	4.7–5.4	2.4	11.1–23.0	0.70
	Na_2O	0.7–1.0	0.5	6.5–9.0	0.06
	P_2O_5	–	0.7	0.03	–
	MgO	~3.9	2.4	13.0–25.5	0.83
	SO_3	–	0.4	12.8–15.0	31.0
	K_2O	1.0–2.0	0.8	0.4–0.5	0.43
Chemical analysis	*f*-CaO	–	–	–	11.4
	$Ca(OH)_2$	–	–	–	3.42
Thermal analysis	$CaCO_3$	<0.1	–	–	–
	LOI	0.6	–	–	–

[a]Victorian brown coal fly ash (VBC-FA); Fly ash (FA)

fresh FA contains a large amount of gypsum ($CaSO_4$) and calcium oxide (free-CaO). Typically, the FA generated from a circular fluidized-bed (CFB) boiler is considered as a pozzolanic constituent. It has been widely used as supplementary cementitious materials (SCMs) or fine aggregates in blended cement [5, 6]. In another approach, the fresh FA can be used as a soil modification material for the following: (1) increasing strength, (2) enhancing load carrying capacity, and (3) reducing the potential volume change of soil [7].

Due to their high chemical activity, stabilization processes for fresh FA are required prior to further utilization in civil engineering [8, 9]. These APC residues, such as fly ash, are currently handled entirely by cement solidification and chemical agent stabilization treatments and then placed in sanitary landfills [10]. Several researches have applied highly pure CO_2 as curing atmosphere for FA-based cement mortar to eliminate free-CaO content and to enhance initial strength development [9, 11, 12].

12.3 Bottom Ash

Bottom ash (BA) is part of the non-combustible residues from combustion in an incinerator or furnace. The portion of the residues that fall by themselves to the bottom hopper of an incinerator or furnace is referred as BA. Usually, the BA was generated in large quantities from a municipal solid waste incinerator (MSWI). Incineration is an environmental-friendly treatment and management option to dispose municipal solid waste (MSW), especially where recycling or reuse is not possible. The MSWI can provide several advantages [1, 13, 14]:

- Mass and volume reductions of original waste can be up to 85 and 95%, respectively.
- Separation in fractions of different residues (e.g., ferrous metals, nonferrous metals, and granulate fractions) and the possibility of waste-to-resource conversion.
- It provides disinfection, reduction of organic matter.
- It can offer possibility of energy recovery.
- Fly ash (FA), or so-called air pollution control (APC) ash and

The combustion of MSW in incinerators would result in gaseous effluent containing approximately 12 vol.% CO_2 and two solid waste streams [15]:

- Fly ash (FA), or so-called air pollution control (APC) ash and
- Bottom ashes (BA).

Despite the significant volume reduction of MSW by incineration, the produced amounts of solid residues are still substantial. For each ton of MSW mass incinerated, approximately 35 kg of FA and 160 kg of BA are produced. In other words, these solid wastes would be up to a mass fraction of 20% from the original MSW [15, 16]. Particularly, the MSWI-BA accounts for 80–90% of the total mass of the MSWI residues [17]. In Taiwan, there are currently 24 large-scale MSW incinerators commercialized, with an annual incinerated MSW around 5.93 million tons in the years of 2002–2009 [10].

12.3.1 Physico-chemical Properties

MSWI-BA is a heterogeneous mixture of slag including ferrous and nonferrous metals, ceramics, and other non-combustible materials. The elemental composition of MSWI-BA depends primarily on the composition of the waste input, which may vary with location, season, and recycling schemes in operation [18, 19]. Table 12.2 presents the physicochemical properties of MSWI-BA. The major elements in MSWI-BA are O, Cl, Ca, Si, Al, Fe, Na, K, Mg, and C, with trace elements including Cu, Zn, S, Pb, Cr, Ni, Sn, Mn, Sb, V, and Co [1, 19]. According to the XRF analysis, the major components in the MSWI-BA are SiO_2, CaO, Al_2O_3, and Fe_2O_3, while the trace compounds are Na_2O, MgO, P_2O_5, TiO_2, and SO_3.

Mineralogical studies indicate that the main crystalline phases of MSWI-BA are typically [18, 19] as follows:

- Silicate:
  - Quartz [SiO_2],
  - Gehlenite [$Ca_2Al_2SiO_7$],
  - Olivine [$(Mg, Fe)_2SiO_4$],
  - Epidote ($Al–Ca–Fe–SiO_2$),

Table 12.2 Summary of physico-chemical properties of MSWI-BA [14]

Properties	Items		Fresh MSWI-BA		
			<125 μm	125–350 μm	350–500 μm
Physical	True density[a] (g cm^{-3})		2.70	2.73	2.78
	Mean diameter (μm)		59.6	175.8	501.5
	Median diameter (μm)		45.1	164.9	513.5
	BET surface area[a] (m^2 g^{-1})		6.63 ± 0.02	4.00 ± 0.02	3.51 ± 0.01
	Langmuir surface area[1] (m^2 g^{-1})		9.31 ± 0.30	5.57 ± 0.20	4.87 ± 0.17
Chemical	XRF	SiO_2 (%)	44.9	50.7	45.0
		CaO (%)	21.1	16.4	18.3
		Al_2O_3 (%)	9.3	8.7	9.5
		Fe_2O_3 (%)	8.5	9.3	11.8
		Na_2O (%)	5.0	5.9	5.2
		P_2O_5 (%)	2.6	2.3	2.6
		MgO (%)	2.1	1.9	2.1
		TiO_2 (%)	1.4	1.2	1.4
		SO_3 (%)	1.3	0.8	0.9
		K_2O (%)	0.9	0.8	0.8
	TGA	$CaCO_3$ (%)	7.07 ± 0.32	2.57 ± 0.02	3.38 ± 0.55
	Chem.[b]	f-CaO (%)	0.07	0.00	0.07

[a]Analyzed by Particulate Technology Laboratory, NTU. [b]Chemical analysis.

  - Augite [(Ca,Na)(Mg,Fe,Al,Ti)$(Si,Al)_2O_6$], and
  - Pigeonite (Al–Ca–Fe–Mg–Mn–Ti–Na–SiO_2).

- Sulfate:
  - Anhydrite [$CaSO_4$],
  - Ettringite [$Ca_6Al_2(SO_4)_3(OH)$], and
  - Gypsum [$CaSO_4 \cdot 2H_2O$].
- Carbonate:
  - Calcite [$CaCO_3$] and
  - Aragonite [$CaCO_3$].
- Iron oxide:
  - Hematite [Fe_2O_3] and
  - Magnetite [Fe_3O_4].
- Others trace amounts:
  - Tassieite (Ca–Fe–Mg–Na–P–OH) and
  - Ktenasite (Cu–H–O–S–Zn).

Quartz was found to be the main component in MSWI-BA. Even though $Ca(OH)_2$ was not identified by the XRD measurements, Ca–Al–Si oxide and Ca–Na–Si oxide were present in significant quantities [14]. Calcium-bearing compounds in fresh MSWI-BA are primarily composed of various types of oxides with silicates that contain other metals, such as Fe and Al metals.

With the specific chemical characterizations of MSWI-BA, it exhibits its potential CO_2 fixation capability due to its calcium content and alkaline properties. In addition, the mineral carbonation of MSWI-BA can immobilize heavy metals and effectively prevent their leaching, especially for Cr, Cu, Pb, Zn, and Sb [1, 17, 18].

12.3.2 Challenges in Treatment, Disposal, and/or Utilization

Bottom ash accounts for 80–90% of the total weight of the MSW in incinerators [17]. The available and common methods for disposal of ashes from the MSW are landfill (82%), recycling or composting (11%), and thermal treatment (7%) [20]. According to the European Waste Catalogue, MSWI-BA is typically classified as a non-hazardous waste. Therefore, it is currently being utilized as an aggregate substitute in road bases and bituminous pavement in European countries [19, 21]. Similarly, in Taiwan, the BA also is classified as a general (non-hazardous) industry waste. Although BA can pass nearly all of the standards of the toxicity characteristic leaching procedure (TCLP), its high chloride content makes its reuse limit [10]. In addition, the potentially high leaching of salts and other elements would increase the pretreatment cost for further utilization [1].

12.4 Electric Arc Furnace Dust (EAFD)

Electric arc furnace dust (EAFD) is considered to be a hazardous waste because of its chemically, physically, and mineralogically complex, and the presence of lead, cadmium, chromium, and zinc [22]. Since EAFD is generated when automobile scrap is remelted in an electric arc furnace, the properties of each dust are site-specific and depend upon the scrap composition and the furnace operating practice [22]. Generally, 10–20 kg of EAFD would be generated for per ton of steel produced [22].

Typically, the major elements in fresh EAFD include zinc, iron, and calcium ions. The zinc element is found in the form of zincite (ZnO), zinc ferrite/franklinite ($ZnFe_2O_4$), zinc chloride ($ZnCl_2$), and diiron zinc tetraoxide (Fe_2ZnO_4). The remaining iron may be present as hematite (Fe_2O_3) and magnetite (Fe_3O_4), or in combination with lime (CaO) as calcium ferrites, or with silica as iron silicates [22]. Other elements are found in low concentrations, such as magnesium (Mg), lead (Pb), silicon (Si), aluminum (Al), manganese (Mn), chromium (Cr), nickel (Ni), copper (Cu), and cadmium (Cd). For example, the lead is typically presented as lead oxide (PbO) and lead chloride ($PbCl_2$). It is noted that the amounts of zinc and lead in the EAFD are economically valuable, while the iron is usually not worth recovering [22].

12.5 Challenges and Perspectives in Ash Utilization

Both the fly ash and bottom ash can be used in various applications, such as concrete production, embankments, cement clinkers, mine reclamation, and road subbase construction. Processes using fresh fine FA as alternative binders and replacement of Portland cement in concrete have been widely developed and deployed, especially in the USA, to reduce the energy and resource consumption, as well as the CO_2 emission for concrete and cement production.

Despite the fact that the FA usually replaces no more than 25% of the Portland cement in concrete, research has successfully demonstrated the use of 100% fly ash concrete with glass aggregate to construct a building by Montana State University [23]. However, the challenges in the substitution using fresh FA or BA in cement and/or concrete are mainly related to early strength development. Engineering experience indicates that, in the case of 50% clinker replacement by fresh FA, the early strength of concrete would be reduced dramatically [24]. To overcome the above challenges, accelerated carbonation process should be an effective method for stabilizing FA and BA; meanwhile, permanently storing CO_2 as solid carbonates [16, 25, 26].

12.5.1 Accelerated Carbonation with Flue Gas CO_2

CO_2 in the flue gas can react with calcium-bearing materials to form stable and insoluble calcium carbonates. Beside the calcium oxide (CaO), magnesium oxide (MgO) also can be the favorable metal oxides in reacting with CO_2. In many countries (such as France and Canada), natural aging and/or ambient air carbonation are the standard practice to stabilize alkaline fly ash and bottom ash [15, 27, 28]. As a result, carbonation of municipal solid waste has recently been receiving more attention due to their availability, low cost, and high CaO and/or MgO contents [16, 29]. Figure 12.1 shows the photographs of MSWI-BA (left) before and (right) after carbonation. The accelerated carbonation can improve the chemical and physical characteristics of solid wastes and facilitate their reuse in a various applications, such as synthesis of construction materials [14, 18]. Since accelerated carbonation is an exothermic process, additional heat input and energy cost could be reduced [30].

After carbonation, both calcite ($CaCO_3$) and quartz (SiO_2) are found to be a major component of the carbonated MSWI-BA, according to the XRD analysis shown in Fig. 12.2 [14]. The calcium-bearing species in MSWI-BA are primarily

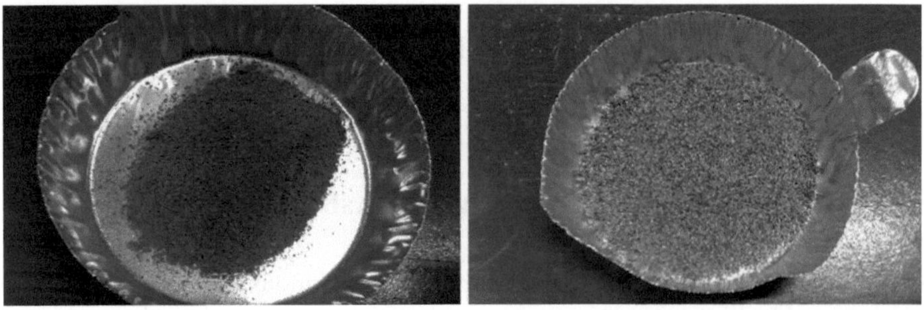

Fig. 12.1 Photograph of municipal solid waste incinerator bottom ash (*left*) before and (*right*) after carbonation

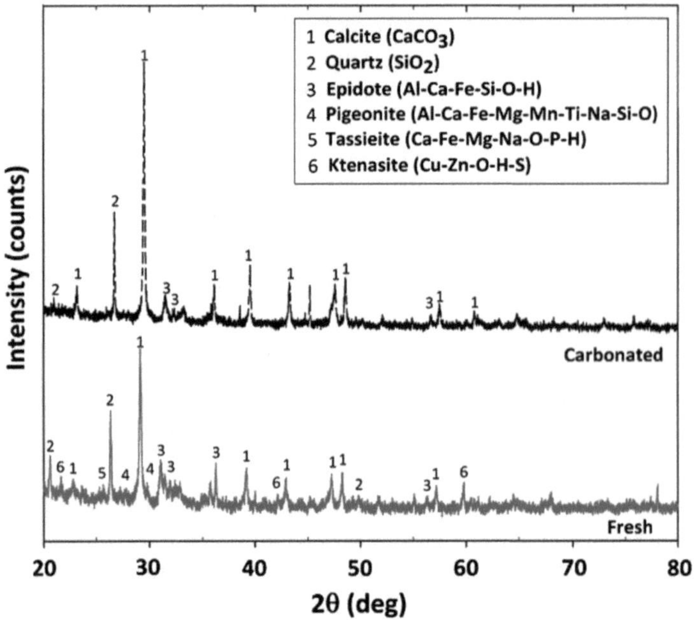

Fig. 12.2 X-ray diffraction pattern of fresh and carbonated municipal solid waste incinerator bottom ashes. Reprinted by permission from Macmillan Publishers Ltd: Ref. [14], copyright 2015

composed of various types of silicates, such as CaO–Al–silicates and CaO–Fe–silicates. The dissociation of calcium-bearing components in the fresh MSWI-BA should occur after introducing the ash into the solution. The calcium ions react with carbonate ions (CO_3^{2-}) under an alkaline condition, leading to the formation of $CaCO_3$ upon introducing of CO_2 into the reactor. Thus, the leaching concentration of calcium ions from MSWI-BA into the solution is an important indicator for accelerated carbonation.

Table 12.3 Comparison of performance and operating conditions using municipal solid waste residues in the literature

Material type	Method[a]	Operating conditions[b]					Performances	
		Liquid types[c]	Pres. (bar)	L/S ratio (mL/g)	Temp. (°C)	Time (min)	Carbonation conversion (%)	Capture capacity per kg waste
Fly ash [31]	AC	DI water	3	0.3	8-42	4320	35.1	100 g CO_2
Fly ash [32]	AC	DI water	1	4	25	14,400	28.8	120 g CO_2
Fly ash [2]	AC	DI water	1	2.5	20	180	–	200 g CO_2
Bottom ash [15]	DC	Moist. = 20%	17	–	25	180	67.7	87 g CO_2
Bottom ash [18]	DC	Moist. = 65%	3	0.3–0.4	25	1440	–	32 g CO_2
Bottom ash [14]	AC	CRW	1	10	25	120	90.7	102 g CO_2

[a] *AC* aqueous carbonation; *DC* dry carbonation. [b] all experiments were conducted in 100% CO_2 gas. [c] *CRW* cold-rolling mill wastewater

Table 12.3 presents the performance of carbonation using MSWI residues, such as APC and bottom ash, in the literature. The dry carbonation process possessed a relatively low capture capacity of 0.08–0.09 kg CO_2/kg MSWI-BA (equivalent to 24 L CO_2/kg BA) [15]. In converse, the aqueous carbonation of MSWI-FA offers a CO_2 fixation capacity of 0.10–0.12 kg per kg ash (corresponding to an average weight gain of 7–12%) under a reaction time of 72–240 h. [31, 32]. Similarly, the aqueous carbonation process using MSWI-APC, a CO_2 capacity of 200 g CO_2 per kg APC can be achieved at a relatively mild operating condition (i.e., 20 °C for 3 h) [2]. This might also be attributed to the active chemical properties of MSWI fly ash.

12.5.2 Heavy Metal Leaching Potential

Several heavy metals, such as lead (Pb), barium (Ba), mercury (Hg) and arsenic (As), can be potentially leached out from the solid matrix with the fresh FA or BA [4]. Therefore, several research has focused on the effect of accelerated carbonation on the leaching behavior of trace heavy metals from FA and/or BA. It indicates that heavy metals can be immobilized effectively by accelerated carbonation, especially Cr, Cu, Pb, Zn, and Sb are the most significant [1, 17, 18]. Cd and Pb have a strong affinity with calcium carbonate and also form complexes with Fe and Al (hydr) oxides [15].

For example, Sb leaching in fresh and carbonated BA (sand fraction) is solubility-controlled by $Ca_{1.13}Sb_2O_6(OH)_{0.26}\cdot 0.74H_2O$ [13]. As carbonation proceeds (i.e., lowering pH), preferential Ca-leaching from romeites and a higher Sb solubility were observed [13]. Accelerated carbonation may thus only be successful at reducing Sb leaching from MSWI-BA if relatively low pH values are obtained where adsorption to iron and aluminum oxides surfaces may again reduce Sb solubility [13]. In addition, immobilization of Sb could also be achieved by combining with other process (e.g., sorbent adding) during carbonation reaction [2]. Furthermore, accelerated carbonation has proved to be an ineffective method for the demobilization of chloride (Cl^-) and sulfate (SO_4^{2-}) ions [1].

References

1. Todorovic J, Ecke H (2006) Demobilisation of critical contaminants in four typical waste-to-energy ashes by carbonation. Waste Manag 26(4):430–441. doi:10.1016/j.wasman.2005.11.011
2. Cappai G, Cara S, Muntoni A, Piredda M (2012) Application of accelerated carbonation on MSW combustion APC residues for metal immobilization and CO_2 sequestration. J Hazard Mater 207–208:159–164. doi:10.1016/j.jhazmat.2011.04.013
3. Paris JM, Roessler JG, Ferraro CC, DeFord HD, Townsend TG (2016) A review of waste products utilized as supplements to Portland cement in concrete. J Clean Prod 121:1–18. doi:10.1016/j.jclepro.2016.02.013

4. Pan S-Y, Hung C-H, Chan Y-W, Kim H, Li P, Chiang P-C (2016) Integrated CO_2 fixation, waste stabilization and product utilization via high-gravity carbonation process exemplified by CFB fly ash. ACS Sustain Chem Eng
5. Snellings R, GIlles M, Elsen J (2012) Supplementary cementitious materials. In: Broekmans MATM, Pollmann H (eds) Reviews in mineralogy and geochemistry, vol 74. Mineralogical society of America, pp 211–278
6. Çiçek T, Çinçin Y (2015) Use of fly ash in production of light-weight building bricks. Constr Build Mater 94:521–527. doi:10.1016/j.conbuildmat.2015.07.029
7. Tsiridis V, Petala M, Samaras P, Sakellaropoulos GP (2015) Evaluation of interactions between soil and coal fly ash leachates using column percolation tests. Waste Manage 43:255–263. doi:10.1016/j.wasman.2015.05.031
8. Hu Y, Zhang P, Li J, Chen D (2015) Stabilization and separation of heavy metals in incineration fly ash during the hydrothermal treatment process. J Hazard Mater 299:149–157. doi:10.1016/j.jhazmat.2015.06.002
9. Morandeau A, Thiéry M, Dangla P (2015) Impact of accelerated carbonation on OPC cement paste blended with fly ash. Cem Concr Res 67:226–236. doi:10.1016/j.cemconres.2014.10.003
10. Yang R, Liao WP, Wu PH (2012) Basic characteristics of leachate produced by various washing processes for MSWI ashes in Taiwan. J Environ Manage 104:67–76. doi:10.1016/j.jenvman.2012.03.008
11. Zhang F, Mo L, Deng M (2015) Mechanical strength and microstructure of mortars prepared with MgO-CaO-Fly ash-Portland cement blends after accelerated carbonation. J Chin Ceram Soc 43(8):1–8. doi:10.14062/j.issn.0454-5648.2015.08.01
12. Mo L, Zhang F, Deng M (2015) Effects of carbonation treatment on the properties of hydrated fly ash-MgO-Portland cement blends. Constr Build Mater 96:147–154. doi:10.1016/j.conbuildmat.2015.07.193
13. Cornelis G, Van Gerven T, Vandecasteele C (2012) Antimony leaching from MSWI bottom ash: modelling of the effect of pH and carbonation. Waste Manage 32(2):278–286. doi:10.1016/j.wasman.2011.09.018
14. Chang EE, Pan SY, Yang L, Chen YH, Kim H, Chiang PC (2015) Accelerated carbonation using municipal solid waste incinerator bottom ash and cold-rolling wastewater: performance evaluation and reaction kinetics. Waste Manage 43:283–292. doi:10.1016/j.wasman.2015.05.001
15. Rendek E, Ducom G, Germain P (2006) Carbon dioxide sequestration in municipal solid waste incinerator (MSWI) bottom ash. J Hazard Mater 128(1):73–79. doi:10.1016/j.jhazmat.2005.07.033
16. Bobicki ER, Liu Q, Xu Z, Zeng H (2012) Carbon capture and storage using alkaline industrial wastes. Prog Energy Combust Sci 38(2):302–320. doi:10.1016/j.pecs.2011.11.002
17. Arickx S, Van Gerven T, Vandecasteele C (2006) Accelerated carbonation for treatment of MSWI bottom ash. J Hazard Mater 137(1):235–243. doi:10.1016/j.jhazmat.2006.01.059
18. Fernandez Bertos M, Li X, Simons SJR, Hills CD, Carey PJ (2004) Investigation of accelerated carbonation for the stabilisation of MSW incinerator ashes and the sequestration of CO_2. Green Chem 6(8):428. doi:10.1039/b401872a
19. Teir S (2008) Fixation of carbon dioxide by producing carbonates from minerals and steelmaking slags. Helsinki University of Technology
20. Fernandez Bertos M, Simons SJ, Hills CD, Carey PJ (2004) A review of accelerated carbonation technology in the treatment of cement-based materials and sequestration of CO_2. J Hazard Mater 112(3):193–205. doi:10.1016/j.jhazmat.2004.04.019
21. Astrup T, Mosbaek H, Christensen TH (2006) Assessment of long-term leaching from waste incineration air-pollution-control residues. Waste Manage 26(8):803–814. doi:10.1016/j.wasman.2005.12.008
22. Pickles CA (2009) Thermodynamic analysis of the selective chlorination of electric arc furnace dust. J Hazard Mater 166(2–3):1030–1042. doi:10.1016/j.jhazmat.2008.11.110

23. Hasanbeigi A, Price L, Lin E (2012) Emerging energy-efficiency and CO_2 emission-reduction technologies for cement and concrete production: A technical review. Renew Sustain Energy Rev 16(8):6220–6238. doi:10.1016/j.rser.2012.07.019
24. Crow JM (2008) The concrete conundrum. Chem World 62–66
25. Pan SY, Chiang PC, Chen YH, Tan CS, Chang EE (2013) Ex Situ CO_2 capture by carbonation of steelmaking slag coupled with metalworking wastewater in a rotating packed bed. Environ Sci Technol 47(7):3308–3315. doi:10.1021/es304975y
26. Lackner KS, Wendt CH, Butt DP, Joyce EL, Sharp DH (1995) Carbon dioxide disposal in carbonate minerals. Los Alamos National Laboratory, Los Alamos, NM, USA
27. Santos RM, Mertens G, Salman M, Cizer O, Van Gerven T (2013) Comparative study of ageing, heat treatment and accelerated carbonation for stabilization of municipal solid waste incineration bottom ash in view of reducing regulated heavy metal/metalloid leaching. J Environ Manage 128:807–821. doi:10.1016/j.jenvman.2013.06.033
28. Assamoi B, Lawryshyn Y (2012) The environmental comparison of landfilling vs. incineration of MSW accounting for waste diversion. Waste Manage 32(5):1019–1030. doi:10.1016/j.wasman.2011.10.023
29. Pan S-Y, Chang EE, Chiang P-C (2012) CO_2 capture by accelerated carbonation of alkaline wastes: a review on its principles and applications. Aerosol Air Qual Res 12:770–791. doi:10.4209/aaqr.2012.06.0149
30. Eloneva S, Teir S, Salminen J, Fogelholm CJ, Zevenhoven R (2008) Fixation of CO_2 by carbonating calcium derived from blast furnace slag. Energy 33(9):1461–1467
31. Li X, Bertos MF, Hills CD, Carey PJ, Simon S (2007) Accelerated carbonation of municipal solid waste incineration fly ashes. Waste Manage 27(9):1200–1206. doi:10.1016/j.wasman.2006.06.011
32. Wang L, Jin Y, Nie Y (2010) Investigation of accelerated and natural carbonation of MSWI fly ash with a high content of Ca. J Hazard Mater 174(1–3):334–343. doi:10.1016/j.jhazmat.2009.09.055

Chapter 13
Paper Industry, Construction, and Mining Process Wastes

Abstract Several alkaline wastes are likely to pose a threat to the ambient environment due to their high alkalinity and fine particles, if disposed improperly. These alkaline wastes include paper pulp and mill waste (lime mud), bauxite mining residue (red mud), and cement and demolition wastes. In this chapter, the characteristics of these alkaline solid wastes from paper industry, construction, and mining process are discussed. In addition, the situations and challenges in conventional utilization of the alkaline wastes are illustrated. After appropriate treatment or activation processes, the solid wastes can be utilized in various novel approaches, such as coagulant and adsorbent for water purification process.

13.1 Introduction

The rapid population growth and industrialization have been generating large amount of solid wastes over the world. Throughout the world, since ongoing urbanization prompted massive urban development and renovation activities, a large quantity of alkaline solid wastes has been produced from the following:

- paper industry;
- construction, demolition, and reconstruction sites; and
- mining process

These alkaline wastes are likely to pose a threat to the ambient environment due to their high alkalinity and fine particles, if disposed improperly, including bauxite residue (red mud), paper pulp and mill wastes, and cement wastes. Although the majority of these wastes are materials that could be reused and recycled, they were usually disposed of by landfilling and dumping, triggering serious environmental impacts [1].

To overcome the critical issues, in this chapter, the characteristics and convention utilization of these alkaline solid wastes from paper industry, construction, and mining process are discussed. Moreover, it is noted that there is a high potential for recycling and reuse of solid wastes from accelerated carbonation with positive

P.-C. Chiang and S.-Y. Pan, *Carbon Dioxide Mineralization and Utilization*, DOI 10.1007/978-981-10-3268-4_13

results in base and subbase layers of roadways, which will be discussed in detail in Chaps. 14–16 (in Part IV).

13.2 Paper and Pulping Mill Waste

Paper and pulping mills produced large amounts of pulp annually. Massive inorganic (such as ashes and dregs) and organic residues are generated in bleached kraft pulp, which would potentially lead to adverse environmental pollution. The types of wastes generated from the paper and pulping industry can be categorized as follows:

- White liquor (cooking liquor):
  - It is an acidic mixture of sulfurous acid (H_2SO_3) and bisulfite ion (HSO_3^-).
  - It also contains various inorganic ions such as Mg, NH_3, Na, and Ca.
- Green liquor:
  - It is a solution of carbonate salts, primarily Na_2S and Na_2CO_3.
  - It contains insoluble unburned carbon and inorganic impurities.
- Black liquor:
  - It is a dilute solution with approximately 12–15% solids.
  - It contains wood lignin, organic materials, oxidized inorganic compounds (Na_2SO_4, Na_2CO_3), and white liquor (Na_2S and NaOH).
- Lime mud:
  - It is a waste of the calcinations or conversion of calcium carbonate to lime for causticizing black liquor.
  - The main component of lime mud is calcium carbonate ($CaCO_3$), typically around 65% [2].
  - Approximately 0.47 m^3 of lime mud can be generated for one ton of pulp [2].

Figure 13.1 shows a process flow diagram of pulp and paper manufacturing. The pulp and drying processes are the major energy consumers in the industry. The main production facilities are either pulp mills or integrated paper and pulp mills [3]. The integrated mills can achieve higher energy efficiency by eliminating intermediate pulp drying, compared to a single pulp mill. It is noted that a paper and pulping factory with a unit production capacity for one-ton pulp will generate around 0.5 ton of lime mud [4].

In this section, the physico-chemical properties of paper and pulping mill waste (lime mud) are illustrated. The challenges in convention utilization of lime mud are also provided.

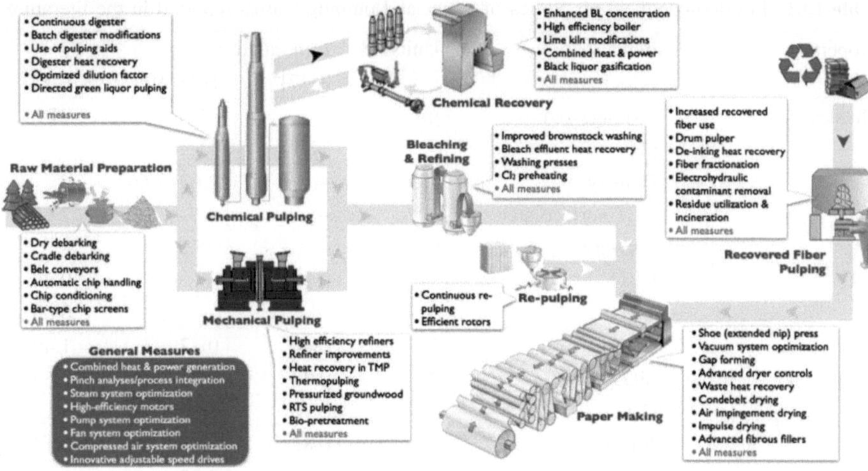

Fig. 13.1 Process flow diagram of pulp and paper manufacturing. Reproduced from Ref. [3] by permission of Institute for Industrial Productivity

13.2.1 Characterization

Table 13.1 presents the physico-chemical properties of paper and pulping wastes in the literature. The regeneration of the cooking liquor results in the formation of several types of portlandite-rich wastes, collectively referred to as paper mill wastes [5]. For instance, lime mud (or calcium mud) is a solid waste originated from the causticization process. Typically, the presence of high concentrations of portlandite ($Ca(OH)_2$) in calcium mud accounts for the high alkalinity [6]. Moreover, the main component of lime mud was found to be $CaCO_3$ [4].

13.2.2 Utilization

In paper and pulping industries, the alkaline by-products are usually sold for the cement manufacture and as alkaline amendment for agricultural soils. Conventional uses such as soil amendments, landfill, and building materials are continually conducted. Land application is one of the several limited methods available to manage solid wastes, which is more economically and ecologically sound than landfill practice [10]. However, it is limited by the presence of chloride and metals (such as Fe, Mn, Cd, Cr, Cu, Ni, Pb, As, and Zn), and its fine particle size and high alkalinity (pH > 12.1) [6, 9, 11].

The lime mud exhibited a promising potential to improve soil fertility by releasing K, Ca, and Mg into soils as an amendment [9]. In addition, applications of lime mud as adsorbents, ameliorants, and additives for treatment of liquid and solid

Table 13.1 Physico-chemical properties of paper and pulping wastes reported in the literature

Properties	Items	Units	Lime mud		
			A [7][a]	B [8, 9]	C [4]
Physical	Average size	μm	–	15	77
	pH	–	12.1	11.96	–
	Density	g/cm^3	–	2.62–2.66	–
	BET surface area	m^2/g	–	2.3–4.7	7.65
	Porosity	%	–	<5.0	–
XRF	Fe_2O_3	%	0.37	–	0.29
	TiO_2	%	–	–	0.06
	Al_2O_3	%	0.40	0.17	1.49
	SiO_2	%	0.37	0.34	2.52
	CaO	%	36.0	83.2	52.4
	Na_2O	%	0.82	0.88	0.14
	MgO	%	1.30	0.35	0.7
	SO_3	%	0.54	2.0	0.31
	K_2O	%	–	0.13	0.01
	P_2O_5	%	0.35	2.4	–
	Free CaO	%	–	–	–
	(O as element)	%	17.85	N.A.	–
Chemical analysis	$Ca(OH)_2$	%	–	55.0	–
	$CaCO_3$	%	39.0	33.0	–
	$Ca_{10}(PO_4)_6(OH)_2$	%	–	12.0	–
Thermal analysis	LOI	%	–	–	41.2

[a]The chemical compositions are presented in element forms

waste present promising effects [9]. On the other hand, CO_2 is generated at pulp mills in both the recovery boiler and lime kiln, which could be used to carbonate the paper mill wastes to produce $CaCO_3$. By the accelerated carbonation, it is reported that approximately 218 kg of CO_2 per ton of paper mill waste could be successfully sequestered into stable calcite [8].

13.3 Cementitious Waste (Construction and Demolition Waste)

Cement-type solid wastes include (1) construction and demolition waste, (2) cement kiln dust, (3) waste concrete, and (4) anorthosite tailings. The physico-chemical properties of these solid wastes are similar to that of Portland cement. Cement and demolition wastes constitute a major portion of total solid waste production in the world. For example, after urban renewal programs or natural disasters (e.g., earthquakes), demolition of older buildings leads to environmental problems particularly in larger urban areas. However, most of them are currently disposal in landfills [12]. They are generally in highly alkaline characteristics and rich in calcium-bearing components. As a result, the utilization of these cement-type solid wastes is limited because of their great concerns on environmental impacts.

13.3.1 Characterization

In this section, the physico-chemical properties of construction and demolition waste are illustrated. Table 13.2 presents the physico-chemical properties of cement-type wastes in the literature. In general, the cement-type wastes are mainly composed of SiO_2, CaO, Al_2O_3, and Fe_2O_3, along with several trace elements such as SO_3, TiO_2, and Na_2O. The chemical composition of metal oxides with an alkaline character of cement-type wastes could be considered as a suitable material for CO_2 mineralization.

Taking the anorthosite tailings as an example, the anorthosite is a rock containing more than 90% of feldspar plagioclase minerals. Feldspar plagioclase is the most abundant mineral in the earth's crust and is considered a solid solution due to

Table 13.2 Physico-chemical properties of cement-type wastes reported in the literature

Properties	Items	Units	Types of samples		
			Cement wastes [13]	Anorthosite tailings [14]	Concrete waste [14]
Physical	Particle size	μm	–	14.0–18.0	21.4–36.7
	pH	–	–	–	–
	Density	g/cm^3	–	–	–
	BET surface area	m^2/g	–	–	–
	Porosity	%	–	–	–
Chemical	Fe_2O_3	%	1.2–1.7	0.72–0.96	2.1–3.2
	TiO_2	%	0.5	0.11	0.34–0.69
	Al_2O_3	%	3.4–6.0	25.6–34.8	6.23–8.47
	SiO_2	%	12.6–25.1	50.0–66.0	37.9–48.5
	CaO	%	35.7–48.3	8.1–11.2	20.7–26.8
	Na_2O	%	2.0	4.8–6.5	1.39–1.63
	MgO	%	3.0–4.8	0.13–0.17	0.63–1.28
	SO_3	%	6.3	–	0.98–1.12
	K_2O	%	0.9–3.4	0.7–0.9	1.74–2.18
	P_2O_5	%	0.3	0.01–0.03	–
	MnO	%	–	0.01	–
	Free CaO	%	–	–	15.4–21.3

its variable composition between two pure poles: anorthite (calcic pole: $CaAl_2Si_2O_8$) and albite (sodic pole: $NaAlSi_3O_8$) [14]. The general process chemistry of carbonation with anorthite can be expressed as Eq. (13.1):

$$CaAl_2Si_2O_{8(s)} + CO_{2(g)} + 2H_2O_{(l)} \rightarrow 4CaCO_{3(s)} + Al_2Si_2O_5(OH)_{4(s)} \quad (13.1)$$

This reaction could naturally occur (so-called weathering), or it can also be accelerated in laboratory under specific operating conditions. According to the results in the literature [14], a total of 45 g CO_2 per kilogram of sample could be fixed within a reaction time of 15 min. In contrast, the waste concrete was found to be much reactive than anorthosite tailings in the aqueous carbonation. This might be attributed to the presence of susceptible phases to dissolving and reacting with CO_2, mainly portlandite ($Ca(OH)_2$) and calcium silicates [14]. In other words, the lower reactivity of andesine could be associated with its structure in framework, which probably inhibits the availability of calcium.

13.3.2 Utilization

The demolition waste consists mainly of concrete, mortar, brick, metal, timber, and plastic [1]. The major source of demolition wastes in rural areas is wastes from building remodeling and structural removal [12]. Typically, the conventional disposal methods for cement and demolition wastes are time-consuming, expensive, and environmentally unsound. Currently, in the member countries in European Union, about 46% of the construction and demolition waste is recycled [15]. It indicates that the most effective way to eliminate the cement-type waste issue is to reuse and recycle the construction materials in construction activities. Several studies have found that the use of cement-treated recycled wastes in the base (or subbase) is feasible for low-volume roads if the mix design is appropriate [12].

13.4 Mining and Mineral Processing Waste

Alumina is an important basic raw material for national economic development. In the aluminum manufacturing industries, a large amount of bauxite residues, so-called the red mud, can be produced from the digestion of bauxite ores with caustic soda (i.e., Bayer process) under elevated heat and high pressure. In general, the production of 1 ton of alumina could generate 0.3–2.5 tons of red mud, depending on the bauxite source and process efficiency of alumina extraction [16–18]. Therefore, for example, it is estimated that over 70 million tons of red mud was generated per year in China [17].

In this section, the physico-chemical properties of mining and mineral processing waste (red mud) are illustrated. In addition, the challenges in convention utilization of red mud are provided.

13.4.1 Characterization

Table 13.3 presents the physico-chemical properties of red mud in the literature. The results indicated that the chemical composition of red mud varied in a wide range, depending on the location and process of manufacturing. In general, the red mud is largely composed of SiO_2, Al_2O_3, Fe_2O_3, and CaO, along with several trace elements such as TiO_2 and Na_2O. The chemical composition of metal oxides with an alkaline character of red mud (i.e., with an average pH value of 13.3 $\pm$ 1.0 [19]) may contribute to a suitable material for CO_2 mineralization. The particle size distribution of red mud is similar to fly ash, having an average particle size of <10 μm. The specific surface area (BET) is typically 10−25 m^2/g [18]. Besides its high alkalinity, bauxite residues usually exhibit a large specific surface area and a high ion-exchange capacity [17].

Characterization of representative samples of red mud by XRD suggests that the mineral phases in red mud include hematite (Fe_2O_3), goethite (α−FeOOH), gibbsite ($Al(OH)_3$), limonite, boehmite, cancrinite, gismondine, goosecreekite, anatase, rutile, and quartz (SiO_2) [16–18,21]. Cancrinite and chantalite are the main calcium-bearing phases in red mud [16]. Since the abundant amount of metals (such as Al, Fe, Ti, and Ga) in red mud, the recycling of the metallic components has been gained great interests, such as metal extraction by using oxalic acid [23].

Table 13.3 Physico-chemical properties of red mud in the literature

Properties	Items	Units	Samples				
			A [16]	B [20]	C [21]	D [22]	E [19]
Physical	Average size	μm	5–50	–	–	–	1.9
	pH	–	7–8	–	–	–	13.3
	density	g/cm^3	1.5–2.2	2.8	2.2	–	2.93
	BET surface area	m^2/g	10.8	22.6	–	–	–
	Porosity	%	0.45	–	–	–	–
Chemical	Fe_2O_3	%	31.9	22.6	2.85	13.7	17.3
	TiO_2	%	21.2	3.37	2.03	2.10	3.43
	Al_2O_3	%	20.1	25.0	40.7	7.02	15.1
	SiO_2	%	8.5	20.2	45.8	18.1	22.8
	CaO	%	2.99	3.83	4.98	42.2	12.2
	Na_2O	%	6.0	8.78	N.D.	2.38	4.37
	MgO	%	–	<0.1	N.D.	2.06	0.27
	SO_3	%	–	N.D.	2.15	–	–
	K_2O	%	–	2.26	0.45	0.24	1.19
	P_2O_5	%	–	–	1.10	–	2.43
	Free CaO	%	–	N.D.	–	–	–

13.4.2 Utilization

The disposal of red mud remains a worldwide issue, in terms of environmental concerns, due to its high alkalinity and leachability of pollutant components. Therefore, utilization of red mud would provide significant benefits in terms of environmental and economic by reducing landfill volume and contamination of soil and groundwater. Great interests have been devoted to various types of utilization pathway for red mud, such as the following:

- coagulant [18],
- adsorbent [18],
- supplementary cementitious materials [20],
- self-compacting concrete [21],
- lightweight aggregate [24],
- soil mixture [25],
- glass-ceramics [22], and
- geopolymers [19].

Table 13.4 presents the performance of various types of utilization pathway for red mud in the literature. The details of the key performance indicators for cement materials can be referred in Chap. 15. For example, coagulants are widely utilized in water purification process, where the commonly used coagulants are Fe^{3+}- and Al^{3+}-based compounds. Due to the high contents of Fe and Al in red mud, it is considered as a promising material for coagulant production. Moreover, red mud can be used as a low-cost adsorbent for phosphorus, fluoride and nitrate ions, and trace heavy metals [18]. However, raw red mud generally presents low adsorption capacity and catalyst sintering. Therefore, several methods are proposed to activate red mud, including acidification, thermal treatment, and carbonation.

Table 13.4 Various types of utilization pathway of red mud in the literature

Pathways	Features	Performance	Reference
Self-compacting concrete	Blended with fly ash	• Pozzolanic activity: 76.6% at 7 d • Pozzolanic activity: 88.5% at 28 d • Reduced drying shrinkage due to internal curing effect	[21]
Soil mixture	Mixed with soil	• No adverse effect on test organisms for at least 10 months • There is an active microflora	[25]
Geopolymers	Mixed with fly ash and alkali activator	• Excellent long-term performance in field engineering applications • Successful formation of amorphous geopolymer gels	[19]
Cement paste	Blended with 3% fresh red mud	• Delayed chloride diffusion and CO_2 penetration in paste • Exhibit good pore refinement	[26]

Several benefits have been associated with the use of carbonated red mud, including soil amendment (removing nitrogen and phosphorus from sewage effluent), fertilizer additive in solid, fillers for plastics, and cement additive. The pozzolanic activity of fresh red mud, i.e., strength activity index (SAI), blended with fly ash in concrete was found to be 76.6 and 88.5% at 7 and 28 d, respectively, indicating that the red mud is roughly equivalent to those of fly ash [21].

References

1. Wu H, Duan H, Zheng L, Wang J, Niu Y, Zhang G (2016) Demolition waste generation and recycling potentials in a rapidly developing flagship megacity of South China: prospective scenarios and implications. Constr Build Mater 113:1007–1016. doi:10.1016/j.conbuildmat.2016.03.130
2. Wirojanagud W, Tantemsapya N, Tantriratna P (2004) Precipitation of heavy metals by lime mud waste of pulp and paper mill. Songklanakarin J Sci Technol 26(1):45–53
3. IETD (2015) Industrial efficiency technology database: pulp and paper. The Institute for Industrial Productivity. http://ietd.iipnetwork.org/content/pulp-and-paper. Accessed 18 May 2016
4. Sun R, Li Y, Liu C, Xie X, Lu C (2013) Utilization of lime mud from paper mill as CO_2 sorbent in calcium looping process. Chem Eng J 221:124–132. doi:10.1016/j.cej.2013.01.068
5. Bobicki ER, Liu Q, Xu Z, Zeng H (2012) Carbon capture and storage using alkaline industrial wastes. Prog Energy Combust Sci 38(2):302–320. doi:10.1016/j.pecs.2011.11.002
6. Perez-Lopez R, Castillo J, Quispe D, Nieto JM (2010) Neutralization of acid mine drainage using the final product from CO_2 emissions capture with alkaline paper mill waste. J Hazard Mater 177(1–3):762–772. doi:10.1016/j.jhazmat.2009.12.097
7. Martins FM, Martins JM, Ferracin LC, da Cunha CJ (2007) Mineral phases of green liquor dregs, slaker grits, lime mud and wood ash of a Kraft pulp and paper mill. J Hazard Mater 147 (1–2):610–617. doi:10.1016/j.jhazmat.2007.01.057
8. Perez-Lopez R, Montes-Hernandez G, Nieto JM, Renard F, Charlet L (2008) Carbonation of alkaline paper mill waste to reduce CO_2 greenhouse gas emissions into the atmosphere. Appl Geochem 23(8):2292–2300. doi:10.1016/j.apgeochem.2008.04.016
9. Zhang J, Zheng P, Wang Q (2015) Lime mud from papermaking process as a potential ameliorant for pollutants at ambient conditions: a review. J Clean Prod 103:828–836. doi:10.1016/j.jclepro.2014.06.052
10. He J, Lange CR, Dougherty M (2009) Laboratory study using paper mill lime mud for agronomic benefit. Process Saf Environ Prot 87(6):401–405. doi:10.1016/j.psep.2009.08.001
11. Sthiannopkao S, Sreesai S (2009) Utilization of pulp and paper industrial wastes to remove heavy metals from metal finishing wastewater. J Environ Manage 90(11):3283–3289. doi:10.1016/j.jenvman.2009.05.006
12. Jia X, Ye F, Huang B (2015) Utilization of construction and demolition wastes in low-volume roads for rural areas in China. Transp Res Rec: Journal of the Transportation Research Board 2474:39–47. doi:10.3141/2474-05
13. Pan S-Y, Chang EE, Chiang P-C (2012) CO_2 capture by accelerated carbonation of alkaline wastes: a review on its principles and applications. Aerosol and Air Qual Res 12:770–791. doi:10.4209/aaqr.2012.06.0149
14. Ben Ghacham A, Cecchi E, Pasquier LC, Blais JF, Mercier G (2015) CO_2 sequestration using waste concrete and anorthosite tailings by direct mineral carbonation in gas-solid-liquid and gas-solid routes. J Environ Manage 163:70–77. doi:10.1016/j.jenvman.2015.08.005

15. Özalp F, Yılmaz HD, Kara M, Kaya Ö, Şahin A (2016) Effects of recycled aggregates from construction and demolition wastes on mechanical and permeability properties of paving stone, kerb and concrete pipes. Constr Build Mater 110:17–23. doi:10.1016/j.conbuildmat.2016.01.030
16. Yadav VS, Prasad M, Khan J, Amritphale SS, Singh M, Raju CB (2010) Sequestration of carbon dioxide (CO_2) using red mud. J Hazard Mater 176(1–3):1044–1050. doi:10.1016/j.jhazmat.2009.11.146
17. Liu W, Yang J, Xiao B (2009) Review on treatment and utilization of bauxite residues in China. Int J Miner Process 93(3–4):220–231. doi:10.1016/j.minpro.2009.08.005
18. Wang S, Ang HM, Tade MO (2008) Novel applications of red mud as coagulant, adsorbent and catalyst for environmentally benign processes. Chemosphere 72(11):1621–1635. doi:10.1016/j.chemosphere.2008.05.013
19. Zhang M, El-Korchi T, Zhang G, Liang J, Tao M (2014) Synthesis factors affecting mechanical properties, microstructure, and chemical composition of red mud–fly ash based geopolymers. Fuel 134:315–325. doi:10.1016/j.fuel.2014.05.058
20. Fujii AL, dos Reis Torres D, de Oliveira Romano RC, Cincotto MA, Pileggi RG (2015) Impact of superplasticizer on the hardening of slag Portland cement blended with red mud. Constr Build Mater 101:432–439. doi:10.1016/j.conbuildmat.2015.10.057
21. Liu R-X, Poon C-S (2016) Utilization of red mud derived from bauxite in self-compacting concrete. J Clean Prod 112:384–391. doi:10.1016/j.jclepro.2015.09.049
22. Yang J, Zhang D, Hou J, He B, Xiao B (2008) Preparation of glass-ceramics from red mud in the aluminium industries. Ceram Int 34(1):125–130. doi:10.1016/j.ceramint.2006.08.013
23. Yang Y, Wang X, Wang M, Wang H, Xian P (2015) Recovery of iron from red mud by selective leach with oxalic acid. Hydrometallurgy 157:239–245. doi:10.1016/j.hydromet.2015.08.021
24. Molineux CJ, Newport DJ, Ayati B, Wang C, Connop SP, Green JE (2016) Bauxite residue (red mud) as a pulverised fuel ash substitute in the manufacture of lightweight aggregate. J Clean Prod 112:401–408. doi:10.1016/j.jclepro.2015.09.024
25. Ujaczki É, Feigl V, Molnár M, Vaszita E, Uzinger N, Erdélyi A, Gruiz K (2016) The potential application of red mud and soil mixture as additive to the surface layer of a landfill cover system. J Environ Sci. doi:10.1016/j.jes.2015.12.014
26. Díaz B, Freire L, Nóvoa XR, Pérez MC (2015) Chloride and CO_2 transport in cement paste containing red mud. Cement Concr Compos 62:178–186. doi:10.1016/j.cemconcomp.2015.02.011

Part IV
Valorization of Carbonation Product as Green Materials

Chapter 14
Utilization of Carbonation Products

Abstract Large-scale utilization of alkaline solid wastes provides a solution to the environmental problems associated with waste dumping, as well as energy and material conservation. However, conventional uses of alkaline solid wastes have encountered several technological barriers, including fatal volume expansion, heavy metal leaching, and low cementitious property of slag. To overcome these barriers of alkaline solid wastes utilization, an accelerated carbonation process is proposed and reviewed in terms of theoretical perspectives and practical considerations. In this chapter, the deployment of accelerated carbonation technologies for simultaneous CO_2 capture and solid waste stabilization is discussed to overcome the barriers, from the perspectives of engineering performance and environmental benefits. The strengths and opportunities of utilizing the carbonate product are comprehensively reviewed in terms of theoretical and practical considerations.

14.1 Introduction

The valorization of industrial alkaline solid wastes is one of the main routes for enhancing resource cycle toward environmental and social sustainability. Application of these "fresh" alkaline solid wastes as an alternative to standard materials has been known for a number of years. Since these materials are generally rich in metal oxides (e.g., calcium oxide, iron oxide, aluminum oxide, and magnesium oxide), a large diversity of utilization can be considered, including the following but not limited to

- Asphalt mixtures, other layers of pavement structure, unbound base courses, and embankments
- Lightweight aggregate [1]
- Supplementary cementitious materials (SCMs) in cement [2]
- CO_2 adsorbent [3]
- Composite [4]
- Soil modification agent [5]

P.-C. Chiang and S.-Y. Pan, *Carbon Dioxide Mineralization and Utilization*, DOI 10.1007/978-981-10-3268-4_14

- Antibacterial cool pigment [6]
- Geopolymer [7, 8]
- Artificial reef [9]

However, some of the industrial solid wastes (see details in Chaps. 11−13) are classified as hazardous materials because they may contain large amounts of heavy metals [10] and cause air particulate pollution [11]. This might result in severe impacts on the environment and human health. Moreover, several barriers, including volume expansion of blended materials and concerns about environmental impacts and social acceptance, have been encountered. Moreover, the use of alkaline solid wastes must comply with strict regulations, consisting of civil-technical and environmental requirements. The above-mentioned barriers hinder these materials from widespread applications in construction engineering.

Accelerated carbonation may provide a solution for multiple waste treatments, i.e., reduction of CO_2 in flue gas, neutralization of alkaline wastewater, and stabilization of alkaline solid waste. Table 14.1 presents the description of alkaline

Table 14.1 Description of alkaline solid wastes for accelerated carbonation and utilization

Waste group	Management issues	Main species involved in carbonation	Application	
			Conventional treatment	Utilization potency after carbonation
Steelmaking slag	Risk of chromium and vanadium release [12]	Larnite; Brownmillerite; Lime; Ettringite; Portlandite [13, 14]	Disposed of in landfills, if not recovered	Aggregates for civil engineering, SCMs, catalyst, etc.
MSWI fly ash	High levels of dioxin, soluble salts, and heavy metals such as cadmium, lead, and zinc (hazardous) [15, 16]	Lime; Portlandite; Ca(OH)Cl; Gehlenite [15]	90% of fly ash produced in UK was sent to landfill [15]	Construction aggregates, SCMs, etc. [16]
MSWI bottom ash	Leaching of heavy metals such as Cu, Zn, and Pb [17]	Gehlenite; Portlandite; Ettringite [18]	More than 50% is used as a secondary building material, road bases, and civil engineering structures in Europe [18]	Replace the sand or gravel fraction in concrete bricks or reuse as secondary raw construction material [17]
Cement kiln dust	Beneficial to agriculture application; however, increased concerns over the health and environmental hazards [19]	Lime; Portlandite; Calcium silicates; Gehlenite [20]	Recycling and reuse as a road base material is an established process	Reuse as road base or construction material. No free-CaO was observed in the reacted samples [20]

solid wastes for accelerated carbonation and utilization. The free-CaO and $Ca(OH)_2$ in alkaline solid wastes can be eliminated after carbonation, thereby preventing the expansion problem of the blended materials. In addition, several studies concluded that the mechanical properties of mortar blended with carbonated solid wastes were superior to those with fresh solid wastes. At the same time, the carbonated materials can meet the standards of construction engineering, providing positive benefits for practical applications. Nonetheless, there are still several important issues for accelerated carbonation, such as the fate of the carbonated products, including calcium carbonates and magnesium carbonates.

Taking steel slags as an example, Table 14.2 presents the characteristics and applications of steel slag associated with their unique physico-chemical properties. The density of steel slag generally ranges between 3.3 and 3.6 g/cm^3 due to its high Fe-oxide content, which results in hardness and wear resistance [21, 22]. The grindability index of steel slag is about 0.7, in contrast to the values of 0.96 and 1.0 for iron slag (i.e., blast furnace slag) and standard sand, respectively [21]. In addition, the chemical compositions of steel slag are found to be highly variable because of the different characteristics of the raw materials used and the types of manufacturing and smelting processes used in steel production [22].

Table 14.2 Characteristics and applications of steel slag, where information was summarized from [22–25]

Categories	Characteristics	Applications
Physical properties	Hard, rough, wear resistance, and adhesive (coarse type)	Aggregates for road and hydraulic construction
	Large surface area and porous	Adsorbents and wastewater treatment
Chemical properties	Cementitious components (C_3S, C_2S, C_4AF)	Cement and concrete production
	Fertilizer components (CaO, SiO_2, MgO, FeO, MnO, P_2O_5)	Agriculture use (e.g., fertilizer and soil improvement)
	CaO, MgO components	CO_2 capture and flue gas desulfurization
	CaO, MgO, MnO, FeO components	Fluxing agent
	FeO, CaO, SiO_2 components	Raw material for cement clinker
	Fe_2O_3, MnO_2 components	Catalyst, e.g., (1) oxidation of humic precursors in humification; (2) used in Fenton reaction
	Fe-oxide components (if high purity)	Pigments and iron ore
	SiO_2 components (if high purity)	Glasses, ceramics, semiconductors, sand, and filler material
	Microsilica	Pozzolan for use with cements

14.2 Routes of Carbonation Product Utilization

With accelerated carbonation, the use of alkaline solid wastes has been considered as an alternative to solve the dilemma of natural resource depletion and solid waste treatment in industries. In this section, the integrated treatment and utilization for alkaline solid wastes, routes of product utilization, and utilization as substitutes in civil engineering are discussed.

14.2.1 Integrated Treatment and Utilization for Alkaline Solid Wastes

Figure 14.1 shows a schematic of integrated alkaline solid waste utilization via accelerated carbonation for iron reclamation, green material production, and CO_2 fixation. After carbonation, the physical, chemical, structural, and mineralogical changes are beneficial to subsequent uses in blended cement. Within the framework, the uncontrollable expansions of cement/concrete due to the hydration of free-CaO and free-MgO in fresh solid waste can be reduced. For use as a SCM in cement, a crushing and grinding process is typically required to obtain relatively small particles of solid waste (except for fly ash). Although the utilization routes for fine solid wastes are limited, accelerated carbonation can be effectively carried out due to the small particle size of solid wastes.

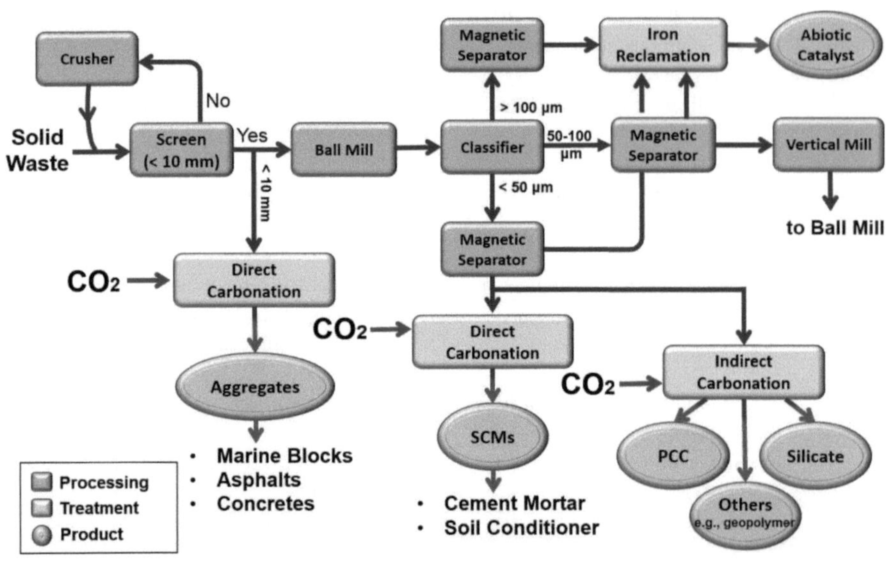

Fig. 14.1 Scheme of integrated alkaline solid waste utilization for iron reclamation, green material production, and CO_2 fixation via accelerated carbonation

Consequently, a multipronged approach to using alkaline solid wastes is suggested as follows:

- Step 1: Using a reclamation process for concentrating iron components, which can be used as pigment and abiotic catalyst.
- Step 2: Applying the carbonation process with flue gas CO_2 to stabilize the active components in slags.
- Step 3: Introducing a ternary blended system using various types of carbonated slags in the case of utilization as supplementary cementitious materials.

14.2.2 Routes of Product Utilization

Figure 14.2 illustrates the categories of utilization route for carbonation products with respect to their particle size distribution and CO_2 capture capacity. As mentioned before, there are numerous utilization routes for carbonation products, such as aggregates, artificial reefs, (abiotic) catalysts, SCM, soil conditioner, and high value-added chemicals such as precipitated calcium carbonate (PCC). The performance of the carbonated solid wastes and/or product can be evaluated via a series of systematic analyses of physico-chemical characteristics followed by comparison with fresh solid wastes. Some of the changes in substantial characteristics are beneficial to the subsequent uses.

Table 14.4 presents the potential applications of products from mineral carbonation processes. Carbonation process is capable of permanently mineralizing

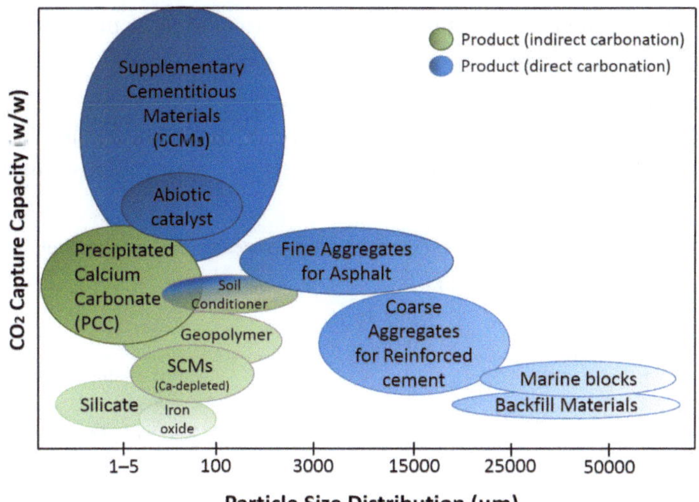

Fig. 14.2 Potential utilization categories of carbonated product with respect to particle size distribution and CO_2 capture capacity

carbon in the form of either fine or coarse aggregates or a SCM to meet the growing green product market as a carbon-negative material. In other words, when alkaline solid wastes are carbonated via direct carbonation, the products can also be used for a broad range of applications, such as construction aggregate (large particles) and cement (fine particles), without the potential presence of free-CaO and free-MgO.

14.2.3 Utilization as Substitutes in Civil Engineering

The carbonated products can be used as various types of substitutes such as aggregates and SCM in civil engineering. In many countries, there is an increasing need to find substitutes for natural sands in concrete to avoid natural resource depletion and reduce energy consumption. Common practice dictates that alkaline solid wastes intended for use as a construction material, either as aggregates or as SCMs, should be weathered for an extended period of time to reduce the amount of free-CaO and its associated hydration expansion in service. With the deployment of accelerated carbonation, the expansive compounds in fresh alkaline solid wastes can be eliminated, enabling the potential use of carbonated slag in civil engineering. Table 14.3 presents the difference in mixture contents for cement paste, mortar, and concrete block. The brief illustration is provided as follows:

- Cement paste: It is formed by the reaction (hydration) of Portland cement clinker with water.
- Mortar compacts: precast products such as masonry units, paving stones, and hollow core slabs.
- Concrete block: a composite material consisting of aggregates (including fine and coarse aggregates), a cementitious paste (including cementitious materials and water), and a binder phase.

The worldwide cement industry is increasingly turning to the use of alkaline solid wastes, such as blast furnace slag and fly ash, as SCMs due to (1) increasing fuel prices and demand for concrete and cement, (2) a lack of raw materials, and

Table 14.3 Difference in mixture contents for cement paste, mortar, and concrete block

Categories	Water	Cement	Fine aggregate (sand)	Gravel	Applications
Cement paste	•	•			Matrix material for the concrete composite
Mortar	•	•	•		Used to hold other hard components together, such as bricks, slate slabs, and stones
Concrete	•	•	•	•	Portland cement concrete, asphalt concrete, and polymer concrete

Table 14.4 Potential applications of products from mineral carbonation

Process	Products	Routes	Applications
Direct carbonation	Carbonates (low-end)	Construction	Cement voids filler, cement additive, $MgCO_3$ boards, SCMs, etc.
		Agriculture	Liming agent, soil amendment in farms, stabilizers, etc.
		Land reclamation	–
Indirect carbonation	Carbonates (high-end)	Industrial chemicals	Chemical applications, gastric antiacid, fillers (PPC), porous filtration coating, etc.
	Silicate	Industrial chemicals	Ceramics, silicate glasses, silicon for semiconductors, circuit boards, refractory materials, etc.
		Sand	Ceramic grade, foundry grade, refractory grade, flux sand, etc.
		Filler materials	Paint, plastic, rubber, etc.
		Microsilica	High-quality pozzolan for use with cements and concrete, etc.
	Iron oxide	Pigments	Paint, ceramics, porcelain, water repellent stains, magnetic ink device, etc.
		Iron ore	Iron and steel industry, etc.

(3) government regulations. However, not every (fresh) alkaline solid wastes could be successfully used in cement and concrete directly.

14.3 Challenges in Utilization of Uncarbonated Solid Wastes

Fresh (untreated or uncarbonated) alkaline solid wastes exhibit good qualities and low costs as substitutes for either aggregates or SCM in civil engineering. However, they pose a number of challenging issues, such as volume instability, heavy metal leaching, and low cementitious activity.

14.3.1 Volume Instability

One of the main obstacles to widespread use of alkaline solid wastes (such as steel slag) in construction is volume instability. The types of volume instability for cement or concrete materials include (1) autoclave expansion, (2) sulfate expansion, and (3) drying shrinkage.

14.3.1.1 Autoclave Expansion Due to Free-CaO or Free-MgO

The active species, such as free-CaO, in cement or concrete containing these solid wastes may gradually absorb moisture in air and cause expansion in the product life cycle. The content of free-CaO in alkaline solid wastes can be divided into two categories [26]:

- Residual free-lime: It is not completely dissolved in a liquid state and grainy or spongy with particle sizes of 2–40 μm.
- Precipitated free-lime: It is found in the grain boundaries of some iron oxide-based compounds, dicalcium ferrite, or R–O phase, with particle sizes of <4 μm.

In some cases, the contents of free-MgO, sulfates, sulfurs, chlorides, and iron oxide (in high amounts) may also contribute [22, 27, 28]. For instance, the reaction of free-CaO and water is presented in Eq. (14.1):

$$CaO_{(s)} + H_2O_{(l)} \leftrightarrow Ca(OH)_{2\,(aq)} + 63.7\,\text{kJ/mol} \tag{14.1}$$

The above reaction proceeds to the right hand under ambient temperature, while to the left hand only above 547 °C. It can induce fatal volume expansion of hardened cement specimen and cause disintegration (segregation), thereby losing the performance and durability of blended cement, especially in the case of electric arc furnace reducing slag and basic oxygen furnace slag [29].

14.3.1.2 Sulfate Expansion

The sulfate resistance is evaluated by measuring sulfate expansion which represented the formation of expansive ettringite and gypsum. Various mechanisms by internal and/or external sulfate attack may contribute to the sulfate expansion of cement and/or concrete. For the internal sulfate ions, ettringite ($C_6A\overline{S}_3H_{32}$) will form in the cement matrix due to the reaction between the C_3A in Portland cement and internal sulfate from gypsum, as expressed in Eq. (14.2):

$$C_3A + 2Ca^{2+} + 2SO_4^{2-} + 26H_2O \rightarrow C_6A\overline{S}_3H_{32} \tag{14.2}$$

At the same time, the remaining C_3A can react with ettringite to form monosulfate ($C_4A\overline{S}H_{12}$).

$$2C_3A + C_6A\overline{S}_3H_{32} + 4H_2O \rightarrow C_6A\overline{S}H_{12} \tag{14.3}$$

By the external sources of sulfate ions outside the cement matrix, ettringite can be also formed as expressed in Eq. (14.4) due to the reaction between the monosulfate and the external sulfate ions, thereby causing the cement to expand.

$$C_4A\overline{S}H_{12} + 2Ca^{2+} + 2SO_4^{2-} + 20H_2O \rightarrow C_6A\overline{S}_3H_{32} \quad (14.4)$$

It is noted that calcium ions are required in this reaction, which can be supplied from $Ca(OH)_2$ (portlandite), or produced by the hydration of C_3S and C_2S in the cement matrix. Therefore, $Ca(OH)_2$ in cement paste can react with external sulfates, forming gypsum ($CaSO_4{\cdot}2H_2O$) in the hardened cement paste. It is generally believed that the formation of gypsum participates results in the expansion and cracking due to the external sulfate attack [30].

Similarly, in the case of the presence of limestone filler in the cement, carbonate hydrates are formed with the consumption of CH, as shown in Eq. (14.5):

$$C_3A + 0.5CaCO_3 + 0.5Ca(OH)_2 + 12H_2O \rightarrow C_4A\overline{C}_{0.5}H_{12} \quad (14.5)$$

14.3.2 Heavy Metal Leaching

Heavy metal leaching is a critical issue in the use of alkaline solid wastes as either aggregates or SCM in civil engineering such as road and hydraulic work. They are usually related to

- Basic oxygen furnace slag [12, 31]: Ba, Cr, Pb, and V ions
- Electric arc furnace slag [32]: Cr, Ba, and V ions
- Fly ash [33, 34]: Hg, Pb, Cr, and Cd ions (other nonmetal ions include Cl^- and dioxin)
- Cement wastes [35]: Pb and Ba ions (other nonmetal ions include SO_4^{2-} and Cl^-)
- Paper sludge [35]: Cr, As, Cu, Mo, Ni, Pb, and Se ions

Basically, the environmental compatibility of alkaline solid wastes as building materials (aggregates) is determined by the potential leaching behavior of heavy metals [36]. Several studies have indicated that the leaching concentrations of heavy metals are low, which may not be relevant under environmental aspects [36], except for the high alkalinity. The alkalinity of solid wastes is affected by the partial solution of the slag lime and the closely related electric conductivity [22].

Another challenge in alkaline solid waste utilization is related to the chromium (Cr) ions, especially for electric arc furnace slag. Hexavalent Cr(VI) compounds are genotoxic carcinogens, while Cr(III) is less toxic and can be used as a pigment. Although Cr as a mineral component may occur in high amounts, the concentrations of hexavalent Cr ions in leachates may not be high since most Cr ions are bound within stable crystalline phases.

14.3.3 Low Cementitious Activity

In Portland cement, both β-C_2S and C_3S are the primary strength-contributing hydraulic phases. Compared to those in cement clinker, however, the amount of cementitious components in several types of alkaline solid wastes is much lower. As a result, blended cements with these solid wastes usually exhibit low strengths, especially at the early stages [37]. For instance, the β-C_2S and C_3S formed in basic oxygen furnace slag are less active because of their large crystal size, leading to a low cementitious activity [38]. In some cases, the γ-C_2S phase is a predominant mineral of C_2S components in alkaline solid wastes [39], which also has no cementitious activity. Although the C_2S phases can develop hydraulic characteristics under suitable conditions [40], the reactions of calcium-bearing minerals (such as larnite, bredigite, ingesonite, or calcium–olivine) with water are not expected to be activated under ambient conditions [41].

14.4 Strategies and Research Needs

14.4.1 Solutions to Overcome Barriers of Conventional Utilization

Usually, appropriate in situ post-treatment of alkaline solid wastes within industrial plants is necessary to achieve desired cementitious activity and volume stability, which are quite different depending on their final application. To guarantee the volume stability and mechanical properties of concrete and/or porous materials, different types of the post-treatment processes and methods can be deployed:

- Weathering procedure [42]
- Immersing and aging treatment [43]
- Microbiologically induced calcite precipitation (MICP) [44] requires bacteria and their nutrients
- Green solvent dimethyl carbonate (DMC) [45]
- Accelerated carbonation [46]

It is noted that the cementitious property of alkaline solid wastes typically increases with their fineness [47]. Also, a combination of inorganic and organic agents in cement paste could improve the cementitious activity of slag, thereby enhancing the early-stage strength of cement without significant influence on the 28-d compressive strength [48]. Aside from the post-treatment processes, an increase in cooling rate and alkalinity of wastes could effectively enhance the activity of cementitious activity, especially for iron and steel slags [47].

In this section, four major post-treatment processes, i.e., (1) weathering procedure, (2) immersing and aging treatment, (3) chemical addition for carbonate precipitates, and (4) accelerated carbonation technology for alkaline solid wastes, are briefly discussed.

14.4.1.1 Weathering Procedure

In weathering procedure, alkaline solid wastes are piled outdoor to react with nature wind and/or rain at the ambient conditions. Weathering procedure of alkaline solid wastes has shown an effective drop of free lime levels to a near-constant nonzero value; however, it requires an extremely long reaction period, e.g., 9–12 months for steel slag [42]. In addition, the alkaline solid waste at the bottom of lump may not be well stabilized due to its limited contacting surface area to the atmosphere.

14.4.1.2 Immersing and Aging Treatment

In steelmaking process, a cooling process called immersing and aging treatment has been developed and deployed to enable in situ rapid elimination of free-CaO to hydrated lime $Ca(OH)_2$ in slag [43]. Therefore, the stabilization and pulverization properties for steel slags can be simultaneously achieved. However, it also generally requires a reaction period of several weeks.

14.4.1.3 Chemical Addition for Carbonate Precipitates

In concrete bricks, green solvents, such as dimethyl carbonate (DMC), can be slowly hydrolyzed to produce carbonate which is used to form calcium carbonate precipitant. The formed calcium carbonate particles can seal and fill the surface cracks or voids in the porous material, thereby enhancing the durability and strength of the material [45]. It was found that the water absorption of the concrete bricks can be significantly reduced and the compressive strength can significantly increase after DMC treatment [45].

14.4.1.4 Accelerated Carbonation Technology

Accelerated carbonation is a novel approach to integrating alkaline solid wastes stabilization with CO_2 fixation (see details in Chaps. 3, 5, 7, and 8). The most valuable benefits of accelerated carbonation for waste stabilization are that the contents of (1) free-CaO and free-MgO in alkaline solid wastes, (2) the leaching potential of heavy metals, and (3) the alkalinity can be reduced in the course of carbonation with flue gas CO_2 [41, 49]. In other words, the carbonated solid wastes can be considered volumetrically stable and environmentally friendly materials. The carbonate product from accelerated carbonation can promote the hydration activity indexes of alkaline solid wastes [50]. As a result, the early strength of blended cement with carbonated slag also could be significantly enhanced over that of uncarbonated alkaline solid wastes.

14.4.2 Strategies on Utilization of Carbonated Alkaline Wastes

Huge amounts of alkaline solid wastes are generated annually worldwide. In the case of iron and steel slags, there are currently two important routes for large-scale utilization: (1) as aggregates in road and hydraulic construction; and (2) as supplementary cementitious materials (SCM) in cement and concrete after reclaiming and grinding. As aforementioned, however, these conventional uses of untreated alkaline solid wastes in civil engineering have encountered several technological barriers, including fatal volume expansion, heavy metal leaching, and low cementitious property of solid wastes. Application of accelerated carbonation for alkaline solid wastes provides a synergetic solution to these challenges.

Significant benefits from introducing the carbonation process as a solid waste pretreatment process before utilization can be realized. For instance, since diluted CO_2 in flue gas can be directly introduced for carbonation, additional environmental and economic benefits such as CO_2 emission reduction are obtained. In addition, the changes in physico-chemical properties of carbonated solid wastes were found to be beneficial to sequential uses in cement and concrete. The potential for heavy metal leaching and uncontrollable expansion can be eliminated. Furthermore, in real operations, the diluted CO_2 in flue gases or fumes can be directly used for accelerated carbonation with alkaline solid wastes generated in the same plant.

Three strategies on the effective use of alkaline solid wastes in civil engineering are suggested as follows:

- Pretreatment processes of alkaline solid wastes are usually necessary to overcome the above potential barriers.
- Sequential tests on key performance indicators of materials are necessary to reliably predict the behavior after use within a reasonable time, such as an expansion test.
- Quality assurance and quality control programs, including calibration, validation, and verification, should be established to ensure data quality and provide a useful distinction between materials that are suitable and those that are not. Since the selection of beneficiation routes for products should take advantage of their mineralogical composition and particle size distribution, the database of steel slag characteristics should be established for slag utilization.

14.4.3 Research Needs

Extensive studies on fundamental science and technology development for accelerated carbonation using alkaline solid wastes have been conducted. For the product utilization, however, only a few efforts have been made on connecting the carbonation process with the subsequent utilization of product. From the research point of view, future work should be focused on:

- The mechanisms of various types of carbonated solid wastes with ordinary Portland cement (OPC) in cement chemistry.
- The balance of carbonation efficiency and cement performance to maximize the overall carbon emission reduction.
- Diversification of utilization routes for carbonation products such as high value-added chemicals.

Although it has been successfully carried out in a laboratory-scale operation, only a few large-scale cases of accelerated carbonation using alkaline solid wastes have been deployed around the world. To realize commercialization and industrialization, future efforts should be focused on:

- Process scale-up such as reactor design, material recycling, and residue treatment.
- Instrumentation control and automation, such as online hardware/software sensors and process control model.
- Management information systems, such as information engineering and operations research.

References

1. Ahmed ZT, Hand DW (2015) Direct adsorption isotherms of AEAs and fly ash: α-olefin sulfonate and combination admixtures. ACS Sustain Chem Eng 3(2):216–223. doi:10.1021/sc500697x
2. Mo L, Zhang F, Deng M (2015) Effects of carbonation treatment on the properties of hydrated fly ash-MgO-portland cement blends. Constr Build Mater 96:147–154. doi:10.1016/j.conbuildmat.2015.07.193
3. Yan F, Jiang J, Li K, Tian S, Zhao M, Chen X (2015) Performance of coal fly ash stabilized, CaO-based sorbents under different carbonation-calcination conditions. ACS Sustain Chem Eng 3(9):2092–2099. doi:10.1021/acssuschemeng.5b00355
4. Sengupta S, Ray D, Mukhopadhyay A (2013) Sustainable materials: value added composites from recycled polypropylene and fly ash using a green coupling agent. ACS Sustain Chem Eng 1(6):574–584. doi:10.1021/sc3000948
5. Tsiridis V, Petala M, Samaras P, Sakellaropoulos GP (2015) Evaluation of interactions between soil and coal fly ash leachates using column percolation tests. Waste Manag 43:255–263. doi:10.1016/j.wasman.2015.05.031
6. Sharma R, Shaw R, Tiwari S, Tiwari S (2015) Nano-titania decorated fly ash as self-cleaning antibacterial cool pigment. ACS Sustain Chem Eng 3(11):2796–2803. doi:10.1021/acssuschemeng.5b00692
7. Huang X, Huang T, Li S, Muhammad F, Xu G, Zhao Z, Yu L, Yan Y, Li D, Jiao B (2016) Immobilization of chromite ore processing residue with alkali-activated blast furnace slag-based geopolymer. Ceram Int 42(8):9538–9549. doi:10.1016/j.ceramint.2016.03.033
8. Perná I, Hanzlíček T (2016) The setting time of a clay-slag geopolymer matrix: the influence of blast-furnace-slag addition and the mixing method. J Clean Prod 112:1150–1155. doi:10.1016/j.jclepro.2015.05.069
9. Oyamada K, Okamoto M, Iwata I (2014) Development of restoration technology for coral reefs using "marine block™". JFE Technical Report. JFE Steel Corporation, Japan

10. Qin J, Cui C, Cui X, Hussain A, Yang C (2015) Preparation and characterization of ceramsite from lime mud and coal fly ash. Constr Build Mater 95:10–17. doi:10.1016/j.conbuildmat.2015.07.106
11. Bai J, Li Y, Ren L, Mao M, Zeng M, Zhao X (2015) Thermal insulation monolith of aluminum tobermorite nanosheets prepared from fly ash. ACS Sustain Chem Eng 3 (11):2866–2873. doi:10.1021/acssuschemeng.5b00808
12. De Windt L, Chaurand P, Rose J (2011) Kinetics of steel slag leaching: batch tests and modeling. Waste Manag 31(2):225–235. doi:10.1016/j.wasman.2010.05.018
13. Huijgen WJJ, Ruijg GJ, Comans RNJ, Witkamp GJ (2006) Energy consumption and net CO_2 sequestration of aqueous mineral carbonation. Ind Eng Chem Res 45(26):9184–9194
14. Eloneva S, Teir S, Revitzer H, Salminen J, Said A, Fogelholm CJ, Zevenhoven R (2009) Reduction of CO_2 emissions from steel plants by using steelmaking slags for production of marketable calcium carbonate. Steel Res Int 80(6):415–421
15. Li X, Bertos MF, Hills CD, Carey PJ, Simon S (2007) Accelerated carbonation of municipal solid waste incineration fly ashes. Waste Manag 27(9):1200–1206. doi:10.1016/j.wasman.2006.06.011
16. Wang L, Jin Y, Nie Y (2010) Investigation of accelerated and natural carbonation of MSWI fly ash with a high content of Ca. J Hazard Mater 174(1–3):334–343. doi:10.1016/j.jhazmat.2009.09.055
17. Arickx S, Van Gerven T, Vandecasteele C (2006) Accelerated carbonation for treatment of MSWI bottom ash. J Hazard Mater 137(1):235–243. doi:10.1016/j.jhazmat.2006.01.059
18. Rendek E, Ducom G, Germain P (2006) Carbon dioxide sequestration in municipal solid waste incinerator (MSWI) bottom ash. J Hazard Mater 128(1):73–79. doi:10.1016/j.jhazmat.2005.07.033
19. Huntzinger DN, Gierke JS, Sutter LL, Kawatra SK, Eisele TC (2009) Mineral carbonation for carbon sequestration in cement kiln dust from waste piles. J Hazard Mater 168(1):31–37. doi:10.1016/j.jhazmat.2009.01.122
20. Huntzinger DN, Gierke JS, Kawatra SK, Eisele TC, Sutter LL (2009) Carbon dioxide sequestration in cement kiln dust through mineral carbonation. Environ Sci Technol 43 (6):1986–1992
21. Meng H, Liu L (2000) Stability processing technology and application prospect of steel slag. Steelmaking 25(6):74–78
22. Yi H, Xu G, Cheng H, Wang J, Wan Y, Chen H (2012) An overview of utilization of steel slag. Procedia Environ Sci 16:791–801. doi:10.1016/j.proenv.2012.10.108
23. Qi G, Yue D, Fukushima M, Fukuchi S, Nishimoto R, Nie Y (2012) Enhanced humification by carbonated basic oxygen furnace steel slag–II. Process characterization and the role of inorganic components in the formation of humic-like substances. Bioresour Technol 114:637–643. doi:10.1016/j.biortech.2012.03.064
24. Chiang Y-W, Santos RM, Elsen J, Meesschaert B, Martens JA, Van Gerven T (2013) Two-way valorization of blast furnace slag into precipitated calcium carbonate and sorbent materials. Paper presented at the Accelerated Carbonation for Environmental and Material Engineering. KU Leuven, Belgium
25. Chiou CS, Chang CF, Chang CT, Shie JL, Chen YH (2006) Mineralization of reactive black 5 in aqueous solution by basic oxygen furnace slag in the presence of hydrogen peroxide. Chemosphere 62(5):788–795. doi:10.1016/j.chemosphere.2005.04.072
26. Arribas I, Santamaría A, Ruiz E, Ortega-López V, Manso JM (2015) Electric arc furnace slag and its use in hydraulic concrete. Constr Build Mater 90:68–79. doi:10.1016/j.conbuildmat.2015.05.003
27. Frías Rojas M, Sánchez de Rojas MI, Uría A (2002) Study of the instability of black slags from electric arc furnace steel industry. Mater Constr 267(52):79–83
28. Juckes LM (2003) The volume stability of modern steelmaking slags. Miner Process Extr Metall 112(3):117–197

29. Shi C, Qian J (2000) High performance cementing materials from industrial slags - a review. Resour Conserv Recy 29(3):195–207
30. Tian B, Cohen MD (2000) Does gypsum formation during sulfate attack on concrete lead to expansion? Cem Concr Res 30:117–123
31. Bodor M, Santos RM, Cristea G, Salman M, Cizer Ö, Iacobescu RI, Chiang YW, van Balen K, Vlad M, van Gerven T (2016) Laboratory investigation of carbonated BOF slag used as partial replacement of natural aggregate in cement mortars. Cement Concr Compos 65: 55–66. doi:10.1016/j.cemconcomp.2015.10.002
32. Mombelli D, Mapelli C, Barella S, Di Cecca C, Le Saout G, Garcia-Diaz E (2016) The effect of chemical composition on the leaching behaviour of electric arc furnace (EAF) carbon steel slag during a standard leaching test. J Environ Chem Eng 4(1):1050–1060. doi:10.1016/j.jece.2015.09.018
33. Yang R, Liao WP, Wu PH (2012) Basic characteristics of leachate produced by various washing processes for MSWI ashes in Taiwan. J Environ Manage 104:67–76. doi:10.1016/j.jenvman.2012.03.008
34. Arickx S, De Borger V, Van Gerven T, Vandecasteele C (2010) Effect of carbonation on the leaching of organic carbon and of copper from MSWI bottom ash. Waste Manag 30(7): 1296–1302. doi:10.1016/j.wasman.2009.10.016
35. Sanna A, Hall MR, Maroto-Valer M (2012) Post-processing pathways in carbon capture and storage by mineral carbonation (CCSM) towards the introduction of carbon neutral materials. Energy Environ Sci 5(7):7781. doi:10.1039/c2ee03455g
36. Motz H, Geiseler J (2001) Products of steel slags an opportunity to save natural resources. Waste Manag 21:285–293
37. Kourounis S, Tsivilis S, Tsakiridis PE, Papadimitriou GD, Tsibouki Z (2007) Properties and hydration of blended cements with steelmaking slag. Cem Concr Res 37(6):815–822. doi:10.1016/j.cemconres.2007.03.008
38. Li J, Yu Q, Wei J, Zhang T (2011) Structural characteristics and hydration kinetics of modified steel slag. Cem Concr Res 41(3):324–329. doi:10.1016/j.cemconres.2010.11.018
39. Zhang T, Yu Q, Wei J, Li J, Zhang P (2011) Preparation of high performance blended cements and reclamation of iron concentrate from basic oxygen furnace steel slag. Resour Conserv Recycl 56(1):48–55. doi:10.1016/j.resconrec.2011.09.003
40. Ghosh S, Rao P, Paul A, Raina K (1979) The chemistry of dicalcium silicate mineral. J Mater Sci 14(7):1554–1566
41. Setién J, Hernández D, González JJ (2009) Characterization of ladle furnace basic slag for use as a construction material. Constr Build Mater 23:1788–1794. doi:10.1016/j.conbuildmat.2008.10.003
42. Das B, Prakash S, Reddy PSR, Misra VN (2007) An overview of utilization of slag and sludge from steel industries. Resour Conserv Recycl 50(1):40–57. doi:10.1016/j.resconrec.2006.05.008
43. Lin HC, Ou MY, Huang CH, Hsueh WH (2007) Water immersing and aging treatment of steel-making slag. Taiwan (ROC) Patent
44. Hammes F, Ae Seka, de Knijf S, Verstraete W (2003) A novel approach to calcium removal from calcium-rich industrial wastewater. Water Res 37:699–704
45. Amidi S, Wang J (2015) Surface treatment of concrete bricks using calcium carbonate precipitation. Constr Build Mater 80:273–278. doi:10.1016/j.conbuildmat.2015.02.001
46. Pan SY, Chen YH, Chen CD, Shen AL, Lin M, Chiang PC (2015) High-gravity carbonation process for enhancing CO_2 fixation and utilization exemplified by the steelmaking industry. Environ Sci Technol 49(20):12380–12387. doi:10.1021/acs.est.5b02210
47. Wanga Q, Yana PY, Feng JW (2011) A discussion on improving hydration activity of steel slag by altering its mineral compositions. J Hazard Mater 186:1070–1075
48. Luo X, Liu JX, Wang B, Zhu GL, Lu ZF (2011) Effect of accelerators on the early strength of steel slag cementitious materials. J Beijing Univ Chem Technol (in Chinese) 38(1):73–75

49. Pan SY, Chiang PC, Chen YH, Tan CS, Chang EE (2013) Ex Situ CO_2 capture by carbonation of steelmaking slag coupled with metalworking wastewater in a rotating packed bed. Environ Sci Technol 47(7):3308–3315. doi:10.1021/es304975y
50. Liang XJ, Ye ZM, Chang J (2012) Early hydration activity of composite with carbonated steel slag. J Chin Ceram Soc (in Chinese) 40(2):228–233

Chapter 15
Supplementary Cementitious Materials (SCMs) in Cement Mortar

Abstract In this chapter, the performance of the blended cement mortar with carbonated solid wastes, including physico-chemical properties, morphology, mineralogy, compressive strength, and autoclave soundness, is illustrated. The specification of performance testing for constriction materials with carbonated solid waste is also provided. In general, a high carbonation conversion of solid waste exhibits a higher mechanical strength in the early stage than pure Portland cement mortar. Moreover, the mortar with carbonated solid waste generally possesses superior soundness to the mortar using fresh solid waste. Since the chemistry of the cement hydrations is complicated and has not been completely clear, the principles and mechanisms of performance enhancement due to the use of carbonated waste in blended cement system are reviewed and discussed.

15.1 Introduction

Cement is a binder substance used in construction that sets and hardens to bind other materials together. The most widely used cement is for producing mortar in masonry and concrete (a combination of cements and aggregates to form a strong building material). Fresh alkaline solid wastes, such as iron and steel slags [1], fly ash [2], and bottom ash [3], have been extensively evaluated for potential use as supplementary cementitious materials (SCMs) in a blended cement and/or concrete block.

According to a report from the Portland Cement Association [4], the use of SCMs in blended cement could increase the later-age strength of the concrete, as compared with the use of only Portland cement. SCMs can be divided into two categories: (1) self-cementing or (2) pozzolanic. Self-cementing materials react in a similar manner to Portland cement. In contrast, the pozzolanic materials (primarily siliceous in composition) does not exhibit cementitious properties in the presence of water [5]. In those alkaline solid wastes, several components are known to contribute the hydration of cement, thereby enhancing the strength development of cement. For example, β-C_2S and C_3S are known as the primary strength-contributing hydraulic phases in Portland cement (see details in Sect. 15.3: Cement Chemistry). However, in some case, it also may reduce the early-age strength of the

P.-C. Chiang and S.-Y. Pan, *Carbon Dioxide Mineralization and Utilization*, DOI 10.1007/978-981-10-3268-4_15

concrete and cement. As a result, in this section, the challenges in the utilization of fresh solid wastes as SCMs are addressed. To overcome the barriers, an alternative is proposed and illustrated by utilizing the carbonated solid wastes (from accelerated carbonation) as SCMs.

15.1.1 Utilization of Fresh Solid Wastes as SCMs

Table 15.1 presents the uses of fresh alkaline solid wastes as SCMs in cement and/or concrete. For instance, basic oxygen furnace slag (BOFS) and fly ash (FA) are accepted as a pozzolanic martial and have been utilized in civil engineering projects for construction. Fresh BOFS or FA is often mandatory material in the production of high-strength concrete as SCMs [6]. The high strength is frequently achieved through an increase in the cementitious material content [4]. For BOFS, however, the following challenges in the utilization as a concrete product or a road base material exist [7–10]:

- It is hard, so grinding to a certain fineness as SCMs is energy-intensive and costly.
- Free-CaO and free-MgO may lead to fatal expansion of hardened cement-BOFS paste.
- The strength of the cement mortar is low, especially for early stage due to their large crystal size in BOFS, although it contains large amounts of β-C_2S and C_3S.
- The predominant minerals in BOFS, e.g., γ-C_2S, generally have no cementitious characteristics.
- Potential environmental impacts of heavy metal leaching and high alkalinity.

Therefore, the utilization of fresh BOFS as a concrete product or a road base material is still restricted.

15.1.2 Utilization of Carbonated Solid Wastes as SCMs

To overcome the above challenges, a viable treatment for alkaline solid wastes is through accelerated carbonation process with flue gas CO_2 prior to utilization as SCMs. The use of carbonated solid wastes as SCMs provides several benefits [8, 11]:

- keep globally available industrial alkaline solid wastes out of landfills,
- provide an economic approach to sequester CO_2 at the same time producing SCMs for construction use,
- create an alternative source, thereby reducing the need to transport suitable natural sands or the energy required to produce manufactured aggregates,
- reduce the amount of leachable metals such as chromium after carbonation, and
- reduce the amount of free-CaO and its associated hydration expansion in service.

Table 15.1 Use of fresh alkaline solid wastes as SCMs

Types	Clinker substitutes	Positive characteristics	Limiting characteristics
Self-cementing	Ground blast furnace slag (GBFS)	Higher long-term strength Improved chemical resistance	Low early strength High electric power demand for grinding
	Class C fly ash	Lower water demand Improved workability Higher long-term strength Better durability (depending on application)	Lower early strength Availability changed with fuel sources by the power sector Heavy metal leaching potential
pozzolan	Natural pozzolan (e.g., volcanic ash), rice husk ash, silica fume	Contributes to strength development Better workability Higher long-term strength Improved chemical resistance	Most natural pozzolan leads to reduced early strength Cement properties may vary significantly
	Limestone	Improved workability Improved mechanical properties	Maintaining strength may require additional power for grinding clinker
	Artificial pozzolan (e.g., calcined clay)	Similar to natural pozzolan	Calcination requires extra thermal energy and Reduces positive CO_2 abatement effect
	Basic oxygen furnace slag (BOFS)	Improved workability Improved chemical resistance	Difficulty in grinding potential heavy metal leaching High alkalinity low early-stage strength (<3 days) Low soundness
	Class F coal fly ash	Lower water demand Improved workability Higher long-term strength Mitigated alkali–silica reaction (lower inclusion rates) Raised sulfate resistance	Availability changed with production from bituminous coal Heavy metal leaching potential

Although reuse potentials of carbonated solid wastes are sometimes limited due to the small particle size of the feedstock needed for efficient carbonation reaction, it could be beneficial for subsequent uses. Most applications for iron and steel slags require that it be crushed and sieved into relatively small particles, which need extremely fine grinding for a cement additive. Therefore, this can avoid additional energy consumption and the cost of grinding specifically for CO_2 sequestration.

Figure 15.1 shows an integrated approach to capturing CO_2 in flue gas and stabilizing solid wastes for utilization as supplementary cementitious materials via a carbonation process. Several researches have investigated the use of carbonated steel slag as SCMs in blended cement [12] and a fine aggregate in concrete [8]. After carbonation, there are both chemical and physical changes in the structure and surface characteristics of alkaline solid waste properties, thereby affecting the performance of blended cement. It is noted that the carbonated product is a solid of lower porosity, lower tortuosity, and lower pore area with calcite infilling the pore space.

Steel slag (such as BOFS) was shown to exhibit less cementitious activity due to the nature of β-C_2S and C_3S formed in BOFS compared to those in cement clinker, thereby resulting in a relatively lower early strength. In comparison, iron slag (such as BFS) possesses great cementitious properties, which have been successfully utilized as SCMs in the cement industry. Accelerated carbonation can offer an overall solution to most of the encountered barriers of steel slag utilization, including low cementitious activity, volume instability, and heavy metal leaching. The hydrations of CaO and MgO in the fresh slag may be undesirable for usage as concrete and asphalt aggregate, road base, and fill materials due to the tendency of

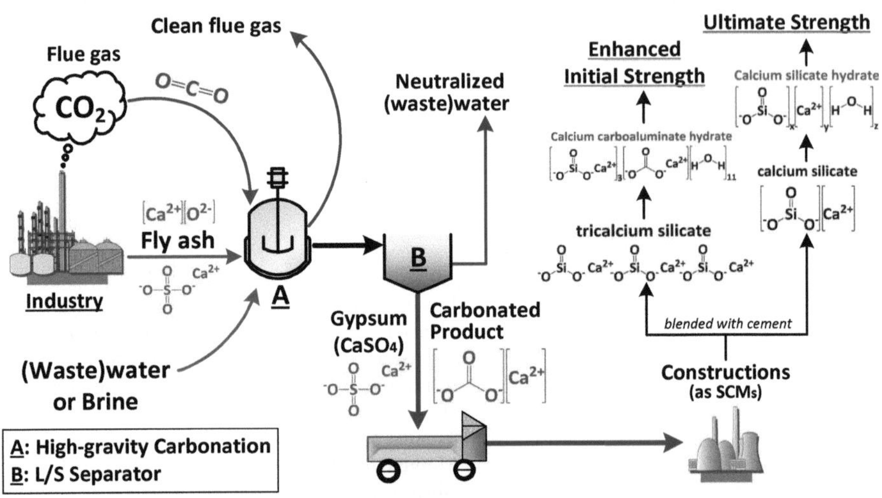

Fig. 15.1 An integrated approach to capturing CO_2 in flue gas and stabilizing solid wastes for utilization as supplementary cementitious materials via a carbonation process. Reprinted with the permission from Ref. [13]. Copyright 2016 American Chemical Society

fresh slag to have high water absorption and expansion properties. This property would be eliminated in the course of carbonation.

15.2 Specification of Performance Testing for Constriction Materials

15.2.1 Workability, Durability, and Mechanical Properties

Table 15.2 presents the most important properties, including bulk density, shape, resistance to fragmentation (i.e., resistance to impact and crushing), strength, water absorption, resistance to freezing and thawing, volume stability, and resistance to abrasion and polishing. After carbonation, the changes in physico-chemical properties of alkaline solid wastes are observed. The details of changes in physico-chemical properties of solid wastes can be referred to Sect. 5.4 in Chap. 5.

Table 15.3 summarizes the important tests and the associated specifications of cement paste, mortar, and concrete, including workability, mechanical properties, and durability. The use of alkaline solid wastes is closely related to their chemical and physical characteristics. They must be able to withstand static and dynamic forces and environmental strains (such as heat, freeze, rain, and thaw) over the long term. Consequently, the technical properties of processed construction materials are of fundamental importance. In this section, technical requirements in both the American Society for Testing and Materials (ASTM) and Chinese National Standards (CNS) are discussed and compared.

Table 15.2 Key physico-chemical properties for the application of alkaline solid waste in civil engineering

Properties	Major effect on material performance	Testing methods
Bulk density	Resistance to fragmentation (i.e., resistance to impact and crushing)	ASTM C29/C29 M-16
Shape		Microscope
Hardness	Resistance to abrasion and polishing	Rockwell harness test
C_3S, C_3A	Early strength	Chemical analysis
C_2S	Late strength	Chemical analysis
Sulfate (gypsum)	Set (hardening) of cement	Chemical analysis
Free-CaO, free-MgO	Volume stability	Immersion expansion ratio (USA and Japan) steel slag stability test method (China)
	Water absorption	
Active components ($Ca(OH)_2$, free-CaO)	Resistance to freezing and thawing	Chemical analysis
Heavy metals	Leaching behavior	Tank leaching test (Germany) TCLP (Taiwan, ROC)

Table 15.3 Important testing items and the associated specifications of cement paste, mortar, and concrete

Categories	Testing items	Specimen	Applicable test method	Descriptions
Workability	Normal consistency	Paste	ASTM C187	Amount of water required
	Setting time	Paste	ASTM C150, C191	Vicat time of setting
	Flow test	Mortar	ASTM C1437	Flow and water content
	Bleeding test	Paste/Mortar	ASTM C243	Bleeding rate/capacity
Mechanical property	Compressive strength	Mortar	ASTM C109	Compressive strength
	Flexural strength	Mortar	ASTM C348	Flexural strength
	Tensile strength	Mortar	ASTM C496	Tensile strength
Durability	Autoclave expansion (soundness)	Paste	ASTM C151	Potential delayed expansion caused by hydration of CaO and MgO
	Drying shrinkage	Mortar	ASTM C596	Change in length on drying of mortar bars
	Sulfate resistance	Mortar	ASTM C452	Potential expansion of mortars exposed to sulfate
		Mortar	ASTM C1012	Length change of mortars exposed to sulfate
		Mortar	ASTM C150	
	Permeability	Concrete	DIN 1048	Related to pore structure
	Frost resistance	Concrete	DIN 1045	Maximum w/c ratio = 0.60
Stability	Marshall stability	Asphalt	ASTM D6927	Monitor the plant process
	Resilient modulus	Bituminous	ASTM D7369	Pavement design, evaluation and analysis

Note MSRC moderate sulfate-resisting cement; *HSRC* high sulfate-resisting cement

15.2.1.1 Workability

Workability is one of the physical parameters of cement mortar or concrete, which indicates the amount of useful internal work to fully compact the mortar or concrete without bleeding or segregation in the finished product. In other words, a cement-based material is said to be "workable" when it is easily placed and compacted homogenously without bleeding or segregation. The workability can be determined by several indicators, including

- normal consistency,
- setting time, and
- flowability.

Numerous factors are largely related to the workability and the cement paste hydration, such as

- water-to-cement (w/c) ratio,
- nature of aggregates or supplementary cementitious materials,
- curing age,
- curing temperature,
- mode of compaction,
- placement and transmission methods, and
- humidity of the environment.

It is noted that the rule of thumb for improving the workability of cement/concrete is to increase the mixing time and temperature. Other approaches to improving workability for concrete are to (1) use well-rounded and smooth aggregate instead of irregular shape and (2) use non-porous and saturated aggregate. Regarding the w/c ratio, a significant increase of hydration was observed with the increase of the w/c ratio from 0.23 to 0.47, while the change of hydration becomes insignificant when the w/c ratio increases beyond 0.47 [14].

To maintain consistent workability in cement pastes, a standard flowability of 110 ± 5%, as specified by the ASTM C 230 [15], should be maintained by adjusting the quantity of mixing water. Similarly, in CNS, the workability of blended cement, including normal consistency (followed CNS 3590 [16]), flow test (followed CNS 786 [17]), and setting time (followed CNS 786 [17]), should be evaluated. For instance, the setting time of the blended cement via the Vicat test should meet the following standards:

- ASTM C595 [15]: no less than 45 min and no more than 7 h.

15.2.1.2 Mechanical Properties

The important mechanical properties for a cement-based material are as follows:

- compressive strength,
- flexural strength, and
- tensile strength.

Concrete can offer relatively high compressive strength but significantly low tensile strength. Therefore, it is usually reinforced with materials that are string in tension such as steel and rebar, so-called reinforced concrete (RC). The major composition of concrete includes cement, water, aggregates, supplementary cementitious materials, and chemical admixture. There are several types of blended cement regulated in the ASTM, such as IS (Portland blast furnace slag cement), IP (Portland-pozzolan cement), IL (Portland-limestone cement), and IT (ternary blended cement). The requirements for minimal compressive strength of cement mortar (comparable to Portland cement type I) are provided as follows:

- ASTM C595 [15] and C109 requirements for IL, IP, IS (<70), and IT (S < 70):
  - 3-days strength >1890 psi (13.0 MPa),
  - 7-days strength > 2900 psi (20.0 MPa), and
  - 28-days strength >3620 psi (25.0 MPa).
- CNS 61-R2001 requirement [18]:
  - 3-days strength >1800 psi (12.4 MPa),
  - 7-days strength >2800 psi (19.3 MPa), and
  - 28-days strength >4000 psi (27.6 MPa).

15.2.1.3 Durability

The volume stability of mortar with alkaline solid wastes may be a key criterion for using them as a construction material [19]. All applications of alkaline solid wastes and carbonated products would be practicable only if there was sufficient volume stability as the decisive criterion. Since the contact of alkaline solid wastes with water (or moisture in air) might cause these mineral phases to react to hydroxides, a volume increase of the solid waste would be observed, thereby resulting in a disintegration of the pieces and a loss of strength.

1. Contents of free-CaO and free-MgO

The contents of free-CaO (lime) and free-MgO in materials are the most important indicator of volume stability. Typically, the MgO content in the cement is limited to less than 6% [20]. In Germany, good engineering practice (GEP) shows that steel slags with a free-lime content up to 7% may be used in unbound layers, and up to 4% in asphaltic layers [19].

2. Autoclave expansion

In Germany, the steam test is commonly conducted to measure that of iron and steel slags for road construction, while the boiling test is used for hydraulic construction [19]. The standards for a maximum autoclave expansion of blended cement mortar can be found as follows:

- ASTM C596 requirement [15]: <0.80%
- CNS 1258 specification [18]: <0.80%

3. Immersion expansion

For iron and steel slags, the test of immersion expansion ratio is commonly used in the USA and Japan to evaluate the volume stability [21]. According to the CNS for "steel slag stability test method" in 2009, the immersion expansion ratio should be limited to

- CNS [22]: <2% for road construction

4. Drying shrinkage

The standards for a maximum drying shrinkage of blended cement mortar are shown below:

- ASTM C157/C596 [15]: <0.15% for blended cement with low heat of hydration.

5. Sulfate resistance

The sulfate resistance is evaluated by measuring sulfate expansion which represents the formation of expansive ettringite and gypsum. The important factors in determining sulfate resistance of blended cement include [23]

- alumina content of SCMs,
- substitution ratio of SCMs in blended cement, and
- C_3A content of clinker.

The standards for sulfate resistance of blended cement mortar are shown below:

- ASTM C-1012 requirement:
  - <0.10% (6-month) for moderate sulfate resistance cements,
  - <0.05% (6-month) for high sulfate resistance cements, and
  - <0.10% (12-month) for high sulfate resistance cements.

Several approaches to increasing the sulfate resistance of blended cement have been reported, such as the addition of extra materials (e.g., calcium sulfate and limestone powder) to control the hydration process of blended cement. Extra additions of limestone filler and calcium sulfate reduce the transformation of monosulfate to ettringite by reaction with external sulfate ions by Hoshino et al. [24].

15.2.2 Effect of Chemical Properties on Material Functions

Figure 15.2 shows the important chemical properties in alkaline solid waste associated with its related properties for the utilization in civil engineering. For example, the physico-chemical properties of steel slags mainly depend on the cooling technique applied [10]. The major mineral compositions contained in steel

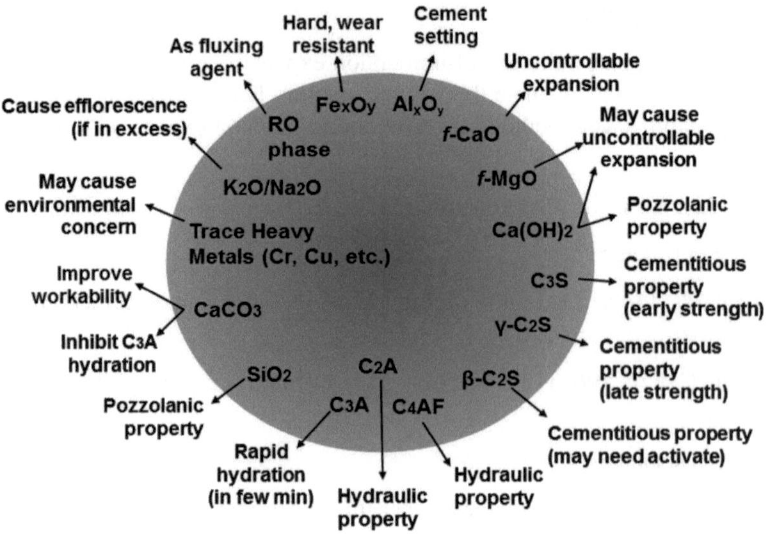

Fig. 15.2 Major chemical compositions in solid wastes (or natural ores) associated with its related properties for utilization in civil engineering

slags include tricalcium silicate (C_3S), dicalcium silicate (C_2S), RO phase (CaO–FeO–MnO–MgO solid solution), tetra-calcium aluminoferrite (C_4AF), dicalcium-ferrite (C_2F), olivine (Mg_2SiO_4), merwinite, wustite (Fe_xO_y), and lime (*f*-CaO) [19, 25]. As mentioned before, the contents of free-CaO (lime) and free-MgO in materials may cause uncontrollable expansion.

Table 15.4 presents the function of various compounds in cement clinker and their associated ratio in cement. According to their compositions and application, there are five types of Portland cement that exist, as regulated in the ASTM C150. Type 1 Portland cement is known as common (or general-purpose) cement. $Ca_3SiO_5(C_3S)$ and $Ca_2SiO_4(C_2S)$ contents are considered to be the main constituents for strength development in cement. The sum of their component percentage varies from 70 to 80%. For type 1 Portland cement, the compound compositions are 55% C_3S, 19% C_2S, 10% C_3A, 7% C_4AF, 2.8% MgO, and 1.0% free-CaO. A limitation on the composition of C_3A is that it should not exceed 15% [26].

15.2.3 Key Evaluation Parameters

Following are the major physico-chemical requirements of ordinary cement and/or supplementary cementitious materials:

Table 15.4 Function of various compounds in cement clinker and their associated ratio in cement. Information was collected from [27–29]

Categories	Items	Features	Specification
Ingredients of cement	CaO	• In excess: (1) cement unsound and (2) to expand and disintegrate • In deficiency: (1) the strength of cement decreases and (2) cement sets quickly	
	SiO_2	• Imparts strength to cement due to the formation of C2S, C3S • In excess: (1) greater strength, (2) but prolongs its setting time	
	Al_3O_2	• Imparts quick setting quality to cement • Acts as a flux to lower clinkering temperature • In excess: strength of cement decreases	<7%
	$CaSO_4$	• Increasing initial setting time of cement • Also added during grinding process for controlling initial setting time	Added by 3–4% (during clinker grinding)
	Fe_2O_3	• Provides color, hardness, and strength to cement • Helps fusion of raw materials during manufacture	<3%
	MgO	• Imparts hardness and color to the cement (if in small amount) • In excess: makes cement unsound and leads cracks after mortar or concrete hardness	< 5%
	SO_3	• Makes cement sound (if present in very small quantity) • In excess: makes cement unsound	2.1–2.8%
	Alkalies (K_2O and Na_2O)	• Should be present in small quantities • In excess: cause efflorescence and lead to the failure of concrete from that cement	<1%
Impurity	Chloride	• May restrict the use of some marine deposits	<0.01%
Composition of cement clinker	C_3S	• Generates more heat and hydrates rapidly than C2S • Possesses less resistance to chemical attack than C2S • Develops initial strength for the first 28 days	
	C_2S	• Hydrates and hardens more slowly • Offers resistance to chemical attack (e.g., sulfate)	

(continued)

Table 15.4 (continued)

Categories	Items	Features	Specification
		• Imparts ultimate strength to the cement (beyond 7 days) • Takes 2 to 3 years for complete hydration	
	C_3A	• Weak against sulfate attack • Reacts fastly, generating a large amount of heat • Does not contribute to development strength • Causes initial setting of cement	<15%
	C_4AF	• Poor cementing value • Reacts slowly, generating small amount of heat • Comparatively inactive • Acts as a flux during manufacturing • Contributes to the color effects that makes cement "Gray"	

1. Blaine fineness

The Blaine fineness of a cement powder, usually expressed in units of m^2/kg, is a single parameter for the characterization of the specific surface area. It is based on the Blaine air permeability test, which is described by ASTM C204. The Blaine fineness of a cement is assumed to be linked to physical and mechanical properties, such as strength, setting time, and flow properties (or rheology) [30]. Table 15.5 presents the Blaine fineness for different materials. For example, the Blaine fineness of OPC is typically in the ranges between 300 and 500 m^2/kg (i.e., from 3000 to 5000 cm^2/g).

2. Loss of ignition (LOI)

The percentage of loss on ignition (LOI, %) can be calculated as follows:

$$\mathrm{LOI}(\%) = (\mathrm{W}/\mathrm{W}') \times 100 \quad (15.1)$$

where W is the loss in mass between 105 and 750 °C (221 and 1382 °F), and W' is the mass of moisture-free sample used. It is noted that a high LOI value may result

Table 15.5 Blaine fineness for different materials

Material	Blaine (m^2/kg)
Ordinary Portland cement (OPC)	300–500
Limestone filler	200–1100
Raw meal	200–800
Fly ash	200–600
Ground clay	100–200

in discoloration, poor air entrainment, segregation, and low compressive strength of the mixed components. Therefore, the LOI of pozzolanic materials containing fly ashes is typically limited by regulations and standards, for example, to less than 6% in accordance with the ASTM C 618 [2].

3. Lime saturation factor (LSF)

The LSF is actually the ratio between the lime (CaO) in the raw material and the maximum amount of lime that can be combined with silica, alumina, and ferrite in the raw meal during burning and cooling. The LSF ranges from 0.90 to 1.04 for ordinary Portland cement (OPC) clinkers, with an average of 0.97 [27].

$$LSF = (CaO - 0.7\ SO_3)/(2.8\ SiO_2 + 1.18\ Al_2O_3 + 0.65\ Fe_2O_3) \tag{15.2}$$

4. Silica ratio (SR)

The SR indicates the solid-to-liquid ratio in the sintering zone of the cement kiln, typically ranging between 1.5 to 4.0, with favorable values between 2.3 and 2.8 [27].

$$SR = SiO_2/(Al_2O_3 + Fe_2O_3) \tag{15.3}$$

5. Alumina ratio (AR)

For clinker of normal composition, the AR lays between 1.5 and 4, with favorable values between 1.4 and 1.6 [27].

$$AQ = Al_2O_3/Fe_2O_3 \tag{15.4}$$

6. Alkalinity

Alkalinity (A) can be used to evaluate the hydration activities of alkaline solid wastes. If alkalinity is greater than 2.4, the solid wastes should be considered as highly reactive [31].

$$A = CaO/(SiO_2 + P_2O_5) \tag{15.5}$$

It is noted that the hydrolysis of BOFS can induce a rapid increase of alkalinity (e.g., the presence of Ca-OH and free-CaO), and the reactivity of steel slag increases with its alkalinity. The contents of free-CaO (as much as 6% in total) in steelmaking slag come from two sources: (1) residual free-lime from the raw material and (2) precipitated lime from the molten slag [32].

7. Pozzolanic activity

According to the ASTM: C311-11b [33], the strength activity index (SAI) is usually used to evaluate the pozzolanic activity of the materials:

$$\text{SAI}\ (\%) = (\text{A}/\text{B}) \times 100 \qquad (15.6)$$

where A is average compressive strength of test mixture cubes (MPa), and B is average compressive strength of control mix cubes (MPa). Normally, the compressive strength of mixture samples at ages of 7 d and 28 d is determined for the pozzolanic activity.

In the ASTM C595, the properties of pozzolanic materials for use in the blended cements are regulated. For example, the SAI index with Portland cement at 28 d should be higher than 75% [15], in accordance with the ASTM C311 test method. Moreover, the LOI of natural pozzolan, fly ash, and silica fume should be lower than 10, 6, and 6%, respectively. Table 15.6 compares the standards of physical properties for ordinary cement (OC) in various regions and/or regulations, such as ASTM, IS, and EN.

15.3 Cement Chemistry: Principles

Cements used in construction can be characterized as either hydraulic or non-hydraulic, depending on the ability of the cement to set in the presence of water. In principle, hydraulic cement hardens by hydration in the presence of water, while non-hydraulic cement (such as slaked lime) hardens by carbonation in the presence of CO_2. Table 15.7 presents the chemical constituents of major oxides in a raw meal for ordinary Portland cement clinker. In the case of hydraulic cement, the components of silicates are responsible for the mechanical properties of the cement. The chemistry of the cement hydration reactions is quite complicated and has not been completely clear.

15.3.1 C_3S and C_2S Hydrations

$Ca_3SiO_5(C_3S)$ and $Ca_2SiO_4(C_2S)$ contents are considered to be the main constituents for strength development in cement, where the sum of their component percentage varies from 70 to 80%. As shown in Eq. (15.7), the C_3S hydrates more rapidly with a higher hydration heat than that of C_2S and would enhance the initial strength development for the first 28 days. Conversely, the C_2S hydrates slowly (long-term hydration), as shown in Eq. (15.8), which is responsible for the ultimate strength development (usually takes 2–3 years for its completion hydration).

$$\text{Rapid hydration}: 2C_3S + 6H \rightarrow C_3S_2H_3 + 3CH, \quad \Delta H = 114.16\ \text{kJ/mol} \qquad (15.7)$$

$$\text{Slow hydration}: 2C_2S + 4H \rightarrow C_3S_2H_3 + CH, \quad \Delta H = 38.75\ \text{kJ/mol} \qquad (15.8)$$

Table 15.6 Properties of ordinary cement (OC) regulated in various standards

Standards	LSF	Percentage of alumina to that of iron oxide	Insoluble residue	Weight of magnesia (MgO)	Sulfur trioxide, SO_3	Loss on ignition, LOI	$SiO_2 + Al_2O_3 + Fe_2O_3$	Standard flow
ASTM C 618 [34], for fly ash	–	–	–	–	<4.0%	<10.0%	>70.0%	110 ± 5%
IS 269-1967/1975 (Indian standard) [28], for OC	0.66–1.02	>0.66	>2%	<5%	<2.75%	<4%	–	–
EN 196-1	–	–	–	<5%	–	–	–	–

Table 15.7 Chemical constituents of major oxides in a raw meal for ordinary Portland cement clinker

Types	Constituents	Units	Values		
			[28]	[27]	[7]
XRF	CaO	%	63.0	63–70	–
	SiO_2	%	22.0	19–24	–
	Al_2O_3	%	6.0	3–7	–
	Fe_2O_3	%	3.0	1–5	–
	MgO	%	2.50	0.7–4.5	–
	SO_3	%	1.75	–	–
Mineralogy	C_3S	%	40	52–85	62.9 ± 1.1
	β-C_2S + γ-C_2S	%	32	0.2–27	15.9 ± 0.4[a]
	C_3A	%	10.5	7–16	–
	C_4AF	%	9	4–16	–
	Free-lime (CaO)	%	–	0.1–5.6	–
Thermal analyses	Loss on ignition	%	1.50	–	1.39
	Insoluble residue	%	0.25	–	0.26

[a]for only β-C_2S

where C, S, and H represent CaO, SiO_2, and H_2O, respectively. $C_3S_2H_3$ represents calcium silicate hydrate (C–S–H colloid), which is in a non-crystalline amorphous solid–liquid phase. CH represents $Ca(OH)_2$. The C_3S hydration is considered to contribute to the initial strength development; however, the hydration reaction will release two molar calcium hydroxide, which is unfavorable for the long-term durability of concrete.

15.3.2 C_3A Hydration

In cement, $3CaO{\cdot}Al_2O_3(C_3A)$ is the first component to react with water and causes the initial set, as shown in Eq. (15.9). Therefore, the setting speed is mainly related to the reaction of C_3A with water, where the reaction immediately hardens the paste (i.e., ~mins) [20].

$$C_3A + 6\ H \rightarrow C_3AH_6 \tag{15.9}$$

where A represents Al_2O_3. The hydration of C_3A phase displays several features for cement development:

- little contribution to the strength of concrete,
- easily affected by sulfates, and
- great heat of hardening (self-heating), leading to a negative effect on the strength after 28 days.

The negative effect on the ultimate strength development is caused by weakening of cement matrix bond with microcrack formation due to high thermal expansion of air and water in the mortar [35]. Therefore, it is rendered ineffective by adding gypsum ($CaSO_4$) during the grinding of clinkers [28]. By doing so, the C_3A component is not available in substantial quantity for hydration reaction when water is added to the cement. In Portland cement, when the ratio of $CaSO_4$ to C_3A is low, ettringite forms during the early hydration (Eqs. (15.10) and (15.11)) and then converts to the calcium aluminate monosulfate (AFm). When the ratio is intermediate, only a portion of the ettringite converts to AFm and both can coexist, as shown in Eq. (15.12). However, at higher ratios, ettringite is unlikely to convert to AFm.

$$C_3A + 3CaSO_4 \rightarrow \text{ettringite(mostly in the form of AFt and AFm)} \quad (15.10)$$

$$C_3A + 3C\check{S} \cdot H_2 + 26H \rightarrow C_3A \cdot 3C\check{S} \cdot 32H \text{ (ettringite)}, \quad \Delta H = 1{,}350 \text{ J/g} \quad (15.11)$$

$$2C_3A + C_3A \cdot 3C\check{S} \cdot H_{32}\text{(ettringite)} \rightarrow 3C_3A \cdot C\check{S} \cdot H_{12}\text{(AFm)} \quad (15.12)$$

where $C\check{S} \cdot H_2$ represents $CaSO_4 \cdot 2H_2O$, $C_3A \cdot 3C\check{S}$ represents ettringite, and $3C_3A \cdot C\check{S}$ represents monosulfidic-type calcium aluminate. Magistri et al. [29] suggested that an optimum SO_3 content in cement mortar should be ranged between 2.1 and 2.8% for early development of compressive strength.

Similarly, using carbonated alkaline solid wastes (with the presence of $CaCO_3$ product) in cement mortar as SCMs would affect the C_3A hydration. It is noted that the hydration of C_3A could be suppressed by $CaCO_3$ due to the formation of the calcium carboaluminate ($C_3A \cdot CaCO_3 \cdot 11H_2O$) on the surface of the C_3A grains [36]. In the absence of $CaSO_4$, the C_3A would react with $CaCO_3$ to form both "hexagonal prism" phase (tricarbonate, $C_3A \cdot 3CaCO_3 \cdot 30H_2O$) and "hexagonal plate" phase (monocarbonate, $C_3A \cdot CaCO_3 \cdot 11H_2O$). The formation of the hexagonal $C_3A \cdot CaCO_3 \cdot 11H_2O$ phase could offer high mechanical strength to the cement samples on the first day of curing [37]. More details are discussed in Sect. 15.4.3.1.

15.3.3 C_4AF Hydration

C_4AF, the main component of ferrite, reacts slowly and generates small amount of reaction heat. It is comparatively inactive and acts as a flux during manufacturing. It contributes to the color effects that makes cement "gray".

$$C_4AF + x \cdot C\check{S} \cdot H_2 + y4H \rightarrow C_3A \cdot 3C\check{S} \cdot H_{32} + C_3A \cdot C\check{S} \cdot H_{12} + \ldots, \quad \Delta H = 460 \text{ J/g} \quad (15.13)$$

15.4 Blended Cement with Carbonated Wastes: Performance and Mechanisms

As aforementioned, various types of carbonated alkaline solid wastes could be considered as a suitable resource in substituting cement as a mortar or concrete block. In this section, the physico-chemical properties of the carbonated solid waste are illustrated, in terms of physical property and morphology, chemical property, and toxicity characteristic leachingprocedure (TCLP). Moreover, the performance of workability, mechanical strength development, and durability of blended cement with carbonated solid wastes is reviewed.

15.4.1 Physico-chemical Properties

15.4.1.1 Physical Property and Morphology

After accelerated carbonation, both the density and particle size distribution of alkaline solid waste may decrease because the reactive calcium species in the fresh solid waste are gradually leached out during carbonation. This might result in a less dense and shrinking matrix of solid waste [12]. On the other hand, the formation of fine $CaCO_3$ precipitates after carbonation would lead to an increase in both fineness and specific surface area. Figure 15.3a, b shows the SEM images of fresh and carbonated steel slag, respectively. Before carbonation, the entire fresh solid waste is smooth without crystallized precipitates on the surface of the particle. After carbonation, the carbonated solid waste exhibited rhombohedral crystals (i.e., $CaCO_3$), with a size of 1–3 μm, formed uniformly on the surface of the carbonated solid waste. In addition, the carbonation conversion of alkaline solid waste should affect the changes in physical properties of alkaline solid waste, such as particle size and specific surface area.

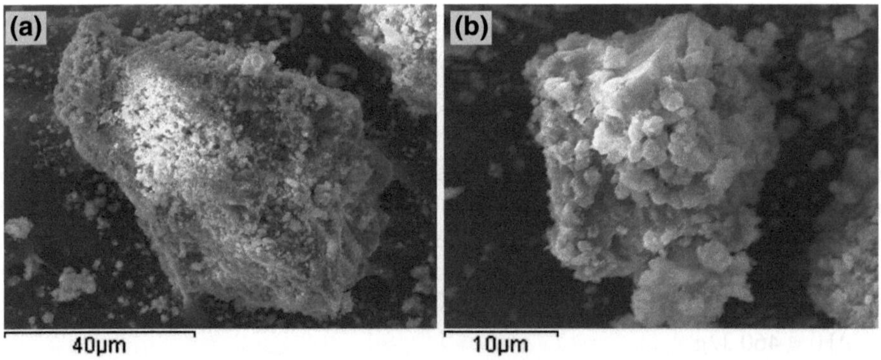

Fig. 15.3 SEM images of (a) fresh and (b) carbonated solid wastes, exemplified by basic oxygen furnace slag

15.4.1.2 Chemical Property

The content of free-CaO was observed to be eliminated after carbonation, which can be beneficial to mitigate the expansion potential of mortar containing solid waste. For example, the crystallography of steel slag was examined by XRD analysis [12], indicating that the formed product was $CaCO_3$ (i.e., calcite). This evidence reveals that the gaseous CO_2 is successfully mineralized as calcium carbonate precipitates. CO_2 was attached onto the surface of the steel slag in the course of carbonation.

15.4.1.3 Toxicity Characteristic Leaching Procedure (TCLP)

One of the purposes of toxicity characteristic leaching procedure (TCLP) tests is to evaluate the heavy metal leachability, in accordance with NIEA R201.14C [38]. The TCLP was originally developed by the United States Environmental Protection Agency (USEPA) to simulate the pollutant leaching for solid waste disposal in a landfill [39]. Depending on the pH of the sample, the TCLP procedure uses acetic acid solution (pH = 2.88 ± 0.05) or acetate buffer solution (pH = 4.92 ± 0.05) to simulate the presence of metals and organic compounds leached by acid rain. The concentrations of different heavy metals in the solution can be analyzed with inductively coupled plasma-optical emission spectroscopy (ICP–OES), or atomic absorption spectroscopy (AAS).

For example, Table 15.8 presents the TCLP results of basic oxygen furnace slag (BOFS) before and after carbonation. The results indicated the fresh BOFS (F-BOFS) might be classified as hazardous materials because of its high leaching concentration of heavy metals, such as total chromium (Cr) metal greater than 5 mg/L. After the HiGCarb process, however, the carbonated BOFS (C-BOFS) can potentially be used as green building materials since both the C-BOFS-1 and C-BOFS-2 were above the related standards. The leaching concentrations of total Cr metal were significantly reduced, by 99.3%, after carbonation [12]. With the increase of carbonation conversion to 48%, the leaching concentrations of Hg, Cr, and Cr(VI) from BOFS were not detected by ICP-OES. It thus suggests that the accelerated carbonation can effectively improve the physico-chemical properties of BOFS and minimize the leaching of heavy metals (such as Hg, Cr, Cr, Ag, and Ba) from solid matrix to become environmentally friendly products.

15.4.2 Workability

Extensive studies have been conducted to evaluate the effect of fresh alkaline solid waste on the performance of workability. The water demand of cement paste largely depends on the physical properties of the SCMs and its portion in the blending. It is noted that blended cement with SCMs having a higher surface area would demand a

Table 15.8 Toxicity characteristic leaching procedure (TCLP) results of fresh and carbonated BOFS comparable to limits by regulations in Taiwan. Reprinted with the permission from Ref. [12]. Copyright 2015 American Chemical Society

Items	Units	Values			Limits by regulations in Taiwan		
		F-BOFS	C-BOFS-1	C-BOFS-2	Utilization Product	Green Building Materials[b]	Hazardous Materials[c]
δ	%	8.8	17.0	47.9	–	–	–
Hg	mg/L	0.0009	0.0005	ND (<0.0003)	0.016	0.005	0.2
Pb	mg/L	ND (<0.029)	ND (<0.029)	ND (<0.029)	4.0	0.3	5.0
Cd	mg/L	ND (<0.0035)	ND (<0.0035)	ND (<0.0035)	0.8	0.3	1.0
Cr	mg/L	6.77	0.065	ND (<0.047)	4.0	–	5.0
Cr^{6+}	mg/L	ND (<0.0035)	0.030	ND (<0.0035)	0.2	1.5	2.5
As	mg/L	ND (<0.0007)	0.001	0.002	0.4	0.3	5.0
Se	mg/L	ND (<0.0010)	ND (<0.0010)	ND (<0.0010)	0.8	–	1.0
Cu	mg/L	ND (<0.012)	ND (<0.012)	ND (<0.012)	12.0	0.15	15.0[d]
Ag	mg/L	0.014	0.007	0.006	–	0.05	5.0[e]
Ba	mg/L	0.249	0.124	0.088	10.0	–	–

[a]δ = carbonation conversion; [b]leaching limit regulated by Taiwan Architecture & Building Center; [c]leaching limit regulated by NIEA R201.14C, EPA, Taiwan; [d]only for bottom and fly ash from incinerator, sludge, dust, catalyst, filter, etc; [e]only for waste solution from photograph and heliograph production

large amount of water. For instance, compared to a pure Portland cement, the blended cements with fresh alkaline solid waste (e.g., steel slag [7]) present higher water requirement for normal consistency and longer setting time. Moreover, compared to that with fresh solid wastes, the blended cements with carbonated solid wastes exhibit higher water demand due to larger surface areas of the carbonated particles.

On the other hand, the setting time increases with the increase of substitution ratios of SCMs in blended cement pastes. The setting speed is mainly related to the reaction of C_3A with water, as shown in Eq. (15.14), where the reaction would immediately harden the paste in few minutes [20].

$$C_3A + 6H \rightarrow C_3A_2H_6 \quad (15.14)$$

In the case of fly ash (FA) [13], the water demands for both fresh and carbonated FA are typically in the range between 0.26 and 0.29 mL-water/g-cement.

Meanwhile, the setting times for the fresh and carbonated FA at a substitution ratio of 20% were 225 and 320 min, respectively. For comparison, the initial and final setting times of pure cement are approximately 150 and 250 min, respectively. As the FA substitution ratios increase, the total content of C_3A in cement pastes decreases since the main compositions of FA are $CaSO_4$ and $CaCO_3$, thereby leading to a delay of exothermic peak. According to the experiences [13], the setting times of blended cement pastes with up to 20% substitution ratio of the fresh or carbonated FA met the ASTM and CNS requirements. Furthermore, several studies indicated that incorporation of fine pure $CaCO_3$ in blended cement could reduce set retardation in ternary blended cement [40–43]. This might be attributed to the fact that the $CaCO_3$ can provide additional nucleation sites for hydration reaction, thereby reducing the setting time.

15.4.3 Compressive Strength

As the substitution ratio of either fresh or carbonated solid waste increases, the compression strength of blended cement generally decreases due to less OPC amount in mortars, thereby reducing the available $Ca_3SiO_5(C_3S)$ and $Ca_2SiO_4(C_2S)$ contents. In addition, after aqueous carbonation, the available calcium silicate contents (e.g., CS, C_2S, and C_3S) may be reduced due to the reaction with CO_2 to form carbonates as shown in the following reaction formulas:

$$CaSiO_{3(s)} + CO_{2(g)} + 2H_2O_{(l)} \rightarrow CaCO_{3(s)} + H_4SiO_{4(aq)} \quad \Delta H = -87.9\ \text{kJ/mol}\ CO_2 \tag{15.15}$$

$$Ca_2SiO_{4(s)} + 2CO_{2(g)} + 2H_2O_{(l)} \rightarrow 2CaCO_{3(s)} + H_4SiO_{4(aq)} \quad \Delta H = -203.5\ \text{kJ/mol}\ CO_2 \tag{15.16}$$

For carbonated BOFS as example, Fig. 15.4 shows the effect of carbonation conversion and substitution ratio on the compressive strength of cement mortar. The percentage numbers (%) presented on each dot represent the relative compressive strength to Portland cement type I mortar. It was found that the cement mortar using carbonated BOFS with a carbonation conversion of 48% exhibited superior 3-day compressive strength to that using pure Portland cement or fresh BOFS. It might be attributed to the fact that the hydration of C_3A could be enhanced by $CaCO_3$ to form calcium carboaluminate, which helps to develop a higher mechanical strength in the early stage.

Figure 15.5 shows the compressive strength of blended cement with different substitution ratios of fresh (F-) or carbonated (C-) FA on 3, 7, 28, and 56 days, comparable to CNS-61 requirement in Taiwan. It was noted that similar observation was found in the case of carbonated FA.

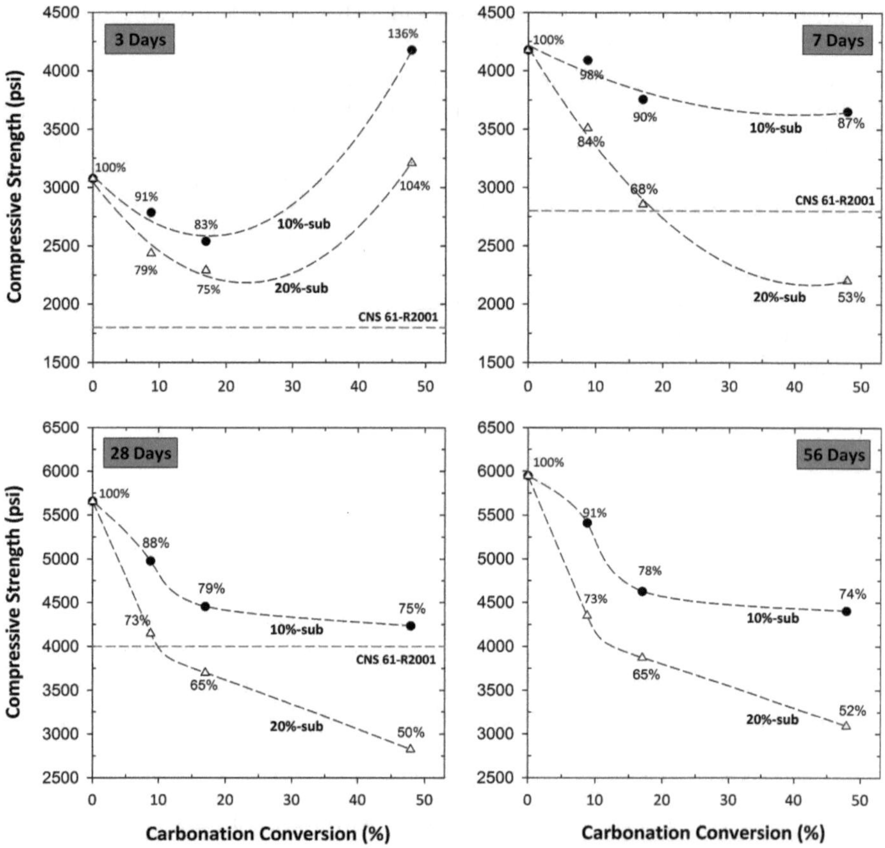

Fig. 15.4 Effect of carbonation conversion and substitution ratio on compressive strength at 3, 7, 28, and 56 days. Carbonation conversion of F-BOFS = 8.8%, carbonation conversion of C-BOFS-1 = 17.0%, and carbonation conversion of C-BOFS-2 = 47.9%. Reprinted with the permission from Ref. [12]. Copyright 2015 American Chemical Society

15.4.3.1 Early Strength Development

The use of carbonated solid wastes as SCMs can be beneficial to the strength development of blended cement, especially the early-age strength. In some case of carbonated steel slag [12] and fly ash [13], at 7 days, mortar with 10% substitution ratio of carbonated solid waste even exhibited higher initial compressive strength than the OPC mortar. Moreover, Pang et al. [44] found that replacing non-carbonated slags with carbonated slags can result in a 20% increase in compressive strength in 28 days, as well as reduce environmental problems such as the leaching of heavy metals. Compared to non-carbonated slag, the hydration activity indexes of carbonated slag on 3 d and 28 d increase by 97 and 16% at the initial water content of 19% [45]. Similar observations were made in several reports [12]. This might be attributed to the enhanced hydration of C_3A by $CaCO_3$ to form stable

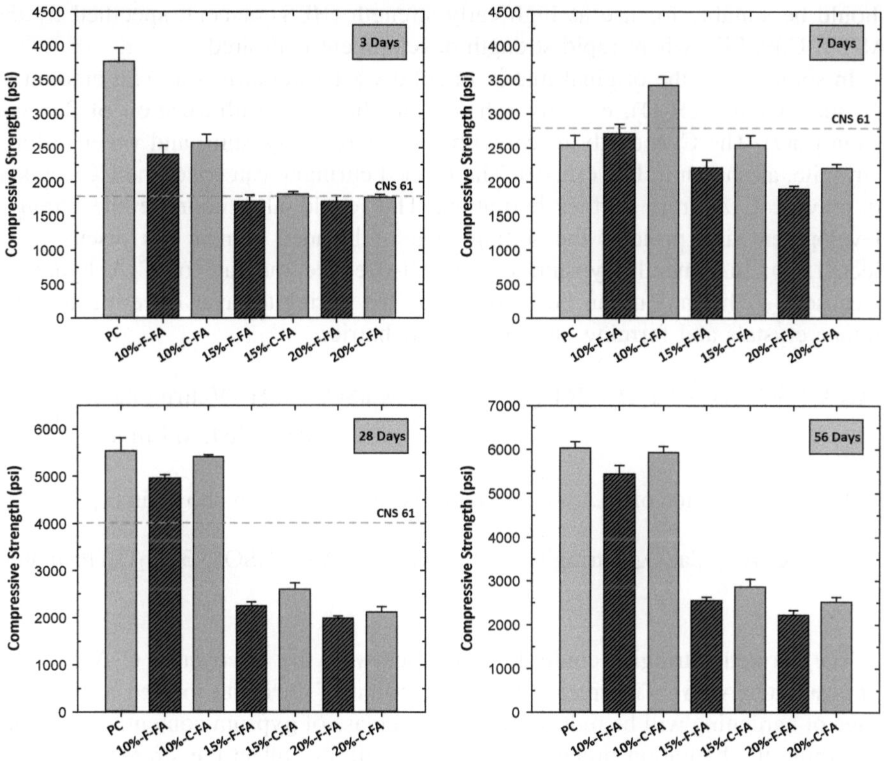

Fig. 15.5 Compressive strength of blended cement with different substitution ratios of fresh (F-) or carbonated (C-) FA on 3, 7, 28, and 56 days, comparable to CNS-61 requirement in Taiwan. Reprinted with the permission from Ref. [13]. Copyright 2016 American Chemical Society

calcium carboaluminate hydrate ($C_3A \cdot CaCO_3 \cdot 11H$), as described in Eq. (15.17). This helps to develop a higher mechanical strength in the early stage [36].

$$\begin{aligned} &2C_3A + 1.5CaCO_3 + 0.5Ca(OH)_2 + 22.5H \\ &\quad \rightarrow C_3A \cdot CaCO_3 \cdot 11H + C_3A \cdot 0.5CaCO_3 \cdot 0.5Ca(OH)_2 \cdot 11.5H \end{aligned} \quad (15.17)$$

The formed unstable compound ($C_3A \cdot 0.5CaCO_3 \cdot 0.5Ca(OH)_2 \cdot 11.5H$) will continuously convert to calcium carboaluminate after 1 d, as shown in Eq. (15.18).

$$2C_3A \cdot 0.5CaCO_3 \cdot 0.5Ca(OH)_2 \cdot 11.5H \rightarrow C_3A \cdot CaCO_3 \cdot 11H + C_3AH_6 \quad (15.18)$$

Although the contents of free-CaO and $Ca(OH)_2$ were reduced in the carbonated solid wastes, the exothermic reaction by the carbonation product ($CaCO_3$) would be faster than C_3S, which would enhance hydration heat and form carboaluminate hydrate for initial strength development [46]. It suggests that the carbonated BOFS should be suitable for use as high early strength (HES) cement, specified by the ASTM C39, [47] where rapid strength development is desired.

In some cases, the original matrix of solid wastes comprises a great amount of gypsum ($CaSO_4{\cdot}2H_2O$), e.g., fly ash (due to the wet desulfurization of flue gas using CaO). The C_3A easily reacts with sulfates (e.g., gypsum) and water to form ettringite, as shown in Eq. (15.19). The formed ettringite can cover the C_3A crystal to prevent C_3A from further hydration. This could also improve the strength development and prolong the setting time of blended cement (as discussed in Sect. 15.2). In general, gypsum is added to reduce the rate of C_3A hydration because C_3A hydration can be hindered by the ettringite layer covering the aluminate crystals and forming an impermeable barrier.

$$C_3A + 3CaSO_4 \cdot 2H_2O + 26H_2O \rightarrow C_3A \cdot 3CaSO_4 \cdot 32H_2O(\text{ettringite}), \quad \Delta H = 364.76\ \text{kJ/mol} \tag{15.19}$$

The direct reaction of C_3A with ettringite is also possible, as shown in Eq. (15.20).

$$C_3A + 3C_3A \cdot 3CaSO_4(\text{ettringite}) + 26H_2O \rightarrow C_3A \cdot 3CaSO_4 \cdot 32H_2O(\text{ettringite}) \tag{15.20}$$

The formed ettringite could further react with the uncovered C_3A to form monosulfate (AFm). This means that the amount of ettringite formed at the early stage of hardening will be reduced with the increase of gypsum content. As a result, disturbing the hydration process at early age will also affect the strength development at later age and may even reduce the strength. This suggests that the optimum gypsum content in blended cement is of crucial importance to achieve maximum strength and minimum drying shrinkage. Therefore, the optimal SO_3 content in blended cement should be in a range between 3.1 and 3.3% [13].

Meanwhile, in the presence of $CaCO_3$ (i.e., carbonated solid wastes), the above reaction can be delayed and reduced, due to the formation of monocarboaluminate, as shown in Eqs. (15.21a) and (15.21b). The carboaluminates are more voluminous and potentially stiffer than sulfoaluminates. Similar observation was also found in the literature [48], which indicated that the formation of ettringite was accelerated by $CaCO_3$ at the very beginning of hydration (e.g., 30 min).

$$C_3A \cdot 3CaSO_4 \cdot 32H_2O(\text{ettringite}) + CaCO_3 \rightarrow C_3A \cdot CaCO_3 \cdot xH_2O(\text{carboaluminate}) + \ldots \tag{15.21a}$$

$$3C_3A \cdot C\check{S} \cdot 12H(\text{AFm}) + 2CaCO_3 + 18H \rightarrow 2C_3A \cdot CaCO_3 \cdot 11H + C_3A \cdot 3C\check{S} \cdot 32H(\text{ettringite}) \tag{15.21b}$$

Tricarbonate is less stable than ettringite at ambient temperature. Sulfate ions are more effective in controlling setting than carbonate ions because sulfate ions enter more readily than carbonate ions into solid solution due to the differences in the stereochemistry [36]. The contribution of $CaCO_3$ in the hydration reactions is only observed primarily on the surface of $CaCO_3$ particle. In other words, the $CaCO_3$ should be considered an inert material.

Aside from the aforementioned chemical enhancement, physical improvement of blended cement mortar with carbonated solid wastes, which is in highly fine particle size, could result in

- High surface area: This provides activate and favorable sites for nucleation and growth of hydration products, such as C–S–H, thereby accelerating hydration reactions [49, 50].
- Microfiller effect: The $CaCO_3$ product can serve as microfiller occupying free spaces between clinker grains, thereby leading to a more compact structure in the early stage. Moreover, liquid cannot intrude into the structure to induce corrosion and further damage to the structure [51].
- Reduced porosity of mortar system: The mechanical properties of cement blended with carbonated solid wastes are also superior to those with uncarbonated solid wastes, especially for early strength [9, 45, 52].

With continuous curing, this effect does not produce additional increases in strength at early stages, unless the blended particle (e.g., cement) is ground finer to compensate. For instance, in the presence of sulfate, the monocarboaluminate can be formed by the hydration of C_3A or C_4AF, if the fineness of $CaCO_3$ particles was greater than 680 m^2/kg [53], revealing an increase in reactivity with finer $CaCO_3$ or particles.

It suggests that the carbonation product, i.e., $CaCO_3$, should be superior to the original CaO or $Ca(OH)_2$ in solid wastes, in terms of physical properties. The volume of $CaCO_3$ product is 11.8% greater than that of $Ca(OH)_2$ [54]. Pang et al. [44] reported that the number of pores smaller than 1 μm in steel slag decreased by 24.4% after carbonation, resulting in a decrease in water absorption and an increase in impermeability. Moreover, $CaCO_3$ is a kind of highly elasticity-resistant material, which can improve early strength when used as SCMs in blended cement [55]. During the carbonation process, the product of micron-sized $CaCO_3$ formed gradually in the C–S–H matrix, building up a denser and more compacted structure, which is beneficial to the strength development of blended cement [56–58].

Similar observations were found in the literature for cement pastes containing $CaCO_3$, either as a chemical reagent or as a limestone constituent [12, 35, 49]. At a low addition level of less than 5%, some modification on the heat of hydration at early ages may occur depending on the fineness of $CaCO_3$ particles. Also the long-term heat flow would be a bit lower than without limestone because of the smaller fraction of hydrating clinker [49].

15.4.3.2 Late Strength Development

Most of the strength-developing properties of cement are controlled by C_3S and C_2S (see discussion in Sect. 15.2); however, these components are partially consumed during carbonation. With the increase of substitution ratio from 10 to 15%, a significant difference in compressive strength after 28 days was observed in the case of carbonated steel slag [12] and fly ash [13]. Therefore, the low pozzolanic (cementitious) activity of carbonate solid waste (such as steel slag) with a relatively higher carbonation conversion caused by the consumption of $Ca(OH)_2$ during carbonation can lead to a low strength development at the late stage.

Although the C_3A content in cement (typically ~10% in clinker) contributes little to the strength of concrete, the C_3A hydration will result in a great amount of hardening heat. The hydration of C_3A provides a decisive effect on the rheological properties of cement paste. A high rate of hydration would result in the saturation of the solution with aluminate and calcium ions and as a consequence to the crystallization of C_4AH_x, leading to the quick stiffening of paste [59]. This might lead to a negative effect on the strength after 28 days by weakening of cement matrix bond with microcrack formation due to its high thermal expansion of air and water in the mortar [35].

In addition, the decomposition of monocarbonate could still occur depending on time, temperature, and humidity of the environment [37]. Partial decomposition of the $C_3A{\cdot}CaCO_3{\cdot}11H$ phase might occur to form C_3AH_6 and $CaCO_3$ on 7 days, as shown in Eq. (15.22).

$$C_3A \cdot CaCO_3 \cdot 11H \rightarrow C_3AH_6 + CaCO_3 + 5H_2O \tag{15.22}$$

This seems to be one of the key factors that lead to a decrease in the compressive strength after 3 days.

15.4.4 Durability

15.4.4.1 Autoclave Expansion

Accelerated carbonation is an effective method to improve the durability of blended cement or concrete block because relatively insoluble $CaCO_3$ is formed from the soluble and active components, e.g., free-CaO and $Ca(OH)_2$, in alkaline solid wastes [60, 61]. After carbonation, free-CaO content in alkaline solid waste could be depleted, even entirely eradicated, as presented in Eq. (15.23), which reduces the risk of fatal fracture from volume expansion [51].

$$CaO_{(s)} + CO_{2(g)} + H_2O_{(l)} \rightarrow CaCO_{3(s)} + H_2O_{(l)}, \Delta H = -178.3 \text{ kJ/mol } CO_2 \tag{15.23}$$

In the case of basic oxygen furnace slag (BOFS), Fig. 15.6 shows the soundness of blended cement mortars with different substitution ratios of F-BOFS (as fresh) or C-BOFS (as carbonated). The expansion capacity of mortar increases with the increases of the substitution ratio of F-BOFS due to higher free-CaO contents in

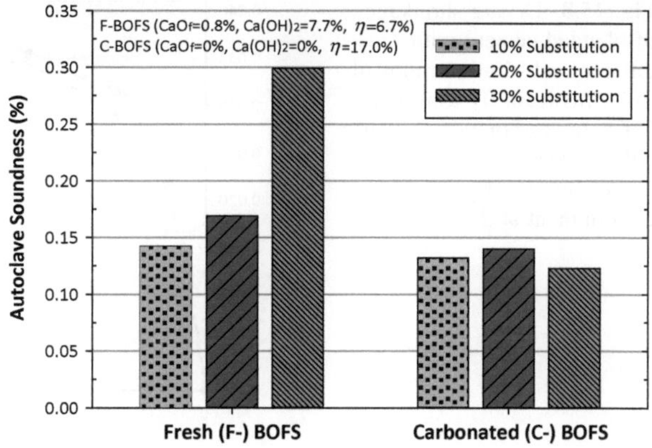

Fig. 15.6 Autoclave soundness expansion of cement mortar with different substitution ratios using fresh BOFS (F-BOFS) or carbonated BOFS (C-BOFS) with a carbonation conversion of 17%

mortar. Using the F-BOFS at a substitution ratio of 20%, the maximal autoclave expansion of mortar was approximately 0.3%. Instead, using C-BOFS in blended cement, the expansion increment could be successfully stabilized at a value of less than 0.15%. This was attributed to the elimination of free-CaO content in C-BOFS with relatively stable compounds (e.g., $CaCO_3$).

In the case of fly ash (FA), Fig. 15.7 shows the autoclave expansion of blended cement mortars with F-FA (as fresh) and C-FA (as carbonated). The results indicated that the autoclave expansion of the blended cement mortars increased with the increase of substitution ratio of the FA because of higher free-CaO contents in mortar. In the case of 20% substitution ratio, the maximal expansions for the F-FA and C-FA were about 0.34 and 0.28%, respectively. In all cases of using the C-FA, the expansion increment could be successfully stabilized. This was attributed to the elimination of reactive free-CaO content in the C-FA with relatively stable compounds ($CaCO_3$).

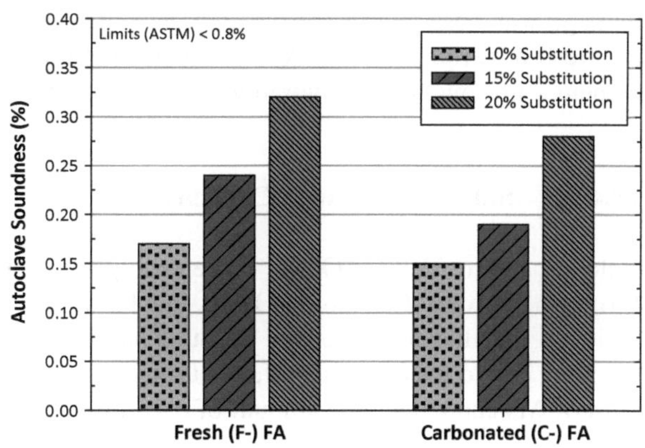

Fig. 15.7 Autoclave expansion ratio (soundness) of cement mortars blended with different substitution ratios of F-FA (as fresh fly ash) and C-FA (as carbonated fly ash)

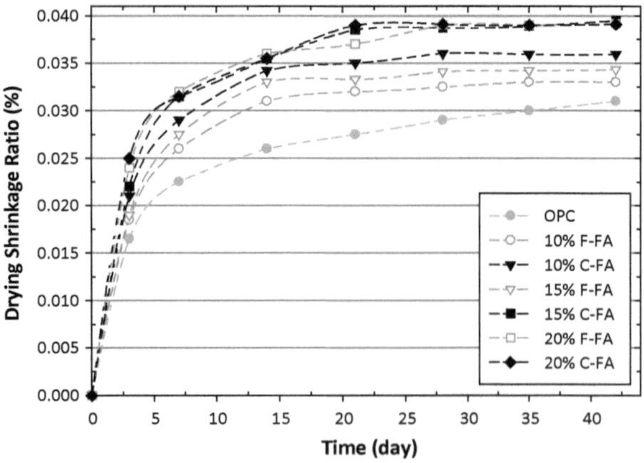

Fig. 15.8 Drying shrinkage of blended cement with different substitution ratios of F-FA (as fresh fly ash) and C-FA (as carbonated fly ash). All samples were cured in air and exposed a dry environment at 23 °C

Similar observations were also found that the durability of cement mixtures could be greatly enhanced by employing either pure limestone powder [62, 63] or carbonated materials [64, 65] in blended cement. As mentioned before, this can improve not only the dimensional stability but also the tensile strength, stiffness, and durability of cement and/or concrete.

15.4.4.2 Dry Shrinkage

Figure 15.8 displays the drying shrinkage of blended cement mortars, which is comparable to the OPC mortar. This is likely caused by the volume change due to water loss by evaporation in a specimen when the blended cement is exposed to a dry environment. The results revealed that the drying shrinkages of all specimens increased rapidly within the first 7 d and gradually reached to a steady state up to 42 days. The drying shrinkages of all the blended cement mortars containing BOFS were higher than those of OPC control within 42 days. However, no significance damage was observed in the blended cement with up to 20% substitution ratio of the fresh or carbonated FA. All specimens met the ASTM requirement for blended cement with low heat of hydration, i.e., a maximal shrinkage of 0.15% [15].

15.4.4.3 Sulfate Resistance Capacity

Calcium hydroxide ($Ca(OH)_2$) in cement paste can react with external sulfates, forming gypsum ($CaSO_4 \cdot 2H_2O$) in the hardened cement paste. It is generally believed that the formation of gypsum participates results in the expansion and cracking due to the external sulfate attack [66]. The replacement of Portland cement with SCMs can reduce the amount of calcium hydroxide in cement paste [67]. This is likely caused by (1) dilution of the Portland cement and (2) consumption of $Ca(OH)_2$ by pozzolanic reaction of the SCMs. Therefore, the replacement of Portland cement with SCMs decreases the availability of $Ca(OH)_2$ to participate in the formation of gypsum with external sulfates.

15.5 Multiple Blended Systems

As described in Sect. 15.4, the encountered barriers of fresh alkaline solid wastes (e.g., F-BOFS) utilization as SCMs in the blended cement, such as instable expansion property and low early-age compressive strength, can be overcome by the carbonation process. However, partial Portland cement replacement by carbonated solid wastes should be limited to a certain ratio (e.g., C-BOFS should be up to 20%), as higher contents would result in a negative impact on the compressive strength of the cement compositions. This is largely due to the reduced amount of the main hydration products which are able to induce bonding properties. Furthermore, the particle size distribution of solid waste as SCMs plays an important role in compressive strength development of blended cement. In the case of both fresh and carbonated steel slag, mortar prepared with either the coarsest (<1.6 mm) or finest fractions (<0.08 mm) of steel slag exhibited significantly lower compressive strengths than the reference OPC mortar [68]. To overcome this challenge, using a mixture formula with different particle sizes of alkaline solid wastes might be a synergetic solution.

The new approach is to introduce the ternary (or multiple) blended system, which might be able to produce high-performance blended cement, with a larger amount of SCMs. A ternary system is a cement mixture that includes two cementitious alternatives into a cement mix, and quaternary mixes utilize three amendments [5]. Due to the variety in characteristics and availabilities of SCMs, it is achievable to blend several types of SCM together in a cement mixture system to meet desired functions. Based on the close packing theory, Zhang et al. [69, 70] proposed a gap-graded blended cement using discontinuous particle size distribution of solid residuals, for example, combined fine and coarse SCMs. This novel method can effectively increase the initial packing density of blended cement paste, thereby enhancing early and late strength development of cement [70]. Moreover, the gap-graded blended cements exhibit a shorter setting time, higher early strength, and superior durability [71].

The use of ternary system and quaternary system can be designed to incorporate less commonly used waste materials and improve the performance of the cement products by combining with different pozzolan [5]. According to the calorimetric and thermogravimetric analyses, a favorable slag–limestone synergy exists that enables high-volume replacement of OPC cement without significant loss in early- and later-age properties [72]. The early-age compressive strengths are beneficially impacted by the presence of limestone, whereas the clinker factor does not play a significant role in later-age strengths in the blended systems. Compared to direct mixture of limestone OPC, the method of interground limestone with OPC had limited the improvement on the mechanical properties and sulfate resistance [67, 72].

References

1. Crossin E (2015) The greenhouse gas implications of using ground granulated blast furnace slag as a cement substitute. J Clean Prod 95:101–108. doi:10.1016/j.jclepro.2015.02.082
2. Mo L, Zhang F, Deng M (2015) Effects of carbonation treatment on the properties of hydrated fly ash-MgO-Portland cement blends. Constr Build Mater 96:147–154. doi:10.1016/j.conbuildmat.2015.07.193
3. Chusilp N, Jaturapitakkul C, Kiattikomol K (2009) Effects of LOI of ground bagasse ash on the compressive strength and sulfate resistance of mortars. Constr Build Mater 23(12):3523–3531. doi:10.1016/j.conbuildmat.2009.06.046
4. Caldarone MA, Taylor PC, Detwiler RJ, Bhidé SB (2005) Guide specification for high performance concrete for bridges, 1st edn. Portland Cement Association, Canada
5. Paris JM, Roessler JG, Ferraro CC, DeFord HD, Townsend TG (2016) A review of waste products utilized as supplements to Portland cement in concrete. J Clean Prod 121:1–18. doi:10.1016/j.jclepro.2016.02.013
6. Wilson ML, Kosmatka SH (2011) Design and control of concrete mixtures. In: High-performance concrete, 15th edn. Portland Cement Association, Washington, DC, p 299
7. Zhang T, Yu Q, Wei J, Li J, Zhang P (2011) Preparation of high performance blended cements and reclamation of iron concentrate from basic oxygen furnace steel slag. Resour Conserv Recycl 56(1):48–55. doi:10.1016/j.resconrec.2011.09.003
8. Monkman S, Shao Y, Shi C (2009) Carbonated ladle slag fines for carbon uptake and sand substitute. J Mater Civ Eng 21:657–665. doi:10.1061//asce/0899-1561/2009/21:11/657
9. Wu HZ, Chang J, Pan ZZ, Cheng X (2009) Carbonate steelmaking slag to manufacture building materials. Adv Mater Res 79–82:1943–1946. doi:10.4028/www.scientific.net/AMR.79-82.1943
10. Pan S-Y, Chang EE, Chiang P-C (2012) CO_2 capture by accelerated carbonation of alkaline wastes: a review on its principles and applications. Aerosol Air Qual Res 12:770–791. doi:10.4209/aaqr.2012.06.0149
11. Pan S-Y, Chiang A, Chang E-E, Lin Y-P, Kim H, Chiang P-C (2015) An innovative approach to integrated carbon mineralization and waste utilization: A review. Aerosol Air Qual Res 15:1072–1091. doi:10.4209/aaqr.2014.10.02
12. Pan SY, Chen YH, Chen CD, Shen AL, Lin M, Chiang PC (2015) High-gravity carbonation process for enhancing CO_2 fixation and utilization exemplified by the steelmaking industry. Environ Sci Technol 49(20):12380–12387. doi:10.1021/acs.est.5b02210
13. Pan S-Y, Hung C-H, Chan Y-W, Kim H, Li P, Chiang P-C (2016) Integrated CO_2 fixation, waste stabilization, and product utilization via high-gravity carbonation process exemplified by circular fluidized bed fly ash. ACS Sustain Chem Eng 4(6):3045–3052. doi:10.1021/acssuschemeng.6b00014
14. Taylor GD (1991) Construction materials. Longman Group, UK
15. ASTM C 595/C 595 M-15 (2015) Standard specification for blended hydraulic cements. Annual Book of ASTM Standards. ASTM American Society for Testing and Materials, West Conshohochen, PA
16. CNS 3590-1988 (1988) Method of test for normal consisteney of hydraulic cement. Chinese National Standards, Taiwan (ROC)

17. CNS 786-1983 (1983) Method of test for time of setting of hydraulic cement by vicat needle. Chinese National Standards, Taiwan (ROC)
18. CNS 61-R2001 (2011) Portland cement. Chinese National Standards, Taiwan (ROC)
19. Motz H, Geiseler J (2001) Products of steel slags an opportunity to save natural resources. Waste Manag 21:285–293
20. Neville AM (2002) Properties of concrete. Pitman Pub
21. Gao JP (2008) Steel slag stability test method-the key technologies of steel slag used in the building materials domain. Metall Stand Qual (in Chinese) 46(6):25–29
22. Yi H, Xu G, Cheng H, Wang J, Wan Y, Chen H (2012) An overview of utilization of steel slag. Procedia Environ Sci 16:791–801. doi:10.1016/j.proenv.2012.10.108
23. Ogawa S, Nozaki T, Yamada K, Hirao H, Hooton RD (2012) Improvement on sulfate resistance of blended cement with high alumina slag. Cem Concr Res 42(2):244–251. doi:10.1016/j.cemconres.2011.09.008
24. Hoshino S, Yamada K, Hirao H (2007) XRD/Rietveld analysis of the hydration and strength development of slag and limestone blended cement. J Adv Concr Technol 4(3):357–367
25. Kourounis S, Tsivilis S, Tsakiridis PE, Papadimitriou GD, Tsibouki Z (2007) Properties and hydration of blended cements with steelmaking slag. Cem Concr Res 37(6):815–822. doi:10.1016/j.cemconres.2007.03.008
26. C150/C150 M A (2015) Standard specification for portland cement. Annual book of ASTM standards. ASTM American Society for Testing and Materials, New York
27. Justnes H (2012) Alternative low-CO_2 "Green" clinkering processes. In: Broekmans MATM, Pollmann H (eds) Reivews in mineralogy & geochemistry, vol 74., Applied mineralogy of cement & concreteMinerological Society of America Virginia, USA, pp 83–99
28. Rajput RK (2007) Cement. In: Engineering material. S. Chand & Company Ltd., Delhi
29. Magistri M, Recchi P, Forni P (2015) Optimization in the use if cement additives: effect of gypsum dehydration on the reactivity of performance enhancers. Mapei SpA, Italy
30. Harrigan ET (2013) Measuring cement particle size and surface area by laser diffraction. Research results digest. National Cooperative Highway Research Program
31. Shi CJ, Qian JS (2000) High performance cementing materials from industrial slags—a review. Resour Conserv Recy 29(3):195–207
32. Birat J-P (2009) Steel and CO_2—the ULCOS program, CCS and mineral carbonation using steelmaking slag. In: 1st International Slag Valorisation Symposium, Leuven
33. ASTM C 311-11b (2011) Standard test methods for sampling and testing fly ash or natural pozzolans for use in portland-cement concrete. Annual Book of ASTM Standards. ASTM American Society for Testing and Materials, West Conshohochen, PA
34. ASTM C 618 (2001) Standard specification for coal fly ash and raw or calcined natural pozzolan for use as a mineral admixture in concrete. Annual Book of ASTM standards, vol ASTM C 618-2001. ASTM American Society for Testing and Materials, West Conshohochen, PA
35. Kurdiwski W (2014) Cement and concrete chemistry. Springer, New York London. doi:10.1007/978-94-007-7945-7
36. Hawkins P, Tennis P, Detwiler R (2003) The use of limestone in portland: a state-of-the-art review. Portland Cement Association, USA
37. Luz AP, Pandolfelli VC (2012) CaCO3 addition effect on the hydration and mechanical strength evolution of calcium aluminate cement for endodontic applications. Ceram Int 38 (2):1417–1425. doi:10.1016/j.ceramint.2011.09.021
38. NIEA R201.14C (2009) Toxicity characteristic leaching procedure. vol 0980070269. Environmental Analysis Laboratory, EPA, Taiwan (ROC)
39. USEPA (1992) Test methods for evaluating solid waste, Physical/Chemical Methods. Government Printing Office, Washington, DC, USA
40. Bentz DP, Sato T, De la Varga I, Weiss WJ (2012) Fine limestone additions to regulate setting in high volume fly ash mixtures. Cement Concr Compos 34(1):11–17
41. Bentz DP (2014) Activation energies of high-volume fly ash ternary blends: hydration and setting. Cement Concr Compos 53:214–223

42. Bentz DP, Ardani A, Barrett T, Jones SZ, Lootens D, Peltz MA, Sato T, Stutzman PE, Tanesi J, Weiss WJ (2015) Multi-scale investigation of the performance of limestone in concrete. Constr Build Mater 75:1–10
43. Gurney LR, Bentz DP, Sato T, Weiss WJ (2012) Reducing set retardation in high-volume fly ash mixtures with the use of limestone. Transp Res Record: J Transp Res Board 2290(1): 139–146
44. Pang B, Zhou Z, Xu H (2015) Utilization of carbonated and granulated steel slag aggregate in concrete. Constr Build Mater 84:454–467
45. Liang XJ, Ye ZM, Chang J (2012) Early hydration activity of composite with carbonated steel slag. J Chin Ceram Soc (in Chinese) 40(2):228–233
46. Zajac M, Rossberg A, Le Saout G, Lothenbach B (2014) Influence of limestone and anhydrite on the hydration of Portland cements. Cement Concr Compos 46:99–108. doi:10.1016/j.cemconcomp.2013.11.007
47. ASTM C 39 (2001) Standard test method for compressive strength of cylindrical concrete specimens. Annual Book of ASTM Standards. ASTM American Society for Testing and Materials, West Conshohochen, PA, USA
48. Tan Z (2012) Chemical reaction of limestone with C_3S and C_3A
49. Lothenbach B, Le Saout G, Gallucci E, Scrivener K (2008) Influence of limestone on the hydration of Portland cements. Cem Concr Res 38:848–860. doi:10.1016/j.cemconres.2008.01.002
50. Thongsanitgarn P, Wongkeo W, Chaipanich A, Poon CS (2014) Heat of hydration of Portland high-calcium fly ash cement incorporating limestone powder: effect of limestone particle size. Constr Build Mater 66:410–417. doi:10.1016/j.conbuildmat.2014.05.060
51. Chi J, Huang R, Yang C (2002) Effects of carbonation on mechanical properties and durability of concrete using accelerated testing method. J Mar Sci Technol 10:14–20
52. Wu HZ, Chang J, Pan ZZ, Cheng X (2011) Effects of carbonation on steel slag products. Adv Mater Res 177:485–488. doi:10.4028/www.scientific.net/AMR.177.485
53. Stark J, Freyburg E, Lohmer K Investigation into the influence of limestone additions to portland cement clinker phases on the early phase of hydration. In: Dhir RK, Dyer TD (eds) International conference on modern concrete materials: binders, additions and admixtures, London, 1999. Thomas Telford
54. Fernandez Bertos M, Simons SJ, Hills CD, Carey PJ (2004) A review of accelerated carbonation technology in the treatment of cement-based materials and sequestration of CO_2. J Hazard Mater 112(3):193–205. doi:10.1016/j.jhazmat.2004.04.019
55. Haecker C-J, Garboczi E, Bullard J, Bohn R, Sun Z, Shah S (2005) Modeling the linear elastic properties of Portland cement paste. Cem Concr Res 35:1948–1960
56. Mahoutian M, Shao Y, Mucci A, Fournier B (2014) Carbonation and hydration behavior of EAF and BOF steel slag binders. Mater Struct 48:3075–3085
57. Matschei T, Glasser FP (2010) Temperature dependence, 0 to 40 °C, of the mineralogy of Portland cement paste in the presence of calcium carbonate. Cem Concr Res 40(5):763–777. doi:10.1016/j.cemconres.2009.11.010
58. Bentz DP, Jones SZ, Peltz MA, Stutzman PE (2015) Influence of internal curing on properties and performance of cement-based repair materials. Nat Inst Stan Technol. doi:10.6028/nist.ir.8076
59. Kurdowski W (2014) Cement hydration. In: Cement and concrete chemistry. Springer, New York. doi:10.1007/978-94-007-7945-7_4
60. Setién J, Hernández D, González JJ (2009) Characterization of ladle furnace basic slag for use as a construction material. Constr Build Mater 23:1788–1794. doi:10.1016/j.conbuildmat.2008.10.003
61. Wu HC, Zhou H, Ding L, Tan WJ, Liu M, Chang J (2010) Performance evaluation of carbonated steel slag blended with cement. Cement 2:6–9
62. Rashad AM, Seleem HEDH (2014) A study on high strength concrete with moderate cement content incorporating limestone powder. Build Res J 61(1):43–58. doi:10.2478/brj-2014-0004

63. Martin LHJ, Winnefeld F, Müller CJ, Lothenbach B (2015) Contribution of limestone to the hydration of calcium sulfoaluminate cement. Cement Concr Compos 62:204–211. doi:10.1016/j.cemconcomp.2015.07.005
64. Pan SY, Chiang PC, Chen YH, Chen CD, Lin HY, Chang EE (2013) Systematic approach to determination of maximum achievable capture capacity via leaching and carbonation processes for alkaline steelmaking wastes in a rotating packed bed. Environ Sci Technol 47 (23):13677–13685. doi:10.1021/es403323x
65. Santos RM, Van Bouwel J, Vandevelde E, Mertens G, Elsen J, Van Gerven T (2013) Accelerated mineral carbonation of stainless steel slags for CO_2 storage and waste valorization: effect of process parameters on geochemical properties. Int J Greenhouse Gas Control 17:32–45. doi:10.1016/j.ijggc.2013.04.004
66. Tian B, Cohen MD (2000) Does gypsum formation during sulfate attack on concrete lead to expansion? Cem Concr Res 30:117–123
67. Hossack AM, Thomas MDA (2015) Varying fly ash and slag contents in Portland limestone cement mortars exposed to external sulfates. Constr Build Mater 78:333–341. doi:10.1016/j.conbuildmat.2015.01.030
68. Bodor M, Santos RM, Cristea G, Salman M, Cizer Ö, Iacobescu RI, Chiang YW, van Balen K, Vlad M, van Gerven T (2016) Laboratory investigation of carbonated BOF slag used as partial replacement of natural aggregate in cement mortars. Cement Concr Compos 65: 55–66. doi:10.1016/j.cemconcomp.2015.10.002
69. Zhang T, Yu Q, Wei J, Zhang P (2011) A new gap-graded particle size distribution and resulting consequences on properties of blended cement. Cement Concr Compos 33(5): 543–550. doi:10.1016/j.cemconcomp.2011.02.013
70. Zhang T, Yu Q, Wei J, Zhang P, Chen P (2011) A gap-graded particle size distribution for blended cements: Analytical approach and experimental validation. Powder Technol 214 (2):259–268. doi:10.1016/j.powtec.2011.08.018
71. Zhang T, Gao P, Luo R, Wei J, Yu Q (2014) Volumetric deformation of gap-graded blended cement pastes with large amount of supplementary cementitious materials. Constr Build Mater 54:339–347. doi:10.1016/j.conbuildmat.2013.12.053
72. Arora A, Sant G, Neithalath N (2016) Ternary blends containing slag and interground/blended limestone: Hydration, strength, and pore structure. Constr Build Mater 102:113–124. doi:10.1016/j.conbuildmat.2015.10.179

Chapter 16
Aggregates and High Value Products

Abstract To the development of technologies for global environmental protection and resource conservation, CO_2 mineralization technologies are considered as an cleaner production in industries for high-level reuse of solid waste and solidification of CO_2 from exhaust gas. This chapter provides the performance and challenges of utilizing fresh solid wastes as aggregates in concrete blocks. To improve the performance of concrete with fresh solid waste, a newly developed process, i.e., accelerated carbonation, was introduced. Through the accelerated carbonation process, additional high value-added chemicals, such as precipitated calcium carbonate, precipitated silica, and adsorbent, can be also produced for further utilizations. Moreover, other applications of carbonated solid wastes, such as (1) marine blocks as artificial reefs for marine forests and (2) abiotic catalyst for enhanced humification, are illustrated.

16.1 Aggregates

16.1.1 Classification

Concrete is made of roughly 70–80% aggregate (sand and gravel), 10–15% cement, and 5–10% additives and water. Construction aggregate can be used to strengthen composite materials, such as concrete and asphalt concrete, for a myriad of uses ranging from railroad bases to housing foundations. In the concrete mixture, the volume of cementitious paste should be large enough to (1) fill up the voids between aggregate particles, and (2) provide excess cementitious paste to form paste films coating the aggregate particles [1]. This can avoid entrapment of air inside the concrete mixture, which might impair the strength and durability. However, an excess ratio of cementitious paste in concrete will form paste films, thereby lubricating the concrete mixture and attaining the required workability.

P.-C. Chiang and S.-Y. Pan, *Carbon Dioxide Mineralization and Utilization*, DOI 10.1007/978-981-10-3268-4_16

The aggregates for construction can be classified as two types:

- coarse aggregate (generally ranging from 9.5 to 37.5 mm), including gravel and crushed stone;
- fine aggregate (usually smaller than 9.5 mm), including sand and crushed stone.

Extensive studies have been conducted to evaluate the performance of the use of limestone as fine aggregates in concrete [2]. It was found that the addition of limestone to reduce the cementitious paste volume would improve not only the dimensional stability but also the tensile strength, stiffness, and durability of concrete [1]. Aside from limestone minerals, many industrial waste materials can potentially be used as economical and environmentally friendly sand substitutes for cementitious building products.

16.1.2 Fresh Solid Waste as Aggregates: Performance and Challenges

Recycled concrete aggregate (RCA), collected from old roads and buildings, has shown promise as an alternative to natural aggregate. Table 16.1 presents the various properties of RCA for replacement of natural aggregate in construction concrete. While RCA has similar gradation to the natural aggregate, RCA particles are more rounded in shape and have more fine particles broken off in L.A. abrasion and crushing tests [3]. This suggests that the use of RCA as a structural concrete should be viable because the performance of RCA concrete beams was still within standard specifications [3–6]. Furthermore, creep can be minimized by incorporating fly ash as either an additive or a replacement in concrete in the case of RCA utilization [6].

16.1.3 Carbonated Solid Wastes as Aggregates

Carbonated alkaline solid waste can function as construction aggregates to partially replace sand, gravel, and crushed stone. Therefore, they can be considered as an economic and environmentally friendly sand substitute for cementitious building products. Several studies have been carried out to evaluate the performance of utilizing carbonated solid wastes as aggregates, such as carbonated steel slag as a fine aggregate in concrete [7].

Fernández-Bertos et al. [8] suggested that the carbonated particles became coarser due to agglomeration, which should be beneficial for use in aggregate manufacturing. The formed $CaCO_3$ crystals in carbonated solid waste appear to reduce the free-CaO content and fill pore spaces of specimens, thereby exhibiting good soundness and enhanced compactness, respectively [9]. In addition,

Table 16.1 Various properties of recycled concrete aggregate for natural aggregate replacement [3–6]

Properties	Unit	Performance of RCA replacement[a]					Remarks
		RCA 100%	RCA 50%	RCA 30%	RCA 15%	NA	
Density	kg/m^3	2394	n.r.	n.r.	n.r.	2890	• RCA is approximately 7–9% lower than NA
Porosity and water absorption	%	4.0–5.2	n.r.	n.r.	n.r.	0.5–2.5	• NA has low water absorption due to low porosity • RCA has greater water absorption due to greater porosity
Crushing test	%	23–24	n.r.	n.r.	n.r.	13–16	• Higher values for RCA were observed
L.A. abrasion test	%	32	n.r.	n.r.	n.r.	11	• Impacted and crushed by steel balls to determine amount of fine particles that break down • Higher values for RCA were observed
Compressive strength	MPa	n.r.	29	31.7	32.7	38.6	• Decreases compressive strength • Greater midspan deflections under service load
Tensile strength (splitting)	MPa	n.r.	2.7	2.7	3.0	3.3	• Concrete with RCA has comparable splitting tensile strength
Modulus of rupture (MOR)	MPa	n.r.	8.9	9.0	9.7	10.2	• MOR of RCA was slightly less than that of conventional concrete
Modulus of elasticity	MPa	15–28	n.r.	n.r.	n.r.	26–33	• Modulus of elasticity of RCA (28-d) is lower up to 45% than NA concrete
Drying shrinkage	Micro strain	540–790	n.r.	n.r.	n.r.	275–590	• Creep can be minimized by incorporating fly ash whether as addition or replacement (in the case of RCA)

Note [a]*RCA* Recycled concrete aggregate; *NA* natural aggregate; *n.r.* not report

Monkman et al. [7] found that the 28-d strengths of the mortars made with the carbonated ladle slag sand were comparable to those with normal river sand mortars. It is revealed that the carbonated ladle slag can be successfully used as fine aggregates to prepare mortar in precast products, such as masonry units, paving stones, and hollow core slabs.

These materials can be further treated by carbonation curing to enhance the strength development (as discussed in Chap. 8). It thus suggests that accelerated carbonation should be used to manufacture alkaline solid wastes as building materials, with high early compressive strength and eligible soundness, along with the benefit of CO_2 capture at the same time.

16.2 High Value-Added Chemicals

16.2.1 Precipitated Calcium Carbonate (PCC)

Recently, research has focused on the production of precipitated calcium carbonate (PCC) from indirect carbonation of alkaline solid wastes because it can be a profitable refining method for solid wastes if the purity requirements for commercial PCC can be achieved. Calcium ($CaCO_3$) and magnesium ($MgCO_3$) carbonates, considered as high value-added chemicals, can be used as coating pigments and as fillers in the pulp, rubber, plastic, paper, and paint industries. They can offer opacity, high brightness, and improved printability due to its good ink receptivity [10–12]. Approximately 75% of the produced PCC can be used in the paper industry as a brightness coating and filler [13, 14]. It can serve as a replacement for more expensive pulp fiber and optical brightening agents to improve the quality and printing characteristics of paper, such as smoothness, gloss, whiteness, opacity, brightness, and color. In addition, PCC can be utilized as an additive and filler in construction materials. Companies in North America, the EU, and Australia are working on developing a similar process, which involves CO_2 capture by bubbling flue gases through saline water to produce solid carbonates as an aggregate material in cement [15].

Indirect carbonation of alkaline solid wastes can produce high-purity PCCs. The application potentials for the PCC produced from the indirect carbonation depend on the obtained purity and variety of the crystal structures. In most applications, the favored crystalline forms of PCC are the rhombohedral calcite type, the orthorhombic acicular aragonite type, and the scalenohedral calcite type [16]. Other physico-chemical properties of PCC, such as particle size distribution, specific surface area, morphology, and polymorphism, also play an important role.

For the process design of carbonation, it suggests that a reaction temperature higher than 80 °C is unfavorable to efficient CO_2 dissolution, although the conversion for the extraction reaction is enhanced at higher temperatures [17]. The precipitation of $CaCO_3$ is endothermic and feasible at temperatures above 45 °C, while the carbonation of Mg^{2+} ions should be possible only at temperatures over 144 °C [18]. Therefore, in the case of 80 °C, the formation of $MgCO_3$ could be neglected. One of the findings in the literature indicated that a maximum carbonation conversion of 70% can be achieved, with a PCC purity of up to 98% [18]. Similarly, a maximum carbonation conversion of 86% can be achieved under a CO_2

concentration of 10 vol.% at 30 °C after acetic acid extraction, with a PCC purity as high as 99.8% [19].

16.2.2 *Precipitated Silica*

Silica (quartz, SiO_2) can be high value-added materials if the purity of precipitates can reach the requirement. Depending on the degree of purity refinement, precipitated silica can be used in various types of applications, such as filler (after mining), smelting plants and gravel (after crushing), glass (after separation), ceramics (after grinding), solar cells (after acid wash), and semiconductors (after advanced processing). The precipitated silica can also be utilized in pesticides, detergents, tires, oral health products, creams, footwear, and plastics. For instance, in the application of tires, the silica can improve wet skid performance and provide lower rolling resistance, thereby saving the fuel usage [20].

Silica is commonly used as a coating pigment together with polyvinyl alcohol for matte inkjet papers [21]. However, current methods of silica production for pigment purposes are quite expensive, i.e., 800–1000 €/t, since the water glass (sodium silicate, Na_2SiO_3) is used as the raw material [21]. During indirect carbonation using acetic acid or ammonium chloride, silica gel can be formed and then separated via filtration before sequential carbonation [18]. It was observed that the produced silica should be qualified for inkjet paper coating, as a replacement for commercially precipitated silica [21].

16.2.3 *Adsorbent*

In indirect carbonation, solid residues (such as steel slag) after extraction step could be converted into sorbent materials via hydrothermal conversion. After the hydrothermal conversion, a material with a larger surface area and micropores can be produced. Therefore, it offers the potential to behave as a sorbent. Mineralogical investigation of the sorbent materials after hydrothermal conversion indicated that two zeolitic phases were formed:

- ferrierite ($Na_2Mg(Si_{15}Al_3)O_{36}{\cdot}9H_2O$) and
- gmelinite ($Na_4(Si_8Al_4)O_{24}{\cdot}11H_2O$).

It can provide the sorption potential for heavy metals and organic compounds [22]. Since aluminum is an essential component of aluminosilicate-based zeolites, the organic acids used for the calcium extraction from solid wastes should prevent the leaching of aluminum ions, which is the key consideration for the synthesis of zeolites.

16.3 Other Applications

16.3.1 Marine Blocks as Artificial Reefs for Marine Forests

Artificial reefs, exhibiting a high stability in seawater, have often been used to recover and develop seaweed beds in at least 40 counties [23–25]. Seaweed beds play several important roles in the following:

- water quality purification,
- biological production,
- sediment stabilization,
- prevention of trawlers from destroying seabed, and
- conservation of the detritus food chain and primary consumers.

The preservation and development of seaweed bed with such functions can contribute to the increase in catches of useful seafood and fishes.

Conventionally, the materials commonly used as artificial reefs are concrete blocks and fieldstones. As an alternative to concrete, large marine blocks (e.g., 1 m^3 in size) have been developed using steel slags and CO_2 without supplying cement. Several studies have been conducted to evaluate the performance of marine blocks on the growth of brown algae, such as *Ecklonia cava* [24]. The brown algae can form a community called "marine forest" in coastal waster at 2–25 depth, which can be used as nursery and breeding spaces for commercial fishes and other animals. Furthermore, JFE Steel Corporation in Japan manufactured marine blocks as artificial reefs for seaweed/coral breeding using carbonated steel slag. The formed $CaCO_3$ components (the same composition to shells and coral) can act as great breeding habitats for seaweed and coral. As a result, the seaweed adhered and grew on the marine block is much more than on normal concrete. It suggests that the marine blocks could form artificial reefs offering the same functions as natural rocky reefs.

16.3.2 Abiotic Catalyst for Enhanced Humification

Humification can occur naturally in soil or in the production of compost. It is considered the primary process to stabilize soil carbon and organic matter in the form of humic substances (i.e., humus). Humus has a characteristic of black or dark brown color. Application of biotic and abiotic humification for organic carbon stabilization has been of great significance in bio-organic waste disposal, particularly in the composting process [26, 27]. Most of the humus in soils has persisted for more than a hundred years, rather than have been decomposed to CO_2. Therefore, humus can be regarded as stable organic matter that has been protected from decomposition by microbial or enzyme. It usually is hidden and occluded inside small aggregates of soil particles or tightly attached to clays.

The humification theory predates a sophisticated understanding of soils. It was noted that Fe(III)- and Mn(IV)-oxide minerals could act as effective oxidants and substantially enhance the polycondensation of humic precursors [28]. It is significant for accelerating the transformation and stabilization of biowastes during composting. For instance, the carbonated steel slag (e.g., basic oxygen furnace slag, BOFS) usually contains high quantities of Fe(III)- and Mn(IV)-oxide minerals. It was found that the carbonated BOFS could substantially promote the formation of humic-like acids (HLAs) from humic precursors, such as amino acids, saccharides, and phenols, under ambient temperature [29]. For humification, it suggests that carbonated BOFS should be used instead of uncarbonated BOFS in the first place since uncarbonated slag has been linked to metal leaching issues. Accelerated carbonation has been shown to reduce the leaching of metals by orders of magnitude to make the slag environmentally benign [27, 30].

References

1. Li LG, Kwan AKH (2015) Adding limestone fines as cementitious paste replacement to improve tensile strength, stiffness and durability of concrete. Cement Concr Compos 60: 17–24. doi:10.1016/j.cemconcomp.2015.02.006
2. Schneider M, Romer M, Tschudin M, Bolio H (2011) Sustainable cement production—present and future. Cem Concr Res 41(7):642–650. doi:10.1016/j.cemconres.2011.03.019
3. McNeil K, Kang THK (2013) Recycled concrete aggregates: a review. Int J Concr Struct Mater 7(1):61–69. doi:10.1007/s40069-013-0032-5
4. Sagoe-Crensil KK, Brown T, Taylor AH (2001) Performance of concrete made with commercially produced coarse recycled concrete aggregate. Cem Concr Res 31(5). doi:10.1016/S0008-8846(00)00476-2
5. Shayan A, Xu A (2004) Value-added utilisation of waste glass in concrete. Cem Concr Res 34 (1):81–89. doi:10.1016/s0008-8846(03)00251-5
6. Behera M, Bhattacharyya SK, Minocha AK, Deoliya R, Maiti S (2014) Recycled aggregate from C&D waste and its use in concrete—a breakthrough towards sustainability in construction sector: a review. Constr Build Mater 68:501–516. doi:10.1016/j.conbuildmat.2014.07.003
7. Monkman S, Shao Y, Shi C (2009) carbonated ladle slag fines for carbon uptake and sand substitute. J Mater Civ Eng 21:657–665. doi:10.1061//asce/0899-1561/2009/21:11/657
8. Fernandez Bertos M, Simons SJ, Hills CD, Carey PJ (2004) A review of accelerated carbonation technology in the treatment of cement-based materials and sequestration of CO_2. J Hazard Mater 112(3):193–205. doi:10.1016/j.jhazmat.2004.04.019
9. Wu HZ, Chang J, Pan ZZ, Cheng X (2009) Carbonate steelmaking slag to manufacture building materials. Adv Mater Res 79–82:1943–1946. doi:10.4028/www.scientific.net/AMR.79-82.1943
10. Azdarpour A, Asadullah M, Mohammadian E, Hamidi H, Junin R, Karaei MA (2015) A review on carbon dioxide mineral carbonation through pH-swing process. Chem Eng J 279:615–630. doi:10.1016/j.cej.2015.05.064
11. Teir S (2008) Fixation of carbon dioxide by producing carbonates from minerals and steelmaking slags. Helsinki University of Technology
12. Eloneva S, Said A, Fogelholm C-J, Zevenhoven R (2012) Preliminary assessment of a method utilizing carbon dioxide and steelmaking slags to produce precipitated calcium carbonate. Appl Energy 90(1):329–334. doi:10.1016/j.apenergy.2011.05.045
13. Teir S, Eloneva S, Zevenhoven R (2005) Co-utilization of CO_2 and calcium silicate-rich slags for precipitated calcium carbonate production (part I). In: Proceedings of the 18th

international conference on efficiency; cost; optimization; simulation and environmental impact of energy systems (ECOS 2005). Trondheim, Norway, 20–22 June 2005
14. Zevenhoven R, Teir S, Eloneva S (2008) Heat optimisation of a staged gas–solid mineral carbonation process for long-term CO_2 storage. Energy 33:362–370. doi:10.1016/j.energy.2007.11.005
15. International Energy Agency (IEA) (2013) Post-combustion CO_2 capture scale-up study
16. Imppola O (2000) Precipitated calcium carbonate-PCC. Pigment coating and surface sizing of paper. Gummerus Printing, Finland
17. Kodama S, Nishimoto T, Yamamoto N, Yogo K, Yamada K (2008) Development of a new pH-swing CO_2 mineralization process with a recyclable reaction solution. Energy 33(5):776–784. doi:10.1016/j.energy.2008.01.005
18. Teir S, Eloneva S, Fogelholm C-J, Zevenhoven R (2007) Dissolution of steelmaking slags in acetic acid for precipitated calcium carbonate production. Energy 32(4):528–539. doi:10.1016/j.energy.2006.06.023
19. Teir S, Eloneva S, Fogelholm C, Zevenhoven R (2009) Fixation of carbon dioxide by producing hydromagnesite from serpentinite. Appl Energy 86(2):214–218. doi:10.1016/j.apenergy.2008.03.013
20. Andresen B (2008) Possibilities for mineral sequestration, can value be created?. Institute for Energy Technology, Kjeller
21. Teir S, Bacher J, Kentta E, Satlin J (2013) Silica produced from olivine for inkjet paper coating. Paper presented at the accelerated carbonation for environmental and material engineering, KU Leuven, Belgium
22. Chiang Y-W, Santos RM, Elsen J, Meesschaert B, Martens JA, Van Gerven T (2013) Two-way valorization of blast furnace slag into precipitated calcium carbonate and sorbent materials. Paper presented at the accelerated carbonation for environmental and material engineering, KU Leuven, Belgium
23. Prasad NT, Sadhu S, Murthy KNVV, Pilli SR, Ramesh S, Phani Kumar SVS, Dharani G, Atmanand MA, Venkata Rao MB, Dey TK, Syamsundar A (2014) Carbon-dioxide fixation by artificial reef development in marine environment using carbonated slag material from steel plant. IEEE
24. Oyamada K, Tsukidate M, Watanabe K, Takahashi T, Isoo T, Terawaki T (2008) A field test of porous carbonated blocks used as artificial reef in seaweed beds of Ecklonia cava. J Appl Phycol 20(5):863–868. doi:10.1007/s10811-008-9332-6
25. Huang X, Wang Z, Liu Y, Hu W, Ni W (2016) On the use of blast furnace slag and steel slag in the preparation of green artificial reef concrete. Constr Build Mater 112:241–246. doi:10.1016/j.conbuildmat.2016.02.088
26. Kanno H, Tachibana N, Fukushima M (2011) Optimization of conditions for thermal treatment of rice bran using an accelerator including an organo-iron compound. Bioresour Technol 102:3430–3436
27. Li P, Pan S-Y, Pei S, Lin YJ, Chiang P-C (2016) Challenges and perspectives on carbon fixation and utilization technologies: an overview. Aerosol Air Quality Res. doi:10.4209/aaqr.2015.12.0698
28. Qi G, Yue D, Fukushima M, Fukuchi S, Nishimoto R, Nie Y (2012) Enhanced humification by carbonated basic oxygen furnace steel slag–II. Process characterization and the role of inorganic components in the formation of humic-like substances. Bioresour Technol 114:637–643. doi:10.1016/j.biortech.2012.03.064
29. Qi G, Yue D, Fukushima M, Fukuchi S, Nie Y (2012) Enhanced humification by carbonated basic oxygen furnace steel slag–I. Characterization of humic-like acids produced from humic precursors. Bioresour Technol 104:497–502. doi:10.1016/j.biortech.2011.11.021
30. Huijgen WJJ, Comans RNJ (2006) Carbonation of steel slag for CO_2 sequestration: leaching of products and reaction mechanisms. Environ Sci Technol 40(8):2790–2796

Part V
An Integral Approach for Waste Treatment and Resource Recovery

Chapter 17
Carbon Capture with Flue Gas Purification

Abstract An integrated approach to applying the concept of process integration and intensification for control of multiple air pollutants, such as sulfur dioxide (SO_2), nitrogen oxides (NO_x), particulate matter (PM), and CO_2, in a flue gas is important. For instance, the high-gravity process has been recognized as a process intensification device with excellent micromixing efficiency and mass transfer rate between the gas and liquid phases. Implementation of the high-gravity technology can significantly improve the efficiency of end-of-pipe air pollution control. This chapter first provides the features, principles, and applications of the high-gravity technology. Then, the mechanism and performance of SO_2, NO_x, and PM removal by the high-gravity process are discussed. Finally, the use of high-gravity technology for removing particulate and gaseous pollutant emissions to achieve the goal of integrated air pollution control in industry is illustrated. To move ahead with the industrial upgrades needed for compliance with the future air quality standards, the high-gravity technology should be considered as a technically and economically feasible alternative to simultaneously reduce more than one type of air pollutants (NO_x, SO_2, PM, and CO_2) in a timely manner.

17.1 Integrated Air Pollution Control at Industry: Challenges, Barriers, and Strategies

17.1.1 Air Quality Control at Industry

Rapid industrial development brings economic growth and social progress; however, it also causes problems, such as severe environmental pollution. A large quantity of air pollutants, such as particulates, biological molecules, and other harmful compounds, are emitted into the atmosphere, exceeding the carrying capacity of the environment. This results in adverse impacts and damage to environmental quality, human health, and crop growth. Air pollution is a significant risk factor for a number of health conditions, such as respiratory infections, heart disease, stroke, and lung cancer [1].

P.-C. Chiang and S.-Y. Pan, *Carbon Dioxide Mineralization and Utilization*, DOI 10.1007/978-981-10-3268-4_17

Energy consumption, electricity generation, and vehicle population are all leading to increases in multiple pollutant emissions. Harmful pollutants in industrial flue gas include sulfur dioxide (SO_2), nitrogen oxides (NO_x), volatile organic compounds (VOCs), carbon monoxide (CO), and particulate matter (PM). A number of air pollutants, including SO_2 and NO_2, exhibit acidic properties, which can cause corrosion to materials, construction, and equipment, as well as result in the formation of acid rain which interferes with the growth of plants and crops and causes acidification of water and soil.

Fine particles can be classified as inhalable particulate matter (PM_{10}, with a diameter of 10 μm or less), fine particulate matter ($PM_{2.5}$, with a diameter of 2.5 μm or less), ultrafine particulate matter, and soot. PM_{10} can reduce the visibility and the amount of sunlight radiation, thereby further increasing the concentration of NO_x, hydrocarbon (HC), and hydrochlorofluorocarbons (HCFC). On the other hand, $PM_{2.5}$ is harmful to the respiratory system and human health, with extensive studies indicating a positive relationship between health impacts and the time of exposure to certain concentrations of fine particles [2]. For instance, in China, most of the PM_{10} in the atmosphere comes from coal-fired power plants and vehicles, which is usually associated with significant toxic matter [3]. Furthermore, due to global climate change and greenhouse gas emissions, the reduction in CO_2 from industries also has recently attracted a great deal of attention around the world. As CO_2 accumulates in the atmosphere, concerns about serious and irreversible damage, such as rising water level and species extinction, are being raised regarding its influence on climate change.

17.1.2 Strategies on Overcoming Challenges and Barriers

Table 17.1 presents the current regulations and standards of ambient air quality and emissions from steel plants around the world. Numerous countries have set aggressive air quality standards for key air pollutants as policy objectives on air quality control. In most cases, air quality standards on SO_2, NO_2, CO, O_3, total suspended particles (TSP), and hydrocarbons were first introduced between 1975 and 1990. Standards on PM_{10} fine particles started to apply between 1990 and 2000, and those on benzene were implemented between 2000 and 2010. Additional standards on $PM_{2.5}$ fine particles have been enacted around the world since 2010. Along the way, these environmental standards have been progressively tightened to achieve higher air quality goals.

Challenges in ambient air quality control and management include (1) coal as the major source of energy in power generation or industrial plants, where the coal-fired combustion process produces a great deal of SO_2; (2) centralized industrial parks, where industrial and mining enterprises are usually the main sources of air pollution, with concentrated and complex emissions resulting in regional air pollution issues; and (3) ineffective governance and management.

Table 17.1 Regulations and standards of ambient air quality and stack emissions from steel plants at different countries

Country	NO_2 (μg/m³)			SO_2 (μg/m³)			TSP (mg/m³)[a]		PM_{10} (μg/m³)		$PM_{2.5}$ (μg/m³)	
	1 h	24 h	AA[a]	1 h	24 h	AA	24 h	AA	24 h	AA	24 h	AA
China [4]	200	80	40	500	150	60	0.3	0.2	150	70	75	35
EU [5]	200	–	40	350	125	–	–	–	50	40	–	25[b]
India [5]	–	80	40	–	80	50	–	–	100	60	60	40
Japan [6]	–	80	–	280	110	–	–	–	100	–	35	15
Korea [7]	185	110	55	390	130	54	–	–	100	50	50	25
Saudi Arabia [8]	660	–	100	730	365	80	–	–	150	50	65	15
Taiwan [9]	465	–	95	650	260	80	0.25	0.13	125	65	35	15
USA [10]	–	–	100	1300[d]	–	–	–	NA	150	–	35	15
WHO [11]	200	–	40	–	20	–	–	–	50	20	25	10

[a]*N.A.* not applicable; *AA* annual; *TSP* total suspended particle. [b]The date of this value goes by January 1, 2015, and will be further limited to 20 μg/m³ by January 1, 2020. [c]*N* new plant; *E* existing plant. [d]The level is for 3 h average

As a result, to overcome the aforementioned challenges and barriers, several strategies on mitigating the regional air pollution issue at industries should be implemented: (1) joint prevention and control of regional air pollution and application of climate-friendly air pollution control measures; (2) a comprehensive control policy focusing on multiple pollutants and emission sources at both the local and regional levels; (3) transformation of the industrial energy structure to a clean coal approach; (4) development of clean energy resources and promotion of clean and efficient coal use; and (5) implementation of synchronous control of multiple pollutants, including SO_2, NO_x, VOC, and PM emissions.

17.1.3 Available Air Pollution Control Technology

A wide variety of air pollution control technologies is available to reduce the emission of air pollutants cost-effectively from industrial plants. Table 17.2 summarizes the available air pollution control technologies with respect to their targeted pollutants in flue gas. Most of these technologies target only one specific type of air pollutant (although some target two or more pollutants) at a time. Therefore, conventional industrial air pollution control equipment is typically installed in series to achieve control of multiple pollutants. The Babcock and Wilcox company developed an advanced air pollution control process, so-called SO_x−NO_x−Rox Box (SNRB™) catalytic baghouse, to simultaneously reduce SO_x, NO_x, and PM emissions at a high temperature from coal-fired boilers. A pilot-scale test of the SNRB™ process was successfully carried out (1) by injecting a calcium-based (or sodium-based) sorbent for SO_x reduction; (2) by injecting ammonia in the presence of a selective catalytic reduction (SCR) for NO_x removal; and (3) through high-temperature baghouse filters for PM collection [12].

Similarly, Kang et al. [14] developed a combined unit for simultaneous removal of SO_2, NO_x, and dust at low temperature. The removal efficiencies of SO_2, NO_x, and dust were 94.2, 92.5, and 99.98%, respectively. The temperature of flue gas plays an important role in determining the reaction rate of absorption, adsorption, and oxidation processes, especially for the flue gas desulphurization (FGD) process. The removal efficiency of SO_x in flue gas decreases as the flue gas temperature increases [15]. On the other hand, the sulfur contents in the flue gas would affect the collection efficiency of PM by an electrostatic precipitator (ESP). Consequently, for conventional air pollution control, the equipment sequence would significantly influence the performance of pollutant reduction (such as energy consumption). The combination of equipment also would largely depend on the characteristics of the flue gas and its permitted emission level.

The high-gravity (HIGEE) technology can fulfill the function of wet collectors, which employ (water) solution phase washing to remove particles directly from a gas stream. Meanwhile, HIGEE also simultaneously removes soluble gaseous pollutants, such as SO_2, NO_x and CO_2. Therefore, an integrated approach to applying the HIGEE process at low temperature for the control of multiple air pollutants in flue gas within an industrial plant could be achieved. As presented in Table 17.2, the HIGEE technology can simultaneously reduce more than one type of air pollutant, such as NO_x, SO_2, PM, and CO_2, in a timely manner. In addition, the size of the HIGEE reactor can be greatly reduced over that of a conventional packed column, leading to a significant reduction in capital and operating costs. Several pilot-scale tests have demonstrated the feasibility of the HIGEE process for meeting the requirements of treatment capacity and time scale in industrial plants [16, 17], suggesting that industries should move ahead with the necessary upgrades

Table 17.2 Best available control technologies (BACT) with respect to their target pollutants (adapted from [13])

Best available control technologies	Pollutants in flue gas[a]				
	HCl	NO_x	SO_2	PM	CO_2
Selective non-catalytic reduction (SNCR)	N.A.	Y	N.A.	N.A.	N.A.
Selective catalytic reduction (SCR)	N.A.	Y	N.A.	N.A.	N.A.
Cyclone	N.A.	N.A.	N.A.	Y	N.A.
Wet/dry scrubber	N.A.	N.A.	Y	C	C
Electrostatic precipitator (ESP)	N.A.	N.A.	N.A.	Y	N.A.
Baghouse filter	N.A.	N.A.	N.A.	Y	N.A.
SNRB™ catalytic baghouse	C	Y	Y	Y	N.A.
High-gravity (HIGEE) technology	Y	Y	Y	Y	Y

[a] *Y* Technology can reduce emissions; *N.A.* not applicable for emission reduction; *C* complementary removal of pollutant emission (not significant)

to comply with strict air quality standards in the future. By doing so, they can demonstrate that better environmental performance is both technically and economically feasible.

17.2 Concept of Process Integration and Intensification

To effectively control the emissions of air pollutants and aerosol at industries, the concept of process integration and intensification for end-of-pipe treatment at industries should be the best achievable approach. This section provides the features, principles, and applications of the HIGEE technology as an example of process intensification. Also, the use of HIGEE technology for aerosol and air pollution control to achieve the goal of integrated air pollution control in industry is illustrated.

17.2.1 High-Gravity (HIGEE) Technology

HIGEE technology, carried out in a rotating packed bed (RPB), was originally developed by Ramshaw and Mallinson [18, 19]. It has been recognized as a process intensification device that can generate a high centrifugal force by spinning a torus-shaped rotor. The HIGEE technology can provide a mean acceleration of 100 or even 1000 times greater than the force of gravity [20–22]. In such a case, liquid can be spread and split into micro/nano droplets and thin films under the HIGEE environment. Therefore, the volumetric gas–liquid mass transfer coefficients are an order of magnitude higher than those in a conventional packed bed [20, 22, 23]. This leads to a dramatic reduction in the reactor size and capital cost, compared to that required for equivalent mass transfer in a gravity flow packed bed.

In general, liquid is injected from the liquid distributors into the inner periphery of the rotor, flows uniformly through the packing of the rotor, and then leaves the rotor onto the machine case at the prevailing peripheral speed, i.e., typically 40–60 m/s [24]. The gas can be forced to flow through the packing zone in various directions to come in contact with the liquid within the rotor, including countercurrent (i.e., radially inwards and leave via the rotor center), co-current (i.e., radially outwards and leave via the outer rotor), and cross-flows (i.e., upwards and leave via the top of rotor), as shown in Fig. 17.1. The contact angles between gas and liquid flows within the packed zone for countercurrent, co-current, and cross types are 180°, 0°, and 90°, respectively. For any types of designs, the packing characteristics are generally uniform throughout the packing zone [24]. In recent years, research has been conducted to determine the mass transfer characteristics of different types of RPB reactors and packing materials [25].

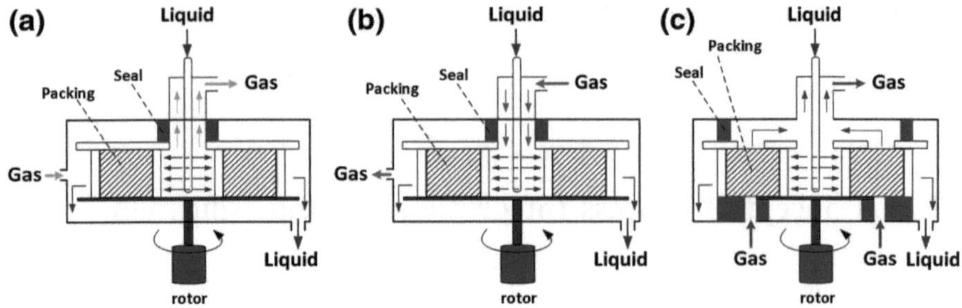

Fig. 17.1 Various types of high-gravity (HIGEE) reactors for reaction between gas and liquid phases: **a** countercurrent flow, **b** co-current flow, and **c** cross-flow

17.2.2 Features and Principles

The HIGEE technology possesses a number of benefits, such as low liquid hold-up, high micromixing efficiency (mixing at the molecular level), high hydraulic and flooding capacity, small reactor size, large mass transfer coefficients, and better self-cleaning of the rotor [26, 27]. Therefore, it has been applied in various fields, such as material synthesis, petroleum products, pollution control, and medical science. The HIGEE process can effectively promote the liquid-film-controlled mass transfer process, such as CO_2 absorption by amine solution [28] and via accelerated carbonation [29]. Since the fluids are sheared dispersion by the rotating bed, the effective interfacial area and the surface update rate can significantly increase, thereby reducing the mass transfer resistance. The relatively smaller size of the reactor allows the RPB to be produced more economically than conventional fixed bed reactors, especially when alloy metals must be used for corrosion resistance or for hazardous materials.

In the HIGEE, contributing factors to the high volumetric mass transfer coefficients are (1) a HIGEE field generated by centrifugal force (10–500 g) to decrease the film thickness of liquid; (2) an increase in the effective gas–liquid contacting efficiency; and (3) a higher hydraulic and flooding capacity, capable of using packing with high porosity (0.90–0.95) and high specific surface areas (2000–5000 m^2/m^3). The centrifugal acceleration (g, m/s^2) generated within a RPB can be determined by Eq. (17.1), which indicates that the centrifugal force increases toward the periphery of the outer rotor.

$$\mathrm{g} = r\omega^2 \tag{17.1}$$

where r (m) is the radius of the packing bed, and ω (s^{-1}) is the rotating speed of the packing bed.

Within the HIGEE reactor, the high-gravity field can be divided into three zones:

- The outer zone
  - The outer zone is the space between the machine shell and the outer periphery of the RPB.
  - Within this zone, the liquid (or solvent) exhaust is thrown from the outer periphery of the packed bed onto the case wall.
  - The formed tiny droplets can contribute to the preliminary purification of high concentration air pollutants if the countercurrent-flow RPB is used.
- The packing zone
  - The packing zone is the main body of the RPB.
  - A highly intensive micromixing between gas and liquid phases (i.e., turbulence state) occurs within this zone, where the main phase mechanisms include collision, interception, and diffusion.
  - The gas–liquid contacting surface updates quickly, thereby contributing to the major reaction and separation.
- The inner zone
  - The inner zone is the area where the liquid is injected from the distributors before it reaches the inner periphery of the rotor.
  - In the case of the countercurrent-flow RPB, this area is also the export channel of gas flow.
  - Since the pollutants, which are in low concentration in the gas phase, can contact the fresh liquid (or solvent), depth purification and/or separation are usually achieved in this area.

17.2.3 Practical Applications

Since the HIGEE process is designed to enhance mass transfer between the gas and liquid phases, the device was originally designed for gas–liquid separation [30]. It consists of a RPB in the shape of a torus or a doughnut, which has a relatively high specific surface area compared to conventional column packing [24]. Recently, HIGEE has been extensively explored for use in a variety of phase contacting processes, such as

- Distillation [31]
- Absorption [32, 33]
- Extraction [34]
- Desulfurization [35, 36]
- Deaeration [37]
- Reactive precipitation [38, 39]
- Chemosynthesis [40, 41]
- Chemical oxidation [42]

- Removal of dust [43]
- Synthesis of nano materials [44, 45]
- Accelerated carbonation [46–48]

From a practical perspective, as the gas flow rate increases, it becomes more difficult for liquid to move through the packing zone. Moreover, a relatively high tangential gas velocity at the center of the packed bed is observed because angular momentum is maintained as the radius decreases. This can disrupt the liquid flow leaving the packing zone or distributors. At its limit, liquid is prevented from any movement and simply accumulates within the column or is removed before it reaches the packing surface, a condition known as flooding. The radial length of the rotor is directly related to the separation duty and removal efficiency, while the hydraulic capacity depends upon the flooding rates at the inner radius and the inner cylindrical surface area. It was noted that operation at approximately 50% of the flooding rates for the HIGEE process can be prudently adopted [24].

The HIGEE technology offers great potential for application to the control of aerosol and air pollutants, such as NO_x, SO_2, and CO_2, due to its excellent micromixing efficiency and mass transfer rate. With the addition of oxidants (e.g., O_3), the NO_x in flue gas can be well mixed with the oxidants and then oxidized into water-soluble nitrate and/or nitric acids. HIGEE also could promote the kinetics of absorption reactions for both SO_2 and CO_2 in the liquid phase, thereby reducing the reactor size. Several successful experiences on the relevant processes are provided and discussed in the following sections.

17.3 Sulfur Oxide (SO_x) Emission Control

The large amount of SO_2 emitted from the burning of fossil fuels, metallic minerals, the preparation of chemical pulp, and the production of acid today leads to severe air pollution and results in great harm to human life and health. Direct SO_2 gas emission into the atmosphere can not only easily lead to the formation of acid rain and urban smog [49] but also support the reactions that create ozone depletion in the stratosphere. To reduce SO_2 emissions from the stack, several desulfurization methods, such as fuel pretreatment, concurrent burning and adsorption, and flue gas desulphurization (FGD), have been proposed [50]. Among these methods, FGD is the most effective technology due to its high efficiency and general availability. A number of FGD methods, such as dry, semidry, and wet scrubbers, have been developed for mitigating the SO_2 emission from fossil fuel boilers in power and industrial plants. FGD wet scrubbers have been applied to coal-fired combustion units in the size range from 5 to 1500 MW, while dry scrubbers and spray scrubbers have generally been applied to units smaller than 300 MW [51].

Wet processes, such as fixed beds with packings [52], spray towers [53], and rotating-stream tray scrubber [54], are widely used for FGD due to their high removal efficiency, low operating cost, and stable operation. For instance, in the

USA, approximately 85% of the FGD units installed are wet scrubbers. Of the remainder, 12% are spray dry systems, and 3% are dry injection systems. However, these traditional wet processes have several disadvantages, such as limited mass transfer rate and large volume of equipment, thereby leading to a huge reactor size and high capital costs. Moreover, these absorption processes are generally not regenerative for SO_2 compounds. Therefore, extensive research has been focused on process intensification for FGD [55–57] and SO_2 recovery [58–60].

17.3.1 Mechanisms and Process Chemistry: Principles

The HIGEE technique can be deployed for SO_2 emission control from a stationary emission point source. The mechanism of SO_2 removal using an RPB is typically by chemical absorption between the gas and liquid phases. With its excellent micromixing ability, the SO_2 in the flue gas can be efficiently dissolved into solutions within a short contact time. Various types of aqueous solutions can be used as the liquid phase, including lime slurry, phosphate buffer solution, and ethylenediamine, as described below:

1. Lime Slurry

Lime (CaO) slurry is the most commonly used solution to remove SO_2 from industrial waste gases by absorption into an aqueous slurry of $Ca(OH)_2$. The process chemistry of CaO hydration can be expressed as Eqs. (17.2) and (17.3). The formed calcium ions can be combined with SO_3^{2-} ions to form $CaSO_3$, as shown in Eq. (17.4), and then further produce the $CaSO_3{\cdot}0.5H_2O$ precipitates, as shown in Eq. (17.5):

$$CaO_{(s)} + H_2O_{(l)} \rightarrow Ca(OH)_{2(aq)} \tag{17.2}$$

$$Ca(OH)_{2(aq)} \leftrightarrow Ca^{2+} + 2\,OH^- \tag{17.3}$$

$$Ca^{2+} + SO_3^{2-} \leftrightarrow CaSO_{3(aq)} \tag{17.4}$$

$$CaSO_{3(aq)} + 0.5\ H_2O \leftrightarrow CaSO_3\ 0.5H_2O_{(s)} \tag{17.5}$$

On the other hand, the SO_2 gas can be easily dissolved in solution to form H_2SO_3, as shown in Eq. (17.6), or $H_2SO_3^-$, as shown in Eq. (17.7):

$$SO_2 + H_2O \rightarrow H_2SO_3 \tag{17.6}$$

$$SO_2 + 2OH^- \rightarrow SO_3^{2-} + H_2O \tag{17.7}$$

2. Phosphate Buffer Solution

The use of phosphate buffers, i.e., alkali metal phosphate (Na_2HPO_4), could significantly increase the capacity of SO_2 absorption in water. The reaction between SO_2 and Na_2HPO_4 solution is reversible and instantaneous, as shown in Eqs. (17.8–17.10):

$$SO_{2(g)} \leftrightarrow SO_{2(aq)} \tag{17.8}$$

$$SO_{2(l)} + Na_2HPO_{4(l)} + H_2O_{(l)} \leftrightarrow NaHSO_{3(l)} + NaH_2PO_{4(l)} \tag{17.9}$$

$$NaHSO_{3(l)} + NaH_2PO_{4(l)} \leftrightarrow SO_{2(l,g)} + Na_2HPO_{4(l)} + Na_2HPO_{4(s)} + H_2O_{(l,g)} \tag{17.10}$$

3. Cobalt Ethylenediamine

Amine species, such as cobalt ethylenediamine $[Co(en)_3]^{3+}$, can be used to enhance the amount of SO_2 dissolved into solution. The $[Co(en)_3]^{3+}$ is formed by adding soluble cobalt salts and ethylenediamine ($H_2NCH_2CH_2NH_2$) into the basic solutions. The reversible hydration reaction of SO_2 in $[Co(en)_3]^{3+}$ aqueous solution can be described as follows:

$$\left[Co(en)_3\right]^{3+} + SO_3^{2-} + H_2O \leftrightarrow \left[Co(en)_2(SO_3^{2-})(H_2O)\right]^{+} + en \tag{17.11}$$

$$[Co(en)_2(SO_3^{2-}) + H_2O]^{+} + 0.5\,O_2 + en \rightarrow SO_4^{2-} + \left[Co(en)_3\right]^{3+} + (H_2O) \tag{17.12}$$

It is noted that $[Co(en)_3]^{3+}$ does not take part in the net reaction and acts as a catalyst to accelerate the oxidation of $SO_3^{2\hat{a}'}$ species [61]. A similar mechanism can be observed for NO_x oxidation and absorption using the oxygen coexisting in the exhaust gas by such a homogeneous catalytic process.

17.3.2 Performance Evaluation: Applications

The SO_2 removal ratio using HIGEE can reach 90.0–99.9%, depending on the type of absorbent used [62, 63]. An increase in the oxygen content in the SO_2 flue gas can promote the oxidation rate of SO_3^{2-} to SO_4^{2-} [62]. In the case of NaOH and a sodium citrate buffer, almost complete removal of SO_2 in the flue gas can be achieved. However, for the sodium citrate buffer, the absorption rate of SO_4^{2-} into the solution using an RPB was found to be approximately 0.136 g/L/h [64], which was lower than that using packing columns (i.e., 0.19–0.20 g/L/h) [65]. This indicates that the RPB may restrain the oxidation reaction of SO_3^{2-} to SO_4^{2-}, probably due to the short retention time in an RPB, reducing the contact time of SO_3^{2-} and O_2 to form SO_4^{2-}. In addition, the HIGEE process can operate at a relatively higher gas–liquid (G/L) ratio, i.e., 140–10,000 (v/v), than the conventional packed bed, i.e., ~10 (v/v). As a result, the treatment capacity of flue gas using the HIGEE process would be greater than that of the same size conventional packed bed reactor.

17.4 Nitrogen Oxide (NO_x) Emission Control

17.4.1 Mechanisms and Process Chemistry: Principles

NO_x is a generic term for the mono-nitrogen oxides, including NO and NO_2. NO_x can be produced from the reaction between nitrogen and oxygen (and even hydrocarbons) during combustion, especially at high temperatures. The solubility of NO gas in water is quite low (i.e., 0.056 g/L at 20 °C), even in the presence of acid and alkaline. However, the nitrogen oxides in the form of NO_2, N_2O_3, and N_2O_4 can be easily dissolved and hydrolyzed in water to form nitrate (HNO_3) and/or nitrite (HNO_2), as shown in Eqs. (17.13–17.16):

$$2\ NO_{2(l)} + H_2O_{(l)} \rightarrow HNO_{2(l)} + HNO_{3(l)} \quad (17.13)$$

$$N_2O_{3(l)} + H_2O_{(l)} \rightarrow 2\ HNO_{2(l)} \quad (17.14)$$

$$N_2O_{4(l)} + H_2O_{(l)} \rightarrow HNO_{2(l)} + HNO_{3(l)} \quad (17.15)$$

$$N_2O_{5(l)} + H_2O_{(l)} \rightarrow 2\ HNO_{3(l)} \quad (17.16)$$

As a result, for the post NO_x capture system in flue gas, it would be much more efficient and effective if the NO gas could be oxidized into NO_2 and N_2O_3, thereby being easily dissolved in water. Ozone is usually used as the oxidant for the NO capture system. It was noted that the reactions between O_3 and NO_x are complicated, as indicated in Fig. 17.2.

When the molar ratio of O_3/NO was less than one, NO was mainly oxidized to NO_2. When the molar ratio was greater than one, the oxidation products were NO_2, N_2O_5, and HNO_3 [66]. Equations (17.17–17.19) show some of the process chemistry of NO_x oxidation.

$$NO_{(g)} + O_{3(g)} \rightarrow NO_{2(g)} + O_{2(g)} \quad (17.17)$$

$$NO_{2(g)} + O_{3(g)} \rightarrow NO_{3(g)} + O_{2(g)} \quad (17.18)$$

$$NO_{2(g)} + NO_{3(g)} \leftrightarrow N_2O_{5(g)} \quad (17.19)$$

Meanwhile, the formed nitrite in water is a relatively unstable compound, which would be self-reacting to form NO gas as shown in Eq. (17.20). However, in the

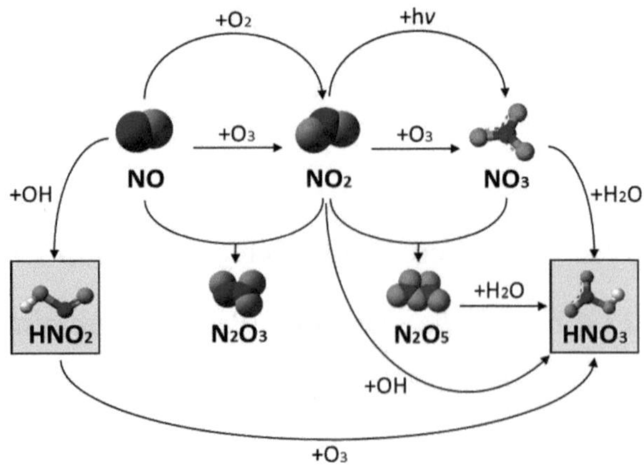

Fig. 17.2 Schematic diagram of main mechanism in NO_x oxidation by ozone

presence of ozone, the decomposition of nitrite in the liquid phase can be inhibited, as expressed in Eq. (17.21):

$$3HNO_{2(l)} \rightarrow 2NO + HNO_{3(l)} + H_2O_{(l)} \tag{17.20}$$

$$3HNO_{2(l)} + O_{3(l)} \rightarrow 3HNO_{3(l)} \tag{17.21}$$

Since integrity of mixing often plays a key role in the removal performance, the HIGEE process can be utilized to enhance the micromixing efficiency, as well as increase the gas–liquid mass transfer rate of the NO_x dissolution, as shown in Eqs. (17.22–17.24):

$$NO_{2(g)} \rightarrow NO_{2(l)} \tag{17.22}$$

$$N_2O_{3(g)} \rightarrow N_2O_{3(l)} \tag{17.23}$$

$$N_2O_{5(g)} \rightarrow N_2O_{5(l)} \tag{17.24}$$

In addition, the absorption reaction between NO_x and water can be promoted by introducing the absorbents, including an alkali component (e.g., NaOH and H_2O_2), a urea solution $[CO(NH_2)_2]$ and a metal chelating agent (e.g., Fe^{II}(EDTA)). The reactions between alkali absorbent and oxidized NO_x gas include the following:

$$NO_2 + 2OH^-(\text{alkali}) \rightarrow NO_2^- + NO_3^- + H_2O \tag{17.25}$$

$$N_2O_3 + 2OH^-(\text{alkali}) \rightarrow 2NO_2^- + H_2O \tag{17.26}$$

$$NO_2^- + H_2O_2(\text{alkali}) \rightarrow NO_3^- + H_2O \tag{17.27}$$

As shown in Eq. (17.28), the reaction between urea solution and NO_x gas can form N_2 and CO_2 gases.

$$CO(NH_2)_2 + NO + NO_2 \rightarrow 2N_2 + CO_2 + 2H_2O \tag{17.28}$$

With the complexing agent such as [Fe^{2+}(EDTA)], NO gas can be converted to a nitroso-acyl complex, where the complex NO is easily oxidized and absorbed in the liquid phase, as described in Eqs. (17.29) and (17.30):

$$NO_{(g)} \rightarrow NO_{(aq)} \tag{17.29}$$

$$NO_{(aq)} + Fe^{2+}L^{n-} \leftrightarrow Fe^{2+}L^{n-}NO \tag{17.30}$$

where L^{n-} is the organic ligand. However, when the solution is under an alkaline condition, ferrous hydroxide will be formed as shown in Eq. (17.31).

$$Fe^{2+} + 2OH^- \leftrightarrow Fe(OH)_2 \tag{17.31}$$

In this case, a severe competition with Fe^{2+} in reacting with EDTA would be observed, leading to less [Fe^{2+}(EDTA)] for combining with NO. This suggests that the pH value of the solution should be maintained at around 7 in the NO removal system to achieve an excellent absorption rate [67].

17.4.2 Performance Evaluation: Applications

The oxidation efficiency of NO_2 to N_2O_5 mainly depended on the flow rate and inlet NO concentration [68]. HIGEE has been successfully introduced as a gas–liquid reactor to improve the NO_x removal ratio [69, 70]. For the HIGEE technique, several key operating factors, such as types of absorbents and oxidants, concentrations of absorbents and oxidants, gas-to-liquid (G/L) ratio, temperature, and rotating speed, play a significant role in the removal efficiency of NO_x in the flue gas. A suitable gravitation field in the HIGEE process, i.e., at around 150 g, can be beneficial to the efficient removal of NO_x [67]. With an excellent micromixing efficiency under such gravitation, an NO_x removal ratio of greater than 96% can be achieved by the HIGEE technique, with an inlet NO_x concentration of 18,000–20,000 mg/m^3 and a urea concentration of 20 wt% [71].

Furthermore, in the presence of a TiO_2 catalyst, the catalytic destruction of O_3 can be expected, thereby creating surface-bound oxygen atoms responsible for the improved oxidation of NO_2 to NO_3. It was found that, at higher temperature (e.g., 100 °C), the presence of a TiO_2 catalyst could considerably improve the efficiency of oxidation of NO_2 to N_2O_5 [68].

17.5 Particulate Matter (PM) Emission Control

The types of anthropogenic or natural PM in the atmosphere include total suspended particulates (TSP), inhalable particulate matter (PM_{10}), fine particulate matter ($PM_{2.5}$), ultrafine particulate matter, and soot. Figure 17.3 shows the relationships between various types of particulate collection equipment and their applicable particle size of collection.

The commonly used particulate collection processes include mechanical collector, air filer, cyclone, wet scrubber, baghouse filter, and ESP. Each type of equipment has been applied based on different removal mechanisms, thereby limiting its appropriate operating window and treatment capacity (space occupation). For instance, the cyclone can work quite well with a particle size greater than 10 μm with a low installation cost; however, it cannot effectively reduce $PM_{2.5}$ and finer PM. In contrast, a baghouse filter can collect even 0.01-μm particulates with an overall removal efficiency of over 99% [72]. However, the baghouse filter has a high maintenance cost due to rapid clogging of the filter. ESP also can achieve a removal efficiency of over 95% with a relatively low maintenance cost. However, the investment and operating costs associated with this technique are high due to the consumption of electricity [73]. For the wet scrubber system, the critical particle sizes (the most difficult to collect) are typically in the range of 0.1–0.5 μm. Due to limits in the collection mechanism, the removal efficiency of particulates with a diameter less than 2.5 μm ($PM_{2.5}$) also was not remarkable, i.e., typically 60–70%

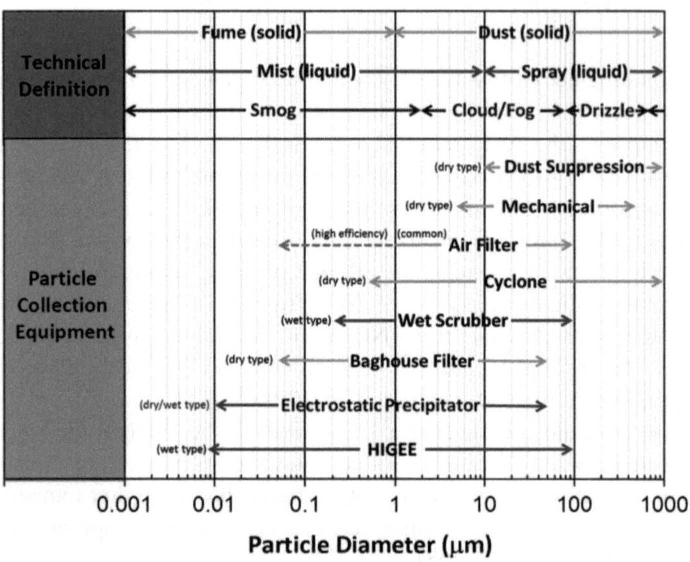

Fig. 17.3 Relationship between various types of particulate control equipment and their applicable particle size of collection (adapted from [13])

for wet scrubber. Therefore, an innovative approach or an integrated process should be developed for effective control of the (ultra) fine particulates such as $PM_{2.5}$. It is also noted that the HIGEE technology could efficiently remove PM in a wide range of particle sizes between 0.01 and 100 μm.

17.5.1 Mechanisms and Principles

Wet collectors (scrubbers) for particulates in a gas stream may be generally grouped into two types [74]:

- Type 1: An array of liquid sprays forms the collecting medium, such as spray towers and venturi scrubbers.
- Type 2: Various types of wetted surfaces constitute the collecting medium, such as packed towers.

Using the HIGEE reactor for particulate collection may consist of both types. In the HIGEE, both thin liquid films (inside packing zone) and tiny droplets (outside packing zone) can be generated due to the shear stress, at the same time continuously changing the direction of the channels for gas and particulate via RPB. The injected liquid also could exhibit a "clean" function to packings in the strong centrifugal field, which prevents them from being blocked, thereby maintaining a high efficiency of dust removal. Figure 17.4 shows a schematic of particle collection mechanisms by the HIGEE system, which include diffusion, interception, and inertial impaction. The relative motion between particles and droplets is important because the collection of large particles mainly occurs by impaction and interception, while diffusion is important for small particles.

1. Diffusion

Diffusion is a function of particle diameter, concentration velocity, and distance between particle and droplet. It is predominant with low gas velocities and small

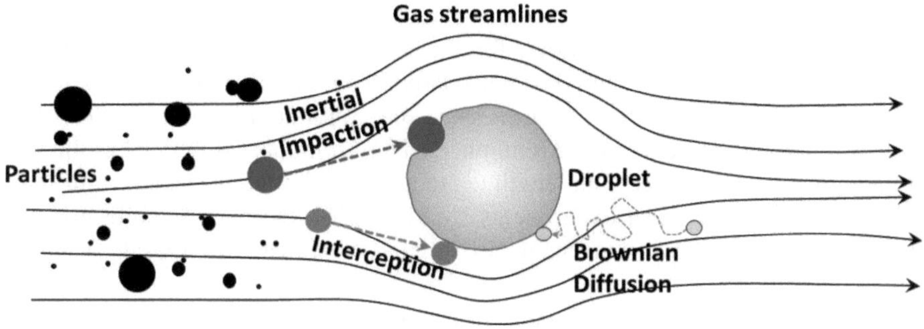

Fig. 17.4 Schematic of particle collection mechanisms in high-gravity (HIGEE) system, including inertial impaction, interception, and Brownian diffusion (adapted from [13])

particles. For the diffusion mechanism, particles with diameters of less than 0.1 μm would experience Brownian motion in an exhaust stream, since small particles attain a high diffusion coefficient. They could possibly diffuse and be collected onto droplets. The rate of particle diffusion depends on the diameters of particles and droplets and on their relative velocity. It was noted that the collection efficiency by diffusion increases as the particle size decreases.

To determine the contribution of the diffusion mechanism, several dimensionless groups should be defined in advance. The Knudsen (*Kn*) number is defined as the ratio of the molecular mean free path length (λ) to a representative physical length scale. The mean free path of air molecules at ambient temperature and pressure is about 0.068 μm [75]. In addition, the diffusion Peclet parameter (*Pe*) of particle can be determined in Eq. (17.32):

$$Pe = \frac{2 \cdot U \cdot D_{\mathrm{d}}}{D} \tag{17.32}$$

where U (m/s) is the droplet face velocity, and D_{d} (m) is the average diameter of the liquid droplets. D (m^2/s) is the Brownian diffusion coefficient of particles, which can be evaluated by the Stokes–Einstein equation as follows:

$$D = \mathrm{k}_B \cdot T \cdot B \tag{17.33}$$

where k_B (J/K) is the Boltzmann's constant, i.e., 1.381×10^{-23}, and T (K) is the temperature. B is the mechanical mobility of a particle, which can be calculated by

$$B = \frac{C_{\mathrm{c}}}{3 \cdot \pi \cdot \mu_{\mathrm{g}} \cdot D_p} \tag{17.34}$$

where μ_{g} (kg/m/s) is the dynamic viscosity of gas, and D_{p} (m) is the particle diameter. The Cunningham correction factor (C_{c}) can be determined in Eq. (17.35):

$$C_{\mathrm{c}} = 1 + \frac{2\lambda}{D_p}\left[a_1 + a_2 \cdot \exp\left(\frac{-a_3 D_{\mathrm{p}}}{2\lambda}\right)\right] \tag{17.35}$$

where the values of a_1, a_2, and a_3 for the air stream are 1.26, 0.40, and 1.10, respectively. The C_{c} value should be considered in determining the St number when particles become smaller than 15 μm, since the continuum assumption in Stokes Law is no longer correct at high *Kn* numbers. The *Kn* number is defined as the ratio of the molecular mean free path length (λ) to a representative physical length scale.

According to analytical experiments by Pitch [76], the particle collection efficiency due to the diffusion mechanism (η_{d}) can be estimated by Eq. (17.36):

$$\eta_{\mathrm{d}} = 2.9 \cdot k_{\mathrm{d}}^{-1/3} \cdot Pe^{-2/3} \cdot (1 + 0.39 \cdot k_{\mathrm{d}} \cdot Pe^{1/3} \cdot Kn) \tag{17.36}$$

where the k_d value can be calculated by Eq. (17.37):

$$k_{\mathrm{d}} = Ku + 1.15 \cdot Kn \tag{17.37}$$

The Ku is the Kuwabara hydrodynamic factor, which can be determined by Eq. (38):

$$Ku = \left(-0.75 + \rho_{\mathrm{L}} - 0.25\rho_{\mathrm{L}}^2\right) - 0.5 \cdot \ln(\rho_{\mathrm{L}}) \tag{17.38}$$

where ρ_{L} (kg/m^3) is the density of liquid droplets.

2. Interception

Interception, a function of particle diameter, occurs when the pathway of a particle comes within one radius of a liquid droplet. The particle touches the liquid droplet and then is captured and thereby removed from the gas stream. Interception is the dominant mechanism for capturing particles with a diameter in the range of 0.1–1 μm. The interception parameter (S) of the particle is defined in Eq. (17.39):

$$S = \frac{D_{\mathrm{p}}}{D_{\mathrm{d}}} \tag{17.39}$$

Based on the Navier-Stokes equations for low Reynold numbers [77], the particle collection efficiency due to interception (η_{s}) can be estimated by Eq. (17.40):

$$\eta_{\mathrm{s}} = \frac{1}{2[2 - \ln(Re)]}\left[2(1+S)\ln(1+S) - (1+S) + \frac{1}{(1+S)}\right] \tag{17.40}$$

3. Inertial Impaction

Inertial impaction, a function of particle mass and velocity, is predominant with high gas velocities and large particles. For the impaction mechanism, dust particles would tend to follow the streamline of the exhaust gas stream in a HIGEE system. However, when tiny liquid droplets are introduced into the system, the particles cannot always follow the original stream, as they diverge around the droplet. This leads to the impaction of particles onto the droplet due to the mass of particles. The particle collection efficiency due to inertial impaction (η_{I}) can be estimated by Eq. (17.41):

$$\eta_{\mathrm{I}} = \frac{f}{2} \cdot \frac{St}{Ku^2} \tag{17.41}$$

where the f value is assumed to be 2 if the S value is greater than 0.4 [78]. Otherwise, the f value should be determined by Eq. (17.42):

$$f = \left(29.6 - \rho_{\mathrm{L}}^{0.62}\right)S^2 - 27.5 \cdot S^{2.8} \tag{17.42}$$

The probability of particle impaction increases with the increase in both the particle diameter and the relative velocity between the particle and droplets and with the decrease in the droplet size. Normally, if the produced droplets are in the range 150–500 μm (in the case of a wet scrubber), particles with diameters great than 1.0 μm could be collected by the impaction mechanism [79]. In contrast, since the minimum diameters of droplets in the HIGEE are estimated to be in the range 0.0013–50 μm [80], the removal efficiency of $PM_{2.5}$ should be significantly enhanced by impaction in the HIGEE.

17.5.2 Key Performance Indicators

Collection efficiency is determined by considering the efficiency of a single spherical collector and then summing over the number of drops per unit volume of gas stream [74]. The collection efficiency (η_{PM}) of PM with a certain particle size can be calculated by Eq. (17.43):

$$\eta_{\mathrm{PM}}(\%) = \frac{C_{\mathrm{in,d}} - C_{\mathrm{out,d}}}{C_{\mathrm{in,d}}} \times 100 \tag{17.43}$$

where $C_{\mathrm{in,d}}$ (mg/m^3) is the concentration of PM with a certain particle size (d) at the inlet of the PM control system, and $C_{\mathrm{out,d}}$ (mg/m^3) is the concentration of PM with a certain particle size (d) at the outlet of the PM control system. The volume of flue gas is calculated under normal temperature (293 K) and pressure (101.325 kPa) conditions. It was also noted that five dimensionless groups are largely related to the collection efficiency, i.e., the Reynolds number of the sphere (*Re*), the Schmidt number of the particle (*Sc*), the Stokes number of the particle (*St*), the diameter ratio of particle and droplet (κ), and the viscosity ratio of water to air (ω), as expressed in Eq. (17.44).

$$\eta_{\mathrm{PM}}(\%) = f(Re, Sc, St, \kappa, \omega) \tag{17.44}$$

The values of *Re*, *Sc*, *St*, κ, and ω can be expressed by Eqs. (17.45–17.49):

$$Re = \frac{\rho_{\mathrm{g}} V_{\mathrm{g}} D_{\mathrm{d}}}{\mu_{\mathrm{g}}} \tag{17.45}$$

$$Sc = \frac{\mu_g}{\rho_g D_d} \quad (17.46)$$

$$St = C \cdot \frac{\rho_p V_g D_p^2}{18 \mu_g D_d} \quad (17.47)$$

$$\kappa = \frac{D_p}{D_d} \quad (17.48)$$

$$\omega = \frac{\mu_l}{\mu_g} \quad (17.49)$$

where ρ_g (kg/m^3) is the gas density; V_g (m/s) is the gas velocity; μ_l (kg/m/s) is the liquid viscosity; and C_c (−) is the Cunningham correction factor.

17.5.3 Performance Evaluation: Applications

Various types of RPB, such as a tapered rotating fluidized bed [81], a cross-flow RPB [82] and a countercurrent-flow RPB [83, 84], were applied for PM control in the flue gas. In general, the dedusting efficiency of countercurrent-flow RPB (i.e., 98.5%) is higher than that of cross-flow RPB (i.e., 97.5%) [85]. The PM removal ratio in a cross-flow RPB increases with increases in the rotating speed, liquid-to-solid ratio and gas flow rate [82]. In addition, the results indicate that the removal ratio of fine PM (~μm) by the RPB should be significantly higher than that by conventional processes such as cyclone, wet scrubber, and ESP [82, 83]. In particular, fine particles larger than 3.5 μm could be almost completely removed by the HIGEE technique [86].

References

1. Jasarevic T, Thomas G, Osseiran N (2014) 7 million premature deaths annually linked to air pollution. WHO
2. Morawska L, Zhang J (2002) Combustion sources of particles: health relevance and source signatures. Chemosphere 49:1045–1058
3. Siegmann K (2000) Soot formation in flames. J Aero Sci 31:S217–S218
4. MEP (2012) Ambient air quality standards, vol GB 3095–2012. Ministry of Environmental Protection, P.R. China
5. CPCB (2009) National ambient quality standards, vol B-29016/20/90/PCI-I. Central Pollution Control Board, India
6. MOE (2009) Environmental quality standards. Ministry of the Environment, Japan
7. MOE (2011) Air quality standards and air pollution level. Ministry of Environment, Korea
8. ECD (2004) Royal commission environmetal regulations. Environmental Control Department, Kingdom of Saudi Arabia

9. Taiwan EPA (2012) Air quality standards, vol 1010038913. EPA Taiwan, Taiwan (ROC)
10. USEPA (2016) Natinoal ambient air quality standards (NAAQS). US Evironmental Protection Agency, USA
11. WHO (2006) WHO air quality guidelines for particulate matter, ozone, nitrogen dioxide and sulfur dioxide. Summary of risk assessment, vol WHO/SDE/PHE/OEH/06.02. World Health Organization, Switzerland
12. Kudlac GA, Farthing GA, Szymanski T, Corbett R (1992) SNRB catalytic baghouse laboratory pilot testing. Environ Prog 11(1):33–38. doi:10.1002/ep.670110115
13. Pan S-Y, Wang P, Chen Q, Jiang W, Chu Y-H, Chiang P-C (2017) Development of high-gravity technology for removing particulate and gaseous pollutant emissions: a review of its principles and applications. J Cleaner Prod
14. Kang YS, Kim SS, Hong SC (2015) Combined process for removal of SO_2, NO_x, and particulates to be applied to a 1.6-MWe pulverized coal boiler. J Ind Eng Chem 30:197–203. doi:10.1016/j.jiec.2015.05.022
15. Gutiérrez Ortiz FJ, Ollero P (2008) Modeling of the in-duct sorbent injection process for flue gas desulfurization. Sep Purif Technol 62(3):571–581. doi:10.1016/j.seppur.2008.03.012
16. Pan S-Y, Chiang A, Chang E-E, Lin Y-P, Kim H, Chiang P-C (2015) An innovative approach to integrated carbon mineralization and waste utilization: A review. Aerosol Air Qual Res 15:1072–1091. doi:10.4209/aaqr.2014.10.02
17. Zhao H, Shao L, Chen J-F (2010) High-gravity process intensification technology and application. Chem Eng J 156(3):588–593. doi:10.1016/j.cej.2009.04.053
18. Ramshaw C, Mallinson R (1979) Mass transfer process. European Patent 2568B Patent
19. Ramshaw C (1983) High distillation—an example of process intensification. Chem Eng 13–14
20. Lin C, Chen B (2008) Characteristics of cross-flow rotating packed beds. J Ind Eng Chem 14 (3):322–327. doi:10.1016/j.jiec.2008.01.004
21. Wang M (2004) Controlling factors and mechanism of preparing needlelike $CaCO_3$ under high-gravity environment. Powder Technol 142(2–3):166–174. doi:10.1016/j.powtec.2004.05.003
22. Kelleher T, Fair JR (1996) Distillation studies in a high-gravity contactor. Ind Eng Chem Res 35:4646–4655
23. Tan C, Chen J (2006) Absorption of carbon dioxide with piperazine and its mixtures in a rotating packed bed. Sep Purif Technol 49(2):174–180. doi:10.1016/j.seppur.2005.10.001
24. Reay DA, Ramshaw C, Harvey A (2013) Intensification of separation processes: Higee. In: Process intensification: engineering for efficiency, sustainability and flexibility. 2nd edn. Elsevier, Amsterdam, p 200–223
25. Liu YZ (2009) Chemical engineering process and technology in high gravity. National Defence Industry Press, Beijing
26. Cheng H-H, Tan C-S (2011) Removal of CO_2 from indoor air by alkanolamine in a rotating packed bed. Sep Purif Technol 82:156–166. doi:10.1016/j.seppur.2011.09.004
27. Xiang Y, Wen LX, Chu GW, Shao L, Xiao GT, Chen JF (2010) Modeling of the precipitation process in a rotating packed bed and its experimental validation. Chin J Chem Eng 18(2): 249–257
28. Yu C-H, Cheng H-H, Tan C-S (2012) CO_2 capture by alkanolamine solutions containing diethylenetriamine and piperazine in a rotating packed bed. Int J Greenhouse Gas Control 9:136–147. doi:10.1016/j.ijggc.2012.03.015
29. Pan SY, Eleazar EG, Chang EE, Lin YP, Kim H, Chiang PC (2015) Systematic approach to determination of optimum gas-phase mass transfer rate for high-gravity carbonation process of steelmaking slags in a rotating packed bed. Appl Energy 148:23–31. doi:10.1016/j.apenergy.2015.03.047
30. Munjal S, Dudukovic MP, Ramachandran P (1989) Mass-transfer in rotating packed beds—I. Dev Gas-Liq Liq-Solid Mass Transf Correl Chem Eng Sci 44(10):2245–2256

31. Li X, Liu Y (2010) Characteristics of fin baffle packing used in rotating packed bed. Chin J Chem Eng 18(1):55–60. doi:10.1016/s1004-9541(08)60323-7
32. Li Y, Liu Y, Zhang L, Su Q, Jin G (2010) Absorption of NO_x into nitric acid solution in rotating packed bed. Chin J Chem Eng 18(2):244–248. doi:10.1016/s1004-9541(08)60349-3
33. Cheng H-H, Tan C-S (2006) Reduction of CO_2 concentration in a zinc/air battery by absorption in a rotating packed bed. J Power Sources 162(2):1431–1436. doi:10.1016/j.jpowsour.2006.07.046
34. Liu YZ, Qi GS, Yang LR (2003) Study on the mass transfer characteristics in impinging stream rotating packed bed extractor. Chem Ind and Eng Pro 22(10):1108–1111
35. Wang YG, Li ZH, Zhang WS, Zeng D, Guo K (2009) Removal of sulfur dioxide from flue gas in rotating packed bed by dual-alkali method (in Chinese). Petro-Chem Tec 38(8):893–896
36. Liu LS, Zhang YH, Liu YZ (2001) Flue gas desulfurization by rotating packed bed (in Chinese). J Basic Sci Eng 9(2–3):292–296
37. Chen JM, Sun YH (2002) Deoxygenation of boiler feed water by high gravity technology (in Chinese). Chem Ind and Eng Pro 21(6):414–416
38. Xiang Y, Wen L, Chu G, Shao L, Xiao G, Chen J (2010) Modeling of the precipitation process in a rotating packed bed and its experimental validation. Chin J Chem Eng 18(2):249–257. doi:10.1016/s1004-9541(08)60350-x
39. Zhao H, Wang J, Zhang H, Shen Z, Yun J, Chen J (2009) Facile preparation of danazol nanoparticles by high-gravity anti-solvent precipitation (HGAP) method. Chin J Chem Eng 17(2):318–323. doi:10.1016/s1004-9541(08)60210-4
40. Zhang D, Zhang P, Zou H, Chu G, Wu W, Zhu Z, Shao L, Chen J (2010) Synthesis of petroleum sulfonate surfactant by different sulfonating agent with application of HIGEE technology. Chin J Chem Eng 18(5):848–855. doi:10.1016/s1004-9541(09)60138-5
41. Zhang QL, Li GM, Liu YZ (2009) Preparation of p-Hydroxybenzaldehyde by combination reactor of rotating packed bed and guide tray tower (in Chinese). Chem Online 10:937–941
42. Chen YH, Chang CY, Su WL, Chen CC, Chiu CY, Yu YH, Chiang PC, Chiang SIM (2004) Modeling ozone contacting process in a rotating packed bed. Ind Eng Chem Res 43(1):228–236
43. Zhang YH, Liu LS, Liu YZ (2003) Experiment study on flue gas dedusting by hypergravity rotary bed (in Chinese). Environ Eng 21(6):42–43
44. Chen J-F, Shao L, Guo F, Wang X-M (2003) Synthesis of nano-fibers of aluminum hydroxide in novel rotating packed bed reactor. Chem Eng Sci 58(3–6):569–575. doi:10.1016/s0009-2509(02)00581-x
45. Shao L, Chen J (2005) Synthesis and application of nanoparticles by a high gravity method. China Particuology 3(1–2):134–135. doi:10.1016/s1672-2515(07)60180-8
46. Pan S-Y, Hung C-H, Chan Y-W, Kim H, Li P, Chiang P-C (2016) Integrated CO_2 fixation, waste stabilization, and product utilization via high-gravity carbonation process exemplified by circular fluidized bed fly ash. ACS Sustain Chem Eng 4(6):3045–3052. doi:10.1021/acssuschemeng.6b00014
47. Pan SY, Chiang PC, Chen YH, Tan CS, Chang EE (2013) Ex Situ CO_2 capture by carbonation of steelmaking slag coupled with metalworking wastewater in a rotating packed bed. Environ Sci Technol 47(7):3308–3315. doi:10.1021/es304975y
48. Chang EE, Pan SY, Chen YH, Tan CS, Chiang PC (2012) Accelerated carbonation of steelmaking slags in a high-gravity rotating packed bed. J Hazard Mater 227–228:97–106. doi:10.1016/j.jhazmat.2012.05.021
49. Mondal MK, Chelluboyana VR, Rao JS (2013) Solubility of SO_2 in aqueous blend of sodium citrate and sodium hydroxide. Fluid Phase Equilib 349:56–60. doi:10.1016/j.fluid.2013.03.025
50. Hao JM, Wang SX, Lu YQ (2009) Handbook on sulfur dioxide pollution control technology in coal combustion. Chemical Industry Press, Beijing
51. USEPA (2003) Air pollution control technology fact sheet. USEPA

52. Kiil S, Michelsen ML, Dam-Johansen K (1998) Experimental investigation and modeling of a wet flue gas desulfurization pilot plant. Ind Eng Chem Res 37(7):2792–2806. doi:10.1021/ie9709446
53. Ruhland F, Kind R, Weiss S (1991) The kinetics of the absorption of sulfur dioxide in calcium hydroxide suspensions. Chem Eng Sci 46(4):939–947. doi:10.1016/0009-2509(91)85087-e
54. Sun WS, Wu ZB, Li Y, Tan TE (2002) Sodium-enhanced limestone WET FGD in rotating-stream tray scrubber. Environ Sci 23(5):105–108
55. Wu Y, Li Q, Li F (2007) Desulfurization in the gas-continuous impinging stream gas–liquid reactor. Chem Eng Sci 62(6):1814–1824. doi:10.1016/j.ces.2006.01.027
56. Berman Y, Tanklevsky A, Oren Y, Tamir A (2000) Modeling and experimental studies of SO_2 absorption in coaxial cylinders with impinging streams: part I. Chem Eng Sci 55 (5):1009–1021. doi:10.1016/s0009-2509(99)00380-2
57. Klaassen R (2003) Achieving flue gas desulphurization with membrane gas absorption. Filtr Sep 40(10):26–28. doi:10.1016/s0015-1882(03)00033-8
58. Erga O (1986) Sulfur dioxide recovery by means of adipic acid buffers. Ind Eng Chem Fund 25(4):692–695. doi:10.1021/i100024a035
59. Akyalçın L, Kaytakoğlu S (2010) Flue gas desulfurization by citrate process and optimization of working parameters. Chem Eng Process 49(2):199–204. doi:10.1016/j.cep.2009.12.008
60. Bekassy-Molnar E, Marki E, Majeed JG (2005) Sulphur dioxide absorption in air-lift-tube absorbers by sodium citrate buffer solution. Chem Eng Process 44(9):1039–1046. doi:10.1016/j.cep.2005.02.001
61. X-l Long, Z-l Xin, M-b Chen, Li W, W-d Xiao, W-k Yuan (2008) Kinetics for the simultaneous removal of NO and SO_2 with cobalt ethylenediamine solution. Sep Purif Technol 58(3):328–334. doi:10.1016/j.seppur.2007.05.004
62. X-p Jiang, Liu Y-z Du, W-f C-l, Mu (2011) Removal of low concentration sulfur dioxide from flue gas by sodium citrate solution scrubbing in high-gravity rotating packed bed (in Chinese). China Acad J 3:49–53
63. Shivhare MK, Rao DP, Kaistha N (2013) Mass transfer studies on split-packing and single-block packing rotating packed beds. Chem Eng Process 71:115–124. doi:10.1016/j.cep.2013.01.009
64. Jiang X, Liu Y, Gu M (2011) Absorption of sulphur dioxide with sodium citrate buffer solution in a rotating packed bed. Chin J Chem Eng 19(4):687–692
65. Hong T, Li LB, Wang CG, Ma GD, Zhao XH, Fu XK, Li WC (2004) Characteristics of SO_2 absorption in flue gas by citrate solution in packing columns (in Chinese). Chem Eng 32 (4):49–52
66. Sun C, Zhao N, Zhuang Z, Wang H, Liu Y, Weng X, Wu Z (2014) Mechanisms and reaction pathways for simultaneous oxidation of NO_x and SO(2) by ozone determined by in situ IR measurements. J Hazard Mater 274:376–383. doi:10.1016/j.jhazmat.2014.04.027
67. Zhang L-L, Wang J-X, Sun Q, Zeng X-F, Chen J-F (2012) Removal of nitric oxide in rotating packed bed by ferrous chelate solution. Chem Eng J 181–182:624–629. doi:10.1016/j.cej.2011.12.027
68. Jõgi I, Erme K, Raud J, Laan M (2016) Oxidation of NO by ozone in the presence of TiO_2 catalyst. Fuel 173:45–51. doi:10.1016/j.fuel.2016.01.039
69. J-h Li, Liu Y-Z (2013) Study on NO_x ozone oxidation and intensifying absorption in high gravity field. Mod Chem Ind 33(5):40–44
70. Chen J, Sun B, Zhao H, XU C, Lo Y (2014) An equipment and technology of high-gravity efficient removal of NO_x. China Patent
71. Liu Y, Li P, Li Y, Kang R, Diao J (2007) Pilot test on treatment of high concentration nitrogen oxides by high gravity technology. Chem Ind Eng Prog 26(7):1058–1061. doi:10.16085/j.issn.1000-6613.2007.07.029
72. Watanabe Y, Sato H, Hirai Y, Kim IS, Hinata S, Kim J (2009) Novel method to evaluate the net wear volume of bag-filter by fly ash. J Hazard Mater 161(2–3):775–780. doi:10.1016/j.jhazmat.2008.04.024

73. Bianchini A, Cento F, Golfera L, Pellegrini M, Saccani C (2016) Performance analysis of different scrubber systems for removal of particulate emissions from a small size biomass boiler. Biomass Bioenergy 92:31–39. doi:10.1016/j.biombioe.2016.06.005
74. Flagan RC, Seinfeld JH (1988) Removal of particles from gas streams. In: Fundamentals of air pollution engineering. Prentice-Hall, Inc., Englewood Cliffs, New Jersey
75. Jennings SG (1988) The mean free path in air. J Aerosol Sci 19(2):159–166. doi:10.1016/0021-8502(88)90219-4
76. Pitch J (1964) Impaction of aerosol particles in the neighborhood of a circular hole. Collect Czech Chem Commun 29:2223–2227
77. Lamb H (1993) Hydrodynamics. 6th edn. Cambridge University Press, Cambridge
78. Hakobyan NA (2015) Introduction to basics of submicron aerosol particles filtration theory via ultrafine fiber media. Armenian J Phys 8(3):140–151
79. Byeon S-H, Lee B-K, Raj Mohan B (2012) Removal of ammonia and particulate matter using a modified turbulent wet scrubbing system. Sep Purif Technol 98:221–229. doi:10.1016/j.seppur.2012.07.014
80. Xu C, Jiao W, Liu Y, Guo L, Yuan Z, Zhang Q (2014) Effects of airflow field on droplets diameter inside the corrugated packing of a rotating packed bed. China Pet Process Petrochem Technol 16(4):38–46
81. Nakamura H, Deguchi N, Watano S (2015) Development of tapered rotating fluidized bed granulator for increasing yield of granules. Adv Powder Technol 26(2):494–499. doi:10.1016/j.apt.2014.12.003
82. Fu J, Qi G, Liu Y, Tian J, Guo Q, Dong M (2015) Research on removal of fine particles by cross-flow rotating packed bed. Chem Ind Eng Prog 34(3):680–694. doi:10.16085/j.issn.1000-6613.2015.03.013
83. Zhang Y, Liu L, Liu Y (2003) Experimental study on flue gas dedusting by hypergravity rotary bed (in Chinese). Environ Eng 21(6):42–58
84. Song Y, Chen J, Fu J, Chen J (2003) Research on particle removal efficiency of the rotating packed bed (in Chinese). Chem Ind Eng Prog 22(5):499–502
85. Fu J, Qi G-S, Liu Y-Z, Tian J-X, Guo Q, Dong M-Y (2015) Removal of fine particles by high gravity wet cleaning (in Chinese). Chem Eng 43(4):6–10
86. Deng X, Tian D, Luo Y (2010) Experimental study of micron dust removal using high-gravity rotating bed. Sulfuric Acid Ind 6:16–20

Chapter 18
Waste-to-Resource (WTR) Green Supply Chain

Abstract Green supply chain has been aggressively constructed in different industrial parks around the world. The win–win benefits in both environmental and economic aspects can be achieved by implementing the waste-to-resource supply chain in the industrial park. Portfolio options of technologies for different types of waste-to-resource supply chains can be considered for achieving circular economy system. In this chapter, the strategies on implementation of waste-to-energy supply chain are proposed to overcome the challenging barriers from the aspects of technology, finance, institution, and regulation. A total of six key task forces are proposed for effectively executing the strategies. In addition, several successful lessons on waste-to-resource supply chains, such as green fuel pellet for heating supply and codigestion of organic wastes for biogas production, are provided.

18.1 Importance and Significance

The global environment and ecosystems have numerous functions including the supplies of food, clean water, and raw material for the mankind. However, human activity has impacted nearly every aspect of the environment. The adverse impacts of human activity on the environment have been known, such as follows:

- The depletion of the ozone layer,
- The destruction of various ecosystems, and
- The formation of increasingly severe weather phenomena.

In response to climate change, several key challenges of the twenty-first century have been identified, such as (1) mitigation of and adaptation to global warming; (2) protection of the population against natural hazards and disasters; and (3) optimization of food, energy and water (FEW) nexus. To prevent these destructive consequences, individuals and groups started to take concerted efforts to protect the environment in the early twentieth century. Environmental protection movement does not necessitate a slowdown of economic development. In this section, the

P.-C. Chiang and S.-Y. Pan, *Carbon Dioxide Mineralization and Utilization*, DOI 10.1007/978-981-10-3268-4_18

concepts of sustainable development, green economy, and circular economy are provided. The definition of a green supply chain is also discussed.

18.1.1 Sustainable Development

Conventionally, the standard of excessive consumption is problematic since it necessitates a trade-off between economic development and environmental sustainability. However, this trade-off approach becomes unnecessary within the framework of a sustainable development goal. The goals of the sustainable development were originally established at the Earth Summit in 1992. It has been further defined by the World Commission on Environment and Development [1]: a sustainable development should be:

> … development that meets the needs of the present without compromising the ability of future generations to meet their own needs.

The three pillars of sustainability include sustainable economy, sustainable environment, and sustainable society. Figure 18.1 shows the three pillars and their key elements for achieving a sustainable development goal by construction of a green supply chain. It suggests that the sustainable development goal should include economic development, environmental protection, and social equity [2].

Fig. 18.1 Conceptual diagram of three aspects for achieving a sustainable development goal

At the United Nations Conference on Sustainable Development (called Rio+20 Summit) in 2012, the green economy has been one of the main themes in the international debates on sustainable development. The achievements of the Rio+20 Summit include the following:

- "Future We Want" outcome document,
- Sustainable development goals (SDGs),
- High-level political forum on sustainable development (HLPF),
- Strengthened UNEP,
- Civil society participation and commitments,
- Green economy, and
- Passed responsibility of "Post-2015 Dev Agenda" to UNEP Governing Council and UN General Assembly.

Therefore, a green economy should be considered as an important tool, both at the global and national as well as at the corporate level, in the context of sustainable development.

18.1.2 Green Economy

As suggested by United Nations Environment Programme (UNEP), the key to sustainable development is to create a green economy. The green economy is defined as an economy system that aims at reducing both environmental risks and ecological scarcities. It should encapsulate three sectors: the industry, the people, and the government. Figure 18.2 shows the history of important international movements on green economy toward a sustainable development goal. Sustainable development without degrading the environment should reply on a green economy. Therefore, it can be achieved if fundamental changes are made to the existing supply chains of energy and material production, especially in industrial parks [3]. In contrast to prior economic regimes, the feature of green economy is the direct valuation of natural capital and ecological services as being valuable in terms of economy.

In 2009, the "Global Green New Deal" report was released by the United Nations Department of Economic and Social Affairs [4]. It addressed several existing barriers such as the difficulty to call for a global effort to "target price supports, establish policy coordination, and create an extension program to ramp up" for the use of green supply chain and renewable energy. Moreover, it provided several strategies for achieving an international green economy that involves a mixture of new policies and public investments. In addition, it specifically mentioned the importance of deploying renewable energy over simply reducing GHG emissions since [4, 5]

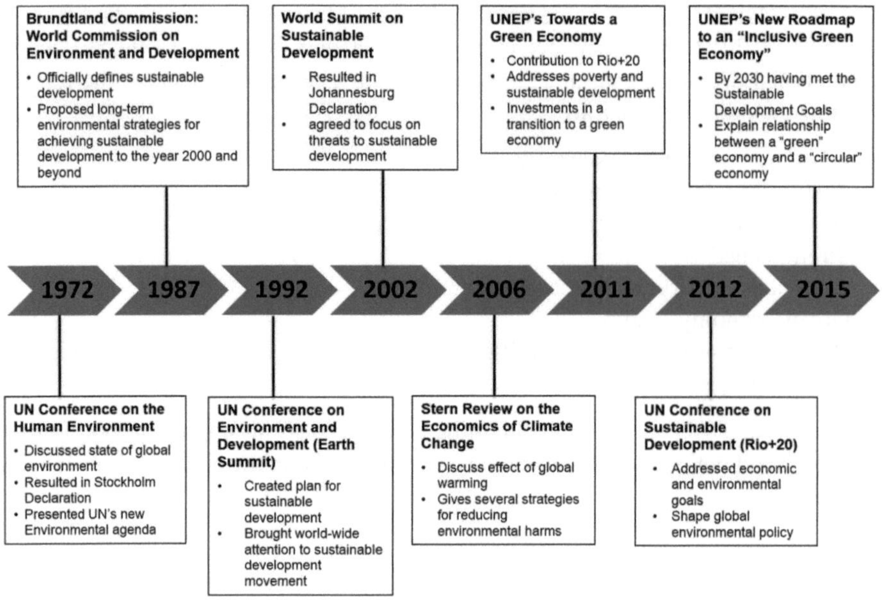

Fig. 18.2 Important international movement on the sustainable development and green economy

> energy is the key to economic development and social wellbeing, and renewable energy is the key to a future without dangerous climate change.

After that, in 2011, the necessity of a green economy, and methods for obtaining one in recent international meetings and publications were discussed. According to the definition suggested by UNEP [6], a green economy should encapsulate all industries, people, and governments, thereby resulting in

> … improving human well-being and social equity, while significantly reducing environmental risks and ecological scarcities.

In addition, a green economy should support the development of green technologies and green infrastructure that

> … reduces carbon dependency, promotes resource and energy efficiency, and lessens environmental degradation.

In 2012, the green economy became one of the main themes in the international debates on sustainable development toward the Rio+20 summit. Also, the thematic consultations of the "Post-2015 Development Agenda" include 11 areas: (1) inequalities, (2) governance, (3) growth and employment, (4) health, (5) education, (6) environmental sustainability, (7) food security and nutrition, (8) conflict and fragility, (9) population dynamics, (10) energy, and (11) water. Therefore, the green economy approach should be an effort to focus sustainable development and poverty reduction on transforming economic activities and economies.

18.1.3 Circular Economy System

The establishment of a waste-to-resource (WTR) supply chain can offer an approach to simultaneously addressing the issues of energy demand, waste management and greenhouse gas (GHG) emissions in order to achieve a circular economy system (CES). Circular economy is a generic term for an industrial economy that is producing zero waste and pollution by innovative design or intention. Therefore, it is based on the "win–win" philosophy that a prosper economy and healthy environment could be coexisted [7]. The CES is contrast to a conventional "linear economy" which is a "take, make, and dispose" model of industrial production. Figure 18.3 shows a conceptual framework of relationships between the environment and WTR supply chain for a CES. The technologies, such as fresh water production and waste production, in an industrial system should be linked together with the environment, energy and GHGs emissions to establish a business model for the CES.

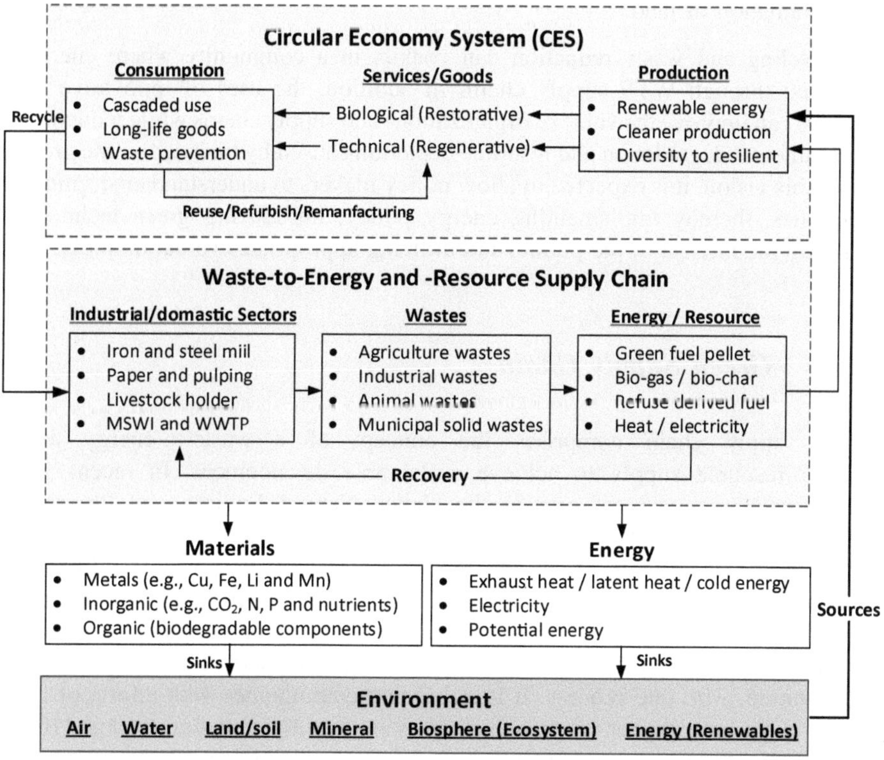

Fig. 18.3 Relationship among the environment, waste-to-energy (WTE) supply chains and circular economy system (CES)

In other words, both the industry system and the environment should be maintained as a circular relationship for facing existing environmental issues and resource scarcity. Within the CES, a closed loop of material flows including biological and technical nutrients exists:

- Biological nutrients: designed to reenter the biosphere safely.
- Technical nutrients: designed to circulate at high quality in the production system without entering the biosphere or being restorative and regenerative by design.

Several reports have studied the role of CES, especially in the developing countries, in establishing a supply chain of services and/or goods using indicator mechanism [8, 9]. The CES is based on the "5R principles" to decouple the economic growth from environmental degradation and build a resource-saving society:

- Reduction,
- Reuse,
- Recycling,
- Recovery of energy, and
- Reclamation of land.

Recycling and waste reduction can coexist in a community where energy is generated through WTR supply chain. In addition, the uses of innovative technologies can improve the value of organizations and supply chains while reducing the environmental degradation and resource depletion caused by their economic growth. Under this vision, it is expected to allow policy makers to understand emerging new techniques, thereby implementing energy policy, introducing green technology, attracting the interest of the public, and utilizing appropriate evaluation tools.

18.1.4 Green Supply Chain

Green supply chain comprises the concept of a waste-to-energy and/or waste-to-resource supply to achieve sustainable development. In recent years, international organizations such as the United Nations Environment Programme (UNEP) have moved to promote a movement toward sustainable development. Since the diverse stream of human waste creates problems when landfill space becomes limited and chemical leachate spills into the environment, WTR supply chain is a feasible method of green material production that revolves two of humanities' environmental issues, i.e., the landfilling and the natural resource conservation, with one process. It is noted that communities with efforts of green supply chain have higher waste recycling ratios than the national average [10]. In other words, there is an urgent need to develop and implement the green technologies into the existing facilities, especially in the developing countries.

The composition and amount of the solid wastes from a municipality and/or industry depend on the level of economic and social developments, energy sources, cultural norms, and geographical conditions (location, climate). For instance, municipal solid wastes (MSW) include commercial waste, medical waste, construction waste, and household waste. Typically, the MSW can be broadly categorized as organic, paper, plastic, glass, metals, and others (such as textiles, leather, rubber, multilaminates, e-waste, appliances, ash, and other inert materials). These solid wastes are usually non-biodegradable and take a long time to transform into natural compounds [11].

18.2 Barriers and Challenges

Typically, barriers from the below aspects could be encountered while implementing the green supply chain:

- Regulatory barrier,
- Institutional barrier,
- Financial barrier, and
- Technological barrier.

These barriers are hard to be distinctly separated because, for instance, policies (or regulations) often act on more than one barrier simultaneously. Similarly, this is especially true for the institutional and financial barriers as they can habitually be closely related.

18.2.1 Regulatory Aspect

Figure 18.4 shows the pyramid of regulatory framework at different levels. Regulatory barriers encompass unclear national vision (corresponding to policy), goal (corresponding to strategy), objective (corresponding to program), target (corresponding to project), and indicator (corresponding to plan).

Political issues, for instance as a consequence to an outdated infrastructure, could act as obstacles to creating an effective green economy. These are the challenges in appropriate policy formulation and government authority allocation. Therefore, regulatory barriers often prevent institutions from efficiently developing technology and processes that are crucial for the green supply chains.

18.2.1.1 Strict Laws and Regulations

Overly strict laws and regulations would prevent development and implementation of green technology. In the financial sector, strict laws may deter investors or

Fig. 18.4 Pyramid of regulatory framework at different levels

insurers from developing green supply chains [6]. In addition, a long time to receive permits and complete the proper procedures for developing and implementing green technology is one of the main impediment factors in the expansion of green supply chain. For instance, in France, it may take as long as eighty months to obtain the proper permits [12].

18.2.1.2 Lack of Efficient Governance and Available Information

Authorities have to identify the waste management needs and energy demands of a community while considering an available budget. Prior to project commencement, the biggest obstacle that almost all involved parties possess is the lack of information. According to the findings from various analysis methods such as analytical hierarchy analysis [13] and stakeholder interviews [14, 15], the needs of informing stakeholders about currently available technology were sometimes neglected.

18.2.1.3 Intellectual Property (IP) Constrains

Policy makers should emphasize the uses of best available technologies (BAT) to ensure environmental safety and facility efficiency. However, making the spread of technology depends on the certain groups controlling the information. Laws regulating intellectual property (IP) rights constrain the share of information among the industries since the laws determine the groups who can control the relevant information and technology. This would lead to a slowdown in technology transformation [16]. As a result, Committee of African Heads of State and Government on Climate Change (CAHOSCC) has called for the removal of restrictions on IP rights to allow African countries to develop clean energy and green infrastructure [17].

18.2.1.4 Approval in the Use of Wastes

In some circumstances, companies may experience obstacles in obtaining governmental approvals to use alternative materials and/or fuels from the recovered and/or recycled wastes. If a by-product is classified as a controlled waste (such as fly ash), strict procedures in transportation and cumbersome documentation are obliged for implementation. Although the by-product synergies appear techno-economically feasible with a positive sustainability, practical implementation has been halted due to the uncertainties of the legislative framework, especially with regard to the final responsibility for the approved reuse options and community concern.

18.2.2 Institutional Aspect

Institutional barriers are pervasive when it comes to creating the green supply chains. Typically, the institutions in play include the (1) enacting authority for policy and regulation, (2) distribution market, (3) project investors, and (4) local community.

18.2.2.1 Lack of Awareness

Decisions should be made with clear awareness of the public risks and economic benefits regarding the financial and technology support. A lack of awareness will lead to misplaced public perception, thereby hindering the progress of policy implementation. Similarly, another major barrier of establishing green supply chains is the lack of awareness of customers about the benefits of green products. Without the demands of green products from customers, the company and/or industry will not replace old technology for innovative green product. As a result, an information exchange platform among the government, industry, and customers should play a crucial role to achieve a successful green supply chain.

18.2.2.2 Unclear Ownerships

For the waste-to-energy supply chain, a critical issue pertaining to the efficiency of supply chain is the ownership of district energy system (DES) center, such as municipal solid waste incinerator (MSWI) plants. The shift of plant ownership and operation from government to a private would generally increase efficiency. Therefore, policy makers should take this into account when deciding the ownerships, operation, and management of the plants [18].

18.2.2.3 Lack of Partnerships

Different motives from political and financial ones could lead to endorsements on opposite sides of a movement. For policy makers, when drafting policy for assisting in the creation of green supply chains, it is difficult to propose a common goal and strategy. Even for people holding similar beliefs and values in a single country, there is also difficulty agreeing on a common direction. For the entire globe to agree on a unified strategy would be more difficult since there are more cultural differences and varying levels of development across continents [17]. This struggle plagues not only the public sector but also the private sector for agreeing on green supply chain strategies.

In addition, the implementation of a climate change strategy is not straightforward or uncontroversial. For instance, South Korea announced its "Four Major Rivers Restoration Project" in 2009 as part of its Green New Deal policy. The ultimate goals of the Four Major Rivers Restoration Project were to (1) combat water scarcity, (2) improve water quality, (3) implement flood control measures, and (4) restore the rivers' ecosystems. However, the opposition decried the project, claiming that it would cause habitat loss, flooding, and a contamination of the water supply. The opposite side argued that their position would benefit the environment while the other would harm it [19].

18.2.2.4 Outdated Infrastructure

An outdated infrastructure, pervasive in both developed and developing countries, would act as obstacles to moving toward construction of green supply chains and creating an effective green economy. Similarly, in the developing countries, without the basic infrastructure such as roads and communication networks, it is difficult to transfer and implement green technologies. As a result, the restructuring of outdated infrastructure is necessary for efficient development and implementation of green supply chains toward green economy. However, governments and/or business may be hesitant to take on such a task since restructuring old infrastructure would require a large input of time and may involve significant costs.

18.2.3 Financial Aspect

For a green supply chain, financial barriers may be embodied by (1) high-capital start-up costs for equipment, (2) inaccurate electricity prices, and (3) pipeline and/or grid interconnection costs. Moreover, the marketplace includes the major challenges of competition with established forms of energy production and appropriate allocation of energy subsidies. In this section, the financial barriers including insufficient incentive, inappropriate allocations of energy subsidies, and inaccurate prices for energy and electricity are illustrated.

18.2.3.1 Insufficient Incentive

Financial incentive for a particular industry and its associated businesses to invest in green technology may not be available at the beginning stage. This may be attributed to a variety of reasons: first, the cost of green technology may be a hard burden, especially in the developing countries and their industries. The high upfront cost of establishing green supply chains may deter institutions from making such a green transition. In the developing countries, the upfront cost may serve as an even greater barrier since they have fewer funds to invest in green technology. As a result, these countries typically continue to be burdened with outdated infrastructures and technologies without sufficient incentive. Second, the payback period for implementing green supply chains, generally between five to ten years, is too long for businesses due to people's natural propensity for risk aversion [6, 20, 21]. Thus, the benefits of green supply chains may not be apparent or immediate enough to incentivize a business, or even the government.

18.2.3.2 Inappropriate Allocations of Energy Subsidies

Subsidies are measures that reduce costs for consumers and producers: (1) keep prices for consumers below market levels, or (2) for producers above market levels. Subsidies are typically provided by the governments to fund popular and mature forms of supply for energy or products. The forms of subsidies include the following:

- Direct regulation and transfers,
- Preferential tax exemptions and rebates,
- Price controls,
- Trade restrictions,
- Public funding, and
- Limits on market access.

However, the material and energy supply industries (such as petroleum and nuclear power) typically obtained a market advantage over other relatively newer industries. In the USA, about half of the government expenditures on energy are from subsidies [22]. Subsidies spread government benefits unevenly and discourage consumers from seeking cleaner alternatives. This also is highly related to the financial and institutional barriers. Eliminations of these subsidies will significantly improve competition in the energy industry and eliminate the unfair advantage given to the nuclear and fossil fuel technologies. Thus, the government should play a central role in the development of new energy industries and green supply chain.

18.2.3.3 Inaccurate Prices for Energy and Electricity

In some countries and their industries, renewable energy prices might be too expensive to be a viable energy option. Research indicated that population with a living budget of even US$10 per day (much higher than the US$1.25 per day that 1.4 billion people in the developing countries) cannot afford renewable energy [4]. As a result, without any policy changes to make renewable energy affordable, these people are forced to use less desirable energy options. Even if the relevant policy is available, there is no guarantee that the developing countries will be able to support these policies. For instance, a major strategy in establishing green supply chain is using subsidies to promote growth in green industries. However, in the least developed and developing countries, governments may not have enough budgets to subsidize to an effective level due to the current high cost of renewable energy and sustainable materials [4].

18.2.4 Technological Aspect

The institutional, regulatory, and financial barriers would further exacerbate technological barriers by preventing the creation of innovative technology. In this section, the technological barriers, including (1) inefficient performance of technology, (2) lack access to green technology, and (3) lack of implementing green practices such as demonstration projects, are illustrated.

18.2.4.1 Inefficient Performance

In particular, in the individual industries and/or power plants, technology barriers often come in the form of implementing the most efficient and environmentally friendly type of technology. For example, for district energy system (DES), certain steam generators used in the incinerator exhibit slow start-up and poor efficiency, thereby generating huge amounts of wastes [23]. This could be overcome by disseminating the state-of-the-art information of innovative technology, and providing appropriate subsidies to the industries.

18.2.4.2 Lack Access to Green Technology

The green technologies can improve efficiency of resource uses and reduce environmental pollutions, leading to a better environment management system toward a green economy [21, 24]. However, local industries and enterprises, especially in the developing countries, still rely on conventional technology and lack access to green technology. Moreover, technological barriers are typically related to the financial and institutional barriers such as resistance of organization to technology advancement adoption due to technological transfer.

18.2.4.3 Lack of Implementing Green Practices and Demonstration Plans

Lack of implementation of innovative green practices and demonstration plans is an important barrier to implement efficient green supply chains. Innovative green practices, such as energy conservation, and reusing and recycling of materials, are essential to achieve a green economy. The innovative green practices are associated with the explicitness of green practices, accumulation of knowledge, organizational encouragement, and quality of human resources. Also, finding appropriate sites for demonstration plans should be a crucial task force to optimize the engineering performance (e.g., overall energy efficiency) and maximize the environmental and economic benefits. For example, district energy systems (DES) should be constructed nearby customers within the region. In constructing eco-industrial parks (EIPs), the steel mill, petrochemical, paper and pulping mill, and cement industries play important roles because of their unique features by utilizing a huge amount of energy and generating a great amount of wastes [25].

18.3 Strategies on Building Green Supply Chain

The barriers, challenges, and strategies for attaining an international green economy have been proposed by many reports, such as the Global Green New Deal [4]. As presented in Table 18.1, the most challenging barriers and strategies for constructing green supply chains are summarized. For instance, eliminations of the unfair subsidies can improve competition in the innovative green industry since unevenly spread of government benefits would discourage consumers from seeking cleaner alternatives and encourage overconsumption of resources.

To overcome the aforementioned barriers in different aspects, it suggests that an effective green supply chain for a green economy should include the following eight key task forces:

- Command and control,
- Economic instruments,
- Information platform,
- Technical assistance,
- Research and development,
- Public and private partnership,
- International collaboration, and
- Environmental education.

For instance, through effective command and control, and environmental education, the National policies could be properly executed under a clear government responsibility. Similarly, the use of command and control, economic instruments, and information platform could internalize the externalities and improve the social

Table 18.1 Potential barriers and overcome strategies for constructing green supply chains

Categories	Barriers and challenges	Strategies
Regulatory	• Loose regulatory laws for green technology allows for greater development and effective implementation • Existing loose environmental regulations and exclusion of CO_2 as regulated pollutants • Long time required for reviewing environmental impact assessment	• Shorten authorization procedures for developing and implementing green technology • New pollutant-targeted regulations (e.g., carbon tax and mandatory energy audits) • Shorter authorization procedures for developing and implementing green technology
Institutional	• Different focuses and concerns between central and local governments • Low level of Bureau of Energy and Environmental Protection Administration in government hierarchy • Information availability of industries due to confidentiality and commercial issues	• Development of networking among central and local governments • Upgrade as Environment and Resource Department • Establishment of networking platform for information exchanges
Financial	• Lack of fund and resource for construction of green supply chain • Low price for utility resources discourages recycling and relatively low costs for waste disposal • Distance between companies inhibits synergies	• Providing economic incentives (e.g., price support, guarantee loans) • Implementation of feed-in tariff (FITs) for green technologies and waste reuse and recycling • Subsidies on development of piping network for renewable energy and district heating and cooling system
Technological	• Lack of own technologies and manufacturing for key components • Existing low energy and material efficiency technologies • Availability of reliable green technologies	• Integration of best available technologies for innovation • Research and development for clean and green technologies • Developments of demonstration plans for providing opportunities for new synergies

acceptance in cooperation with a sound public–private partnership. On the other hand, to achieve the vision and goals, research and development should be enforced with sufficient economic supports using appropriate economic instruments. At the same time, a comprehensive performance evaluation (CPE) program should be established with the support of research and development, technical assistance and international collaboration to assess the performance of green supply chains and promote the environmental education.

To effectively deploy the green supply chains, policy mechanisms can tackle multiple barriers from the aspects of regulatory, institution, finance, and technology at once. In this section, the most important strategies on implementing green supply chains for a green economy, including (1) implementation of National sustainable policy; (2) establishment of government responsibility; (3) provision of economic incentives and price supports; (4) internalization of externalities, social acceptance and investor mobilization; (5) integration of best available technologies for innovation; and (6) development of comprehensive performance evaluation program, are illustrated.

18.3.1 Implementation of National Sustainable Policy

The goal of sustainable economic development is to ensure the daily needs of the people while maximizing the net benefits of economic activities. However, in most developing countries, predominant emphasis was given to achieving rapid economic growth and prioritizing industrial development. Since no countries can be forced to participate in international regulation, two strategies are suggested to make the regulation more globally acceptable:

- Involvement of a regulation context and
- Implementation of green industries by technology-forcing, guaranteed market and economies of scale

On the other hand, to meet the major prerequisite of pursuing sustainable development, the National sustainable policy should be implemented at both the central and local governments. Two important task forces, i.e., (1) establishment of clear visions and missions on sustainable development and (2) promotion of green technology at private sectors and industries, are illustrated as follows.

18.3.1.1 Establishment of Clear Visions and Missions on Sustainable Development

A policy with uncertain goals can result in the negative consequences of a collapsed project. Therefore, well-defined goals and measures are important to make projects feasible. Figure 18.5 illustrates the visions, goals, and strategies of building green supply chains toward a green economy. Governments should put efforts on promoting the development of green industries, cleaner production, and green consumption [26]. Also, the government should place much greater emphasis upon achieving both economic development and environment protection. In addition, the industries themselves should pursue a more balanced economic development, where raising quality takes precedence over expanding quantity. In seeking to satisfy the basic living needs, people should also abide by the moral imperative to coexist and coprosper with other forms of life to maintain the biological diversity [2].

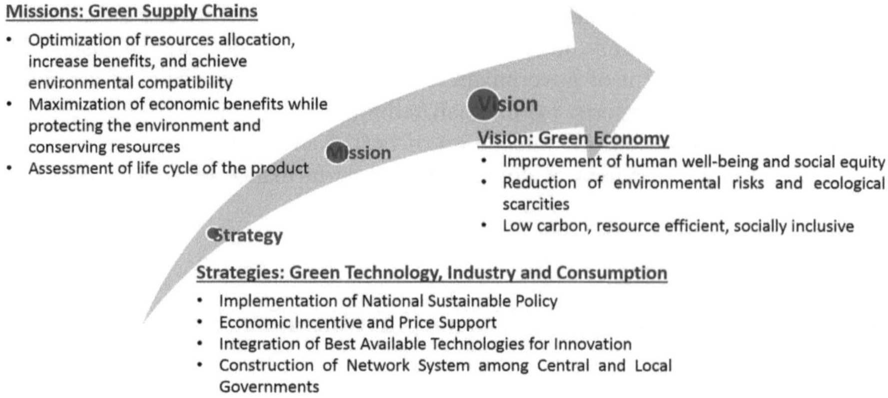

Fig. 18.5 Visions, missions and strategies of a green economy toward sustainable development

Comprehensiveness, consistency, and flexibility must be maintained to obtain success in the policy enforcement. A comprehensive policy should incorporate insight into a wide range of disciplines and account for externalities. Also, a policy needs to be formulated with precautions to handle changes in the political and economic situation. Clear goals and evaluation measures can ensure the consistency of a policy. For policy makers, the concept of consistency is needed to make changes as minimal and as infrequent as possible. A constantly shifting policy would lead to reluctant investors and limited progress during production. Furthermore, in the past, policies were given favorably to the categorical labels, such as financial policy, administrative policy, or social policy. Nowadays, the integrated policy should encompass measures formulated with a multidisciplinary approach. This could be the successful way to accomplish effective green supply chain while paying attention to the multitude of actors in play.

18.3.1.2 Promotion of Green Technology at Private Sectors and Industries

A successful green supply chain among plants can demonstrate its environmental and economic benefits [27]. Development of a sustainable economy seeks to preserve the gains from industrial capitals, including man-made capital, natural capital, and human capital. For the private sectors and industries, the concept of eco-industrial park (EIP) is imperative to facilitate the development of innovative products and green services for upgrading industrial technologies. With a view to promote the marketability of regenerated products, the promotion of green supply chains in industrial parks helps firms to publicize the sale of such products, promote green purchasing, and develop marketing channels for green products. Appropriate policies should be established to foster industrial symbiosis, thereby accelerating the development of green technologies for effectively material reuse and waste recycling [28, 29].

18.3.2 Establishment of Government Responsibility

Mostly, the government is the entity that brings all stakeholders, such as energy companies and the local communities, together through policy formulation. To overcome different barriers in an organized way, a national government would often create a department and equip it with the proper authority. Depending on the scale of the program, a governing body can range from a local government to an international organization (such as the European Union). In the following part, the main responsibilities of government, such as (1) appropriate policy with effective governance, (2) cultivation of green market and enterprise culture, and (3) involvement of stakeholders in policy-making system, are illustrated.

18.3.2.1 Appropriate Policy with Effective Governance

An appropriate policy with governance at both the city and national levels must be adopted to overcome financial, technical, and social barriers. Governance should be steered to direct cities' significant resources of physical, human, natural, and intellectual capital toward a green economy [30]. In the context of green supply chain, actions including agenda in policy development, formulation, adoption, and evaluation should be implemented to reduce the amount of GHG emissions and wastes while generating green products in a profitable manner.

On the other hand, to produce a shift to environmentally cleaner forms of renewable energy, several government measures can be implemented including the following:

- Demand-side management,
- Eliminating conventional subsidies,
- Pricing electricity more accurately,
- Enacting a national feed-in tariff (FIT) mechanism,
- Taxes on pollution, and
- Energy service companies (ESCO).

Municipal authorities can lower costs by linking public investment with the ESCO. The ESCO can establish special funds, credit lines loan guarantee programs, market transformations, and/or grants to address barriers in investments. Several full-scale ESCO models have been established in the developed countries, such as North America. However, no successful ESCO model was found in the developing countries, which might be attributed to the lack of legal and financial policies in place to enforce complex contracts [31].

18.3.2.2 Cultivation of Green Market and Enterprise Culture

It is not appropriate for governments to take on official pricing strategies or policy measures for institutional green transition. Rather, the unofficial measures could be equally effective. As shown in Fig. 18.6, the governments can assist in cultivating green enterprise culture and market environments of green consumption, and encouraging green technological innovation. By publicizing green consumption through publicity, schools, and media outlets, it is possible to widely spread the green knowledge to consumers and push businesses to a greener production. Therefore, environmental education on green economy and green supply chain should be critically promoted.

The ultimate goal of the investment and policy changes is to create a "virtuous cycle." It is noted that all subsidies could be removed in the following decade without hindering the development of green supply chains and a green economy [4]. An initial series of investment and policy changes would facilitate industrial scaling-up, expand markets for green products, and accelerate growth rates in cleaner production. Finally, technological improvements will further accelerate industrial scaling-up.

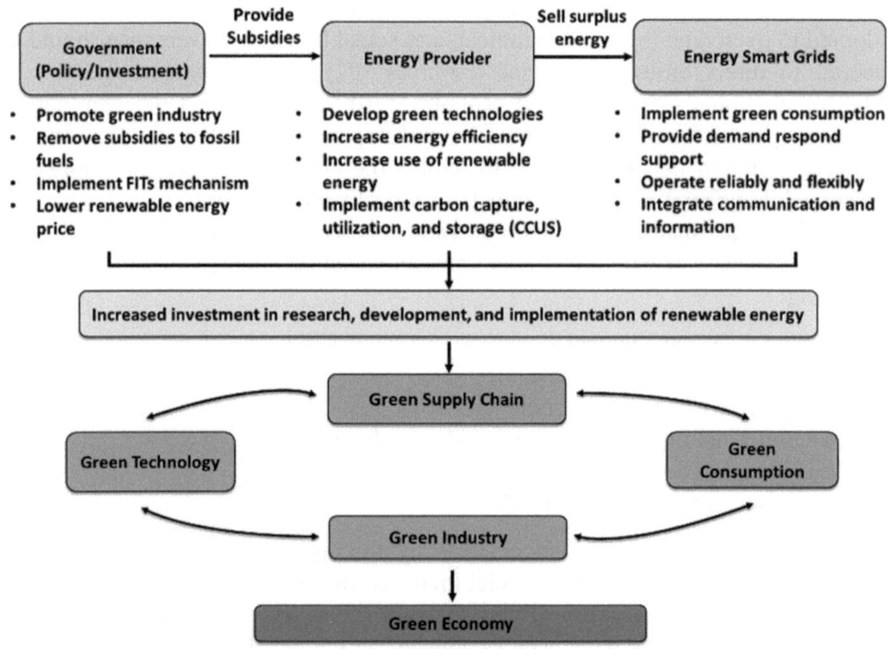

Fig. 18.6 Schematic diagram of price support and governance in cities for building green supply chains toward a green economy. Reprinted by permission from Macmillan Publishers Ltd: Ref. [32], copyright 2015

18.3.2.3 Involvement of Stakeholders in Policy-Making System

The development and implementation of a green supply chain generally involve different stakeholders, such as the owners of plants, a governing body (or competent authorities of plants), local government, industrial sectors, energy supply/distribution companies, communities, and citizens. Social familiarity because of common cultures and trusts among the stakeholders is considered as the vital factor for minimizing the risks and uncertainties of failure. A central governing body (such as the competent authorities of industrial park) should link up all the stakeholders to provide information on regulatory and financial support [29, 33]. Describing and outreaching policy information along with the coordinating collective actions are imperative for a long-term mutually beneficial success among the newly developed projects [34]. An efficient strategic decision-making system for green supply chains should include the following [23]:

- Supply and demand contracts;
- Network configuration such as sourcing, location and capacity of energy production facilities, locations of storage facilities, and network design;
- Ensuring sustainability.

On the other hand, policy makers should identify the specific concerns and issues and even recognize the need for intervention when the market fails [35]. Therefore, appropriate policy measures should be used to ensure compatibility and viability of a project. In some cases, a policy just needs to simply overcome a few local barriers rather than enacting large overbearing financial mechanisms.

18.3.3 Provision of Economic Incentives and Price Supports

Renewable energy plays an inevitable role in a circular industrial economy. In many countries, however, the price of renewable energy is much too high for it to be a viable option of green energy. This is likely due to excess subsidies that lower the prices of fossil fuels so that they are much cheaper than the cost of the renewable energy. These issues can be solved by simply removing those uneven subsidies to allow the international market to determine the price of fossil fuels. It is estimated that eliminating fossil fuel subsidies would reduce global GHG emissions by 6% and increase GDP by 0.1% [36].

Beyond policy in the form of price supports, a series of policy changes will bolster the price supports to be effective. This assistance of pricing strategies would solve the institutional and financial barriers in the developing countries. To create a cost-effective green certificate market, many economic incentives and price support tools can be used, such as (1) feed-in tariff (FIT) mechanism, (2) emission trading scheme, and (3) tax exemptions and rebates.

18.3.3.1 Feed-in Tariff Mechanism

To decrease the price of renewable energy, the primary method suggested by the "Global Green New Deal" report is to provide price supports through a "feed-in tariff (FIT)" mechanism. The FIT mechanism can increase investment in developing renewable energy, thereby increasing the installed capacity of renewable energy. Aside from that, the FIT can offer several benefits including the following [22]:

- Ensuring a stable investment stream for project developers,
- Suppliers getting paid immediately,
- Quickly expanding renewable power, and
- Providing a predictable industry to produce new high-paying jobs.

The FIT mechanism forces electric utilities to purchase renewable power in a nearby service area at a fixed price above market rates for a specific period of time. Germany is one of the greatest success countries for implementing FIT around the world. Its FIT covers the costs of electricity grid interconnection and metering by spreading it across all electricity customers, and then slowly declining the tariff over time.

18.3.3.2 Carbon Pricing System

Aside from the FIT mechanism, the most widely used tool should be the carbon pricing system. Carbon pricing system is to put a cost on the negative externalities generated by non-green technology and makes using those technologies undesirable. Therefore, it could spur on the research and development of green technologies. Normally, carbon pricing system can take the form of a trading scheme or a tax on emissions wherein the right to certain levels of GHG emissions is traded.

1. Emission Trading Scheme (ETS)
   The implementation of the emission trading scheme (ETS) is based on the "cap and trade" principle, where a maximum (cap) is set on the total amount of GHG that can be emitted by all participating installations. "Allowances (or permission)" for GHG emissions will be allocated for free or auctioned off by the authorities and subsequently can be traded among all the participants. Under the "measurable, reportable, and verifiable (MRV)" principle, installations must monitor and report their GHG emissions. After verifying their GHG emissions by third party, installations should hand in enough allowances to authorities to cover their GHG emissions. If emission exceeds the permission, a company must purchase allowances from other participants. Conversely, a company having well performance on emission reduction can sell its leftover credits. The European Union Emissions Trading System (EU-ETS), launched in 2005, was the first large-scale ETS for GHG emissions in the world. However, an issue

with the emission trading is that it lowers the net emissions of an industry but could increase the emissions at a single site [37]. This individual issue should be considered by the policy presiding over that single facility.

2. Tax Exemptions and Rebates
   Tax exemptions and rebates can be implemented in many forms, such as fuel taxes and land development taxes. Carbon taxes can provide large fiscal revenues while lowering carbon emissions by putting a price on pollution emission [38]. Taxes on pollution account for externalities that have traditionally remained unpaid for. Due to the rising price of carbon emissions worldwide, companies would be forced to adopt cleaner technologies. Moreover, a company with overtaxed from pollution costs may be forced to reevaluate its overall efficiency on the use of energy and materials.

18.3.3.3 Other Measures

To promote the renewable energy, electricity can be priced more accurately by other measures:

- Abolishment of "price ceilings": This would allow electricity rates to reflect current market prices instead of fueling excessive consumption, inhibiting investment, and undervaluing energy efficiency.
- Elimination of "declining block-rate pricing (i.e., the per unit price of energy decreases as the energy consumption increases, which is offered to large-scale energy consumers)": this would promote energy efficiency and reduce the consumption of electricity.
- Reflection of "time use (i.e., how the electricity usage varies throughout the day)" in electricity bills: This would adjust customers' consumption with respect to peak and off-peak consumption hours. A consumer will buy electricity in a more efficient manner [22].

18.3.4 Internalization of Externalities, Social Acceptance, and Investor Mobilization

Global environmental issues can be classified into three parts: (1) pollution, (2) biodiversity, and (3) trade-related. Global public environmental issues, such as GHG emissions and ozone depletion, should be addressed by collective actions and binding agreements to avoid "free-rider" problems. With a view to improve social welfare, the concept of environmental externalities, social acceptance, and investor mobilization are introduced in this section.

18.3.4.1 Internalization of Externalities

There are two types of environmental market failure: (1) environmental externalities and (2) environmental degradation. Externalities can cause divergence between social cost (or benefits) and private costs (or benefits). Externalities arise when certain actions of a producer or a consumer have unintended external effects on others. In general, negative externalities (such as pollution) arise when a producer imposes cost on other producers and/or consumers, where the imposer is not charged. Mostly, the probability of human exposure to pollution such as solid residues and wastewater can be effectively mitigated through proper operating procedures. However, airborne emissions are typically no longer controllable once they are released. There has been consensus among public health officials that airborne pollutants from incineration would lead to premature mortality. Airborne emissions comprise a large number of substances that are environmental persistence, long half-life, and inherent toxicity. Even at a low level, they would exhibit severe impacts on environment, ecosystem quality, and human health [39]. As a result, facilities should deploy up-to-date green technologies with appropriate flue gas controls to reduce airborne emissions.

18.3.4.2 Social Acceptance

Since a community is made up of a spectrum of different viewpoints, it cannot be treated as a collective whole. People with different viewpoints may react differently to certain political decisions and scientific information. As a result, these variances should be considered in the decision-making process to complement the policy measures. In general, "cultural theory" can be applied as a heuristic tool to evaluate the public opinions and acceptance on a certain issue. It is noted that even when the information is available to the public, certain factions will remain skeptical about the need to implement green technologies [40]. A study indicated that local attitudes were surprisingly in favor of green supply chains. However, the development of green supply chains was still limited due to the absence of public information, insufficient technology information, incomplete legal framework, and inadequate political decision [41].

18.3.4.3 Investor Mobilization

Prior to green facility installation, investor behavior and mobilization can be evaluated by several factors. Two of the most important parameters affecting the decision making of stakeholders are as follows:

- Awareness of investor [42] and
- Payback period [43].

1. Awareness
   For investors and stakeholders, awareness is the amount of information regarding a certain technology and its associated market [44]. It should be established through the understanding of governance, objectives, targets, business models, technical knowledge, risks, and rewards [45]. Making investors aware of the costs and benefits of a successful green supply chain is a tipping point since lack of awareness could result in a market failure. In particular, in the developing countries, there is lack of consensus on green practices such as promotion of energy efficiency.
2. Payback Period
   Payback period refers to the period of time required to recoup the investments or to reach the break-even point. It is correlated to the probability of an investment being made [46], e.g., a shorter payback time will yield a higher probability of investment. Interestingly, a larger number of recommendations in a preliminary assessment means more work for stakeholders further down the road, thereby leading to a negative influence on the probability of implementation. The neglect of the stakeholders in following through with recommendations can be attributed to a lack of economic incentives.

18.3.5 Integration of Best Available Technologies for Innovation

Technological barriers require comprehensive and integrated strategies that include solutions from the institutional, regulatory, and financial aspects. Simple technologies are available with limited resources, which should make significant improvements in economic efficiency, resource use, and human well-being [30]. The complexity of various technologies for green supply chains, however, may hinder the formation of green supply chains. It thus suggests that the innovation centers should adapt the relevant knowledge to localize the experience on implementation of green technologies. The technology knowledge should be available to policy makers, investors, and communities to support national institutions and serve as a link to international experts and knowledge base.

For optimization of WTE supply chain, several approaches such as bioethanol supply model [47], taxonomy criterion [48], mathematical programming [49], and multiobjective decision making [50] have been employed. Integration of best available technologies (BAT) for innovation can provide opportunities of green technologies and products. With consideration to the life cycle of the production process, the recycling-based technologies should be implemented in industrial parks. In industrial manufacturing processes, the integrated approaches include the following [28, 33]:

- Waste-to-resource and -energy technologies;
- Energy conservation technologies, such as waste heat recovery;
- Cleaner productions for energy, water, and materials;
- Energy-efficient and water-efficient technologies; and
- Carbon capture and utilization (CCU) technologies.

Several important technologies for the constructions of green supply chains are illustrated as follows.

18.3.5.1 Waste-to-Energy (WTE) Technologies

The biosolids from the wastewater treatment plant can be converted to biogas for electricity and heat. Biosolids gasification has been receiving the most attention as viable options for waste-to-energy (WTE), which is capable of providing a clean and manageable process with the possibility of net energy gains [51]. The WTE technologies can convert the biobased wastes into a form of biobased chemicals or energy, which can be used for heating and energy supplies of a district. For building the WTE supply chains, the commonly used technologies used in industrial park are as follows:

- Green fuel pellet [52],
- Bioheating [53],
- Combustion [54] or incineration [55],
- Gasification [51, 56], and
- Anaerobic codigestion [57].

By using the proper technologies, different types of biomass can be converted into various types of bioenergy products, such as biogas, biofuel, and biochar. The suitable feedstock for the WTE supply chain includes the following:

- Agriculture and forestry wastes,
- Energy crops,
- Domestic and household wastes,
- Animal residues, and
- Industrial residues.

As shown in Fig. 18.7, the WTE techniques can be divided into four categories: physical, thermal, chemical, and biological methods. As suggested by USEPA [56], there is significant interest around the globe in developing this technology to commercial scale based on the quantity of research data pertaining to sludge gasification. However, the pulp and paper mill sludge may not be a suitable candidate for gasification due to the high moisture and mineral contents, resulting in low energy values and uneconomical even for a full-scale operation [56].

In this section, the commonly used processes in the WTE supply chains, including green fuel pellet, combustion, gasification, and anaerobic digestion processes, are briefly illustrated.

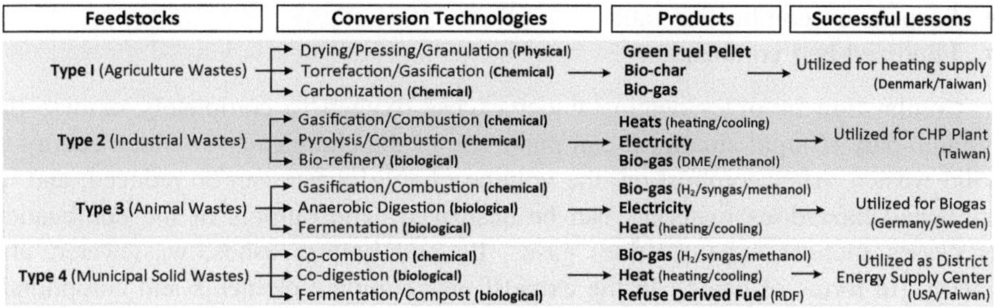

Fig. 18.7 Waste-to-energy (WTE) supply chain for bioenergy utilization. Reprinted by permission from Macmillan Publishers Ltd: ref. [3], copyright 2015

1. Green Fuel Pellet

Green pellet fuels are biofuels made from compressed organic matter of biomass. They are considered as environmentally friendly fuels due to their lower sulfur content and lower pollutant emission than heavy fuel oil in the course of combustion. Wood pellet fuels are the most common type of pellet fuels, generally made from compacted sawdust and related industrial wastes such as lumber, furniture, and construction wastes. The advantages of using green pellet fuels as alternative sources of heating and power include as follows:

- Substantial increase in low heating value (LHV) compared with green chips.
- Reduction in transportation costs.
- Simplified transportation and handling.
- Reduction of biological activity and stable storing.
- Homogeneous manageable fuel for power plants.

According to the life cycle assessment, the energy consumption of wood pellet production was mainly on the manufacturing process ($\sim$71%), followed by its transportation ($\sim$23%) [58]. On the other hand, the solid residues generated from wood pellet combustion (such as bottom ashes) can be used as farmland fertilizers and soil conditioners due to the high contents of calcium, potassium, magnesium, and phosphorus [3, 52]. Another pellet fuel produced by the physical method is called refuse-derived fuel (RDF). The RDF is made from materials that have been sorted out of municipal solid waste streams to exclude non-combustible materials such as glass and metals.

2. Combustion

Combustion is referred as a thermal treatment or an incineration (in the case of municipal solid wastes treatment). The commonly used combustion technologies can be categorized into

- Pile combustion,
- Stoker combustion,

- Suspension combustion, and
- Fluidized-bed combustion.

Combustion involves heating under excess oxygen to completely oxidize the organic part of input stream. It can make use of the chemically bounded energy in solid wastes. After combustion, the volume of solid waste can be reduced, and its contained hazardous materials can be destroyed. The outputs of the combustion processes include exhaust (flue) gases, fly and bottom ashes, wastewater, and energy (in terms of heats). In the exhaust gas, complex elements and compounds can be found: such as N_2, CO, CO_2, NO_x, SO_x, polychlorinated di-benzodioxine, furan, methane, ammonia, hydrochloric acid, and hydrogen fluoride [59]. The emissions of air pollutants could be reduced by various methods such as follows:

- Modifying fuel composition,
- Modifying moisture content of fuel,
- Modifying particle size of fuel, and
- Improving construction chamber shape and incineration application.

Prior to combustion, input solid wastes are often physically altered to increase energy efficiency and decrease emissions. Since the moisture content in municipal solid waste directly affects the efficiency of combustion, the solid waste stream is often processed to ensure an optimal level of moisture content.

3. Gasification

Conventionally, the treatments for biosolids, such as paper and pulp mill sludge and municipal sewage sludge, were landfill, incineration, or land application. Aside from combustion, the sludge can be sent to a gasification process to generate biogas. Gasification can convert organic part of input stream into methane (CH_4), syngas (CO and H_2), and CO_2. It typically can be achieved by reacting the materials at high temperatures (e.g., >700 °C) with a controlled amount of combination of steam, oxygen, and/or nitrogen. The advantages of gasification for sludge treatment include [56]:

- Higher value of versatile end products.
- Availability of the feedstock.
- High efficiency of gasification system.
- Low costs for syngas conversion process.

Appropriate pretreatments on sludge are required if the gasification process is applied. For example, the appropriate moisture content in sludge should be typically between 10 and 20%, which is much lower than those in raw sludge, i.e., 40–99%. After gasification, the syngas, if purified and cleaned, can be further converted to liquid fuels via a catalytic Fischer-Tropsch (FT) process. The produced liquid fuels can be applied in various applications [56] such as follows:

- Feed into an internal combustion engine as transport fuels,
- Feed into generator for electricity production,

- Combusted for heat recovery,
- Used in fuel cell applications, and
- Production of a variety of chemicals.

4. Anaerobic (co-)digestion

Anaerobic digestion is a series of biological processes where microorganisms break down biodegradable components in the absence of oxygen. It is a versatile technology by which a renewable energy in the form of biogas can be produced in the course of microbial decomposition of biosolids. As a result, it can significantly reduce the costs for treating wastes and pollution. After anaerobic digestion, the reacted residues have a fairly homogeneous content with respect to major nutrients such as sodium, phosphate, and potassium, which is beneficial to using as a fertilizer.

Anaerobic digestion of a certain biomass (such as manure) as a sole substrate might not be profitable because of low biogas production and some exploitation problems. To overcome this barrier, codigestion of various complimentary feedstocks has been developed and implemented as a good engineering practice (GEP). The codigestion process could avoid the probabilities of ammonia and lipids from inhibiting the process due to a better nutritional balance [52, 61]. For stable anaerobic digestion operation, the carbon-to-nitrogen (C/N) ratio should range between 20 and 30. In this case, the anaerobic codigestion of municipal sewage sludge with swine manure and poultry manure can achieve a high biogas yield of 400 dm^3 per kg VS [60].

18.3.5.2 Waste Heat Recovery

The heat recovery in the incineration or manufacturing processes not only enhances the use of district heating but also reduces the energy consumption with a better valorization of the waste. The exhaust (waste) heat can be classified into various levels:

- High quality: higher than 500 °C,
- Medium quality: 250–500 °C, and
- Low quality: lower than 250 °C.

The process waste heat could be further utilized to generate electricity and/or steam by various technologies, such as heat exchanger, adsorption chiller, transcritical CO_2 heat pump, refrigeration cycles, and organic Rankine cycle (ORC).

1. Multiple Energy Production System

For waste heat recovery, several mature technologies regarding multiple energy production system can be used in district energy supply [61, 62]:

- Combined heat and power (CHP): known as "cogeneration" and
- Combined cooling, heating and power (CCHP): known as "tri-generation."

By utilizing exhaust heat, both CHP and CCHP boost system efficiency and decrease CO_2 emissions. The principles of CHP and CCHP are similar since they derive energy from a single source. CHP utilizes a heat engine and/or power station to simultaneously generate electricity and available heat. In CHP, the high-temperature heat or steam first drives a gas or steam turbine-powered generator, and then, the resulting low-temperature exhaust heat is used for water or space heating. The moderate temperatures of outlet steam after the CHP process were typically at 100–180 °C, which can be used by the adsorption chillers, and/or refrigerators for cooling demand such as air conditioner. It is noted that a well-designed CHP system could offer an energy efficiency of over 80% [54, 61].

The main difference between CHP and CCHP is that, for CCHP, cooling is one of the desired end products for the customers. Cooling can be generated by a heat pump or absorption chiller using the exhaust heat from process or heat delivered to buildings. Moreover, a great advantage of deploying CCHP systems for energy supply is the flexibility of the system. For instance, in winter, the CCHP can be seen as CHP since there is no demand for air-conditioning in building.

2. Organic Rankine Cycle (ORC) System

Organic Rankine cycle (ORC) power facility has been recently used for exhaust (waste) heat recovery from the flue gas in various industrial processes because of its simplicity, reliability, low maintenance, and easy remote monitoring [63]. The ORC facility can effectively extract low- to medium-grade thermal heat (typically at temperatures of $66 - 260$ °C) in the flue gas for power generation [64–66]. Moreover, it can be operated at low pressures (less than 1380 kPa or 200 psig) [63]. To evaluate the thermodynamic and economic performances, the thermal efficiency and net power output index are frequently used, respectively [67]. In the ORC, the R245fa has been commonly used as the organic working fluid because of its relatively high latent heat of gasification and heat exchange efficiency, and relatively low environmental impacts on ozone depletion and GHG emissions [64]. The boiling point and specific heat of R245fa are 15.1 °C at 1 atm and 0.9369 kJ/(kg·°C) at 30 °C, respectively. The density of R245fa at 30 °C is about 1324.6 kg/m^3.

18.3.5.3 Carbon Capture and Utilization (CCU) Technologies

The implementations of carbon capture and utilization (CCU) technologies should combat the environmental and energy issues in industries for security and sustainability. With the CCU technologies, appropriate value of carbon management mechanism is added into fossil fuel, biomass, and renewable energy. Furthermore, updates in the main infrastructures, such as road and land accessibility, water availability, solid waste disposal, and an electrical grid, should be required for construction of WTE supply chain.

18.3.6 Development of Comprehensive Performance Evaluation Program

To evaluate the performance of a green supply chain, two methods, i.e., cost–benefit analysis (CBA) and life cycle assessment (LCA), are commonly utilized. CBA can be used to estimate the costs and profits associated with a project, while LCA can quantify the environmental impacts and benefits. Aside from the CBA and LCA, several concepts are imperative to carried out a comprehensive performance evaluation (CPE) for establishing green supply chains, including determination of plan-do-check-action (PDCA) principle, key performance indicators (KPIs), and demand-side management (DSM), which are illustrated as follows.

18.3.6.1 Plan-Do-Check-Action (PDCA) Principle

Plan-do-check-action (PDCA) principle was originally suggested by Shewhart [68] and could be implemented in the design of comprehensive performance evaluation as illustrated in Fig. 18.8. Establishing commercialized (or business) models should be essential to demonstrate and evaluate the performance of innovative green technologies. For example, the incineration plant can be integrated with the steam cooking system to form a district energy supply center. Biosolid wastes from the

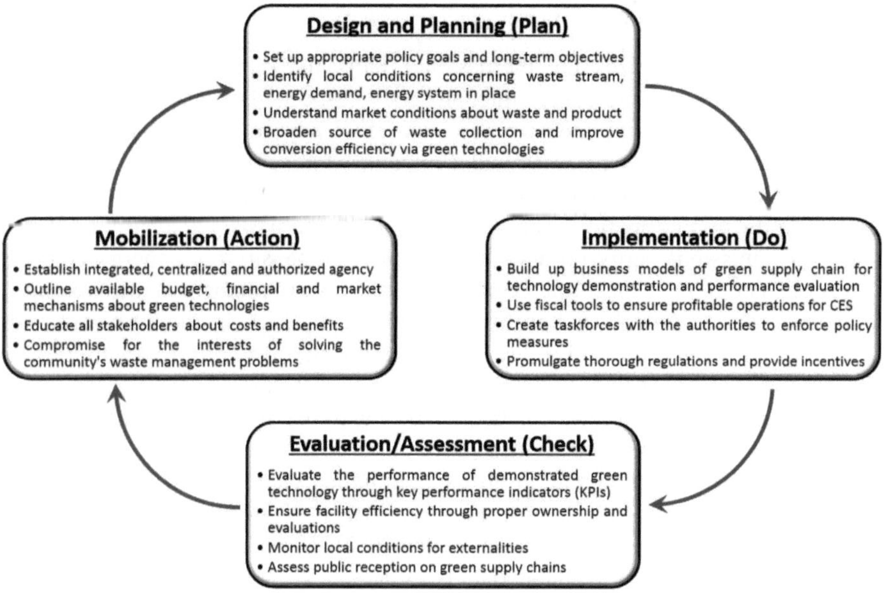

Fig. 18.8 Policy-making cycle for establishment and implementation of green supply chains

large and/or small industrial plants could be utilized as the energy source for power generation and/or heating purpose, respectively. After performance evaluations, policy makers and stakeholders can compare the operation results with their original goals and properly revise it if needed. Through revisions and amendments, a policy can be adjusted in light of new information. The above procedure is consistent with the management method of PDCA cycle.

18.3.6.2 Key Performance Indicators (KPIs)

The CPE programs are required to assess whether the policy factors are successful at achieving preset goals. Since a policy acts on a community in a multifaceted way, performance evaluations should be conducted in a diverse approach. The CPEs should be performed for at least three times: (1) before a policy is enacted to ensure proper planning, (2) during the process of implementation to ensure optimal function, and (3) after goals have been met to ensure insight into future improvements. For the CPE, key performance indicators (KPIs) must be established for evaluating the progress toward the implementation of green supply chains. The KPIs are quantifiable measures of an institution's ability to accomplish their set goals. As suggested by UNEP, three primary areas act as the most beneficial KPIs when measuring various aspects of green economies [1]:

- Indicators of resource efficiency (engineering aspect),
- Economic transformation (economic aspect), and
- Human progress and well-being (social aspect).

Table 18.2 summarizes the themes and KPIs for establishing green supply chains from environmental, economic, and social aspects. For instance, the KPIs in the environmental aspect measure resource efficiency of a green supply chain. Because the scales of land areas and companies for various industrial parks are quite different, the performance of each industrial park should be compared via annual production of energy or economic values as the basis, e.g., carbon intensity (i.e., CO_2 emission/energy production) and energy intensity (i.e., energy/GDP).

For the economic aspect, the economic transformation indicators often assign a monetary value to the cost and profits of greening strategies, including investments, jobs, and industrial growth. The levels of investments made in green activities can be compared with that in environmentally harmful activities. Moreover, economic performance of green supply chains can be measured by the growths of goods, services, and jobs in green activities [69].

For the social aspect, human progress, community development, and well-being indicators are suitable for gauging the performance of green supply chains since they consider if the economic development goal of sustainable development is fulfilled. However, when available data is sufficient, the indicators are often seen as

Table 18.2 Themes and key performance indicators (KPIs) from environmental, economic, and social aspects

Aspects	Themes	Indicators
Environmental	Pollution prevention and control (per unit output value increase)	Air pollutant emissions such as VOCs, NO_x, SO_2, and particles
		Wastewater discharge
		COD emissions
		Solid waste generation
		Target for CO_2 reduction
	Energy and resource consumption (per GDP)	Land consumption
		Energy consumption
		Freshwater consumption
		Chemicals consumption
	Energy and resource recycling	Energy-saving efficiency
		Ratio of reclaimed industrial wastewater
		Water consumption per unit output value
		Material consumption per unit output value
		Waste recycling ratio per unit output value
	Environmental planning and management	Averaged pollution standard index (PSI)
		Ratio of green land
		Green building indicators
		Environmental management system (EMS)
		Sustainable material management (SMM) system
Economic	Cost reduction and clean production	Measures for promoting pollution prevention and resource recovery
		Cost reduction of CO_2 emission control by waste recycling
		Ratio of material shipping expense in the total output value
	Profit increase and green consumption	Gross domestic production (GDP)
		Gross industrial output value (GIOV)
		Industrial added value (IAV)
		Discounted cash flow
	Tax	Carbon tax
		Fuel tax
		Pollution tax
	Incentive and pricing support	Feed-in tariff on renewable energy (or green technology)
		Government subsidy on construction
		Credit lines loan guarantee
	Corporate image promotion and green industry	Budget/expenditure of environmental protection
		Total investment for pollution control

(continued)

Table 18.2 (continued)

Aspects	Themes	Indicators
Social	Public participation and acceptance	Number of visitors in open house events per year
		Completeness of message platform
		Publication of environmental report
		Public satisfaction of environment
		Public cognition of eco-industrial park
	Community development	Interchange plan for public transportation system
		Plan for biking and walking route
		Social familiarity
		Betweenness centralization
		Density average distance
	Fairness and justice	Green park area per capital
		Number of pleaded environmental pollution events
		Compliance with laws and regulations
	Population and health	Safety nets
		Health status

less legitimate than the key economic indicators such as GDP for making policy decisions.

18.3.6.3 Demand-Side Management (DSM)

Demand-side management (DSM), or demand-side response, programs are implemented to change the consumption pattern of consumers, such as the behavior of a household. The goal of the DSM is to encourage the consumer to use less energy during peak hours, or to move the time of energy use to off-peak times (such as nighttime and weekends). A successful DSM program should comprise marketing strategies with multiple approaches, such as follows [22]:

- Programs targeting specific audiences,
- Technical assistance for customers,
- Simple program procedures for customers to estimate potential benefits, and
- Financial incentives to attract attention and reduce initial costs.

Similarly, large-scale energy plants also should be located near their source of heat demand to maximize the overall energy efficiency [70]. Also when changes occur, the performance of a plant and its associated energy distribution network should be able to predict [35]. In catering to green supply chain, a cost–benefit analysis can provide an estimate on the direct costs of operating and maintenance and the fixed costs to optimize the design of green supply chain [10].

18.3.6.4 Life Cycle Assessment (LCA)

Life cycle assessment (LCA) is an evaluation of potential environmental impacts for a production system or service throughout its life cycle by compiling all inputs and outputs (e.g., material, energy, and pollutants). LCA was originally used in product analysis, and it recently has been widespread in analyses of pollution control facility and/or environmental engineering areas, such as follows:

- Waste management system [31, 71],
- Incineration facility [39, 72], and
- Carbon footprint in industrial park [73].

Environmental impacts evaluated through LCA can be associated with analysis of material and energy consumptions [74] and stakeholder involvement [14]. On the other hand, the decisions from human health risk assessment need to be supported by strong scientific evidences, and the extent of uncertainties in assessments should be carefully determined. Scenario evaluation by LCA can be used to estimate exposure levels in humans, with the consideration to the time of contact and the sources of hazardous materials. Another tool for approximating actual human exposure levels to a pollutant of interest is biomonitoring.

18.4 Implementation of WTR Green Supply Chains: Case Study

To achieve the goals of being environmentally bearable, economic viable, and social equitable, "building a green supply chain within industrial park" should be extensively promoted to make traditional industries around the world. It is noted that development of eco-industrial parks (EIPs) can simultaneously achieve the environmental protection, economic development, and social equity.

18.4.1 Eco-Industrial Parks (EIPs) as a Business Model

To meet the demands of a circular economy, eco-industrial parks (EIPs) have been extensively established in different regions, such as Australia [75], Denmark [76], Europe [12, 77], USA [34], Japan [29], Mainland China [28, 78, 79], Korea [33], and Taiwan [26, 80]. The definition of the EIPs can be found in the literature [81]:

> community of manufacturing and services companies seeking enhanced environmental and economic performance through collaboration in managing environmental and resources issues including energy, water, and materials.

The EIPs are to promote energy conservation/efficiency, carbon reduction, and green production by the implementation of green supply chains. The objectives of EIPs are to

- Establish an integrated framework that embraces economic development, environmental quality, and social equity,
- Stimulate investments in the private sector, increasing employment opportunities related to resources recycling and encouraging rural and urban community developments,
- Build up a recycling-based sustainable society to achieve the goals of zero emissions,
- Manage waste reduction and reuse technologies to achieve goals of total recovery and zero waste, and
- Build recycling-based ecocities and/or ecovillages, raise resource-recycling ratios, and reduce water and energy consumptions.

To meet the objectives of EIPs, five strategies are suggested as follows:

- Policy makers should create policy for simultaneously reducing GHG emissions and improving energy efficiency.
- Action plans should increase manufacturing efficiency while seeking synergetic cooperation between all manufacturers in the industrial park.
- Creation of a cost-effective integrated green certificate market by implementation of pricing instruments, such as tax exemptions and carbon credits.
- Crucial information should be made easily accessible including the following:
  - Updated manufacturing processes,
  - Supply and demand of materials and energy,
  - Resources for assistance, and
  - Human training resources.
- Life cycle analysis (LCA) should be utilized as a structured basis for evaluating the performance of environmental impacts and benefits in EIPs.

18.4.2 Iron and Steel Industry

In the case of iron and steel industry, China Steel Corp. (CSC) in Lin-Hai Industrial Park (Kaohsiung, Taiwan) has successfully established the business model and served as the center of green supply chain since 2008. Figure 18.9 shows the schematic diagram of construction of the green supply chains in the Lin-Hai Industrial Park. Lin-Hai Industrial Park consists of a total of 482 manufacturers in the fields of mechatronics, steel manufacturing, chemical engineering, and transportation. By the end of 2012, a total of 15 green supply chains including steam, hydrogen, nitrogen, waste alkaline solution, incinerator bottom ash, and electric arc furnace dust were established. For instance, the alkaline solid wastes can be used for carbonation process to react with flue gas CO_2 to form stable carbonate precipitates [82]. Meanwhile, the physicochemical properties of the carbonated solid waste can be upgraded since the free-CaO content is eliminated, which is beneficial to the application as construction materials [25, 83]. Moreover, the alkaline

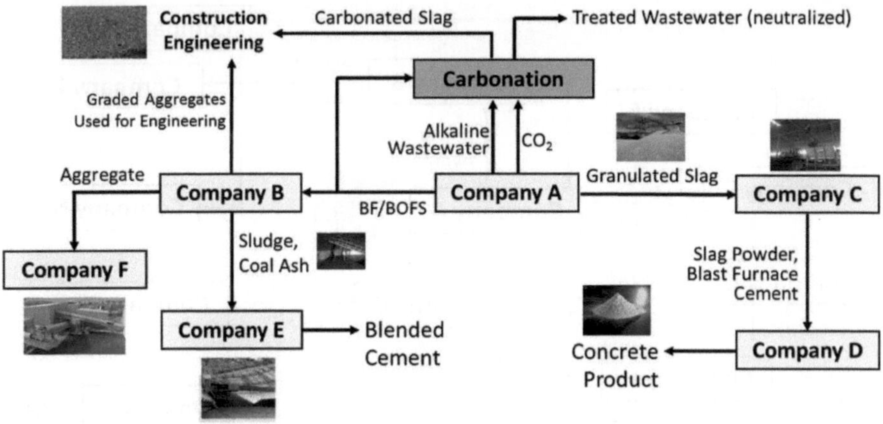

Fig. 18.9 Conceptual diagram of green supply chains in the case of alkaline solid wastes in the Lin-Hai Industrial Park (Taiwan)

wastewater, if available, can be introduced in the carbonation reaction, and the wastewater can be neutralized after reaction.

From this green supply chain model, the total amount of steam supply was estimated at about 2.5 Mt/y. As a result, the environmental benefits of steam supply include the following: (1) a CO_2 reduction of 574,000 t/y, (2) a SO_x reduction of 1830 t/y, (3) a NO_x reduction of 1270 t/y, and (4) particle matter (PM) reduction of 180 t/y. On the other hand, the total amount of recycling wastes was determined at 0.67 Mt/y, corresponding to a waste utilization ratio of 84.7%. Accordingly, the total economic profits attributed by the green supply chains was estimated to be US$ 100 million per year.

18.4.3 Petrochemical Industry

Lin-Yuan Industrial Park (Kaohsiung, Taiwan) comprises a total of 30 industries, where 27 of them are petrochemical related industries including Formosa Plastic Corp. and China Petroleum Corp. Since 1992, Formosa Plastic Corp. had been served as the district energy supply (DES) center for many companies in the industrial park. Figure 18.10 shows the schematic diagram of construction of the green supply chains in the Lin-Yuan Industrial Park. For example, the Formosa Plastic Corp. utilized the CHP technology to generate electricity and heat (i.e., steam), where four boilers with a steam capacity of 200 tons/h were installed. The exhaust heat with different qualities was used to simultaneously generate electricity, steam, and hot water. The high-quality steam ($\sim$3.5 kg/cm^2) was utilized to drive the steam turbine for electricity generation and median-quality steam supply. Similarly, the low quality steam (1 kg/cm^2) was used not only to recycle the chilled water for air-cooled heat exchanger but also generate hot gas ($\sim$105 °C) delivering

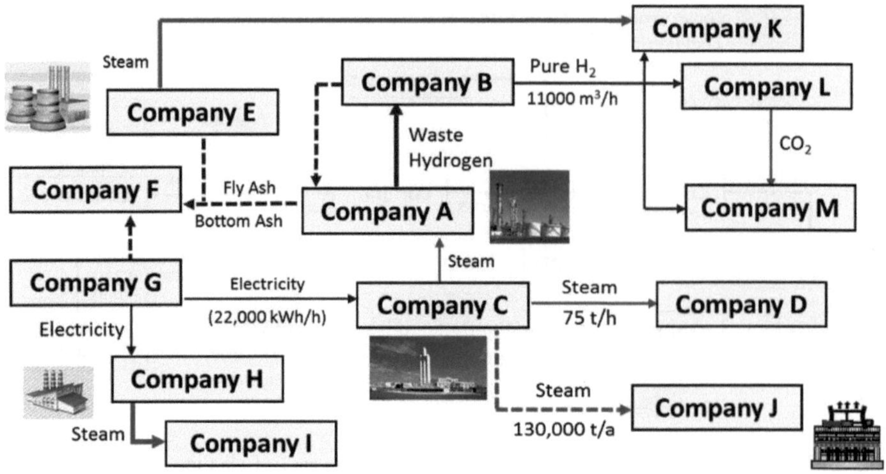

Fig. 18.10 Schematic diagram of construction of the green supply chains in the Lin-Yuan Industrial Park (Taiwan)

to another plant for high-density polyethylene (HDPE) production. With the aforementioned energy integration system, the overall heat efficiency increased up to 60.5% because of the district steam supply. Furthermore, the industry waste gas supply (such as hydrogen) was estimated at 8600 ton per year.

In this industrial park, both the rainwater and wastewater are recycled into the manufacturing process for reuse. The water recycling technologies, including the membrane bioreactor (MBR), ultrafiltration (UF) and reverse osmosis (RO) processes, can be used to purify the wastewater to the acceptable levels of turbidity (<0.2 NTU) and total suspended solid (<1 mg/L). Also the large molecules and ions in the wastewater can be removed though the RO process.

Since 2012, a total of seven green supply chains (including electricity, steam, hot water, hydrogen and bottom ash) have been established with a potential amount of 38,000 tons per year. The environmental benefits of steam supply include (1) a CO_2 reduction of 32,300 ton per year, (2) a SO_x reduction of 370 ton per year, and (3) a NO_x reduction of 160 ton per year. Accordingly, the economic profits in the Lin-Yuan Industrial Park were estimated to be US$5.3 million per year.

References

1. UNEP (2012) Measuring progress towards an Inclusive green economy. UNEP
2. CEPD (2004) Taiwan agenda 21: vision and strategies for national sustainable development. Council for Economic Planning and Development (CEPD), Executive Yuan, Taiwan
3. Pan S-Y, Du MA, Huang IT, Liu IH, Chang EE, Chiang P-C (2015) Strategies on implementation of waste-to-energy (WTE) supply chain for circular economy system: a review. J Clean Prod 108:409–421. doi:10.1016/j.jclepro.2015.06.124

4. UNDESA (2009) A global green new deal for climate, energy, and development. United Nations Department of Economic and Social Affairs
5. Jupesta J, Boer R, Parayil G, Harayama Y, Yarime M, Oliveira JAPd, Subramanian SM (2011) Managing the transition to sustainability in an emerging economy: evaluating green growth policies in Indonesia. Environ Innov Societal Transitions 1(2):187–191. doi:10.1016/j.eist.2011.08.001
6. UNEP (2011) Towards a green economy: pathways to sustainable development and poverty eradication—a synthesis for policy makers
7. Tukker A (2013) Product services for a resource-efficient and circular economy—a review. J Clean Prod. doi:10.1016/j.jclepro.2013.11.049
8. Su B, Heshmati A, Geng Y, Yu X (2013) A review of the circular economy in China: moving from rhetoric to implementation. J Clean Prod 42:215–227. doi:10.1016/j.jclepro.2012.11.020
9. Geng Y, Fu J, Sarkis J, Xue B (2012) Towards a national circular economy indicator system in China: an evaluation and critical analysis. J Clean Prod 23(1):216–224. doi:10.1016/j.jclepro.2011.07.005
10. Jamasb T, Nepal R (2010) Issues and options in waste management: a social cost–benefit analysis of waste-to-energy in the UK. Resour Conserv Recycl 54(12):1341–1352. doi:10.1016/j.resconrec.2010.05.004
11. Saha S, Roy TB (2011) Assessment of the status of solid waste management in mega cities in India: an overview. Int J Agric Environ Biotechnol 4(4):305–315
12. ECORYS (2010) Assessment of non-cost barriers to renewable energy growth in EU Member. Rep. Rotterdam
13. De Lange WJ, Stafford WH, Forsyth GG, Le Maitre DC (2012) Incorporating stakeholder preferences in the selection of technologies for using invasive alien plants as a bio-energy feedstock: applying the analytical hierarchy process. J Environ Manage 99:76–83. doi:10.1016/j.jenvman.2012.01.014
14. Huttunen S, Manninen K, Leskinen P (2014) Combining biogas LCA reviews with stakeholder interviews to analyse life cycle impacts at a practical level. J Clean Prod 80:5–16. doi:10.1016/j.jclepro.2014.05.081
15. Matos S, Silvestre BS (2013) Managing stakeholder relations when developing sustainable business models: the case of the Brazilian energy sector. J Clean Prod 45:61–73. doi:10.1016/j.jclepro.2012.04.023
16. Steiner A (2010) Eleventh annual grotius lecture series: focusing on the good or the bad: what can international environmental law do to accelerate the transition towards a green economy? Am Univ Int Law Rev 25(5):843–875
17. Keating M (2009) Africa and climate change: with one voice. World Today 65(10):10–11
18. Chen YT, Chen CC (2012) The privatization effect of MSW incineration services by using data envelopment analysis. Waste Manag 32(3):595–602. doi:10.1016/j.wasman.2011.11.007
19. Normile D (2010) Restoration or devastation? Science 26:1556–1570
20. Leichenko RM, O'Brien KL, Solecki WD (2010) Climate change and the global financial crisis: a case of double exposure. Ann Assoc Am Geogr 100(4):963–972. doi:10.1080/00045608.2010.497340
21. van Loon-Steensma JM, Schelfhout HA, Vellinga P (2014) Green adaptation by innovative dike concepts along the Dutch Wadden Sea coast. Environ Sci Policy 44:108–125. doi:10.1016/j.envsci.2014.06.009
22. Sovacool BK (2009) The importance of comprehensiveness in renewable electricity and energy-efficiency policy. Energy Policy 37(4):1529–1541. doi:10.1016/j.enpol.2008.12.016
23. Iakovou E, Karagiannidis A, Vlachos D, Toka A, Malamakis A (2010) Waste biomass-to-energy supply chain management: a critical synthesis. Waste Manag 30(10):1860–1870. doi:10.1016/j.wasman.2010.02.030
24. Styles D, O'Brien P, O'Boyle S, Cunningham P, Donlon B, Jones MB (2009) Measuring the environmental performance of IPPC industry: I. Devising a quantitative science-based and policy-weighted environmental emissions index. Environ Sci Policy 12(3):226–242. doi:10.1016/j.envsci.2009.02.003

25. Pan S-Y, Chang EE, Chiang P-C (2012) CO_2 capture by accelerated carbonation of alkaline wastes: a review on its principles and applications. Aerosol Air Qual Res 12:770–791. doi:10.4209/aaqr.2012.06.0149
26. Huang Y-H, Wu J-H (2013) Analyzing the driving forces behind CO_2 emissions and reduction strategies for energy-intensive sectors in Taiwan, 1996–2006. Energy 57:402–411. doi:10.1016/j.energy.2013.05.030
27. Shih SM, Wang S, Zhang CC, Wang TF, Chiang PC, Ji CJ (2004) Taiwan 21 agenda. Taiwan (ROC)
28. Liu W, Tian J, Chen L (2014) Greenhouse gas emissions in China's eco-industrial parks: a case study of the Beijing economic technological development area. J Clean Prod 66:384–391. doi:10.1016/j.jclepro.2013.11.010
29. Dong H, Ohnishi S, Fujita T, Geng Y, Fujii M, Dong L (2014) Achieving carbon emission reduction through industrial & urban symbiosis: a case of Kawasaki. Energy 64:277–286. doi:10.1016/j.energy.2013.11.005
30. Puppim de Oliveira JA, Doll CNH, Balaban O, Jiang P, Dreyfus M, Suwa A, Moreno-Peñaranda R, Dirgahayani P (2013) Green economy and governance in cities: assessing good governance in key urban economic processes. J Clean Prod 58:138–152. doi:10.1016/j.jclepro.2013.07.043
31. Assamoi B, Lawryshyn Y (2012) The environmental comparison of landfilling vs. incineration of MSW accounting for waste diversion. Waste Manag 32(5):1019–1030. doi:10.1016/j.wasman.2011.10.023
32. Li J, Pan SY, Kim H, Linn JH, Chiang PC (2015) Building green supply chains in eco-industrial parks towards a green economy: Barriers and strategies. J Environ Manage 162:158–170. doi:10.1016/j.jenvman.2015.07.030
33. Behera SK, Kim J-H, Lee S-Y, Suh S, Park H-S (2012) Evolution of 'designed' industrial symbiosis networks in the Ulsan Eco-industrial Park: 'research and development into business' as the enabling framework. J Clean Prod 29–30:103–112. doi:10.1016/j.jclepro.2012.02.009
34. Yi H (2014) Green businesses in a clean energy economy: Analyzing drivers of green business growth in U.S. states. Energy 68:922–929. doi:10.1016/j.energy.2014.02.044
35. White W, Lunnan A, Nybakk E, Kulisic B (2013) The role of governments in renewable energy: the importance of policy consistency. Biomass Bioenergy 57:97–105. doi:10.1016/j.biombioe.2012.12.035
36. Steiner A (2010) Focusing on the good or the bad: what can international environmental law do to accelerate the transition towards a green economy. Am U Int'l L Rev 843:843–875
37. IEA (2009) Cogeneration and district energy: sustainable energy technologies for today and tomorrow. OECD, International Energy Agency, France
38. Carraro C, Favero A, Massetti E (2012) Investments and public finance in a green, low carbon, economy. Energy Econ 34:S15–S28. doi:10.1016/j.eneco.2012.08.036
39. Reis MF (2011) Solid waste incinerators: health impacts. Institute of Preventive Medicine, University of Lisbon
40. West J, Bailey I, Winter M (2010) Renewable energy policy and public perceptions of renewable energy: a cultural theory approach. Energy Policy 38(10):5739–5748. doi:10.1016/j.enpol.2010.05.024
41. Achillas C, Vlachokostas C, Moussiopoulos N, Banias G, Kafetzopoulos G, Karagiannidis A (2011) Social acceptance for the development of a waste-to-energy plant in an urban area. Resour Conserv Recycl 55(9–10):857–863. doi:10.1016/j.resconrec.2011.04.012
42. Afroz R, Masud MM, Akhtar R, Duasa JB (2013) Survey and analysis of public knowledge, awareness and willingness to pay in Kuala Lumpur, Malaysia—a case study on household WEEE management. Jf Cleaner Prod 52:185–193. doi:10.1016/j.jclepro.2013.02.004
43. Abdelaziz EA, Saidur R, Mekhilef S (2011) A review on energy saving strategies in industrial sector. Renew Sustain Energy Rev 15(1):150–168. doi:10.1016/j.rser.2010.09.003

44. Polanec B, Aberšek B, Glodež S (2013) Informal education and awareness of the public in the field of waste management. Proc Soc Behav Sci 83:107–111. doi:10.1016/j.sbspro.2013.06.021
45. Hawkey D, Webb J, Winskel M (2013) Organisation and governance of urban energy systems: district heating and cooling in the UK. J Clean Prod 50:22–31. doi:10.1016/j.jclepro.2012.11.018
46. Abadie LM, Ortiz RA, Galarraga I (2012) Determinants of energy efficiency investments in the US. Energy Policy 45:551–566. doi:10.1016/j.enpol.2012.03.002
47. Avami A (2013) Assessment of optimal biofuel supply chain planning in Iran: technical, economic, and agricultural perspectives. Renew Sustain Energy Rev 26:761–768. doi:10.1016/j.rser.2013.06.028
48. Sharma B, Ingalls RG, Jones CL, Khanchi A (2013) Biomass supply chain design and analysis: basis, overview, modeling, challenges, and future. Renew Sustain Energy Rev 24:608–627. doi:10.1016/j.rser.2013.03.049
49. De Meyer A, Cattrysse D, Rasinmäki J, Van Orshoven J (2014) Methods to optimise the design and management of biomass-for-bioenergy supply chains: a review. Renew Sustain Energy Rev 31:657–670. doi:10.1016/j.rser.2013.12.036
50. Cambero C, Sowlati T (2014) Assessment and optimization of forest biomass supply chains from economic, social and environmental perspectives—a review of literature. Renew Sustain Energy Rev 36:62–73. doi:10.1016/j.rser.2014.04.041
51. Adams PWR, McManus MC (2014) Small-scale biomass gasification CHP utilisation in industry: energy and environmental evaluation. Sustain Energy Technol Assessments 6:129–140. doi:10.1016/j.seta.2014.02.002
52. Pawelzik PF, Zhang Q (2012) Evaluation of environmental impacts of cellulosic ethanol using life cycle assessment with technological advances over time. Biomass Bioenergy 40:162–173. doi:10.1016/j.biombioe.2012.02.014
53. Hagos DA, Gebremedhin A, Zethraeus B (2014) Towards a flexible energy system—a case study for Inland Norway. Appl Energy 130:41–50. doi:10.1016/j.apenergy.2014.05.022
54. Maraver D, Sin A, Sebastián F, Royo J (2013) Environmental assessment of CCHP (combined cooling heating and power) systems based on biomass combustion in comparison to conventional generation. Energy 57:17–23. doi:10.1016/j.energy.2013.02.014
55. Pavlas M, Touš M, Klimek P, Bébar L (2011) Waste incineration with production of clean and reliable energy. Clean Technol Environ Policy 13(4):595–605. doi:10.1007/s10098-011-0353-5
56. USEPA (2012) Aqueous sludge gasification technologies. Greenhouse Gas Technology Center, U.S. Environmental Protection Agency, USA
57. Pagés Díaz J, Pereda-Reyes I, Taherzadeh MJ, Sárvari-Horváth I, Lundin M (2014) Anaerobic co-digestion of solid slaughterhouse wastes with agro-residues: synergistic and antagonistic interactions determined in batch digestion assays. Chem Eng J 245:89–98. doi:10.1016/j.cej.2014.02.008
58. Mahalle L (2013) Comparative life cycle assessment of pellet, natural gas and heavy fuel oil as heat energy sources. FPInnovations, British Columbia
59. Tabasová A, Kropáč J, Kermes V, Nemet A, Stehlík P (2012) Waste-to-energy technologies: impact on environment. Energy 44(1):146–155. doi:10.1016/j.energy.2012.01.014
60. Borowski S, Domanski J, Weatherley L (2014) Anaerobic co-digestion of swine and poultry manure with municipal sewage sludge. Waste Manag 34(2):513–521. doi:10.1016/j.wasman.2013.10.022
61. Rezaie B, Rosen MA (2012) District heating and cooling: review of technology and potential enhancements. Appl Energy 93:2–10. doi:10.1016/j.apenergy.2011.04.020
62. Wu DW, Wang RZ (2006) Combined cooling, heating and power: a review. Prog Energy Combust Sci 32(5–6):459–495. doi:10.1016/j.pecs.2006.02.001

63. Chambers T, Raush J, Russo B (2014) Installation and operation of parabolic trough organic Rankine cycle solar thermal power plant in South Louisiana. Energy Proc 49:1107–1116. doi:10.1016/j.egypro.2014.03.120
64. Peris B, Navarro-Esbrí J, Molés F, Collado R, Mota-Babiloni A (2015) Performance evaluation of an organic Rankine cycle (ORC) for power applications from low grade heat sources. Appl Thermal Eng 75:763–769. doi:10.1016/j.applthermaleng.2014.10.034
65. Guo C, Du X, Yang L, Yang Y (2015) Organic Rankine cycle for power recovery of exhaust flue gas. Appl Thermal Eng 75:135–144. doi:10.1016/j.applthermaleng.2014.09.080
66. Imran M, Park BS, Kim HJ, Lee DH, Usman M, Heo M (2014) Thermo-economic optimization of regenerative organic Rankine cycle for waste heat recovery applications. Energy Convers Manag 87:107–118. doi:10.1016/j.enconman.2014.06.091
67. Yang M-H, Yeh R-H (2015) Thermodynamic and economic performances optimization of an organic Rankine cycle system utilizing exhaust gas of a large marine diesel engine. Appl Energy 149:1–12. doi:10.1016/j.apenergy.2015.03.083
68. Shewhart WA (1930) Economic control of quality of manufactured product/50th anniversary commemorative issue. Paper presented at the American Society for Quality December 1980
69. UNEP (2012) Green economy: metrics and indicators
70. Longden D, Brammer J, Bastin L, Cooper N (2007) Distributed or centralised energy-from-waste policy? Implications of technology and scale at municipal level. Energy Policy 35(4):2622–2634. doi:10.1016/j.enpol.2006.09.013
71. Kirkeby JT, Birgisdottir H, Hansen TL, Christensen TH, Bhander GS, Hauschild M (2006) Environmental assessment of solid waste systems and technologies: EASEWASTE. Waste Manag Res 24(1):3–15. doi:10.1177/0734242x06062580
72. Riber C, Bhander GS, Christensen TH (2008) Environmental assessment of waste incineration in a life-cycle-perspective (EASEWASTE). Waste Manag Res 26(1):96–103. doi:10.1177/0734242x08088583
73. Dong H, Geng Y, Xi F, Fujita T (2013) Carbon footprint evaluation at industrial park level: a hybrid life cycle assessment approach. Energy Policy 57:298–307. doi:10.1016/j.enpol.2013.01.057
74. Damgaard A, Riber C, Fruergaard T, Hulgaard T, Christensen TH (2010) Life-cycle-assessment of the historical development of air pollution control and energy recovery in waste incineration. Waste Manag 30(7):1244–1250. doi:10.1016/j.wasman.2010.03.025
75. Giurco D, Bossilkov A, Patterson J, Kazaglis A (2011) Developing industrial water reuse synergies in Port Melbourne: cost effectiveness, barriers and opportunities. J Clean Prod 19 (8):867–876. doi:10.1016/j.jclepro.2010.07.001
76. Münster M, Morthorst PE, Larsen HV, Bregnbæk L, Werling J, Lindboe HH, Ravn H (2012) The role of district heating in the future Danish energy system. Energy 48(1):47–55. doi:10.1016/j.energy.2012.06.011
77. Pardo N, Vatopoulos K, Riekkola AK, Perez A (2013) Methodology to estimate the energy flows of the European Union heating and cooling market. Energy 52:339–352. doi:10.1016/j.energy.2013.01.062
78. Jung S, Dodbiba G, Chae SH, Fujita T (2013) A novel approach for evaluating the performance of eco-industrial park pilot projects. J Clean Prod 39:50–59. doi:10.1016/j.jclepro.2012.08.030
79. Tian J, Liu W, Lai B, Li X, Chen L (2014) Study of the performance of eco-industrial park development in China. J Clean Prod 64:486–494. doi:10.1016/j.jclepro.2013.08.005
80. Cheng Loong Corp (2012) Establishment of green supply chains in Cheng Loong Corp. In: Chiang P-C (ed) Forum on promoting integrated energy and resource supply chain. Ministry of Economic Affairs (MOEA), Taiwan (R.O.C.)
81. Côté RP, Cohen-Rosenthal E (1998) Designing eco-industrial parks: a synthesis of some experiences. J Clean Prod 6:181–188

82. Pan S-Y, Chiang A, Chang E-E, Lin Y-P, Kim H, Chiang P-C (2015) An innovative approach to integrated carbon mineralization and waste utilization: a review. Aerosol Air Qual Res 15:1072–1091. doi:10.4209/aaqr.2014.10.02
83. Chang EE, Pan SY, Chen YH, Tan CS, Chiang PC (2012) Accelerated carbonation of steelmaking slags in a high-gravity rotating packed bed. J Hazard Mater 227–228:97–106. doi:10.1016/j.jhazmat.2012.05.021

Chapter 19
System Optimization

Abstract Since the accelerated mineralization process would consume additional energy and generate more CO_2 emissions, it should be critically assessed from the view point of 3E (Engineering, Environmental, and Economic) aspects. From the process design point of view, dissolution of reactive species (e.g., calcium ions) and the water solubility of CO_2 exhibit contradictory performances in the limiting step. Therefore, the optimum operating conditions of the accelerated carbonation process should be proposed to compromise the above conflicting phenomena. In this chapter, two different approaches, i.e., (1) mathematical programming and (2) graphical presentation, are illustrated for evaluating the engineering performance of process. To provide a holistic assessment, the 3E triangle analysis is discussed with a case study of high-gravity carbonation using steel slag. Furthermore, in this chapter, the pilot studies and demonstration projects of accelerated carbonation around the world are reviewed. The performance of accelerated carbonation is illustrated from the engineering (CO_2 capture scale and efficiency, and product utilization), economic (energy consumption and operating cost), and environmental (impacts and benefits) aspects.

19.1 Mathematical Programming Approach

19.1.1 Principles

A nonlinear program can be formulated to estimate the maximum (or minimum) objective function, e.g., carbonation conversion of alkaline solid waste. For example, the effect of different operating factors, such as reaction temperature (noted as A) and liquid-to-solid ratio (noted as B), on the carbonation conversion of alkaline solid waste (noted as δ) can be expressed as follows:

$$\delta = f(A, B) \tag{19.1}$$

P.-C. Chiang and S.-Y. Pan, *Carbon Dioxide Mineralization and Utilization*, DOI 10.1007/978-981-10-3268-4_19

Since the δ is a function of various operating factors with interaction terms, the optimality of δ can be determined by a nonlinear program with several equations for the operating factors.

According to the Weierstrass theorem [1, 2], δ has a global optimality in set S only if

- δ is continuous on feasible S.
- S is closed and bounded.

Considering the objective of maximizing δ (A, B) subject to equality constraint h_i (A, B) converting from inequality constants s_i^2, it is convenient to write these conditions in terms of a Lagrange function defined as L (A, B, u) in Eq. (19.2):

$$L(A, B, u, s) = \delta(A, B) + \sum_{i=1} u_i h_i(A, B, s) \tag{19.2}$$

where u_i is the Lagrange multipliers, which can be either positive, negative, or zero. In Eq. (19.2), the maximum value of $\delta(A, B)$ should be equal to maximum $L(A, B, u, s)$, if the constraints were satisfied. To meet the necessary conditions of the Lagrange Multiplier Theorem ($L'(x^*) = 0$ and $L''(x^*) < 0$), the gradients of $L(A, B, u, s)$ function can be determined as follows:

$$L_A(x)|_{x=x*} = \frac{\partial \delta(x^*)}{\partial A} + \sum_{i=1} u_i^* \frac{\partial h_i(x^*)}{\partial A} = 0 \tag{19.3}$$

$$\nabla L_B(x)|_{x=x*} = \frac{\partial \delta(x^*)}{\partial B} + \sum_{i=1} u_i^* \frac{\partial h_i(x^*)}{\partial B} = 0 \tag{19.4}$$

$$\nabla L_u(x)|_{x=x*} = h_i(x^*) = 0 \tag{19.5}$$

$$\nabla L_s(x)|_{x=x*} = u_i^* \frac{\partial h_i(x^*)}{\partial s} = 0 \tag{19.6}$$

where x^* is the optimal solution for the carbonation of alkaline solid waste. Any point that does not satisfy the conditions of the Lagrange multiplier theorem cannot be a maximum point.

In addition, the obtained results should meet the sufficient conditions of optimality by taking the Hessian matrix of Eq. (19.2), as shown in Eq. (19.7):

$$H(x)|_{x=x*} = \begin{bmatrix} \frac{\partial^2 L}{\partial A^2} & \frac{\partial^2 L}{\partial A \partial B} \\ \frac{\partial^2 L}{\partial B \partial A} & \frac{\partial^2 L}{\partial B^2} \end{bmatrix} \tag{19.7}$$

19.1.2 Application: Case Study of Carbonation in a Slurry Reactor

From the statistical point of view, the response surface methodology (RSM), as illustrated in Chap. 9, can be used to evaluate the effect of the relating operational parameters, including reaction time, temperature, and L/S ratio on the carbonation conversion of alkaline solid wastes. For example, the carbonation conversion of steel slag (i.e., basic oxygen furnace slag, BOFS) in a slurry reactor can be expressed as Eq. (19.8) [3]:

$$\delta(\%) = 45.56 + 8.97A + 6.45B - 2.55AB - 5.36A^2 - 0.91B^2 \tag{19.8}$$

where A is the reaction temperature and B is the L/S ratio. By solving Eqs. (19.3)–(19.6), eight optimal candidate points may be theoretically obtained as a solution. In the case of $h_1 = 0$, for instance, the conditions can be determined by Eqs. (19.9)–(19.11):

$$\nabla L_{A,B}(\mathrm{x})\big|_{x=x*} = \begin{bmatrix} \frac{\partial L}{\partial A} \\ \frac{\partial L}{\partial B} \end{bmatrix}_{x=x*} = \begin{bmatrix} -10.72A - 2.55B + 8.97 \\ -2.55A - 1.82B + 6.45 + u_1 \end{bmatrix}_{x=x*} = \begin{bmatrix} 0 \\ 0 \end{bmatrix} \tag{19.9}$$

$$\nabla L_u(x)\big|_{x=x*} = \frac{\partial \mathrm{L}}{\partial u_1} = B - 1 + s_1^2 = 0 \tag{19.10}$$

$$\nabla L_s(x)\big|_{x=x*} = \frac{\partial L}{\partial s_1} = 2u_1 s_2 = 0 \tag{19.11}$$

In this case, the temperature and L/S ratio can be determined to be 62 °C (i.e., A = 0.6) and 20 mL g^{-1} (i.e., B = 1.0), respectively. The s_1 and u_1 values were 0 and −3.1, respectively. Moreover, the obtained results should meet the sufficient conditions of optimality by taking the Hessian matrix of Eq. (19.2), as shown in Eq. (19.12):

$$H(x)\big|_{x=x*} = \begin{bmatrix} \frac{\partial^2 L}{\partial A^2} & \frac{\partial^2 L}{\partial A \partial B} \\ \frac{\partial^2 L}{\partial B \partial A} & \frac{\partial^2 L}{\partial B^2} \end{bmatrix} = \begin{bmatrix} -1.82 & -2.55 \\ -2.55 & -10.72 \end{bmatrix} < 0 \tag{19.12}$$

Since the function of carbonation conversion should be negative definite, the determined condition of Eq. (19.9) results in the maximal value. As shown in Fig. 19.1, it was predicted that the maximum carbonation conversion of BOFS in a slurry reactor for 120 min should be 53.0%. It was noted that the carbonation reaction of steel slag can be enhanced if the alkaline wastewater was introduced as a liquid agent due to the presence of sodium and chloride ions in the wastewater [4, 5].

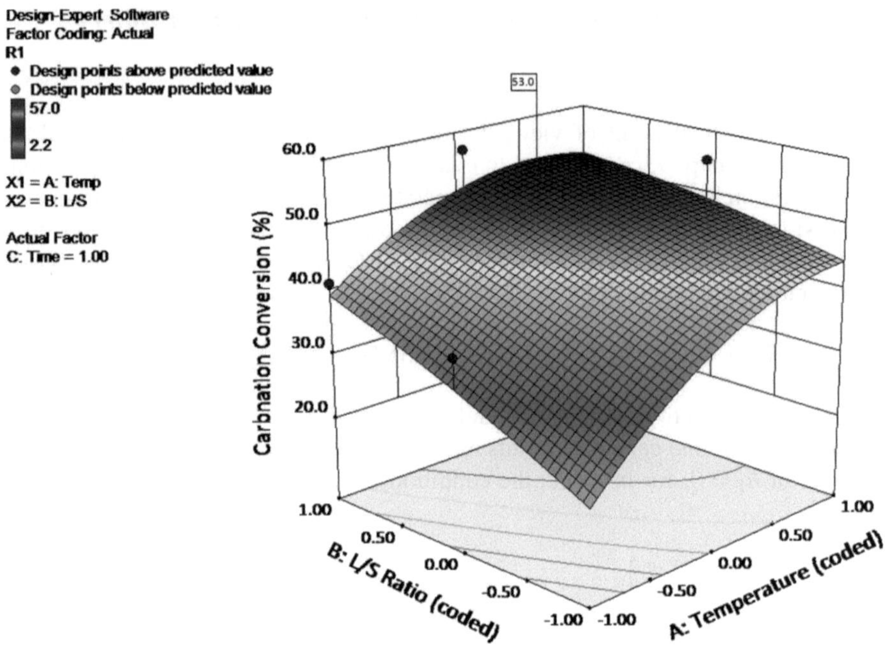

Fig. 19.1 3D response surface plot of various operating factors and conversion of steel slag in a slurry reactor

19.2 Graphical Presentation for Optimization

19.2.1 Maximum Achievable Capture Capacity (MACC)

The maximum achievable capture capacity (MACC) of solid wastes after accelerated carbonation can be obtained by considering both the carbonation conversion and energy consumption. It is noted that the carbonation kinetics can be expressed by an "exponential growth to maximum" model due to the formation of product layer, during the carbonation reaction, as reported in the above-mentioned results. However, the overall energy consumption of batch carbonation process was found to increase linearly with the increase in reaction time. Therefore, the MACC of solid waste can be systematically determined and graphically presented in Fig. 19.2 by balancing "exponential growth of capture capacity (positive capture)" and "linear increase of energy consumption (negative capture)."

The major unit operation processes for accelerated carbonation typically include grinding, transportation, stirring, pumps, blowers, reactor, and liquid–solid separation. The power consumption for most of the unit operations can be determined by multiplying the operating voltage to the operating amplitude of the existing equipment, while the power consumption (*W*) for grinding (crushing) can be

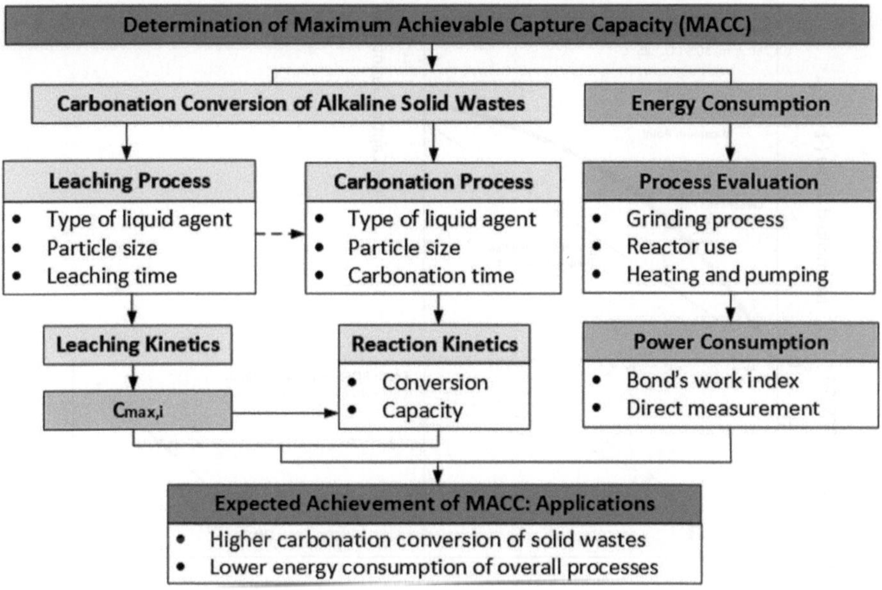

Fig. 19.2 Systematic approach to determining maximum achievable capture capacity (MACC) via leaching and carbonation processes

calculated by Bond's equation as shown in Eq. (19.13) [6]. The Bond's equation has been widely used in the literature [7, 8]:

$$W = W_i\left(\frac{10}{\sqrt{W_{P80}}} - \frac{10}{\sqrt{W_{F80}}}\right) \tag{19.13}$$

where W (kWh/ton) is the power consumption, W_{F80} (μm) and W_{P80} (μm) are the 80% passing size of the feed and the product, respectively, and W_i (kWh/ton) refers to the work index of ground material. For example, the work index for the slag obtained from the ball-mill test is approximately 30.4 kWh/ton [8].

Since the energy consumption of processes increases as the operating time increases, the overall MACC of solid wastes should be achieved at the "maximum point" by considering both the carbonation rate and energy consumption. As shown in Fig. 19.3, for example, in the case of steel slag using alkaline wastewater in a high-gravity carbonation, the operating time for reaching the MACC should be 8.5 min (from Point 1 to Point 2, and then we can obtain the Point A). In that case, the required amount of steel slag for capturing 1 ton of CO_2 by the carbonation process was estimated to be 5.13 ton (from Point 3 to Point 5, and then we can get the Point B), under which the MACC was approximately 0.20 ton CO_2 per ton steel slag. The carbonation of steel slag coupled with alkaline wastewater (i.e., cold-rolling wastewater) with a particle size less than 125 μm exhibits a relatively

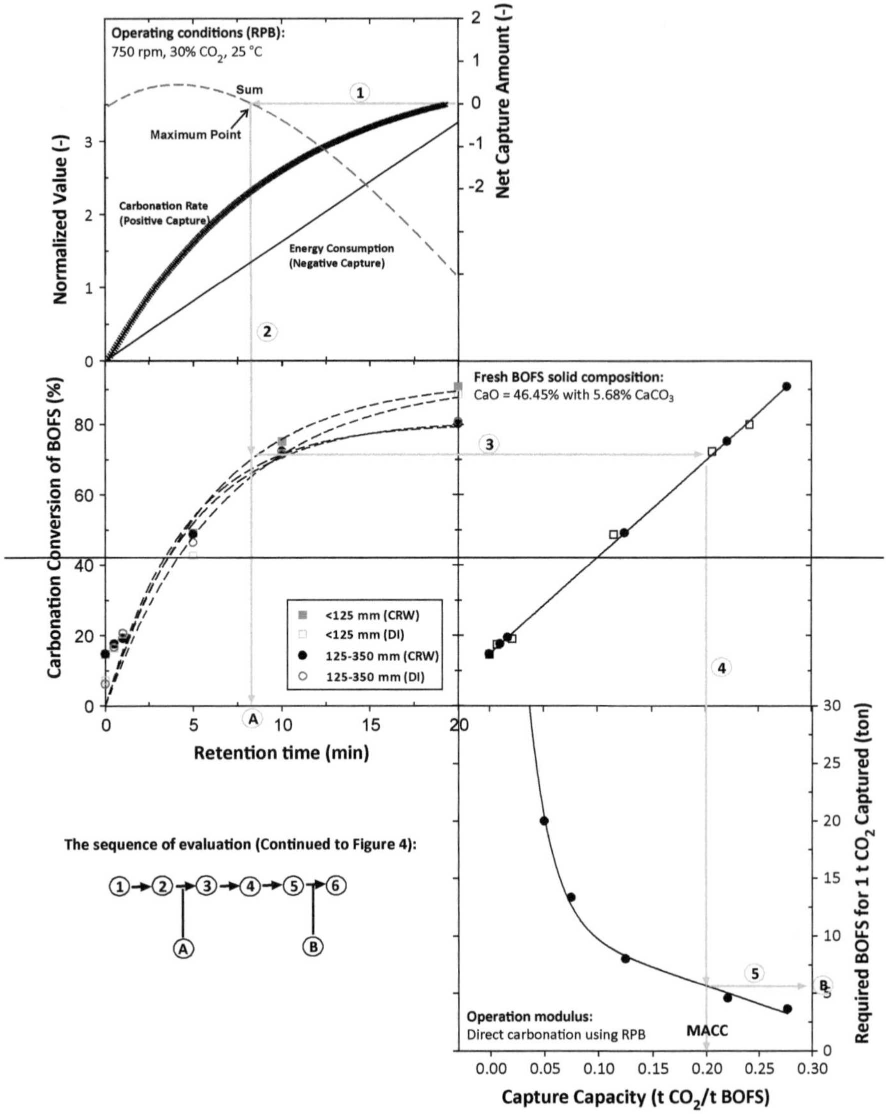

Fig. 19.3 Maximum achievable capture capacity (MACC) for steel slag in high-gravity carbonation, as indicated by *red line*. *CRMW* cold-rolling mill wastewater; *DI* deionized water. Reprinted with the permission from Ref. [4]. Copyright 2013 American Chemical Society

higher performance to achieve the lower energy consumption with higher CO_2 capture capacity.

Accordingly, the energy consumption of the carbonation process using direct carbonation of steel slag with various particle sizes under different reaction times can be determined from Fig. 19.4. As mentioned before, the capture capacity of

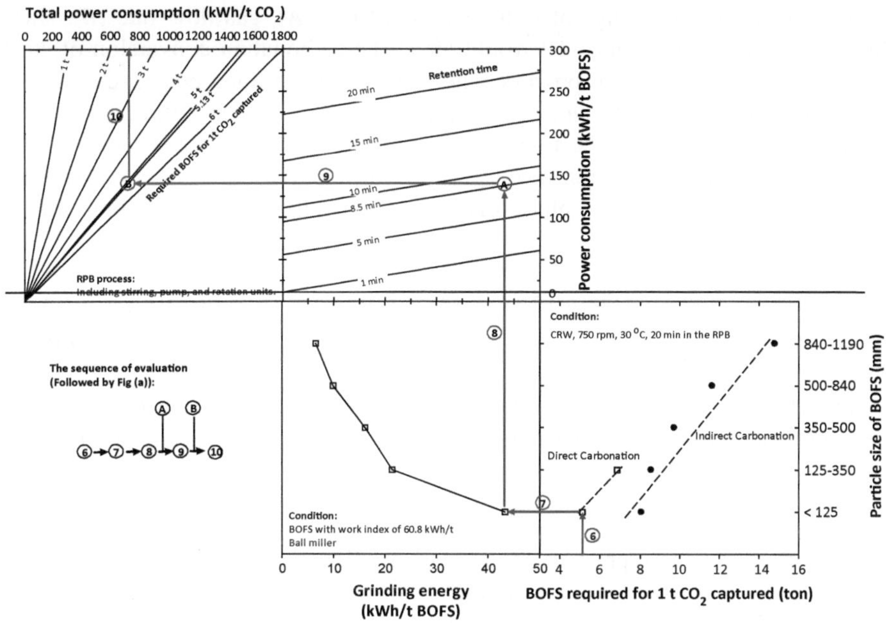

Fig. 19.4 Estimation of energy consumption for accelerated carbonation of steel slag coupled with alkaline wastewater in high-gravity carbonation, as indicated by *red line*. Reprinted with the permission from Ref. [4]. Copyright 2013 American Chemical Society

steel slag at its maximum reaction rate (operated for 8.5 min using alkaline wastewater) was estimated to be 0.195 ton CO_2 per ton steel slag (from Point 6 to Point 8, and then combined with the results of points A and B). The rotation of the packed bed is found to be the most energy-intensive process (i.e., 47.5% of total). Meanwhile, the fraction of energy consumption for the pumps and the grinding process to the total energy consumption is 18.7% and 24.3%, respectively. Although the carbonation process might require additional electricity, it could effectively neutralize the alkaline wastewater (down to a pH value of 6.3) and improve the properties of steel slag for further utilization, since the free CaO and Ca $(OH)_2$ in fresh steel slag could be totally eliminated after carbonation. The treatment cost for waste stabilization is expected to decrease because the carbonation process does not need to introduce additional chemicals or steam; it only needs to utilize waste CO_2 as a reaction agent.

19.2.2 Balancing Mass Transfer Rate and Energy Consumption

To optimize the mass transfer rate in the high-gravity carbonation process, a statistical K_Ga model should be developed according to the experimental data. The important factors including rotation speed (coded as A), gas flow rate (B), slurry flow rate (C), and L/S ratio (D) would affect the carbonation conversion of steelmaking slag in carbonation process. The analysis of the fitted response surface is generally equivalent to the analysis of the actual system if the fitted surface is a satisfactory estimation of the true response function. For instance, the developed model associated with rotation speed (A), gas flow rate (B), slurry flow rate (C), and L/S ratio (D) on overall gas-phase mass transfer coefficient (K_Ga) is presented in Eq. (19.14).

$$K_Ga\,(\text{coded}) = -0.13 + 1.32^*A - 0.71^*B - 0.16^*C - 0.14^*D + 2.04^*AB - 0.13^*A^2 + 0.11^*C^2 \tag{19.14}$$

In addition, a correlation between CO_2 removal efficiency (η) and K_Ga value can be expressed, based on the definition of K_Ga in two-film theory, as Eq. (19.15):

$$\eta(\%) = \eta_{\max}[1 - \exp(-t' \times K_Ga)] \tag{19.15}$$

where the $\eta_{\max}$ value represents the maximum CO_2 removal efficiency of the carbonation process, which has a theoretically maximum value of 100%. The exponential coefficient (t') is the characteristic time of carbonation.

In the case of high-gravity carbonation, the estimated t' value was 7.10 ± 0.45 (s), with a determination coefficient (r^2) of 0.955 [9]. It is noted that the average residence time of liquid flow in a high-gravity reactor should be approximately 0.2–0.8 s [10]. Although the carbonation reaction rate is generally fast enough compared to the retention time of gas in a high-gravity reactor, an appropriate level of gas-to-liquid (G/L) ratio should be maintained for the operation of high-gravity carbonation. It is noted that an increase in gas flow rate, corresponding to a greater G/L ratio, would reduce the retention time of gas in the packed bed zone.

To optimize the gas-phase mass transfer rate and energy consumption, the favorable operating modulus could be systematically determined via graphical presentation, as shown in Fig. 19.5. The energy consumptions could be expressed in terms of kWh per ton CO_2 capture by the process. In the case of the high-gravity carbonation process, a centrifugal acceleration should be maintained at 475 m/s^2 for a relatively lower energy consumption ($L1 \rightarrow L2$) and greater K_Ga value ($L3 \rightarrow L4$). The favorable G/L ratio should range between 40 and 55 for high-gravity carbonation (determined by both $L5 \rightarrow L6$ and $R1 \rightarrow R2 \rightarrow R3$). A further increase in G/L ratio up to 80 would lead to a low K_Ga value and high energy consumption for rotation and pumps, resulting in a poor CO_2 removal efficiency and capacity.

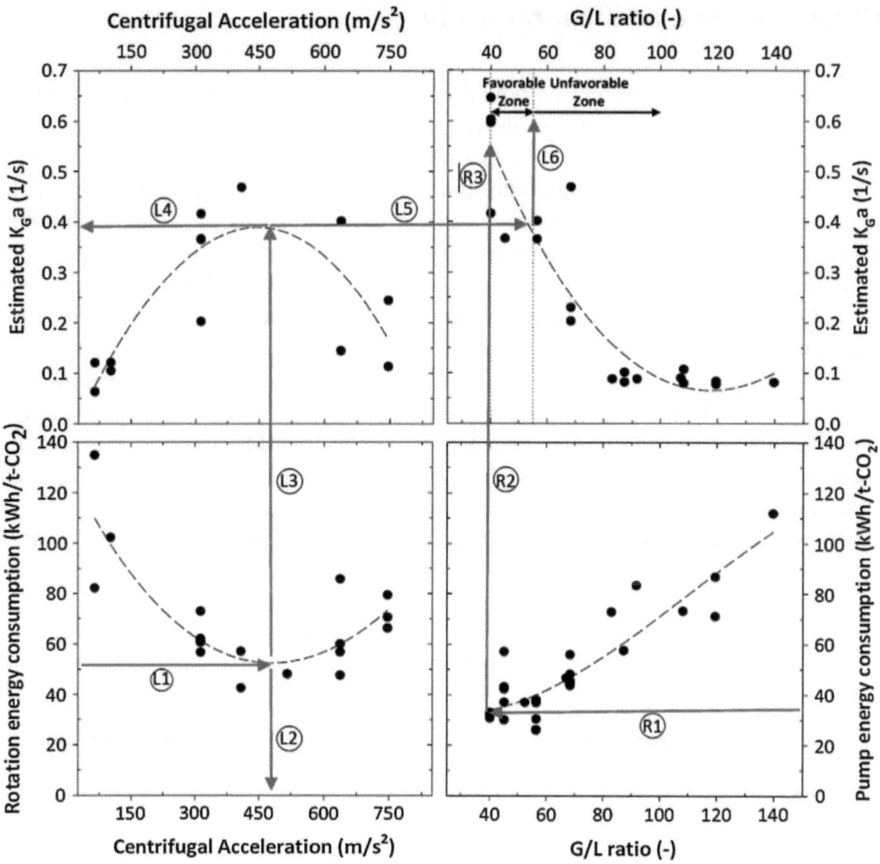

Fig. 19.5 Determination of optimal K_Ga value associated with favorable centrifugal acceleration (i.e., rotation speed) and G/L ratio via graphical presentation for high-gravity carbonation process (as indicated by *red line*). Reprinted by permission from Macmillan Publishers Ltd.: Ref. [9], copyright 2015

19.3 Comprehensive Performance Evaluation via 3E Triangle Model: A Case Study

Accelerated carbonation technologies are a feasible approach to integrating alkaline solid waste treatment with CO_2 fixation. In this section, a case study of the high-gravity carbonation (HiGCarb) process is assessed from the perspectives of engineering, environment, and economy (3E) using a 3E triangle model. The principles, key performance indicators, and data analyses and interpretation for the 3E triangle model are illustrated in Chap. 9. Several methods, such as response surface methodology, life cycle assessment and cost benefit analysis (as discussed in Chap. 9), can be utilized in the 3E triangle analysis for system optimization.

19.3.1 Scope and Scenario Setup

The performance of different carbonation processes should be critically evaluated from the perspectives of process design, energy consumption, and environmental benefits. Figure 19.6 shows the evaluation framework for the performance before (i.e., business-as-usual case) and after integration of HiGCarb process. To critically evaluate the benefits of integrating the HiGCarb process in the steelmaking industry, a systematically evaluation from the 3E aspects is quite important.

Table 19.1 presents a comparison of business-as-usual (BAU) and integration of the HiGCarb process in the steel industry from the 3E aspects. In the BAU case, three existing waste sources are separately operated or treated: (1) CO_2 emissions from hot-stove stack without CO_2 capture or fixation; (2) cold-rolling mill wastewater (CRMW) is neutralized by chemical agents at a wastewater treatment plant discharge; and (3) basic oxygen furnace slag (BOFS) stabilization and disposal at a landfill plant. In contrast, in the integration of HiGCarb process case, the CO_2 emitted from the steel industry is directly used to neutralize alkaline CRMW. At the same time, the contents of free-CaO and $Ca(OH)_2$ in BOFS can be

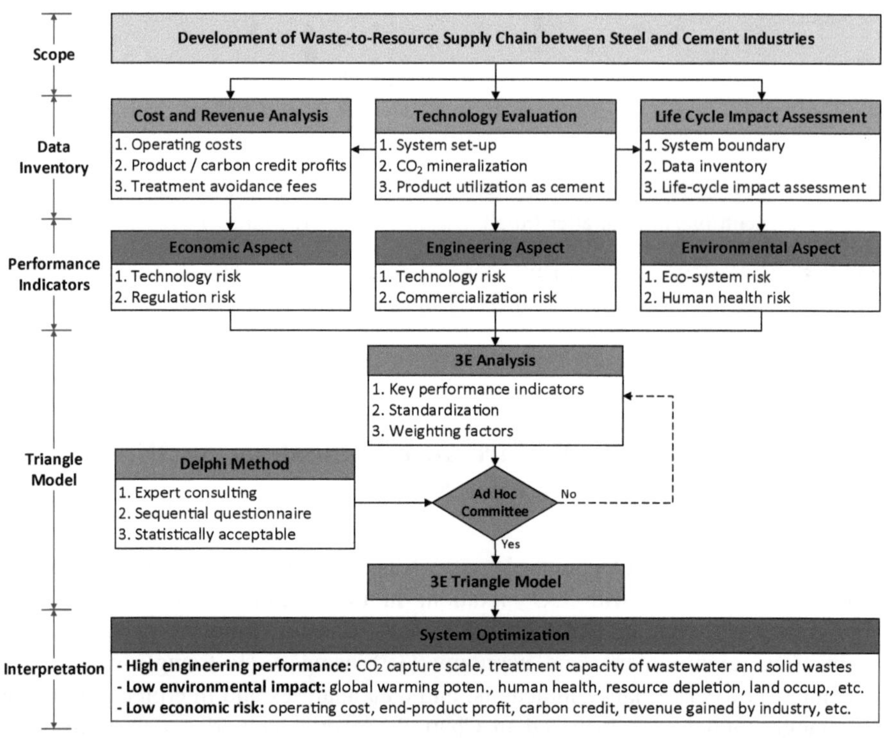

Fig. 19.6 Research framework of 3E assessment for high-gravity carbonation (HiGCarb) process using a triangle model to obtain an optimal operating modulus

Table 19.1 Remarks of business-as-usual (BAU) case and high-gravity carbonation (HiGCarb) process in steel industry

Aspects	System boundary	
	Business-as-usual (BAU)	Integration of HiGCarb process
Engineering	• No need to modify spec or features of process	• Manufacture of RPB reactor • Integration of RPB into industry • Additional energy consumption for HiGCarb process operation
Environmental	• CO_2 emission • CRMW discharge • BOFS landfill	• Reduction in CO_2 emission • Reduction in the use of chemicals for waste treatment • Increase in air pollutant emissions due to electricity use • Reduce cement production in cement industry (carbonated product utilization)
Economic	• Pay for BOFS treatment fee • Pay for CRMW treatment fee • Pay for CO_2 emission fee	• Capital and O&M costs for RPB reactor • Gain direct/indirect carbon credits • Gain profits from product sales • Avoidance of BOFS treatment fee • Avoidance of CRMW treatment fee

eliminated. As a result, the physico-chemical properties of BOFS can be utilized as cement replacement materials.

Furthermore, the carbonated product can be used as supplementary cementitious materials (SCM) in a blended cement mortar or concrete block. As cement manufacturing is an energy- and material-intensive process, with high annual production, significant indirect environmental benefits by the accelerated carbonation process can be realized.

19.3.2 Key Performance Indicators and Data Inventory

A 3E triangle model can provide a holistic assessment from the viewpoint of 3E aspects using a graphical presentation. As exemplified in Table 19.2, a total of 15 key performance indicators (KPIs) are presented for evaluating the HiGCarb process used in a 3E triangle model. Engineering performance (EP) is calculated using three technology indicators (engineering aspect), where EP_1, EP_2, and EP_3 represent the HiGCarb capacities for CO_2 removal, wastewater neutralization, and carbonated product, respectively. To calculate the life cycle environmental impact (LCEI), eight environmental indicators are selected from ReCiPe midpoint and endpoint assessment because of their expected relevance for the HiGCarb process. The rest of impact categories in ReCiPe were excluded for the 3E analysis since they did not exhibit significant difference among the scenarios. Economic cost (EC) is determined using four economic indicators, including operating costs for capturing, end product sale profit, waste treatment free avoided, and carbon credit profit.

Table 19.2 Key performance indicators (KPIs) for the evaluation of accelerated carbonation process

Aspects	Key performance indicators (KPI)[b]		Units	Weighting factors (W_i)	Remarks
Engineering	EP_1	Specific CO_2 capture capacity of BOFS by HiGCarb	t CO_2/t-BOFS	0.70	Technology risk
	EP_2	Wastewater neutralization by HiGCarb	m^3/t-BOFS	0.15	Commercialization risk
	EP_3	End product production by HiGCarb	t/t-BOFS	0.15	Commercialization risk
Environmental[a]	$LCEI_1$	Global warming potential (GWP100) reduction	kg CO_2-Eq/t-BOFS	0.30	Ecosystem risk
	$LCEI_2$	Freshwater ecotoxicity reduction	kg 1,4-DCB/t-BOFS	0.10	Ecosystem risk
	$LCEI_3$	Particulate matter formation reduction	kg PM_{10}-eq/t-BOFS	0.10	Human health risk
	$LCEI_4$	Marine eutrophication potential reduction	kg N-eq/t-BOFS	0.10	Ecosystem risk
	$LCEI_5$	Urban land occupation avoidance	m^2a/t-BOFS	0.10	Ecosystem risk
	$LCEI_6$	Ecosystem quality improvement	points/t-BOFS	0.10	Ecosystem risk
	$LCEI_7$	Human health protection	points/t-BOFS	0.10	Human health risk
	$LCEI_8$	Resource depletion reduction	points/t-BOFS	0.10	Ecosystem risk
Economic	EC_1	Operating costs for capturing	US$/t-BOFS	0.30	Economic risk
	EC_2	End product sale profit	US$/t-BOFS	0.20	Economic risk
	EC_3	Waste treatment free avoided	US$/t-BOFS	0.15	Economic risk
	EC_4	Carbon credit profit	US$/t-BOFS	0.35	Regulation risk

[a]Functional unit = per ton of BOFS input for HiGCarb process. [b]*LCEI* life cycle environmental impact; *EP* engineering performance; *EC* economic cost

The weighting factors (W_i) of each KPI can be determined via Delphi method. The Delphi study is typically conducted over a period of few months and comprised at least two rounds. The participating experts, so-called ad hoc committee, could be consulted through roundtable discussion or sequential (online) questionnaires. The ad hoc committee should comprise highly informed academic and industrial experts from diverse backgrounds as well as government officials from the region.

Table 19.3 presents the examples of operating information and life cycle data inventory, including main material inputs and energy consumption, for nine scenarios. In this analysis, the CO_2 used for the HiGCarb process was introduced from the hot-stove gas at China Steel Corporation (Kaohsiung, Taiwan), where the average CO_2 concentration was 28–32%. Both the BOFS and alkaline CRMW were used directly from the manufacturing process. The gas—(Q_G) and slurry—(Q_S) flow rates were 0.38 m^3/min and 0.33–0.56 $m^{3/}$h, respectively. For these conditions, the capture scale of the HiGCarb process ranged between 75 and 170 kg CO_2 per day.

In this case study, data inventory is obtained from the experiment in field tests, and the scale factor is assumed to be 0.8. Due to different operating conditions, a total of nine scenarios are established based on various levels of CO_2 removal ratio:

- Low level: <70% (noted as L)
- Medium level: 70–90% (noted as M)
- High level: >90% (noted as H)

The choice of functional unit reveals several issues such as net CO_2 fixation amounts per unit weight of BOFS within a certain operating period. The 3E performance can be performed directly from the amounts of the materials and energy used such as electricity. In the 3E triangle model, the total scores for the 3E aspects are summed up after multiplying each KPI by its corresponding weighting factors (W_i). The W_i can be determined by the ad hoc committee using the Delphi method.

19.3.2.1 Engineering Consideration

The mass transfer rate of carbonation in the HiGCarb process could be significantly greater than that using a fix packed bed. As a result, the HiGCarb process can offer a high CO_2 capture efficiency of greater than 98% with a relatively short reaction time at ambient temperature and pressure [11]. The CO_2 removal ratio (η, %) from the flue gas via the carbonation process can be experimentally determined by Eq. (19.16):

$$\eta = \frac{\left(\rho_{CO_2,i}\, Q_{g,i}\, C_{g,i} - \rho_{CO_2,o}\, Q_{g,o}\, C_{g,o}\right)}{\rho_{CO_2,i}\, Q_{g,i}\, C_{g,i}} \times \% \qquad (19.16)$$

Table 19.3 Data inventory including main material inputs and energy consumption for nine scenarios

ID	CO_2 removal ratio	Performance and operation conditions[a]				Inventory (per t-CO_2 captured)			Engineering performance (EP)		
		η (%)	ω (rpm)	Q_G (m^3/min)	Q_S (m^3/h)	CRMW inputs (t)[b]	BOFS inputs (t)	total energy (kWh)	EP_1 (t CO_2/t-BOFS)	EP_2 (m^3/t-BOFS)	EP_3 (t/t-BOFS)
L1	Low level (<70%)	42.3	158	0.38	0.33	105.2	7.01	441.4	0.143	15.0	1.142
L2		51.4	158	0.38	0.50	140.2	7.01	432.7	0.143	20.0	1.144
L3		65.1	200	0.38	0.40	81.3	6.11	356.1	0.164	13.3	1.164
M1	Medium level (70–90%)	71.3	350	0.38	0.40	78.7	3.93	263.7	0.254	20.0	1.256
M2		77.0	500	0.38	0.50	91.0	4.55	290.6	0.220	20.0	1.218
M3		86.4	450	0.38	0.33	52.2	3.48	226.4	0.287	15.0	1.287
H1	High level (>90%)	95.8	350	0.38	0.40	55.8	4.19	247.7	0.238	13.3	1.240
H2		98.3	400	0.38	0.33	47.5	3.16	204.7	0.316	15.0	1.318
H3		99.5	200	0.38	0.56	86.3	6.49	354.6	0.154	13.3	1.155

[a]η is CO_2 removal ratio as determined by Eq. (19.16), ω is rotating speed, Q_G is gas flow rate, Q_s is slurry flow rate, and L/S is liquid-to-solid ratio. [b]*CRMW* cold-rolling wastewater

where $\rho_{CO_{2,i}}$ and $\rho_{CO_{2,o}}$ (g/L) are the CO_2 mass densities at the temperature of inflow and outflow gas streams, respectively. $Q_{g,i}$ (L/min) and $Q_{g,o}$ (L/min) are the volumetric flow rate of the inlet and outlet gas streams, respectively. $C_{g,i}$ (%) and $C_{g,o}$ (%) are the CO_2 volume concentration in the inlet and exhaust gas, respectively.

The CO_2 capture process may involve energy-intensive units, such as stirring, heating, blowers, air compressors, pumps, liquid–solid separation, and material grinding. The power consumption of process could be directly determined by multiplying the operating voltage by the operating current of the existing equipment. Moreover, the energy consumption of material grinding (E_G) can be estimated by Bond equation [6], as shown in Eq. (19.17). It is noted that the Bond equation should give the most accurate estimation of grinding energy requirement within the conventional grinding range of 25,000 to 20 μm [12].

$$E_G = w_i\left(\frac{10}{\sqrt{D_{P80}}} - \frac{10}{\sqrt{D_{F80}}}\right) \quad (19.17)$$

where D_{F80} (μm) and D_{P80} (μm) are the 80% passing size of feed and product, respectively. The w_i (kWh/ton) is the work index of ground material, which expresses the resistance of the material to crushing and grinding. It is noted that the work index is subject to variations because of variations in the inherent properties of materials, variations in the grinding environment, and variations in the mechanism of energy transfer from the grinding equipment to its charge [13].

19.3.2.2 Environmental Consideration

Figure 19.7 shows the LCA system boundaries of the BAU case and the integration of the HiGCarb process into industry. The environmental impacts of the BAU (without the HiGCarb process) and the HiGCarb process are compared by means of LCA, including manufacturing and operation of the reactor and end product use as green materials. The functional equivalent (unit) is assumed to be one ton of fresh BOFS produced from steel industry or delivered to the carbonation process.

In the HiGCarb process, the unit operation processes include slag grinding, stirring machines, blowers, air compressors, pumps, RPB reactor, and electricity generation. Also, for the boundary system in the LCA, the stages of both raw material extraction (i.e., RPB reactor manufacturing) and product use (i.e., substitution in CEM I/42.5 Portland cement) are included. The life time of the system is assumed to be 20 years. As suggested by the LCA method described in the ISO 14040:2006 and ISO 14044:2006 [15, 16], the environmental impacts of the process could be quantified by Umberto 5.6 using the ReCiPe Midpont (E) and Endpoint (E, A) methodology [17].

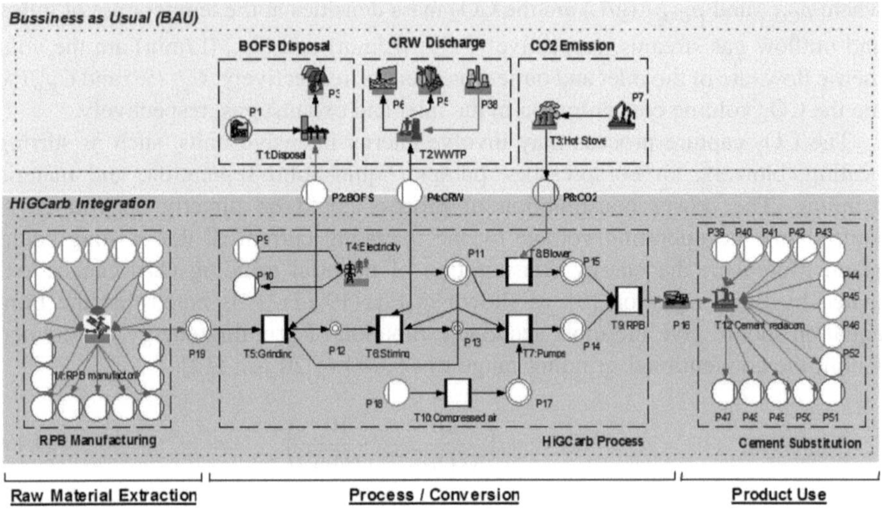

Fig. 19.7 System boundaries of business-as-usual (*top part* of the figure) and HiGCarb (*bottom*) process, which includes manufacturing of the reactor and substitution for cement. Reprinted by permission from Macmillan Publishers Ltd: Ref. [14], copyright 2016

19.3.2.3 Economic Consideration

The amount of revenue gained (RG), in terms of USD/t-BOFS, could be calculated by Eq. (19.18):

$$\mathrm{RG} = \left(P_{\mathrm{cc,dir}} + P_{\mathrm{cc,ind}}\right) + P_{\mathrm{ta}} + P_{\mathrm{ep}} - C_{\mathrm{op}} \quad (19.18)$$

where $P_{\mathrm{cc,dir}}$ is the profit of direct carbon credit by the HiGCarb process, $P_{\mathrm{cc,ind}}$ is the profit of indirect CO_2 avoidance credit by-product use, P_{ta} is the profit of BOFS treatment avoidance, P_{ep} is the profit of end product sale, and C_{op} is the operating cost.

The price of stabilized BOFS was approximately US$6.0/ton [18], which could be considered as the profit of carbonated BOFS product sales. Because the physico-chemical properties of BOFS can be upgraded after the HiGCarb process, no additional treatment of carbonated BOFS such as grinding and stabilizing processes within the steelmaking industry is required [4]. The treatment fee of BOFS was approximately US$10/ton [19], which can be saved in the case of HiGCarb process. In addition, the price of carbon credit in the emission reduction unit (ERU) market was approximately US$8.1/t-$CO_2$ in 2014 [20].

19.3.3 Performance in 3E Perspectives

19.3.3.1 Engineering Performance

Energy consumption is one of the major concerns to the engineering performance and cost effectiveness of a CO_2 capture process. It is noted that the accelerated carbonation by the HiGCarb is conducted at ambient temperature and pressure. Therefore, no additional energy is required to maintain the reaction temperature and pressure, in contrast to autoclave or slurry reactors [21–24]. Figure 19.8 shows the energy consumption of unit processes among the nine scenarios, in terms of a functional unit of "per t-CO_2". The results indicated that with a capture scale of 75–170 kg CO_2 per day, the total energy consumption of the HiGCarb process ranged from 205 to 440 kWh/t CO_2. The preprocessing of material such as feedstock grinding was the most energy-intensive process, contributing 65–79% of total energy consumption. Although the grinding process was energy-intensive, it was required for effective carbonation reaction, as well as for subsequent utilization of carbonated BOFS as SCM in Portland cement. In addition, the energy consumption of the rotating packed bed reactor was the second highest, corresponding to 11–21% of total energy consumption. Furthermore, scenarios with low CO_2 removal ratios (such as L1, L2, and L3) required a longer operating time to achieve the same CO_2 capture scale than other scenarios with high CO_2 removal ratios. Scenario H2 exhibited the lowest energy consumption of about 205 kWh for capturing one ton of CO_2. In this case, it met the criteria suggested by the US Department of Energy (US-DOE): A cost-effective CO_2 capture facility should achieve a CO_2 removal ratio of 90%, while maintaining less than 35% impact on the cost of electricity [25]. In other words, this criterion corresponds to a maximal energy consumption of 420 kWh/t-CO_2 [26].

Regarding the engineering performance, the specific capture capacity of BOFS (EP_1) mainly depends on the operating conditions, but not directly correlated with

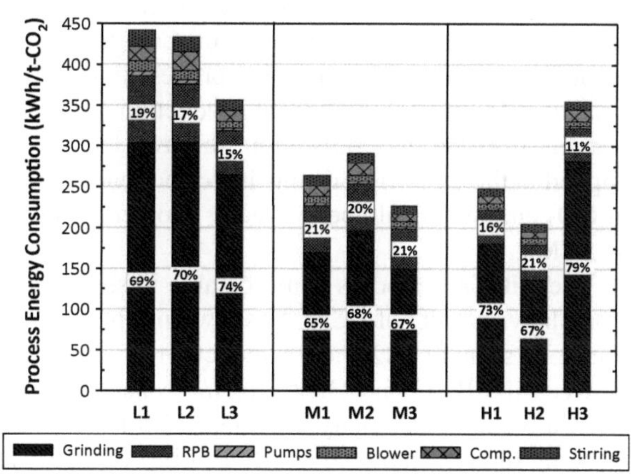

Fig. 19.8 Process energy consumption of HiGCarb for different scenarios, with their corresponding contribution in percentage. *Numbers* within each bar stack represent the contribution percentage

CO_2 removal ratio. Scenario H2 has the highest EP_1 value, corresponding to 316 kg CO_2 per ton BOFS. Among the nine scenarios, although CO_2 removal ratio in scenario H3 was the highest (i.e., 99.5%), the EP_1 of scenario H3 was the second lowest (i.e., 154 kg CO_2/t-BOFS). On the other hand, the CO_2 capture scale is mainly related to the amount of BOFS introduced into HiGCarb system per unit time period. In scenario H3, a relatively low L/S ratio (i.e., greater amount of BOFS) and higher slurry flow rate were used, the CO_2 capture scale of the entire HiGCarb system was promoted, i.e., 56.8 t-CO_2/year, becoming the top three high in the nine scenarios.

For the amount of wastewater neutralization (EP_2), scenario H3 exhibited the highest treatment capacity, followed by scenarios L2 and M2. The EP_2 was not related to CO_2 removal ratio but replied on the slurry flow rate and L/S ratio. Also, for the amount of carbonated product (EP_3), scenario 3 exhibited the highest capacity, followed by scenario H1 and then scenario L3. The carbonated product can be used as SCMs in blended cement mortar, where several mortar properties, such as early-stage compressive strength and soundness could be enhanced [27].

19.3.3.2 Environmental Performance: Impacts and Benefits

From the environmental aspect, the global warming potential (GWP), for instance, can be calculated by considering all the CO_2-equivalent emissions of each element or equipment from the life cycle point of view for the entire HiGCarb process. Figure 19.9 shows the GWP of each scenario as determined by LCA. The impact of the reactor manufacturing and its maintenance could be neglected since the magnitude of environmental impacts by RPB production is 10^4 to 10^5 times less than that of operating processes. The actual CO_2 capture amounts could be offset by the energy consumption due to the manufacturing and operation of equipment, causing additional CO_2 emissions. According to the direct measurement of CO_2 reduction in the flue gas, the capture capacity per ton of BOFS by carbonation reaction ranged between 140 and 320 kg CO_2. On the other hand, the carbonated BOFS product from the HiGCarb process can be used as SCMs in blended cement, thereby resulting in additional avoidance of CO_2 (i.e., indirect CO_2 reduction). Cement manufacturing is a CO_2-intensive process, where 0.73–0.99 tons of CO_2 would be generated for one ton of cement production [28]. To account for the environmental benefits from product utilization, the avoided burden approach [15, 16] has been applied in the LCA. The results indicated that a significant amount of 0.87–1.00 ton of CO_2 emission could be indirectly avoided by utilization of the carbonated BOFS as SCMs.

The HiGCarb process could be able to serve as a "real" CO_2 fixation technique from the viewpoint of LCA. As shown in Fig. 19.9, the highest GWP reduction was found in scenario H2, i.e., a reduction of roughly 1.28 ton CO_2-eq per t-BOFS between the BAU and HiGCarb. In scenario H2, additional CO_2 emissions from HiGCarb due to electricity uses (i.e., 33.8 kg CO_2-eq) were much lower than those of being directly captured by HiGCarb (i.e., 316.7 kg CO_2-eq).

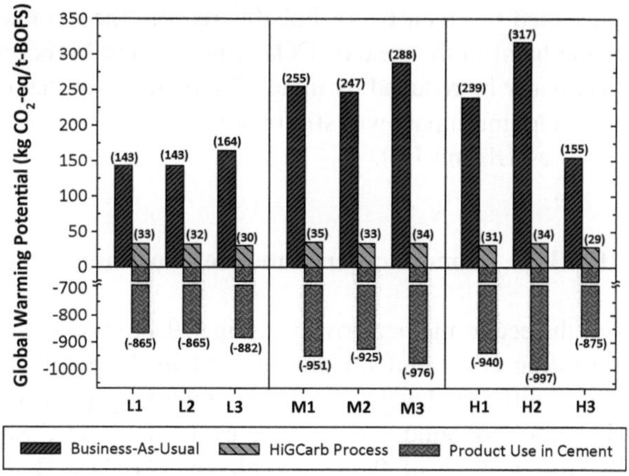

Fig. 19.9 Direct reduction by CO_2 capture and indirect CO_2 avoidance, in terms of global warming potential (GWP), in each scenario for one ton BOFS input to HiGCarb process

According to the LCA results, the HiGCarb process could reduce not only GHG emission but also environmental impacts on ecosystem quality, human health, and resource depletion. Figure 19.10 shows the results of the endpoint impact assessment for different scenarios by the ReCiPemethodology. Because of the various initial material flows such as wastewater discharge and CO_2 emission, the endpoint impacts of BAU (as indicated by the red bars) among the nine scenarios are quite different. For the HiGCarb (as indicated by the yellow bars), the particulate formation (PM) potential was found to be significantly higher than that of the BAU. It was attributed to the fact that the HiGCarb process would consume additional electricity, thereby resulting in a greater human health impact for all scenarios. However, the adverse impacts on human health due to the process of electricity usage could be compensated by the utilization of carbonated BOFS as SCM (as

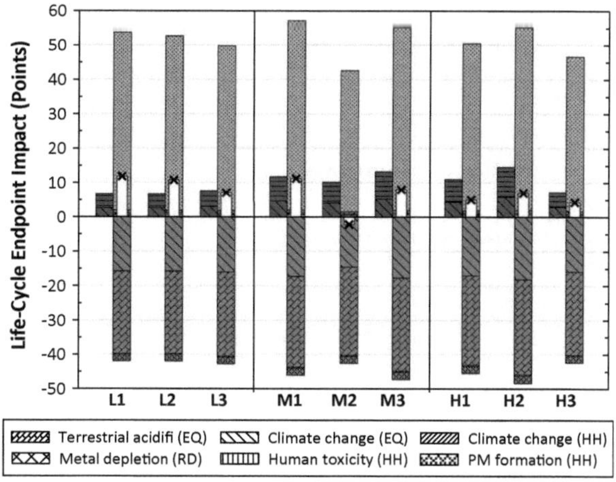

Fig. 19.10 Endpoint impact assessment on different scenarios. *Red bars* represent the BAU scenario; *yellow bars* represent the HiGCarb process; and *green bars* represent the product use as cement substitutes. *White bars with dots* represent the net impact between *yellow* and *green bars* in each scenario

presented in green bars). For this reason, the net endpoint impact (as presented in white bars) in the case of CO_2 removal ratio higher than 75% (scenario M2) could eventually be reduced by up to 12.4 points over that of BAU case. Nevertheless, the net endpoint impact was still relatively higher in scenarios of low CO_2 removal ratio (such as L1 and L2).

19.3.3.3 Economic Performance: Cost and Benefits Analysis

For the economic performance, Fig. 19.11 shows the effect of CO_2 removal ratio on operating costs and revenue gained in three different cases of electricity prices. Three different levels of average electricity price for industrial use in 2013 were used for economic performance evaluation: (1) Case A represents a low industrial electricity price of 0.091 USD/kWh; (2) Case B represents a medium industrial electricity price of 0.168 USD/kWh; and (3) Case C represents a high industrial electricity price of 0.319 USD/kWh. As shown in Fig. 19.11, the operating cost for processing one ton of BOFS was roughly 5.4–5.9 USD in Case A, while increasing to 19.0–20.8 USD per one ton of BOFS input in Case C. On the other hand, the profits from direct and indirect carbon credits were estimated to be 8.3–10.1 USD per ton of BOFS input to the HiGCarb process. Also, no additional CO_2 storage cost is needed for the HiGCarb because the CO_2-based mineral product can be directly used as SCMs for Portland cement in cement industry. As a result, the total profits returned including carbon credit and BOFS-related returns were approximately 25.8–29.0 USD per ton of BOFS input to HiGCarb. According to the above analysis, the highest revenue was gained with a CO_2 removal ratio greater than 93%. In Case A, the revenue gained was estimated to be 20.2–23.2 USD per ton of BOFS input.

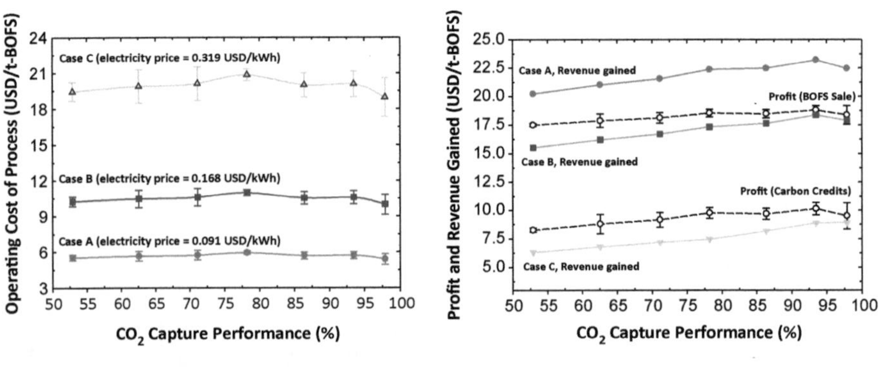

Fig. 19.11 Effect of CO_2 removal ratio on operating costs and revenue gained. *Case A* represents the one with low electricity price; *Case B* represents the one with medium electricity price; and *Case C* represents the one with high electricity price. *Note* Revenue gained is calculated by Eq. (19.18)

19.3.4 Optimization Using 3E Triangle Model

Figure 19.12 shows the results of performance evaluation for different scenarios of HiGCarb process via 3E triangle model. The results indicated that the effects of CO_2 removal ratio (η) on economic costs were not significant since the LCC scores typically ranged between 0.25 and 0.40. However, a poor engineering performance (i.e., scenarios L1, L2, and L3) is typically accompanied by severe environment impacts, with the LCEI scores ranging between 0.52 and 0.60. In other words, an increase in CO_2 removal ratio should effectively reduce the environmental impacts and make integration of the HiGCarb process into the steel industry more environmentally friendly. Among nine scenarios, scenario H2 exhibits a superior engineering performance (as indicated by line 1) with a relatively lower environmental impacts (as indicated by line 2) and a relatively lower economic costs (as indicated by line 3). Although the CO_2 removal ratio of scenario H3 was the highest (i.e., 99%) among all scenarios, the large quantity of BOFS input eventually resulted in medium environmental impacts and economic costs.

To evaluate the significance of HiGCarb process in an industry, data from China Steel Corp (CSC) is used and combined with the results of 3E triangle model. The annual production of BOFS in CSC is assumed to be 1.2 Mt [29], which should be treated and/or utilized. By applying the scenario H2, the annual direct CO_2 fixation by HiGCarb process is estimated to be 0.33 Mt, corresponding to a reduction potential of 1.5% in total CO_2 emission from the studied industry. Meanwhile,

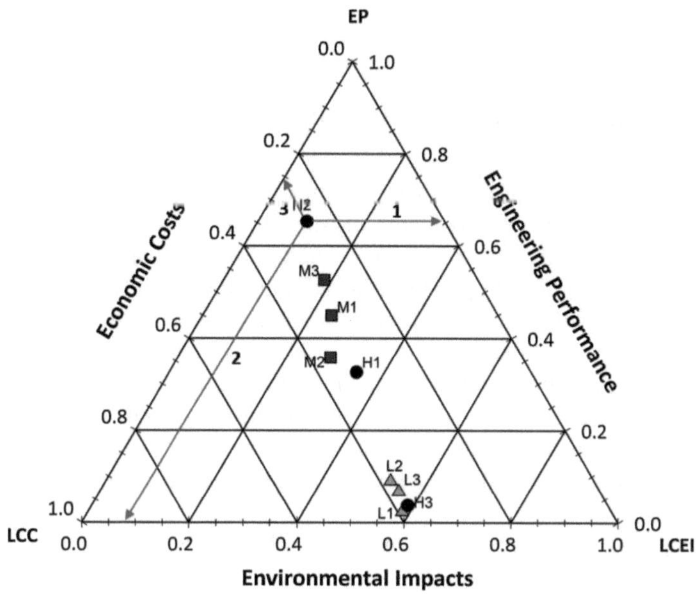

Fig. 19.12 Comprehensive performance evaluation of HiGCarb process for different scenarios via 3E triangle model

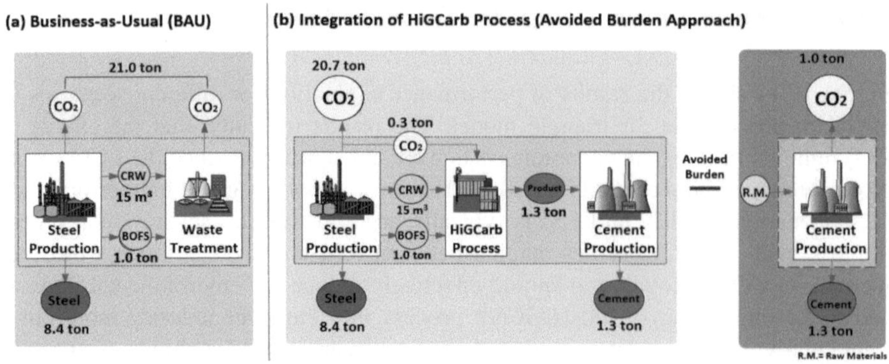

Fig. 19.13 Comparison of performance for business-as-usual and integrated HiGCarb process suggested by the results from 3E triangle model

based on the avoided burden approach, the annual indirect CO_2 reduction from the substitution of carbonated BOFS as cementitious materials is about 1.05 Mt, accounting for ~5% in total CO_2 emission from the steel industry.

As a result, a reduction potential of up to 6.5% in total CO_2 emission from the steel industry could be realized, as shown in Fig. 19.13. The HiGCarb process can establish a waste-to-resource supply chain between the steel and cement industries, thereby reducing the overall CO_2 emissions and resource consumption. At the same time, the alkaline wastes from steel industry could be stabilized and converted into valuable products such as green cement.

19.4 Technology Demonstration and Commercialization

Large-scale CO_2 separation from flue gases in power and/or industrial plants would make a huge volume of CO_2 available on-site. As a result, the subsequent fate of captured CO_2 would be either storage in natural geological structures or direct utilization and conversion. In practice, the suitable storage sites are few in number, and the procedures still involve high energy and economic costs with high risk and uncertainty in terms of long-term storage. Therefore, carbon utilization and conversion technologies are preferred, which should be directly integrated with the associated capture unit. Rather than the amount of CO_2 used, the most important consideration in CO_2 utilization is the development of innovative technologies for cleaner production, thereby directly leading to a reduction in the use of energy and materials.

Industrial wastes (by-products) are generally produced near places of CO_2 emissions. Accelerated carbonation using industrial alkaline wastes is attractive since an integrated approach to combining CO_2 capture and utilization with waste stabilization can be achieved. The potential environmental impacts caused by utilization of those untreated wastes, such as (1) highly alkaline and active properties and (2) heavy metal leaching, could be avoided. Therefore, using emitted CO_2 to carbonate industrial waste (i.e., accelerated carbonation) offers an improvement over existing methods because it does not require the CO_2 or the industrial waste to be transported. It also allows better monitoring of total pollution emissions.

19.4.1 Worldwide Demonstration Plans

Accelerated carbonation may involve several energy-intensive processes, such as material grinding, reactor heating, and gas pressurization. Energy and cost penalties largely depend on plant scale, operation conditions, and operation modulus (such as pretreatment and post-treatment processes). Several pilot studies and demonstration projects of accelerated carbonation using alkaline wastes can be found around the world, as summarized in Table 19.4. For example, since 2007, an accelerated carbonation plant at the Rocks (Wyoming, the USA) has been demonstrated in a 2120-MW coal-fired power plant using fly ash [30]. Another pilot study in the USA has been developed by Calera Corp. The Calera technology can fix approximately 30,000 t CO_2 per year from fossil fuel power plants and other industrial sources. The captured CO_2 is sequestered in geologically stable substances suitable for disposal, storage, and/or use as building materials. In 2009, Calera identified another ideal demonstration site at a brown-coal power plant in the Latrobe Valley, Victoria, Australia, for further demonstration [31].

In France, the Carmex project was initiated in 2007 and launched in 2009 to utilize various materials (such as harzburgite, wehrlite, iherzolite, olivine, and slags) through direct carbonation with and/or without organic ligand and mechanical exfoliation. The Carmex experiences indicate that the use of mineral carbonation is feasible for industries [34]. In this project, the accessible alkaline wastes are matched to large CO_2 emitters through a dedicated geographic information system (GIS). From the technical point of view, a high carbonation conversion of 70–90% can be achieved without additional heat activation of feedstock.

In Australia, the MCi project has been carried out to transform CO_2 into carbonates for use in building or non-fired products, such as bricks, pavers, and plasterboard replacements [35]. A total investment of US$9 million over four years was provided to establish the pilot plant at the University of Newcastle. The serpentine was used as the feedstock to mineralize CO_2 from the Kooragang Island plant.

In San Antonio (Texas, the USA), the Capitol SkyMine® plant was under construction by Skyonic on September 2013, and had been launched since October 2014. This plant can directly remove CO_2 at a scale of 83,000 t-CO_2 annually from industrial waste streams. The carbonate and/or bicarbonate material products can be

Table 19.4 Pilot-scale demonstration projects of accelerated carbonation around the world

Project name	Scale	Launch	Feedstock	Description of features	Performance	References
Rocks, Wyoming, the U.S.	Pilot scale	2007	Coal fly ash	• Conducted at Jim Bridger Power Plant • Direct mineralization of CO_2, SO_2, and Hg in flue gas	• CO_2 removal efficiency: 90% • SO_2 removal efficiency: > 85% • $HgCO_3$ formation: 0.4 mg/kg (25–58 °C)	[30, 32]
Calera (Moss Landing, CA, the U.S.)	Pilot scale	2007	Naturally occurring brines, waste materials	• Stabilize a number of pollutants (e.g., mercury, SO_2) in addition to CO_2 • Potentially low energy penalty compared to other carbon capture processes • Produces calcareous material and HCl	• CO_2 capture capacity: 30,000 t/y • CO_2 capture efficiency: > 50% • SO_2 removal efficiency: > 95% • Replace 40% of clinker in cement production	[31, 33]
Carmex (New Caledonia, France)	Pilot scale	2009	Mafic wastes (mine tailings)	• Ex situ mineral carbonation without heat activation • Matching accessible wastes to large CO_2 emitters through dedicated GIS • Environmental assessment through LCA	• Carbonation yield: 70–90% • 23% of initial Mg was carbonated in Ni-pyrometallurgical slag	[34]
MCi (New South Wales, Australia)	Pilot scale	2013	Serpentine	• Capture CO_2 emissions at its Kooragang Island plant • Solid product can be turned into various green products including building materials (bricks and pavers)	• Total investment of US$9 million over four years • Bricks were used as construction materials, acting as physical carbon sinks	[35]

(continued)

Table 19.4 (continued)

Project name	Scale	Launch	Feedstock	Description of features	Performance	References
HiGCarb process (CSC, Kaohsiung, Taiwan)	Small scale	2014	Basic oxygen furnace slag, cold-rolling mill wastewater	• High-gravity carbonation process • Alkaline wastewater was introduced • Carbonated BOFS was utilized as partial replacement of cement	• CO_2 capture capacity: ~ 60 t/y • CO_2 removal efficiency: > 95% • Wastewater is neutralized to pH 7 • Free CaO and $Ca(OH)_2$ in BOFS are eliminated	[11, 27]
Capitol SkyMine (San Antonio, TX, the USA)	Business model	2014	Brine solution	• Groundbreaking on September 2013 • Facility opened on October 2014 • Produce green chemicals such as HCl, bleach, Cl, and H_2	• CO_2 capture capacity: 83,000 t/y • CO_2 removal anywhere from 15%–99% • Remove CO_2, SO_x, and NO_2 from flue gas • Removes heavy metals (e.g., Hg)	[36]
HiGCarb process (FPCC, Yunlin, Taiwan)	Small scale	2016	By-product lime	• High-gravity carbonation process • Carbonated product is used as SCM	• CO_2 capture capacity: ~ 200 t/y • CO_2 removal efficiency: > 95% • Remove CO_2, SO_x, NO_x, and PM from flue gas	–

cogenerated for use in bioalgae applications to become a profitable process. Aside from mineralizing CO_2, the process can remove SO_x, NO_2, and heavy metals such as mercury from existing power plants and/or industrial plants that can be retrofitted with SkyMine®.

In Taiwan, the first small-scale high-gravity carbonation (HiGCarb) process was launched at China Steel Corporation (CSC) in 2013 to stabilize basic oxygen furnace slag (BOFS) and alkaline wastewater. The CO_2 removal efficiency of hot-stove gas was greater than 95%, with total elimination of CaO_f and $Ca(OH)_2$ content in the BOFS. The annual capture scale was ~60 tons CO_2 at a gas inflow rate of 0.9 m^3/min. Moreover, the carbonated BOFS was used as green cement substitutes in mortar. In 2016, the second HiGCarb process was established at Formosa Petrochemical Corporation (FPCC) using by-product lime to capture CO_2 in the flue gas. The capture scale of this plant was about 0.6 tons CO_2 per day, at a gas inflow rate of 1.8 m^3/min. Moreover, the carbonated by-product could be used as supplementary cementitious materials (SCM) in cement mortar or concrete.

19.4.2 Engineering Performance

19.4.2.1 CO_2 Capture Scale and Efficiency

Scale-up of the post-combustion CO_2 capture process is possible without significant developments or costs [37]. However, deployment of the post-combustion carbon capture process in industries is still challenging because the CO_2 emissions normally come from "multiple" sources. For example, in the steel industry, an integrated steelmaking process is composed of numerous facilities from the entire life cycle of iron ore to steel products including raw material preparation (such as coke production, ore agglomerating plant, and lime production), iron-making (such as blast furnace, hot metal desulphurization), steelmaking (such as basic oxygen furnace, ladle metallurgy), casting and finishing mills. The largest part of direct CO_2 emissions in steel mills is from power plants (about 48% in total CO_2 emissions), followed by blast furnaces at around 30% [38].

In large-scale tests, process integration should be considered to improve the performance of the entire industrial plant, thereby reducing the energy requirement for the capture process. Also, the selection of appropriate site for CO_2 capture should be referred to the management of the use of by-product gases, as well as on the definition of boundary limits.

19.4.2.2 Product Utilization

Accelerated carbonation technology of alkaline wastes could be moved toward commercialization, only if the produced carbonates can be used as a valuable product such as a substitute for components of cement [37]. In addition to the product utilization, the current challenges in the application of the process still include the following:

- Effect of impurities on removal performance
- Acceptance of the product by the cement industry
- Ability to capture large amounts of CO_2
- Energy requirements
- Finding an appropriate water source
- Production of alkalinity
- Having sufficient demand for the end product

For the product utilization, it was reported that both cement kiln dust (CKD) and fly ash (FA) have been successfully used to produce a green Portland ash [31, 39]. The suitability of the calcareous material as a partial replacement for cement clinker in cement has been documented in some non-structural applications in the USA, but the suitability of the calcareous material as a cement ingredient in concrete applications has not yet been demonstrated publicly [31].

19.4.2.3 Integrated Approach to CO_2 Fixation and Solid Waste Utilization

Figure 19.14 shows an integrated approach to deploying the high-gravity carbonation (HiGCarb) process, which could be considered for CO_2 fixation in flue gas and solid waste utilization within an industrial plant. It is noted that the CO_2 removal rate by the HiGCarb process could meet the timescale in industrial plants.

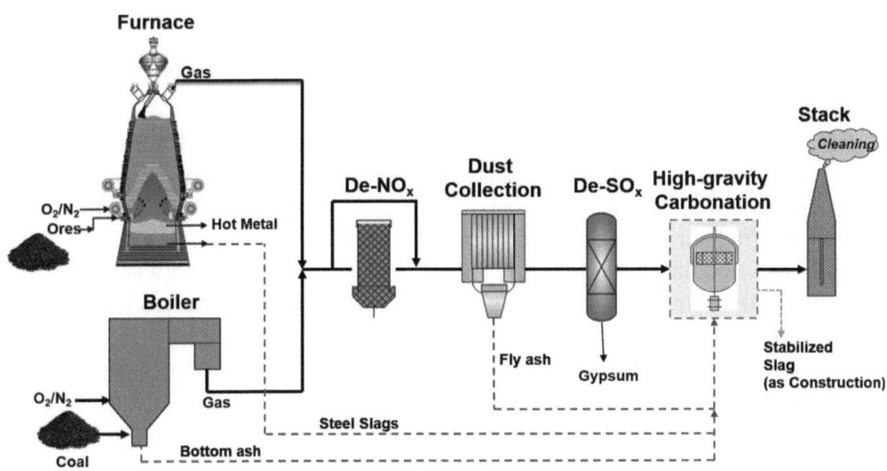

Fig. 19.14 Integrated approach to applying high-gravity carbonation (HiGCarb) process for CO_2 capture in flue gas and solid waste utilization within an industrial plant

19.4.3 Economic Perspectives

19.4.3.1 Energy Consumption

Accelerated carbonation might involve several energy-intensive processes, such as grinding of solid wastes, reactor heating, and gas pressurization. From the perspective of energy consumption, the extensive uses of electricity for any unit process in carbonation could easily diminish the credits from carbon fixation. A life cycle assessment for different types of direct carbonation processes indicates that energy consumption is responsible for the increase in additional CO_2 emission and offsets the overall CO_2 capture efficiency Xiao et al. [40]. In comparison, although the energy consumption of indirect carbonation is typically less than that of direct carbonation, the manufacturing of chemicals for the extraction step may generate additional CO_2 emission and lead to other environmental issues. This suggests that the recovery of the extractants (valuable chemicals) with low energy consumption should be included for implementing indirect carbonation [40, 41].

As suggested by the US Department of Energy (DOE) [25], a cost-effective CO_2 capture facility should meet the following criteria:

- Achieve a removal efficiency (η) of 90%
- Maintain <35% impact on the cost of electricity (COE)

Consequently, heat recovery is an important unit process for accelerated carbonation (exothermic reaction), which could not only improve carbonation performance but also reduce energy loss. For example, the temperature of flue gas streams is high enough for carbonation since it is usually above the dew point. Therefore, heat can be directly obtained from the gas streams or other heat-regenerating systems. To achieve this goal, Santos et al. [42] developed an integrated process where the high pressure is obtained by pumping liquid in an autoclave reactor through a long reaction chamber. Moreover, the heat generated by the carbonation reaction (i.e., exothermal) can be recovered via the integrated process.

Pan et al. [4, 43] introduced a high-gravity carbonation (HiGCarb) process for direct carbonation, where high micromixing between the slurry and gas phases could enhance the overall mass transfer, thereby improving the carbonation conversion and reducing the residence time. In the HiGCarb process, the slurry was first pumped into the center of the reactor, after which it flowed outward motivated by centrifugation. In the meantime, the CO_2 gas entered the reactor from the tangent direction and moved inward due to the pressure gradient.

Figure 19.15 shows the effect of CO_2 removal efficiency on energy consumption and CO_2 capture capacity of the HiGCarb process. Both steel slag grinding and HiGCarb (air compressors, stirring machines, blowers, pumps, and rotating packed bed reactor) processes were considered in energy consumption calculation. The scale of the HiGCarb process was operated at a capture capacity of $\sim$170 kg CO_2 per day, producing $\sim$690 kg of C-BOFS per day. The energy consumption for the

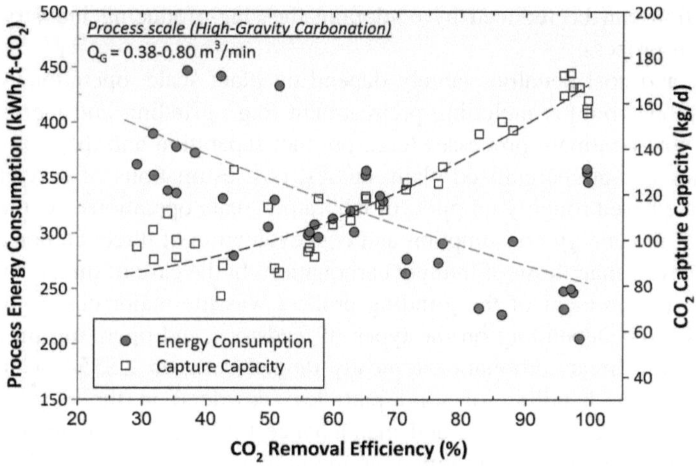

Fig. 19.15 Influence of CO_2 removal efficiency (η) on energy consumption and capture capacity in the case of high-gravity carbonation (HiGCarb) process. Adaption with the permission from Ref. [27]. Copyright 2015 American Chemical Society

high-gravity carbonation include grinding, pumps, blowers, stirring machines, air compressors, and rotation packed bed. The results indicate that the overall energy consumption of the HiGCarb process increases with the decrease in CO_2 capture efficiency. The total energy consumption of the HiGCarb process with CO_2 removal efficiency (η) > 90% was estimated to be 270 ± 60 kWh/t-CO_2 captured (with a 95% confidence interval). It is noted that this value was lower than the DOE requirement, i.e., 420 kWh/t-CO_2 [26]. In addition, the corresponding capture capacity was estimated at 150 kg of CO_2 per day. In this case, the COE of the HiGCarb process was estimated to be 22.4 ± 0.1%, which met the goal of maintaining <35% impact on COE set by the US DOE [25].

19.4.3.2 Operating Cost

Accelerated carbonation using industrial alkaline wastes should be an important part to the reduction in CO_2 from the industrial sector and the use of industrial wastes as cement replacement. Technology may not be the only barrier to the deployment of accelerated carbonation process in the industrial sector. Market competitiveness and the global nature of some of these industries are important issues that should be addressed. From an economic perspective, costs for grinding feedstock can be reduced using the slag in the as-received condition. If the grinding process is needed to improve efficiency of carbonation, it should be done only to the optimum particle size for carbonation to save energy costs. Similarly, the costs of CO_2 pressure can be reduced by using it at atmospheric pressure, and the costs of transportation can be reduced by equipping the slag-producing industry with the carbonation process.

Energy and cost penalties largely depend on plant scale, operation conditions, and operation modulus including pretreatment (e.g., grinding and thermal activation) and post-treatment processes (e.g., product separation and disposal) [44, 45]. Due to lack of commercialized plant studies, cost estimations of accelerated carbonation are based roughly on pilot- or laboratory-scale operations. As presented in Table 19.5, the energy consumption and cost evaluation of direct carbonation were relatively lower than those of indirect carbonation. In the case of direct carbonation, the energy requirement of the grinding process was the major cost in the overall process [4, 46]. Depending on the types of feedstock and operating modulus, the cost of *ex situ* direct carbonation typically ranged between US$54/t-$CO_2$ and US $133/t-$CO_2$. The handling of solid particles (powders) in the process has the potential to raise the operation and maintenance (O&M) costs, compared to CO_2 absorption technologies using ammonia and amine [47].

In contrast, for indirect carbonation using chemical extraction (such as CH_3COOH, HCl, HNO_3, and NaOH) without regeneration of chemicals, a fairly high cost of US$600–4500 would be required for capturing one ton of CO_2 [52]. However, with the regeneration of the chemicals, the recovery process would generate more than 2.5 times the amount of CO_2 fixation in the carbonation process [53]. The operating costs depend largely on the purity of the precipitated calcium carbonate (PCC) product. An average cost of US$80 is required per ton of the PCC production from two-stage indirect carbonation using cement wastes at 50 °C and 30 bar. The major energy consumption processes include pulverization, carbonation, CO_2 separation, CO_2 pressurization, and stirring process for both extraction and carbonation, which are considered at a total of 52.8 MW [54].

To make *ex situ* carbonation more economically feasible, a breakthrough on the use of carbonated solid wastes or products should be sought in the aspects of technology, regulation, institution, and finance. The global cement market is large: with ~3.5 billion metric tons used in 2011 at the processing cost of ~US$100 per ton [37]. It is noted that the carbonated solid waste could potentially be used as partial cement replacement materials (i.e., SCM) [55, 56]. As a result, the benefits returned from carbonation product utilization should be considered in the fiscal analysis of the overall process. From the viewpoint of energy consumption, fine fly ash (FA) should be a good candidate for low-cost carbonation since no grinding process is needed in advance. Moreover, waste heat from manufacturing processes could be integrated instead of electrical heating to reduce the overall energy requirement and operating cost [57].

In comparison, without taking into account long-term monitoring costs, it is estimated that the total cost of *in situ* carbonation should be at US$72–129 per t-$CO_2$, if transportation and storage cost was assumed to be ~US$17 per t-$CO_2$ in basaltic rocks [58]. All of these costs are by far greater than the recent carbon price in European carbon market, i.e., ~US$7 per t-$CO_2$ in 2014 [52]. However, it is noted that the CO_2 price may increase to US$35–90 per t-$CO_2$ by 2040 [59]. In

Table 19.5 Energy consumption and cost evaluation of various processes for direct carbonation

Process descriptions	Evaluation consideration	Feedstock	Performance	Energy consumption (kWh/t-CO_2)	Cost[b] (US$/t-$CO_2$)	References
Carbonation with pretreatment	• Mechanical (grinding to 38–75 μm) and thermal (630 °C for 2 h) pretreatment • Carbonation reaction	Olivine	• Mineral grade: 70%	633–653	54–55	[46]
		Wollastonite	• Mineral grade: 100% (lizardite/antigorite)	429	91	
Carbonation with thermal heat	• Heat activation pretreatment • Partial dehydroxylation of lizarite	Serpentine	• Heat integration • 63% of decrease in energy requirement	n.a.[a]	70	[48]
Carbonation in a fluidized bed	• Pilot-scale plant (at a 532 MW coal-fired power plant) • 70 °C for 120 min	Fly ash (No grinding was required)	• Efficiency: 90% • Capacity: 0.21 t-CO_2/t-ash	n.a.	7.4–27.3	[32, 49]
Carbonation in a spraying glass chamber	• Laboratory scale • Grinding: ~ 45 μm	Mixed steel slag and concrete wastes	• Efficiency: > 50% • Operating (pumps, reactor, sprayers) and labor costs • Cost of water consumption included	15.6 (laboratory scale)	8	[50]
Autoclave reactor (20 bar and 200 °C)	• Depreciation of investments • Viable costs (feedstock, electricity, cooling water) • Fixed costs (labor, supervision, maintenance, insurance, and laboratory)	Wollastonite	• Feedstock costs: US$70 • Grinding (from 100 mm to < 38 μm) • Compression costs: US$34	296 (power) -295 (Heat) (laboratory scale)	133	[51]
		Steel slag	• Fixed operating cost: US$35 • High depreciation cost: US$30	337 (power) -167 (Heat) (laboratory scale)	100	
High-gravity carbonation (HiGCarb)	• Grinding of slag • Stirring, Pumps, Carbonation Reactor (1 atm, 25 °C) • Scale factor: 0.7	Basic oxygen furnace slag	• Grinding costs: US$17.7 • Operating costs: US$24.3	485 (laboratory scale)	57	[4]

[a]*n.a.* Not Available. [b]Assumed EUR€1.0 = US$1.3

another scenario estimated by the International Panel on Climate Change (IPCC), the carbon price would give ~US$55 per t-$CO_2$ as a lower bound estimate.

19.4.4 Environmental Impacts and Benefits

The effect of accelerated carbonation should be carefully weighed and compared according to changes in the environmental impacts. A life cycle assessment (LCA) of the accelerated carbonation process is of particular importance to maximize CO_2 capture capacity while minimizing additional CO_2 emissions due to the process energy consumption. Several operation units, such as material grinding, sieving, and heating, for accelerated carbonation are energy-intensive processes, thereby leading to additional CO_2 emissions. In particular, more than half of the process power consumption may come from the material grinding [8]. These unit processes may also increase other environmental impacts, such as eutrophication (midpoint), acidification (midpoint), and resource depletion (endpoint), due to increases in the concentrations of other pollutants.

19.4.4.1 Reduction in Greenhouse Gas Emission

Deployment of accelerated carbonation in industries and/or power plants will contribute to greenhouse gas (GHG) emission reduction and create additional environmental benefits. For instance, conventionally, precipitated calcium carbonate (PCC) is manufactured by carbonating calcined limestone; therefore, the produced CO_2 is greater than that bound during the carbonation process [60]. Traditional PCC manufacturing resulted in an additional 0.21 kg of CO_2 emissions per kg of PCC [61], mainly caused by oil combustion for lime calcinations. In the indirect carbonation process, the PCC can be produced from a carbon-free feedstock, being a more environmentally sustainable method for producing PCC since no calcination step is required. In this case using acetic acid with wollastonite, a net fixation of 0.34 kg CO_2 per kg of PCC can be achieved, indicating a substantial reduction in GHG emission. Since PCC can be used in the paper industry, a paper mill plant integrated with the indirect carbonation process can transform its CO_2 emissions into PCC toward carbon neutrality.

Similarly, the direct carbonation process can attain huge environmental benefits by taking the use of carbonated solid wastes as SCMs into account. The demand of cement could be reduced if the carbonated solid waste is used as substitutes to replace Portland cement in cement mortar or concrete. Cement production is energy- and material-intensive, which accounts for 4–5% annual CO_2 emission around the world [62]. China accounts for more than 60% of global cement production, where the carbon footprint of cement production in China in 2011 was 0.545 ton-CO_2/ton-cement [63]. The main contributors to CO_2 emission from cement production are

- Clinker process: The production of raw cement material lime (CaO) from limestone ($CaCO_3$), which produces ~0.5 ton CO_2 per ton cement [62].
- Intensive heat required during the cement production.

In addition, a huge amount of natural resources is used for cement production. For instance, Kumar et al. [64] estimated that 1.3–1.5 tons of limestone, 0.2–0.4 tons of clay, and 0.11–0.13 tons of coal are used per ton of cement clinker production.

19.4.4.2 Leaching Behavior of Heavy Metals from Alkaline Solid Wastes

The leaching potential of heavy metals from alkaline solid wastes is of great concern, in terms of human health and environmental impact. Extensive studies have been carried out to evaluate the effect of pH of the solution, carbonation reaction, and mineral structure on the leaching behavior of heavy metals from the solid wastes [55, 65, 66]. Also, different oxidant states of heavy metals in alkaline solid wastes would result in different leaching behaviors, especially for Cr species [67]. It has been demonstrated that accelerated carbonation could effectively reduce the leaching behaviors of most heavy metals from various types of solid wastes, such as steel slag [27], bottom ash [68], and fly ash [69].

As indicated in a few reports [65], the dissolution of calcium silicate minerals (e.g., C_2S) might break down the mineral structure during carbonation, as shown in Eq. (19.19), thereby potentially releasing heavy metals (e.g., vanadium), chlorine and fluoride ions into solution.

$$(CaO)_2{\cdot}SiO_2 + mH_2O \rightarrow CaO \cdot SiO_2 \cdot H_2O_{m-1} + Ca(OH)_2 \qquad (19.19)$$

Calcium-bearing components are the major alkalinity contributors in the alkaline solid wastes. For instance, the pH of the solution containing steel slag typically ranges between 11 and 13. Conversion of the CaO species in solid wastes into $CaCO_3$ with CO_2 gas could effectively decrease the pH of the solution to about 6–8. In parallel with the decrease in pH, the leaching potential of most heavy metals from wastes, such as Pb, Cr, Cu, Zn, Cd, and V ions, could be properly restricted due to the formation of insoluble carbonates [65, 66].

References

1. Stone MH (1948) The generalized weierstrass approximation theorem. Math Mag 21(4): 167–184
2. Weierstrass K (1885) Über die analytische Darstellbarkeit sogenannter willkürlicher Functionen einer reellen Veränderlichen. Erste Mitteilung 633–639
3. Pan SY, Liu HL, Chang EE, Kim H, Chen YH, Chiang PC (2016) Multiple model approach to evaluation of accelerated carbonation for steelmaking slag in a slurry reactor. Chemosphere 154:63–71. doi:10.1016/j.chemosphere.2016.03.093

4. Pan SY, Chiang PC, Chen YH, Chen CD, Lin HY, Chang EE (2013) Systematic Approach to determination of maximum achievable capture capacity via leaching and carbonation processes for alkaline steelmaking wastes in a rotating packed bed. Environ Sci Technol 47 (23):13677–13685. doi:10.1021/es403323x
5. Krevor SCM, Lackner KS (2011) Enhancing serpentine dissolution kinetics for mineral carbon dioxide sequestration. Int J Greenhouse Gas Control 5(4):1073–1080. doi:10.1016/j.ijggc.2011.01.006
6. Bond FC (1961) Crushing and grinding calculations. Part 1 Br Chem Eng 6:378–385
7. Huijgen WJJ, Ruijg GJ, Comans RNJ, Witkamp GJ (2006) Energy consumption and net CO_2 sequestration of aqueous mineral carbonation. Ind Eng Chem Res 45(26):9184–9194
8. Kodama S, Nishimoto T, Yamamoto N, Yogo K, Yamada K (2008) Development of a new pH-swing CO_2 mineralization process with a recyclable reaction solution. Energy 33(5): 776–784. doi:10.1016/j.energy.2008.01.005
9. Pan SY, Eleazar EG, Chang EE, Lin YP, Kim H, Chiang PC (2015) Systematic approach to determination of optimum gas-phase mass transfer rate for high-gravity carbonation process of steelmaking slags in a rotating packed bed. Appl Energy 148:23–31. doi:10.1016/j.apenergy.2015.03.047
10. Guo K, Guo F, Feng Y, Chen J, Zheng C, Gardner NC (2000) Synchronous visual and RTD study on liquid flow in rotating packed-bed contactor. Chem Eng Sci 55:1699–1706
11. Pan SY, Chiang PC, Chen YH, Tan CS, Chang EE (2013) Ex situ CO_2 capture by carbonation of steelmaking slag coupled with metalworking wastewater in a rotating packed bed. Environ Sci Technol 47(7):3308–3315. doi:10.1021/es304975y
12. Hukki RT (1961) Proposal for a solomonic settlement between the theories of von Rittinger, Kick, and Bond. Transl Soc Mining Eng AIME 220:403–408
13. Ipek H, Ucbas Y, Hosten C (2005) The bond work index of mixtures of ceramic raw materials. Min Eng 18(9):981–983. doi:10.1016/j.mineng.2004.12.014
14. Pan S-Y, Lorente Lafuente AM, Chiang P-C (2016) Engineering, environmental and economic performance evaluation of high-gravity carbonation process for carbon capture and utilization. Appl Energy 170:269–277. doi:10.1016/j.apenergy.2016.02.103
15. International Organization for Standardization (2006) ISO 14040—environmental management—life cycle assessment: principle and framework. Management environnemental — Exigences. Switzerland. ISO 14044:2006(E)
16. International Organization for Standardization (2006) ISO 14044—environmental management—life cycle assessment: requirements and guidelines. management environnemental — principes, vol ISO 14040:2006(E). Switzerland. doi:ISO Store order #:783769/Downloaded:2006–11-06
17. Goedkoop M, Heijungs R, Huijbregts M, De Schryver A, Struijs J, van Zelm R (2013) ReCiPe 2008: A life cycle impact assessment method which comprises harmonised category indicators at the midpoint and the endpoint level. Ruimte en Milieu, The Netherlands
18. Zhang T, Yu Q, Wei J, Li J, Zhang P (2011) Preparation of high performance blended cements and reclamation of iron concentrate from basic oxygen furnace steel slag. Resources, Conservation and Recycling 56(1):48–55. doi:10.1016/j.resconrec.2011.09.003
19. Chen C-S (2014) Price fight of product. Taiwan Environmental Information Center. http://boot-topping5.rssing.com/browser.php?indx=3994084&item=3708. Accessed 28 Sept 2014
20. Carbon Place (2014) http://www.carbonplace.eu/
21. Chang EE, Chiu A-C, Pan S-Y, Chen Y-H, Tan C-S, Chiang P-C (2013) Carbonation of basic oxygen furnace slag with metalworking wastewater in a slurry reactor. Int J Greenhouse Gas Control 12:382–389. doi:10.1016/j.ijggc.2012.11.026
22. Chang EE, Pan SY, Chen YH, Tan CS, Chiang PC (2012) Accelerated carbonation of steelmaking slags in a high-gravity rotating packed bed. J Hazard Mater 227–228:97–106. doi:10.1016/j.jhazmat.2012.05.021
23. Chang EE, Pan S-Y, Chen Y-H, Chu H-W, Wang C-F, Chiang P-C (2011) CO_2 sequestration by carbonation of steelmaking slags in an autoclave reactor. J Hazard Mater 195:107–114. doi:10.1016/j.jhazmat.2011.08.006

24. Chang EE, Wang Y-C, Pan S-Y, Chen Y-H, Chiang P-C (2012) CO_2 Capture by using blended hydraulic slag cement via a slurry reactor. Aerosol Air Q Res 12:1433–1443. doi:10.4209/aaqr.2012.08.0210
25. Matuszewski M, Ciferno J, Marano JJ, Chen S (2011) Research and development goals for CO_2 capture technology. U.S. Department of Energy, Washington, DC
26. Datta S, Henry MP, Lin YJ, Fracaro AT, Millard CS, Snyder SW, Stiles RL, Shah J, Yuan J, Wesoloski L, Dorner RW, Carlson WM (2013) Electrochemical CO_2 capture using resin-wafer electrodeionization. Ind Eng Chem Res 52(43):15177–15186. doi:10.1021/ie402538d
27. Pan SY, Chen YH, Chen CD, Shen AL, Lin M, Chiang PC (2015) High-gravity carbonation process for enhancing CO_2 fixation and utilization exemplified by the steelmaking industry. Environ Sci Technol 49(20):12380–12387. doi:10.1021/acs.est.5b02210
28. Hasanbeigi A, Price L, Lin E (2012) Emerging energy-efficiency and CO_2 emission-reduction technologies for cement and concrete production: A technical review. Renew Sustain Energy Rev 16(8):6220–6238. doi:10.1016/j.rser.2012.07.019
29. China Steel Corp (2014) Corporate sustainability report. CSR, CSC, Kaohsiung, Taiwan
30. Reynolds B, Reddy K, Argyle M (2014) Field application of accelerated mineral carbonation. Minerals 4(2):191–207. doi:10.3390/min4020191
31. Zaelke D, Young O, Andersen SO (2011) Scientific synthesis of calera carbon sequestration and carbonaceous by-product applications. University of California, Donald Bren School of Environmental Science and Management, Santa Barbara
32. Reddy KJ, Reddy KJ, Weber H, Bhattacharyya P, Morris A, Taylor D, Christensen M, Foulke T, Fahlsing P (2010) Instantaneous capture and mineralization of flue gas carbon dioxide: pilot scale study. nature preceding available from nature preceding. http://dx.doi.org/10.1038/npre.2010.5404.1. doi:10.1038/npre.2010.5404.1
33. Calera (2011) Calera process: green cement for a blue planet
34. Bodénan F, Bourgeois F, Petiot C, Augé T, Bonfils B, Julcour-Lebigue C, Guyot F, Boukary A, Tremosa J, Lassin A, Gaucher EC, Chiquet P (2014) Ex situ mineral carbonation for CO_2 mitigation: evaluation of mining waste resources, aqueous carbonation processability and life cycle assessment (Carmex project). Min Eng 59:52–63. doi:10.1016/j.mineng.2014.01.011
35. Mineral Carbonation International (2013) http://mineralcarbonation.com/
36. Skyonic (2014) Technology: capitol SkyMine. Skyonic. http://skyonic.com/skymine/
37. IEA (2013) Post-combustion CO_2 capture scale-up study. International Energy Agency
38. Santos S (2013) Challenges to the development of CCS in the energy intensive industries. Paper presented at the 7th IEAGHG International Summer School, Cheltenham, UK
39. Shah SPW, Kejin (2004) Development of "green" cement for sustainable concrete using cement kiln dust and fly ash 15–23
40. Xiao L-S, Wang R, Chiang P-C, Pan S-Y, Guo Q-H, Chang EE (2014) Comparative life cycle assessment (LCA) of accelerated carbonation processes using steelmaking slag for CO_2 fixation. Aerosol Air Q Res 14(3):892–904. doi:10.4209/aaqr.2013.04.012
41. Azdarpour A, Asadullah M, Junin R, Manan M, Hamidi H, Mohammadian E (2014) Direct carbonation of red gypsum to produce solid carbonates. Fuel Process Technol 126:429–434. doi:10.1016/j.fuproc.2014.05.028
42. Santos RM, Verbeeck W, Knops P, Rijnsburger K, Pontikes Y, Van Gerven T (2013) Integrated mineral carbonation reactor technology for sustainable carbon dioxide sequestration: 'CO_2 energy reactor'. Energy Procedia 37:5884–5891. doi:10.1016/j.egypro.2013.06.513
43. Pan S-Y, Chiang P-C, Chen Y-H, Tan C-S, Chang EE (2014) Kinetics of carbonation reaction of basic oxygen furnace slags in a rotating packed bed using the surface coverage model: maximization of carbonation conversion. Appl Energy 113:267–276. doi:10.1016/j.apenergy.2013.07.035

44. Sanna A, Dri M, Hall MR, Maroto-Valer M (2012) Waste materials for carbon capture and storage by mineralisation (CCSM)—a UK perspective. Appl Energy 99:545–554. doi:10.1016/j.apenergy.2012.06.049
45. Pan S-Y, Chang EE, Chiang P-C (2012) CO_2 capture by accelerated carbonation of alkaline wastes: a review on its principles and applications. Aerosol Air Q Res 12:770–791. doi:10.4209/aaqr.2012.06.0149
46. Gerdemann SJ, O'Connor WK, Dahlin DC, Penner LR, Rush H (2007) Ex situ aqueous mineral carbonation. Environ Sci Technol 41(7):2587–2593
47. Yu C-H, Huang C-H, Tan C-S (2012) A review of CO_2 capture by absorption and adsorption. Aerosol Air Q Res 12:745–769. doi:10.4209/aaqr.2012.05.0132
48. Rayson M, Magill M, Sault R, Ryan G, Swanson M (2008) Mineral sequestration of CO_2. Chemical Engineering. The University of Newcastle, Newcastle, NSW
49. Christensen MH (2010) An economic analysis of the Jim Bridger Power Plant CO_2 mineralization process. University of Wyoming, Laramie, Wyoming
50. Stolaroff J, Lowry G, Keith D (2005) Using CaO- and MgO-rich industrial waste streams for carbon sequestration. Energy Convers Manage 46(5):687–699. doi:10.1016/j.enconman.2004.05.009
51. Huijgen WJJ, Comans RNJ, Witkamp G-J (2007) Cost evaluation of CO_2 sequestration by aqueous mineral carbonation. Energy Convers Manage 48(7):1923–1935. doi:10.1016/j.enconman.2007.01.035
52. Sanna A, Uibu M, Caramanna G, Kuusik R, Maroto-Valer MM (2014) A review of mineral carbonation technologies to sequester CO_2. Chem Soc Rev 43(23):8049–8080. doi:10.1039/c4cs00035h
53. Teir S, Eloneva S, Fogelholm C, Zevenhoven R (2009) Fixation of carbon dioxide by producing hydromagnesite from serpentinite. Appl Energy 86(2):214–218. doi:10.1016/j.apenergy.2008.03.013
54. Katsuyama Y, Iizaka A, Yamasaki A, Fujii M, Kumagai K, Yangagisawa Y (2005) Development of a new treatment process of wastes concrete for CO_2 reduction in cement industry. Greenhouse Gas Control Technol 2:1433–1439
55. Salman M, Cizer Ö, Pontikes Y, Santos RM, Snellings R, Vandewalle L, Blanpain B, Van Balen K (2014) Effect of accelerated carbonation on AOD stainless steel slag for its valorisation as a CO2-sequestering construction material. Chem Eng J 246:39–52. doi:10.1016/j.cej.2014.02.051
56. Liang XJ, Ye ZM, Chang J (2012) Early hydration activity of composite with carbonated steel slag. J Chinese Ceram Soc 40(2):228–233 (in Chinese)
57. Balucan RD, Dlugogorski BZ, Kennedy EM, Belova IV, Murch GE (2013) Energy cost of heat activating serpentinites for CO2 storage by mineralisation. Int J Greenhouse Gas Control 17:225–239. doi:10.1016/j.ijggc.2013.05.004
58. Gislason SR, Oelkers EH (2014) Carbon storage in Basalt. Science 344:373–374. doi:10.1126/
59. Wilson R, Luckow P, Biewald B, Ackerman F, Hausman E (2012) 2012 Carbon dioxide price forecast. Synapse Energy Economics, Inc., Cambridge
60. Teir S, Eloneva S, Fogelholm C-J, Zevenhoven R (2007) Dissolution of steelmaking slags in acetic acid for precipitated calcium carbonate production. Energy 32(4):528–539. doi:10.1016/j.energy.2006.06.023
61. Teir S, Eloneva S, Zevenhoven R Co-utilization of CO_2 and calcium silicate-rich slags for precipitated calcium carbonate production (part I). In: Proceedings of the 18th international conference on efficiency; cost; optimization; simulation and environmental impact of energy systems (ECOS 2005), Trondheim, Norway, 20–22 June 2005
62. Gibbs MJ, Soyka P, Conneely D (2001) CO_2 emissions from cement production. good practice guidance and uncertainty management, National Greenhouse Gas Inventories. Intergovernmental Panel on Climate Change (IPCC)
63. Shen W, Cao L, Li Q, Zhang W, Wang G, Li C (2015) Quantifying CO_2 emissions from China's cement industry. Renew Sustain Energy Rev 50:1004–1012

64. Kumar S, Kumar R, Bandopadhyay A (2006) Innovative methodologies for the utilisation of wastes from metallurgical and allied industries. Resour Conserv Recycl 48(4):301–314. doi:10.1016/j.resconrec.2006.03.003
65. van Zomeren A, Van der Laan S, Kobesen H, Huijgen W, Comans R (2011) Changes in mineralogical and leaching properties of converter steel slag resulting from accelerated carbonation at low CO_2 pressure. Waste Manage 31:2236–2244
66. Baciocchi R, Corti A, Costa G, Lombardi L, Zingaretti D (2011) Storage of carbon dioxide captured in a pilot-scale biogas upgrading plant by accelerated carbonation of industrial residues. Energy Procedia 4:4985–4992. doi:10.1016/j.egypro.2011.02.469
67. Baciocchi R, Costa G, Di Bartolomeo E, Polettini A, Pomi R (2010) Carbonation of stainless steel slag as a process for CO_2 storage and slag valorization. Waste Biomass Valorization 1:467–477
68. Chang EE, Pan SY, Yang L, Chen YH, Kim H, Chiang PC (2015) Accelerated carbonation using municipal solid waste incinerator bottom ash and cold-rolling wastewater: performance evaluation and reaction kinetics. Waste Manage 43:283–292. doi:10.1016/j.wasman.2015.05.001
69. Pan S-Y, Hung C-H, Chan Y-W, Kim H, Li P, Chiang P-C (2016) Integrated CO_2 fixation, waste stabilization, and product utilization via high-gravity carbonation process exemplified by circular fluidized bed fly ash. ACS Sustain Chem Eng 4(6):3045–3052. doi:10.1021/acssuschemeng.6b00014

Chapter 20
Prospective and Perspective

Abstract This chapter provides the prospective and perspective of key strategies for effective CO_2 mineralization and utilization, including (1) implementation of national sustainable policy; (2) recovery of valuable elements from solid wastes; (3) enhanced removal of various air pollutants in flue gas; (4) generation of high value-added products for diversified applications; (5) integrated approach to multiwaste treatment as green solutions; and (6) eco-industrial parks as a business model. In addition, research needs on CO_2 mineralization and utilization using alkaline wastes are proposed to achieve the goal of zero waste.

20.1 Strategies Toward "Zero" Waste for Sustainability

Integrated alkaline waste treatment with CO_2 capture and utilization is an attractive approach to achieving direct and indirect reduction in greenhouse gas (GHG) emissions in industries. The accelerated carbonation can not only stabilize alkaline wastes but also fix CO_2 in flue gas from industries as a safe and stable carbonate precipitate. On the other hand, the amount of CO_2 reduction by carbonation could be considered as certified emission reduction (CER) credits, which could be used in the emission trading scheme (ETS) under the clean development mechanism (CDM) issued by the Kyoto Protocol.

Barriers from the aspects of regulation, institution, finance, and technology can be encountered while implementing the green supply chain. These barriers are hard to be distinctly separated because, for instance, policies (or regulations) often act on more than one barrier simultaneously. Similarly, this is especially true for the institutional and financial barriers as they can habitually be closely related. From the viewpoint of treatment and utilization for various alkaline solid wastes, several strategies should be considered:

- implementation of national sustainable policy,
- recovery of valuable elements from solid wastes,
- enhanced removal of various air pollutants in flue gas,

P.-C. Chiang and S.-Y. Pan, *Carbon Dioxide Mineralization and Utilization*, DOI 10.1007/978-981-10-3268-4_20

- generation of high value-added products for diversified applications,
- integrated approach to multiwaste treatment as green solutions, and
- eco-industrial parks as a business model.

20.1.1 Implementation of National Sustainable Policy

For the national sustainable policy, predominant emphasis should be given to achieving rapid economic growth and prioritizing industrial development. Industries should pursue a more balanced economic development in which raising quality takes precedence over expanding quantity, placing much greater emphasis upon achieving both economic development and environment protection. Therefore, industries should engage in "cleaner production" that may not seriously pollute the environment while promoting green consumption. From the government point of view, promotion of eco-industrial parks (EIP) establishment is essential to spur the development of new "green-tech" products and services for upgrading industrial technologies. Appropriate policies should be established to foster industrial symbiosis, thereby facilitating the green technologies for effective material reuse and waste recycling. Several rules are suggested to make the regulation more globally acceptable, including (1) involvement of a regulation context and (2) implementation of green industries by technology-forcing, guaranteed market, and economies of scale.

20.1.2 Recovery of Valuable Elements from Solid Wastes

The valuable elements, such as aluminum (Al), sodium (Na), and iron (Fe), in solid wastes can be extracted and recovered prior to accelerated carbonation. For example, a reclamation process for concentrating iron components from solid wastes (especially for steel slags) should be considered. The concentrated iron materials can be used as pigment and abiotic catalyst.

20.1.3 Enhanced Removal of Various Air Pollutants in Flue Gas

In real operations, the flue gases or fumes (with diluted CO_2) can be directly used for accelerated carbonation with alkaline solid wastes generated in the same plant. In this case, accelerated carbonation process can further reduce several air pollutants, such as sulfur dioxide (SO_2) and particulate matter (PM) in flue gas by different mechanism. Therefore, the additional environmental benefits could be gained due to coremoval of air pollutants using a single process. For example, the SO_2 gas can be combined with calcium ions from solid wastes to form calcium sulfate ($CaSO_4$) precipitates. In addition, the PM can be reduced via different mechanisms, such as impaction, interception, and diffusion by tiny water droplets.

20.1.4 Generation of High Value-Added Products for Diversified Applications

Alkaline solid wastes can be used to sequester great amount of CO_2, especially if the wastes are generated nearby the source of CO_2, for achieving both the environmental and economic benefits. After carbonation, the treated solid wastes could be converted into several high value-added materials, such as glass ceramics (with red mud from the sintering process), supplementary cementitious materials (SCM), and precipitated calcium carbonate (PCC). In the case of utilization as SCM, a ternary blended system using various types of carbonated wastes (such as combined fly ash and steel slag) should be considered. Sequential performance tests on material functions are necessary to reliably predict the behavior after use within a reasonable time, such as an expansion test when it comes to be utilized as SCM. In addition, quality assurance and quality control (QA/QC) programs, including calibration, validation, and verification, should be established to ensure data quality objectives (DQO) and provide a useful distinction between materials that are suitable and those that are not.

20.1.5 Integrated Approach to Multiwaste Treatment as Green Solutions

Pretreatment processes of alkaline solid wastes are usually necessary to overcome the potential barriers in conventional utilization and application. Portfolio solutions can be provided along with the accelerated carbonation. In Chap. 8, an integrated approach to multiwaste treatment via accelerated carbonation has been proposed using CO_2 in flue gas as a chemical to stabilize active components in alkaline solid wastes. For example, after indirect carbonation, the extracted residues can be used in other processes as raw materials, such as accelerated carbonation with flue gas CO_2, rare elements’ extraction, and building materials production. In addition, the brine wastes from water reuse processes (such as electrodeionization and reverse osmosis) can be utilized in carbonation process as liquid agents. From the economic point of view, accelerated carbonation can reduce the treatment cost of wastewater and increase the added value of alkaline solid. Therefore, aqueous accelerated carbonation of alkaline solid wastes is suggested as a link to utilize wastewater for large-scale application.

20.1.6 Eco-industrial Parks (EIP) as a Business Model

Building a waste-to-resource (WTR) green supply chain within industrial park should be extensively promoted to make traditional industries around the world being environmentally bearable, economic viable, and social equitable. Policy makers should create policy for reducing GHG emissions while improving energy efficiency. Therefore, action plans should increase manufacturing efficiency while seeking synergetic cooperation between all manufacturers in industrial park. To ensure the effective implementation, creation of a cost-effective integrated green certificate market by the implementation of pricing instruments, such as tax exemptions and carbon credits, is quite important. Moreover, life cycle analysis (LCA) should be utilized as a structured basis for evaluating the performance of environmental impacts and benefits in the WTR green supply chain.

20.2 Research Needs

Concerns about serious and irreversible damages caused by rapid CO_2 accumulation in the atmosphere are raised. Therefore, effective approaches to controlling CO_2 emissions are required to achieve the goal of constraining global CO_2 concentration below 550 ppm over the next 100 years. Ex situ carbonation of alkaline wastes, which combines the treatment of industrial wastes that are readily available near a CO_2 emission point, could be part of an integrated approach to reducing CO_2 emissions for industrial plants. Although a great amount of alkaline waste is available for CO_2 mineralization, the costs of accelerated carbonation are too high for large-scale industrial deployment.

Another challenge may be the issue of cost allocation in the case of multiple emission reductions. For instance, CO_2 emission in an industrial plant is lower than that in a coal-fired power plant. The management of a variety of CO_2 sources within a single industrial plant and the selection of appropriate process and technology for CO_2 capture should be the major concern for lowering the industrial capture costs. Thus, it may not be a complete solution to carbon capture and utilization. To promote industrialization of CO_2 mineralization, future work should be focused on.

20.2.1 Technology Improvement and Breakthrough

Significant technological breakthroughs are needed before deployment can be considered. A well process design and build-up of full-scale plant with efficient mass transfer among gas–liquid–solid phases (high CO_2 fixation capacity with low energy consumption) are required. It suggests that a carbonation conversion of higher than 85% for solid wastes should be acceptable to achieve waste stabilization and CO_2 fixation.

20.2.2 *Material Function Evaluation for Multiple Products*

Diversified applications of carbonated products as high value-added chemicals or green materials can be achieved. For instance, the chemistry and mechanism of utilizing the carbonated product as SCM in blended cement should be systematically determined, in terms of the physicochemical properties of the product. From the research point of view, future work should be focused on (1) the mechanisms of various types of carbonated solid wastes with ordinary Portland cement (OPC) in cement chemistry; (2) the balance of carbonation efficiency and cement performance to maximize the overall carbon emission reduction; and (3) the diversification of utilization routes for carbonation products such as high value-added chemicals.

20.2.3 *Process Integration for Innovation*

The integration of different existing (or urgent) unit operations, such as water reuse process, with CO_2 capture and utilization is an effective way to achieve green solutions. For instance, accelerated carbonation of natural ores and/or alkaline solid waste is normally suffered from slow kinetics and high energy demand. Research and implementation of these technologies require new collaborative efforts among the crushed stone and cement industries, electric utilities, and the science and engineering communities. Moreover, the procedures of feedstock crushing (in the case of steel slag), process heating, and slurry stirring are normally energy-intensive processes. It needs to be compensated by the exothermic carbonation process to make the process economically viable in an industrial context.

20.2.4 *Demonstration and Action Plans*

The selection of appropriate processes for each case of operation site within an achievable plant size is important. For example, the management of material recycle and post-treatment of residue should be systematically considered in the design phase. This includes the waste-to-resource (WTR) supply chain and system optimization from the 3E (Engineering, Environmental, and Economic) aspects. In the case of developing a demonstration program, all activities that will be carried out to construct, operate, and close a large-scale facility and all factors that could be affected by the above activities should be considered in the strategic environmental assessment (SEA), as illustrated in Chap. 4. Accordingly, the environmental impacts of different alternatives would be analyzed and quantified. In the valuation step, preferable technical alternatives would be identified and weighted. To realize commercialization and industrialization, future efforts should be focused on

(1) process scale-up such as reactor design, material recycling, and residue treatment; (2) implementation of instrumentation control and automation (ICA), such as online hardware/software sensors and process control model; and (3) establishment of management information systems (MIS), such as information engineering and operations research.

20.3 The Future We Want

Reduction in CO_2 emission in industries and/or power plants should be a portfolio option. For instance, CO_2 capture and alkaline solid waste treatment can be combined through an integrated approach, i.e., accelerated carbonation. Gaseous CO_2 is fixed as thermodynamically stable solid precipitates, which are rarely released after mineralization. On the other hand, in many regions, there is an increasing need to find substitutes for natural sands in concrete to reduce material and energy consumption. Within the framework of accelerated carbonation, the use of alkaline solid wastes has been considered as an alternative to simultaneously address the issues of solid waste treatment and natural resource depletion. The waste-to-resource (WTR) supply chain between steelmaking and cement industries can be established by deploying the innovative accelerated carbonation technology, which can simultaneously treat the CO_2, wastewater, and solid wastes generated from the industry. The carbonated products utilized as green materials including cement, aggregate, and precipitate calcium carbonate are illustrated. Therefore, it is concluded that the establishment of a WTR supply chain should provide a method of simultaneously addressing the problems of energy demand, waste management, and GHG emissions to achieve a circular economy system (CES). The "win-win" philosophy that is a prosper economy and healthy environment can be coexisted.

Index

P.-C. Chiang and S.-Y. Pan, *Carbon Dioxide Mineralization and Utilization*, DOI 10.1007/978-981-10-3268-4

D

MIX
Papier aus verantwortungsvollen Quellen
Paper from responsible sources
FSC® C105338

If you have any concerns about our products,
you can contact us on
ProductSafety@springernature.com

In case Publisher is established outside the EU,
the EU authorized representative is:
Springer Nature Customer Service Center GmbH
Europaplatz 3, 69115 Heidelberg, Germany

Printed by Libri Plureos GmbH
in Hamburg, Germany